AIRCRAFT BASIC SCIENCE

● SIXTH EDITION ●

MICHAEL J. KROES
JAMES R. RARDON
RALPH D. BENT
JAMES L. McKINLEY

GREGG DIVISION
GLENCOE

Macmillan/McGraw-Hill

Lake Forest, Illinois   Columbus, Ohio

Mission Hills, California   Peoria, Illinois

Sponsoring Editor: **D. Eugene Gilmore**
Editing Supervisors: **Ira C. Roberts and Larry Goldberg**
Design and Art Supervisor: **Annette Mastrolia-Tynan**
Production Supervisor: **Catherine Bokman**

Text Designer: **Phyllis Lerner**
Cover Photographer: COMSTOCK, INC./Tom Grill

Library of Congress Cataloging-in-Publication Data

Aircraft basic science.

   (Aviation technology series)
   Rev. ed. of: Aircraft basic science/Ralph D. Bent.
5th ed. c1980.
   Includes index.
   1. Airplanes—Design and construction.   I. Kroes,
Michael J. II. Bent, Ralph D. Aircraft basic science.
III. Series.
TL671.2.A3735 1988   629.13   87-22623
ISBN 0-07-004799-5

Send all inquiries to: Glencoe Division, Macmillan/McGraw-
Hill, 936 Eastwind Drive, Westerville, Ohio 43081.

4 5 6 7 8 9 10 11 12 13 14 15   SEM   00 99 98 97 96 95 94 93 92 91

ISBN 0-07-004799-5

CONTENTS

Preface   vii
Acknowledgments   viii
1. Fundamentals of Mathematics   1
2. Science Fundamentals   36
3. Basic Aerodynamics   59
4. Airfoils and Their Applications   74
5. Aircraft in Flight   94
6. Aircraft Drawings   122
7. Weight and Balance   142
8. Aircraft Materials   169
9. Fabrication Techniques and Processes   191
10. Aircraft Hardware   211
11. Hand Tools and Their Application   233
12. Aircraft Fluid Lines and Fittings   264
13. Federal Aviation Regulations and Publications   288
14. Ground Handling and Safety   307
15. Aircraft Inspection and Servicing   328
16. Maintenance Shop Requirements and Practices   349
Appendix   359
Glossary   380
Index   383

PREFACE

Aircraft Basic Science is one of the four textbooks in the Aviation Technology Series. Its purpose is to provide the general technical information needed as a foundation for work as a technician in the field of aviation maintenance. The subjects covered in this text are those that are usually applicable to both airframes and powerplants and their associated systems. The material is of a general nature since in addition to applying to the maintenance technician, it is also applicable to many related fields of study in aviation.

This edition has been updated to include many recent advances in aviation technology and, with the exception of electricity, has been revised to meet the requirements for general subject material specified in FAR Part 147. Electricity and electronics are not covered but are treated in detail in another text in this series.

The aviation technician must be able to read and interpret many drawings found in aircraft technical reference material. The ability to read drawings is not synonymous with the ability to make drawings. Because of the development of computer-aided drafting, the chapter "Fundamentals of Drafting" has been deleted in this revision. Those students wishing more knowledge about drafting will undoubtedly obtain it from specialized courses. The material dealing with the reading and interpretation of drawings and blueprints has been revised and expanded in this edition.

The development and use of new materials in aircraft production have resulted in the previous chapter "Materials and Processes" being expanded into two new chapters. The "Aircraft Materials" chapter emphasizes material properties and identification, leaving aircraft fabrication techniques and processes for a separate chapter.

The text has been expanded by the addition of a chapter to make room for more information in the areas of maintenance publications, safety practices, ground handling, servicing, and aircraft inspections. This book, when used along with the other books in the series and in conjunction with classroom and shop instruction, will provide the student with the technical information needed to qualify for certification as an airframe and powerplant technician.

The book is intended as a text and general reference book for those individuals involved in aircraft maintenance and operation. The material presented is for use in understanding aircraft materials, processes, practices, and operations. Technical information contained in this book should not be substituted for that provided by the manufacturer.

Michael J. Kroes
James R. Rardon
Ralph D. Bent
James L. McKinley

ACKNOWLEDGMENTS

The authors wish to express appreciation to the following organizations for their generous assistance in providing illustrations and technical information for this text: Aircraft Spruce and Specialty, Fullerton, California; Air France, New York, New York; Aeroquip Corporation, Jackson, Michigan; Atlas Copco, Inc., Orange, California; Aluminum Association, Inc., Washington, D.C.; Aluminum Company of America, Los Angeles, California; Avtec Corp., San Pablo, California; Beechcraft Corp., Wichita, Kansas; Bell Helicopter Textron, Division of Textron, Inc., Fort Worth, Texas; Boeing Company, Seattle, Washington; Boeing Vertol, Division of Boeing Company, Philadelphia, Pennsylvania; Cessna Aircraft Co., Wichita, Kansas; Cherry Fasteners, Townsend Division, Textron, Inc., Santa Ana, California; Cleveland Twist Drill, an Acme Cleveland Company, Cleveland, Ohio; Deutsch Metal Components Division, Los Angeles, California; Douglas Aircraft Company, Division of McDonnell Douglas Corporation, Long Beach, California; Evergreen Weigh, Inc., Lynwood, Washington; Federal Aviation Administration, Washington, D.C.; Gates Learjet Corp., Tucson, Arizona; General Electric Corp., Lynn, Massachusetts; Grumman Aircraft Corp., Bethpage, New York; Hi-Shear Corporation, Torrance, California; Hobart Brothers Co., Troy, Ohio; International Aviation Publishers, Riverton, Wyoming; International Nickel Corporation, New York, New York; Lockheed Corp., Burbank, California; Lufkin Division, Cooper Group, Apex, North Carolina; National Aeronautics and Space Administration, Washington, D.C.; Nicholson File Division, Cooper Group, Apex, North Carolina; Northrop Corporation, Hawthorne, California; Northrop University, Inglewood, California; Peterson Publishing Co., Los Angeles, California; Piper Aircraft Corporation, Vero Beach, Florida; Proto Tool Company, Los Angeles, California; Purdue University, West Lafayette, Indiana; Resistoflex Corporation, Roseland, New Jersey; Rocketdyne Division, Rockwell International Corp., Canoga Park, California; Sikorsky Aircraft Division, United Technologies, Stratford, Connecticut; Snap-on Tools, Kenosha, Wisconsin; Stanley Tools, New Britain, Connecticut; Stellite Division of the Cabot Corp., Kokomo, Indiana; Stratoflex, Inc., Fort Worth, Texas; VOI-SHAN Division, VSI Corporation, Culver City, California.

In addition, the authors wish to thank the many aviation schools and instructors who provided valuable suggestions, recommendations, and technical information for the revision of this text.

1 FUNDAMENTALS OF MATHEMATICS

● INTRODUCTION

The science of mathematics, so important to the modern age of technology, had its beginnings in the dim ages of the past. It is probable that prehistoric people recognized the differences in quantities at an early age and therefore devised methods for keeping track of numbers and quantities. In the earliest efforts at trade it was necessary for the traders to figure quantities. For example, someone might have traded 10 sheep for 2 cows. To do this the trader had to understand the numbers involved.

As time progressed, the ancient Babylonians and Egyptians developed the use of mathematics to the extent that they could perform marvelous engineering feats. Later the Greeks developed some of the fundamental laws which are still in use today. One of the great Greek mathematicians was a philosopher named Euclid who prepared a work called *Elements of Geometry.* This text was used by students of mathematics for almost 2000 years. Another mathematician, also a Greek, was Archimedes, who is considered one of the greatest mathematicians of all time. One of his most important discoveries was the value of π (pi), which is obtained by dividing the circumference of a circle by its diameter. Archimedes discovered many other important mathematical relationships and also invented the beginnings of calculus. Modern differential and integral calculus was discovered by Sir Isaac Newton in the seventeenth century. These discoveries are considered some of the most important in the history of mathematics.

Today's modern technology, including aircraft maintenance, is greatly dependent upon mathematics. Computing weight and balance of an aircraft, designing a structural repair, or determining serviceability of an engine part are but a few examples of an aviation maintenance technician's need for mathematics. Electronic calculators and computers have made mathematical calculations more rapid and probably more accurate. However, these devices are only as good as the information input and do not excuse the technician from learning the fundamentals of mathematics.

It is expected that the aviation technician student will have taken, or be taking, mathematics courses that go beyond the material in this chapter. The purpose of this chapter is to refresh the student's understanding of fundamental mathematical processes. Emphasis is placed on those mathematical terms or problems that they will encounter in portions of their technical studies or in employment.

● ARITHMETIC

Numbers

The ten single number characters, or **numerals**—1, 2, 3, 4, 5, 6, 7, 8, 9, and 0—are called **digits.** With these digits in various combinations we can express any number we wish. The arrangement of the digits and the number of digits used determine the value of the number which we are expressing.

Our number system is called a **decimal** system, the name being derived from the Latin word *decem,* meaning ten. In the decimal system the digits are arranged in columns which are powers of 10. The column in which a certain digit is placed determines its expressed value. When we examine the number 3 235 467, we indicate the column positions as follows:

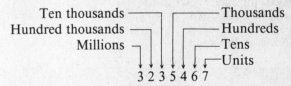

We may analyze the total number by considering the values expressed by each column, thus:

Units	7	7
Tens	6	60
Hundreds	4	400
Thousands	5	5 000
Ten thousands	3	30 000
Hundred thousands	2	200 000
Millions	3	3 000 000

We may now observe that the total number consists of 3 millions, 2 hundred thousands, 3 ten thousands, 5 thousands, 4 hundreds, 6 tens, and 7 units. The total number is read "three million, two hundred thirty-five thousand, four hundred sixty-seven."

There are several classes of numbers which should be understood. **Whole numbers,** also called **integers,** are those which contain no fractions. Examples of such numbers are 3, 10, 250, and 435. A **fraction** is a part of a unit. A **mixed number** contains a whole number and a fraction. An **even number** is one which is divisible by 2. The numbers 2, 4, 6, 8, 10, 48, and 62 are even. **Odd numbers** are those which are not divisible by 2. The numbers 3, 5, 11, 13, 53, and 61 are odd.

Addition and Subtraction

Addition and subtraction may be considered the simplest of mathematical operations; however, it requires practice to do these operations quickly and accurately.

Addition. The sign for addition is the **plus sign** (+). This sign placed between numbers indicates that they are to be added. Numbers to be added may be arranged horizontally or vertically in columns as shown below.

$$324 + 25 + 78 = 427 \qquad \begin{array}{r} 324 \\ 25 \\ +\ 78 \\ \hline 427 \end{array}$$

In arithmetic, numbers to be added are usually arranged in columns for more speed and convenience in performing the addition.

$$\begin{array}{rrr} & 32 & 4382 \\ 7 & 420 & 276 \\ 6 & 8 & 1820 \\ 3 & 19 & 2753 \\ 8 & 26 & 47 \\ +\ 5 & +248 & +\ 238 \\ \hline 29 & 753 & 9516 \end{array}$$

If one wishes to become proficient in addition, it is well to practice adding by sight. It is quite easy to learn to add by sight when the numbers to be added contain only one digit, and this provides a starting point. Examine the following additions and quickly record the answer for each.

Practice Problems

1. 6 + 5 =   4. 7 + 6 =   7. 7 + 4 =
2. 4 + 9 =   5. 9 + 2 =   8. 3 + 2 =
3. 8 + 3 =   6. 5 + 8 =

Add the following by sight:

$$\begin{array}{llllllll} 1. & 25 & 3. & 34 & 5. & 75 & 7. & 59 \\ & \underline{45} & & \underline{64} & & \underline{44} & & \underline{99} \\ \\ 2. & 16 & 4. & 63 & 6. & 54 \\ & \underline{63} & & \underline{35} & & \underline{88} \end{array}$$

Practice is one of the surest ways to learn to add accurately and rapidly. If it is desired to attain proficiency, the student should take time to make up problems or procure problems already prepared and then practice solving the problems until the desired results are obtained.

After a little practice, individuals can reach the point where the sight of any two digits will immediately bring the sum to mind. Thus when they see the digits 6 and 5, for example, they will immediately think "11," or when they see 9 and 7, they will think "16" instantly.

When it is desired to add two-digit numbers by sight, it is merely necessary to add the units and then the tens. Suppose that the numbers 45 and 23 are presented for addition. The units are 5 and 3, so we immediately think 8 units. The tens are 4 and 2, so we think 6 tens. The sum of 6 tens and 8 units is obviously 68. If the units in an addition add to a sum greater than 9, then we must remember to add the ten or tens to the sum of the tens. If we wish to add 36 and 57, we see that the units add to 13, or 1 ten and 3 units. We therefore note the 3 units and carry the ten and add it to the 3 tens and 5 tens. The result is 9 tens and 3 units, or 93.

Subtraction. **Subtraction** is the reverse of addition. The sign for subtraction is the minus sign (−). In ordinary arithmetic a smaller number is always subtracted from a larger number.

In subtraction the number from which another is to be subtracted is called the **minuend,** the number being subtracted from the other is called the **subtrahend,** and the result is called the **difference.**

$$\begin{array}{rl} 675 & \text{minuend} \\ -342 & \text{subtrahend} \\ \hline 333 & \text{difference} \end{array}$$

In subtraction it is well to remember the components of a number, that is, units, tens, hundreds, etc. This will make it easier to perform the necessary operations with clear understanding. In the preceding example, the digits in the subtrahend are smaller than the corresponding numbers in the minuend, and the operation is simple. If a number in the minuend is smaller than the corresponding number in the subtrahend, it is necessary to borrow from the next column. For example,

$$\begin{array}{r} 853 \\ -675 \\ \hline 178 \end{array}$$

In the first column we find the 3 smaller than the 5, and so we must borrow a ten from the next column. We then subtract 5 from 13 to obtain 8. We must then remember that there are only 4 tens left in the second column, and we have to borrow 1 hundred from the next column to make 140. We subtract 70 from 140 and obtain 70 and so we place a 7 in the tens column of the answer. Since we have borrowed 1 hundred from the 8 hundreds of the third column, only 7 hundreds are left. We subtract 6 hundreds from 7 hundreds, thus leaving 1 hundred. We therefore place a 1 in the hundreds column of the answer.

Multiplication

Multiplication may be considered multiple addition. It may be defined as the process of finding the total number or quantity obtained by repeating a specified number or quantity a specified number of times. If we add 2 + 2 to obtain 4, we have multiplied 2 by 2, because we have taken 2 two times. Likewise, if we add 2 + 2 + 2 + 2 to obtain 8, we have multiplied 2 by 4, because we have taken 2 four times.

In multiplication the number to be multiplied is called the **multiplicand,** and the number of times the multiplicand is to be taken is called the **multiplier.** The answer obtained from a multiplication is the **product.** The following problem illustrates these terms:

```
    425     multiplicand
×    62     multiplier
   850
 25 50
 26 350     product
```

It will be observed that the terms *multiplicand* and *multiplier* may be interchanged. For example, 2 × 4 is the same as 4 × 2; hence the terms are actually reversed.

When we employ multiplication to solve a specific problem, the names of the terms have more significance. For example, if we wish to find the total weight of 12 bags of apples and each bag weighs 25 pounds (lb), then the multiplicand is 25 and the multiplier is 12. We then say 12 times 25 lb is 300 lb, or 12 × 25 = 300.

We may obtain an understanding of a multiplication process by analyzing a typical but simple problem such as multiplying 328 by 6.

```
      328
×       6
       48  = 6 × 8
      120  = 6 × 20
     1800  = 6 × 300
     1968
```

In actual practice we do not write down each separate operation of the multiplication as shown in the foregoing problem, but we shorten the process by "carrying" figures to the next column. In the problem shown it can be seen that 6 × 8 is 48 and that the 4 goes into the tens column. Therefore, when we multiply, we merely carry the 4 over and add it to the next multiplication, which is in the tens column. When we use this method, the operation is as follows:

```
     14
    328
      6
   1968
```

$$6 \times 8 = 48$$

Record the 8 (units) and carry the 4 (tens).

$$6 \times 2 = 12$$

Add the 4 to obtain 16. Record the 6 (tens) and carry the 1 (hundreds).

$$6 \times 3 = 18$$

Add the 1 and obtain 19. Record the 19.

When there is more than one digit in the multiplier, we repeat the process for each digit but we must shift one column to the left for each digit. This is because the right-hand digit of the multiplier is **units,** the next digit to the left is **tens,** the next is **hundreds,** etc. If we multiply 328 by 246, we proceed as follows:

```
      328          328
×     246        ×  246
    1 968         1 968
   13 120        13 12
   65 600        65 6
   80 688        80 688
```

Note that we have placed zeros at the end of the second and third multiplications in the first example to show that we were multiplying by 40 and 200, respectively. In actual practice the zeros are not usually recorded. In the above multiplication we multiplied 328 first by 6, then by 40, and finally by 200. When we added these products, we obtained the answer, 80 688.

Accurate multiplication requires great care. First, it is important to know the multiplication tables. Second, care must be taken in recording answers in the correct column. Third, the additions required must be made carefully and accurately. In order to acquire proficiency in accurate multiplying, practice is essential. The following problems will provide a sample of the type of practice needed to improve one's multiplication ability:

Practice Problems

1.	9 ×7	11.	73 × 7	21.	648 ×612
2.	7 ×8	12.	87 × 6	22.	789 ×876
3.	9 ×6	13.	86 × 9	23.	3428 ×6123
4.	11 ×12	14.	24 ×12	24.	5967 ×7298
5.	13 ×11	15.	38 ×14	25.	7642 ×3850
6.	13 ×17	16.	39 ×26	26.	9743 ×3006
7.	26 × 6	17.	93 ×67	27.	9537 ×4060
8.	38 × 4	18.	357 × 78	28.	8457 ×3000
9.	49 × 7	19.	429 × 67	29.	39 999 × 8 009
10.	57 × 8	20.	567 ×329	30.	90 400 × 9 090

In any mathematical problem it is desirable to check the answer for accuracy. There are a number of methods for checking multiplication, and the most obvious is to divide the product by either the multiplicand or the multiplier. If the product is divided by the multiplicand, the quotient (answer) should be the multiplier.

Another method for checking multiplication is to repeat the problem, reversing the multiplicand and multiplier. If the product is the same in each case, the answer is most likely to be correct.

Division

Division may be considered the reverse of multiplication; that is, division is the separating or dividing of a number into a certain number of equal parts. The symbol for division is ÷ and it is read "divided by." For example, 98 ÷ 4 is read "98 divided by 4."

In arithmetic there are two commonly used methods for the division of whole numbers. These are **short division** and **long division.** The terms used to describe the elements of a division problem are **dividend,** which is the number to be divided; **divisor,** the number of times the dividend is to be divided; and **quotient,** the number of times the divisor goes into the dividend. In the problem 235 ÷ 5 = 47, the number 235 is the dividend, 5 is the divisor, and 47 is the quotient.

The process of short division is used most often when it is desired to divide a number by a divisor having only one digit. This is accomplished as follows:

$$\begin{array}{r} 3 \\ 7\overline{)3\ 8\ 5\ 7} \\ 5\ 5\ 1 \end{array}$$

The first step is to divide 38 by 7. Since $7 \times 5 = 35$, it is obvious that after the division of 38 by 7 there will be a remainder of 3. This 3 is held over in the hundreds column and becomes the first digit of the next number to be divided. This number is then 35, and 7 goes into 35 five times without leaving a remainder. The only number left to divide is the 7, into which the divisor goes once. The quotient is then 551. The process of division as explained above may be understood more thoroughly if we analyze the numbers involved. The dividend 3857 may be expressed as $3500 + 350 + 7$. These numbers divided separately by 7 produce the quotients 500, 50, and 1. Adding these together gives 551, which is the quotient obtained from the first division.

Long division is employed most often when the dividend and the divisor both contain more than one digit. The process is somewhat more complex than that of short division, but with a little practice long division may be accomplished easily and accurately.

To solve the problem 18 116 ÷ 28 we arrange the terms of the problem as shown below.

$$\begin{array}{r} 647 \\ 28\overline{)18\ 116} \\ 16\ 8x \\ \hline 1\ 31 \\ 1\ 12x \\ \hline 196 \\ 196 \\ \hline 0 \end{array}$$

The first step in solving the problem is to divide 181 by 28, because 181 is the smallest part of the dividend into which 28 can go. It is found that 28 will go into 181 six times with a remainder of 13. The number 168 (6×28) is placed under the digits 181 and is subtracted. The number 13, which is the difference between 168 and 181, is placed directly below the 6 and 8 as shown, and then the figure 1 is brought down from the dividend to make the number 131. The divisor 28 will go into 131 four times with a remainder of 19. The final digit 6 of the dividend is brought down to make the number 196. The divisor 28 will go into 196 exactly seven times. The quotient of the entire division is then 647.

If we study the division shown in the foregoing example, we will find that the dividend is composed of $28 \times 600 = 16\ 800$, $28 \times 40 = 1120$, and $28 \times 7 = 196$. Then by adding $16\ 800 + 1120 + 196$, we find the sum, which is 18116, the original dividend. We could divide each part of the dividend by 28 separately to obtain 600, 40, and 7 and then add these quotients together; however, it is quicker and simpler to combine the divisions as shown.

If a divisor does not go into a dividend an even number of times, there will be a remainder. This remainder may be expressed as a whole number, a fraction, or a decimal. Fractions and decimals will be discussed later in this chapter.

In the following example the divisor will not go into the dividend an even number of times, so it is necessary to indicate a remainder.

$$\begin{array}{r} 223\tfrac{10}{16} \\ 16\overline{)3578} \\ 32 \\ \hline 37 \\ 32 \\ \hline 58 \\ 48 \\ \hline 10 \end{array}$$

Practice Problems

1. 465 ÷ 15 =
2. 320 ÷ 16 =
3. 575 ÷ 25 =
4. 492 ÷ 12 =
5. 288 ÷ 24 =
6. 6095 ÷ 265 =
7. 4000 ÷ 125 =
8. 33 231 ÷ 627 =
9. 4283 ÷ 252 =
10. 37 985 ÷ 423 =

Fractions

A **fraction** may be defined as a part of a quantity, unit, or object. For example, if a number is divided into four equal parts, each part is one-fourth ($\frac{1}{4}$) of the whole number. The parts of a fraction are the **numerator** and the **denominator,** separated by a line indicating division.

In the illustration of Fig. 1-1 a rectangular block is cut into four equal parts. Each single part is $\frac{1}{4}$ of the total. Two of the parts make $\frac{1}{2}$ the total, and three of the parts make the fraction $\frac{3}{4}$ of the total.

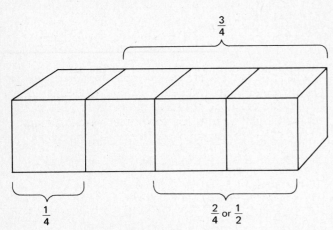

FIG. 1-1 Fractions of a whole.

A fraction may be considered an indication of a division. For example, the fraction $\frac{3}{4}$ indicates that the numerator 3 is to be divided by the denominator 4. One may wonder how a smaller number such as 3 can be divided by a larger number such as 4. It is actually a relatively simple matter to accomplish such a division when we apply it to a practical problem. Suppose we wish to divide 3 gallons (gal) of water into four equal parts. Since there are 4 quarts (qt) in a gallon, we know that there are 12 qt in 3 gal. We can then divide the 12 qt into four equal parts of 3 qt each. Three quarts is $\frac{3}{4}$ gal; thus we see that 3 divided by 4 is equal to $\frac{3}{4}$. The principal fact to remember concerning fractions is that a fraction indicates a division. The fraction $\frac{1}{2}$ means that 1 is to be divided by 2, or that the whole is to be cut in half.

A fraction whose numerator is less than its denominator is called a **proper fraction.** Its value is less than 1. If the numerator is greater than the denominator, the fraction is called an **improper fraction.**

A **mixed number** is a combination of a whole number and a fraction, such as $32\frac{2}{5}$ and $325\frac{23}{35}$, which mean $32 + \frac{2}{5}$ and $325 + \frac{23}{35}$.

Fractions may be changed in form without changing their values. If the numerator and the denominator of a fraction are both multiplied by the same number, the value of the fraction remains unchanged.

For example:

$$\frac{3 \times 3}{4 \times 3} = \frac{9}{12}$$

The value of $\frac{9}{12}$ is the same as $\frac{3}{4}$. In a similar manner, the value of a fraction is not changed if both the numerator and the denominator are divided by the same number.

Hence:

$$\frac{24 \div 12}{36 \div 12} = \frac{2}{3}$$

Thus we see that a large fraction may be simplified in some cases. This process is called **reducing the fraction.** To reduce a fraction to its lowest terms, we divide both the numerator and the denominator by the largest number that will go into each without leaving a remainder. This is accomplished as follows:

$$\frac{36 \div 4}{40 \div 4} = \frac{9}{10} \quad \text{and} \quad \frac{525 \div 25}{650 \div 25} = \frac{21}{26}$$

Addition and Subtraction of Fractions. In order to add or subtract fractions, it is necessary that the denominators of the fractions have equal values. For example, it is not possible to add $\frac{1}{3}$ to $\frac{2}{5}$ until the denominators of the fractions have been changed to equal values. Since 3 and 5 will both go evenly into 15, we can change $\frac{1}{3}$ to $\frac{5}{15}$ and $\frac{2}{5}$ to $\frac{6}{15}$. In this case, 15 is called the **lowest common denominator** (LCD) of the fractions being considered. It is now a simple matter to add the fractions.

$$\frac{5}{15} + \frac{6}{15} = \frac{11}{15}$$

We can see that the above addition makes sense because $5 + 6 = 11$. Since both the 5 and the 6 denote a specific number of fifteenths, we add them to obtain the total number of fifteenths.

The foregoing principle may be understood more easily if we apply it to a practical problem. Suppose we wish to add 3 gal and 5 qt and 1 pint (pt) of gasoline. The most logical method is to convert all quantities to pints. In 3 gal of gasoline there are 3×8 or 24 pt; in 5 qt there are 10 pt. Then we add 24 pt + 10 pt + 1 pt. Obviously the answer is 35 pt. If we wish to convert this quantity to gallons, we must divide the 35 by 8. We then find that we have 4 gal and 3 pt, or $4\frac{3}{8}$ gal.

To prepare fractions for adding or subtracting we proceed as follows:

1. Find the LCD.
2. Divide the LCD by each denominator.
3. Multiply the numerator and denominator of each fraction by the quotient obtained when the LCD was divided by the denominator.

Example. *Perform the following addition:*

$$\frac{3}{4} + \frac{7}{8} + \frac{5}{6}$$

The LCD is 24.

For the first fraction:

$$24 \div 4 = 6$$
$$\frac{3 \times 6}{4 \times 6} = \frac{18}{24}$$

For the second fraction:

$$24 \div 8 = 3$$
$$\frac{7 \times 3}{8 \times 3} = \frac{21}{24}$$

For the third fraction:

$$24 \div 6 = 4$$
$$\frac{5 \times 4}{6 \times 4} = \frac{20}{24}$$

Then:

$$\frac{18}{24} + \frac{21}{24} + \frac{20}{24} = \frac{59}{24} = 2\frac{11}{24}$$

Practice Problems (Addition)

1. $\frac{2}{3} + \frac{1}{2} + \frac{5}{6} =$

3. $\frac{1}{3} + \frac{3}{8} + \frac{4}{7} + \frac{3}{4} =$

2. $\frac{3}{4} + \frac{5}{8} + \frac{11}{16} =$

4. $\frac{5}{6} + \frac{4}{9} + \frac{7}{12} + \frac{1}{4} =$

Practice Problems (Subtraction)

1. $\frac{3}{4} - \frac{5}{9} =$ 3. $\frac{17}{30} - \frac{23}{45} =$ 5. $\frac{24}{71} - \frac{31}{211} =$

2. $\frac{9}{10} - \frac{7}{15} =$ 4. $\frac{28}{58} - \frac{18}{36} =$ 6. $\frac{17}{51} - \frac{8}{48} =$

Adding and Subtracting Mixed Numbers. When adding and subtracting mixed numbers, we must consider both the whole numbers and the fractions. To add $5\frac{3}{8} + 7\frac{2}{3}$, we may add 5 and 7 to obtain 12, and then we must add the fractions. We find that $\frac{3}{8} = \frac{9}{24}$ and $\frac{2}{3} = \frac{16}{24}$ and that $\frac{9}{24} + \frac{16}{24} = \frac{25}{24}$ or $1\frac{1}{24}$. Then $12 + 1\frac{1}{24} = 13\frac{1}{24}$, the total sum of the mixed numbers.

Subtraction of mixed numbers is accomplished by subtracting the whole numbers and then the fractions. For example, subtract $8\frac{2}{3}$ from $12\frac{3}{4}$.

$$\begin{aligned} 12\frac{3}{4} &= 12\frac{9}{12} \\ - 8\frac{2}{3} &= - 8\frac{8}{12} \\ \hline & \quad\; 4\frac{1}{12} \end{aligned}$$

If the fraction of the subtrahend is greater than the fraction of the minuend, it is necessary to borrow 1 from the whole number in the minuend to increase the fraction of the minuend. If we wish to subtract $5\frac{7}{8}$ from $9\frac{1}{3}$, we must increase the $\frac{1}{3}$ to a value greater than $\frac{7}{8}$. The LCD of the fractions is 24, and so $5\frac{7}{8}$ becomes $5\frac{21}{24}$ and $9\frac{1}{3}$ becomes $9\frac{8}{24}$. We must then borrow 1 from 9 and add the 1 to $\frac{8}{24}$. The minuend then becomes $8\frac{32}{24}$. The final form of the problem is then

$$\begin{aligned} 8\frac{32}{24} \\ - 5\frac{21}{24} \\ \hline 3\frac{11}{24} \end{aligned}$$

Multiplication of Fractions. Multiplication of fractions is accomplished by obtaining the product of the numerators and placing this product over the product

of the denominators. This result is then reduced to lowest terms. For example,

$$\frac{2}{5} \times \frac{1}{2} \times \frac{3}{4} = \frac{6}{40} = \frac{3}{20}$$

Where possible in the multiplication of fractions, **cancellation** is employed to simplify the fractions before final multiplication takes place.

$$\frac{\cancel{5}}{\cancel{8}} \times \frac{\cancel{2}}{\cancel{3}} \times \frac{\cancel{9}^{3}}{10} \times \frac{\cancel{4}}{\cancel{3}} = \frac{3}{10}$$

In the foregoing problem we have canceled all values except the 3 in the numerator and the 10 in the denominator. First we canceled the 5s, next we divided the 8 in the denominator by the 2 in the numerator. The 4 that was left in the denominator was then canceled by the 4 in the numerator. The 9 in the numerator was divided by the 3 in the denominator to leave a 3 in the numerator. The product, $\frac{3}{10}$, contains the product of the numerators and the product of the denominators reduced to their lowest terms. This may be proved as follows.

Multiplying numerators:

$$5 \times 2 \times 9 \times 4 = 360$$

Multiplying denominators:

$$8 \times 3 \times 10 \times 5 = 1200$$

Then:

$$\frac{360}{1200} = \frac{3}{10}$$

The reduction of the fraction was accomplished by dividing both the numerator and denominator by 120.

A problem involving a few more operations than that given previously is

$$\frac{\cancel{25}^{5}}{\cancel{8}} \times \frac{\cancel{36}^{\cancel{6}3}}{7} \times \frac{11}{\cancel{20}_{4}} \times \frac{9}{\cancel{44}_{4}} = \frac{135}{56} = 2\frac{23}{56}$$

In this problem it will be noted that the 25 in the numerator and the 20 in the denominator were both divided by 5 to obtain a 5 in the numerator and a 4 in the denominator.

Dividing Fractions. The division of fractions is simply accomplished by inverting the divisor and multiplying. Inverting a fraction means to turn it over; for example, if we invert $\frac{3}{4}$, it becomes $\frac{4}{3}$. It is also of interest to note that $\frac{4}{3}$ is the **reciprocal** of $\frac{3}{4}$. If we invert a whole number, we merely place a 1 above it. Hence, 3 becomes $\frac{1}{3}$ when it is inverted.

Divide: $\frac{5}{8}$ by $\frac{7}{15}$.

$$\frac{5}{8} \div \frac{7}{15} = \frac{5}{8} \times \frac{15}{7} = \frac{75}{56} = 1\frac{19}{56}$$

Multiplying and Dividing Mixed Numbers. Mixed numbers may be multiplied or divided by changing the mixed numbers to improper fractions and then proceeding as with fractions.

Multiply:

$$5\frac{7}{8} \text{ by } 3\frac{2}{3} \qquad 5\frac{7}{8} = \frac{47}{8} \quad \text{and} \quad 3\frac{2}{3} = \frac{11}{3}$$

Then:

$$\frac{47}{8} \times \frac{11}{3} = \frac{517}{24} = 21\frac{13}{24}$$

Divide:

$$9\frac{3}{4} \text{ by } 4\frac{2}{3} \qquad 9\frac{3}{4} = \frac{39}{4} \quad \text{and} \quad 4\frac{2}{3} = \frac{14}{3}$$

Then:

$$\frac{39}{4} \div \frac{14}{3} = \frac{39}{4} \times \frac{3}{14} = \frac{117}{56} = 2\frac{5}{56}$$

It must be pointed out that the procedures explained in the preceding sections do not represent the only possible methods for obtaining the same results; but those given are commonly used and will aid students in refreshing their ability to perform computations with fractions. The following problems will enable students to test their ability with the operations explained in this section.

Practice Problems

1. $\frac{3}{4} + \frac{5}{7} =$

2. $\frac{9}{16} + \frac{11}{24} =$

3. $\frac{7}{8} - \frac{3}{4} =$

4. $\frac{18}{24} - \frac{7}{9} =$

5. $7\frac{9}{16} + 12\frac{14}{24} =$

6. $\frac{25}{27} + 7\frac{15}{42} =$

7. $28\frac{31}{36} - 15\frac{24}{54} =$

8. $\frac{5}{6} \times \frac{12}{15} =$

9. $\frac{16}{25} \times \frac{30}{24} \times \frac{42}{70} =$

10. $\frac{8}{9} \div \frac{4}{27} =$

11. $34\frac{5}{6} \times 12\frac{3}{8} =$

12. $16\frac{3}{10} \div 6\frac{4}{15} =$

Decimals

Decimals, or decimal fractions, provide a means of performing mathematical operations without the necessity of using the time-consuming and complex methods of common fractions. A decimal fraction is a common fraction converted to tenths, hundredths, thousandths, etc. For example, if we convert the common fraction $\frac{3}{4}$ to a decimal, we find that it becomes 0.75. This is accomplished by dividing the numerator by the denominator:

```
     0.75
  4)3.00
    2 8
    ___
     20
     20
     __
```

Any fraction may be converted to a decimal by this same process. Let us assume that we wish to convert the fraction $\frac{28}{35}$ to a decimal.

```
      0.80
  35)28.00
     28.00
```

The decimal 0.80 is the same as 0.8 and may be read "eighty hundredths" or "eight tenths."

If it is desired to convert a fraction to a decimal when the denominator will not go evenly into the numerator, the decimal will be carried to the nearest tenth, hundredth, thousandth, or ten-thousandth according to the degree of accuracy required for the problem. For example, we may wish to convert the fraction $\frac{25}{33}$ to a decimal carried to the nearest ten-thousandth. We proceed as follows:

```
        0.7576
  33)25.0000
     23 1
     ____
      1 90
      1 65
      ____
       250
       231
       ___
       190
       198
```

Rounding Off Decimals. In the problem above, the answer would be alternately 7 and 5 indefinitely if we continued to carry the division onward. Instead we **round off** the answer to the degree of accuracy required. The accuracy required will be a function of the equipment being used. To calculate a sheet-metal layout to "one thousandth" (0.001) of an inch is not necessary if the scale being used can only measure in tenths (0.1) of an inch. To round off to tenths of an inch the calculation would be carried to two decimal places. If the last, or second, digit is less than 5 it would be dropped. If it is 5 or more, 1 is added to the preceding number. To round off the above example to one decimal place the calculation would be 0.75. The second digit is 5, so 1 is added to the first digit resulting in a rounded-off answer of 0.8. If an accuracy of one-hundredth (0.01), or two decimal places, were required, the calculation out to three places would be 0.757. Since the third digit (7) is greater than 5, the rounded-off answer is 0.76. If the calculation resulted in a figure of 0.6336, the third digit being less than 5 would be dropped. The answer, accurate to two decimal places, would be 0.63.

Multiplication of Decimals. The multiplication of decimals is performed in the same manner as the multiplication of whole numbers except that we must use care in placing the decimal point in the product. Let us assume that we wish to multiply 37.5 by 24.2.

```
      37.5
    ×24.2
      750
     1500
     750
    _____
    907.50
```

Having completed the multiplication of numbers containing decimals, we count the number of decimal places in the multiplicand and multiplier and point off this many places in the product. In the foregoing example there is one decimal place in the multiplicand and one in the multiplier. We therefore point off two places in the answer. After the answer is obtained, we may drop any zeros at the right-hand end of the answer. The answer of the foregoing problem would then be 907.5 and be read "nine hundred seven and five tenths." If the zero were left in the answer, the decimal portion would be read "fifty hundredths."

It is often necessary to multiply decimals in which there are no whole numbers. For example,

$$
\begin{array}{r}
0.056 \quad \text{(fifty-six thousandths)} \\
\times 0.325 \quad \text{(three hundred twenty-five thousandths)} \\
\hline
0280 \\
0112 \\
0168 \\
0\,000 \\
\hline
0.018200 \quad \text{(one hundred eighty-two ten-thousandths)}
\end{array}
$$

Since there is a total of six decimal places in the multiplicand and multiplier, we must point off six places in the product. This makes the answer 0.018 200, which would be read "eighteen thousand two hundred millionths." In order to simplify the answer, we drop the two zeros at the right and read the answer "one hundred eighty-two ten-thousandths."

Note: It is customary with some writers to omit the zero ahead of a decimal point not preceded by a whole number. Thus 0.04 would be written .04. In either case the value is the same and the decimal is read "four hundredths." It is not necessary to use the zero before the decimal point, but it may aid in preventing mistakes.

Addition of Decimals. The addition of decimals is a simple matter provided the decimals are properly placed. In adding a column of numbers with decimals, the decimal points should be kept in line in a column as shown below.

$$
\begin{array}{r}
23.065 \\
2.5 \\
354.2 \\
0.637 \\
\hline
380.402
\end{array}
$$

In the foregoing problem observe that the digits in the first column to the right of the decimal points add up to more than 10. When this occurs, we carry the 1 over into the units column. If the column should add up to 20 or more, we would carry the 2 over to the units column.

Subtraction of Decimals. Subtraction of decimals is almost as easy as subtracting whole numbers. It is necessary, however, that we use care to avoid mistakes in the placing of decimal points. This is illustrated in the following problems:

$$
\begin{array}{r}
652.25 \\
-\ 28.64 \\
\hline
623.61
\end{array}
\qquad
\begin{array}{r}
2568.2300 \\
-\ 376.4532 \\
\hline
2191.7768
\end{array}
\qquad
\begin{array}{r}
320.000 \\
-215.365 \\
\hline
104.635
\end{array}
$$

Observe in the foregoing problems that where there are fewer decimal places in the minuend than in the subtrahend, we add zeros to fill the spaces. This aids in avoiding mistakes which could otherwise occur. The addition of the zeros does not affect the value of the decimals.

Division of Decimals. The division of decimals requires much more care than the addition, subtraction, or multiplication. This is because it is easier to misplace the decimal point in the quotient. The principal rule to remember in dividing decimals is to place the decimal point of the quotient directly above the decimal point of the dividend. This is illustrated in the following problems:

$$
\begin{array}{r}
3.32 \\
28{\overline{)92.96}} \\
84 \\
\hline
8\,9 \\
8\,4 \\
\hline
56 \\
56 \\
\hline
\end{array}
\qquad
\begin{array}{r}
0.476 \\
34{\overline{)16.184}} \\
13\,6 \\
\hline
2\,58 \\
2\,38 \\
\hline
204 \\
204 \\
\hline
\end{array}
\qquad
\begin{array}{r}
0.0002268 \\
435{\overline{)0.0986500}} \\
870 \\
\hline
1165 \\
870 \\
\hline
2950 \\
2610 \\
\hline
3400 \\
3480 \\
\hline
\end{array}
$$

In the third problem illustrated above the division does not come out evenly, and so the answer is "rounded off" with an 8 to provide an accuracy to the nearest ten-millionth.

When the divisor contains decimals, we move the decimal point to the right until the divisor is a whole number. We then move the decimal in the dividend the same number of points to the right. This is equivalent to multiplying both the dividend and the divisor by the same number, and so the quotient remains the same.

Divide: 34.026 by 4.538.

$$
4.538{\overline{)34.026}}
$$

$$
\begin{array}{r}
7.49\ + \\
4538{\overline{)34026.00}} \\
31766 \\
\hline
22600 \\
18152 \\
\hline
44480 \\
40842 \\
\hline
3538
\end{array}
$$

Divide: 20.583 by 3.06

$$
3.06{\overline{)20.58\ 3}}
$$

```
              6.726  +
    306)2058.300
        1836
         222 3
         214 2
           8 10
           6 12
           1 980
           1 836
             144
```

Divide: 23.42 by 4.3867

```
    4.3867)23.4200
```

```
              5.34  —
   43867)234200.00
         219335
          14865 0
          13160 1
           1704 90
           1754 68
```

The small + and − signs placed after the quotients in the above examples indicate that a small amount is to be added or subtracted if the number is to be made exact; that is, the exact answer is a little more or a little less than the answer shown.

Converting Decimals to Common Fractions. It has been stated that a decimal is a fraction, and of course this is true. A decimal fraction is a fraction that has 10, 100, 1000, etc., for the denominator. The decimal 0.34 is read "34 hundredths" and may be shown as $\frac{34}{100}$. Also, the decimal fraction 0.005 may be written as $\frac{5}{1000}$. To convert a decimal fraction to a common fraction, we merely write it in the fraction form and then reduce it to its lowest terms by dividing the numerator and denominator by the same number. To convert 0.325 to a common fraction, we write $\frac{325}{1000}$ and then divide the numerator and denominator by 25:

$$\frac{325 \div 25}{1000 \div 25} = \frac{13}{40}$$

To convert 0.625 to a fraction, we divide as follows:

$$\frac{625 \div 125}{1000 \div 125} = \frac{5}{8}$$

It is obvious that many decimals cannot be converted to small common fractions because the numerator and denominator may not have common factors. However, it may be possible to arrive at an approximate fraction which is within the accuracy limits required. For example, 0.3342 may be converted to approximately $\frac{1}{3}$.

Practice Problems

1. 4.025 + 340 =	*7.* 38.2 × 33.02 =
2. 567.2 + 0.056 =	*8.* 324.11 × 0.08 =
3. 0.0067 + 0.1267 =	*9.* 1.065 × 0.005 =
4. 236 − 0.98 =	*10.* 8.56 ÷ 2.34 =
5. 24.2 − 5.965 =	*11.* 57.2 ÷ 0.05 =
6. 350.01 − 45.004 =	*12.* 2.357 ÷ 54.88 =

Convert to decimals:

13. $\frac{5}{8}$ *14.* $\frac{9}{10}$ *15.* $\frac{12}{66}$

Convert to fractions:

16. 0.50   *17.* 0.75   *18.* 0.275

Percentage

Percentage is the term used to indicate a certain number of hundredths of a whole. The expression 5% means $\frac{5}{100}$ or 0.05. To find a certain percentage of a number we multiply the number by the number of percent and then move the decimal point two places to the left. For example, to find 6% of 325 we multiply 325 by 6 to obtain 1950, and then we move the decimal two places to the left and find the answer 19.50, or 19.5. We could just as easily multiply by 0.06 to obtain the same answer.

Certain percentages are equal to commonly used fractions, and it is well to be familiar with these: 25% = $\frac{1}{4}$, 50% = $\frac{1}{2}$, 75% = $\frac{3}{4}$, $12\frac{1}{2}$% = $\frac{1}{8}$, and $33\frac{1}{3}$% = $\frac{1}{3}$. Familiarity with these fractions and their equivalent percentages is helpful in many computations.

If we wish to find what percent one number is of another, we divide the first number by the second. For example, 26 is what percent of 65?

```
                    0.40
    Divide:    65)26.00
                   26 0
```

Since we change a decimal to a percentage by moving the decimal point two places to the right, 0.04 becomes 40%. Thus 26 is 40% of 65.

Practice Problems

Find the following percentages:

1. 30% of 65

2. 25% of 400

3. 88% of 990

4. 125% of 8

5. What percent of 25 is 5?

6. What percent of 320 is 96?

7. What percent of 452 is 63.28?

8. 60 is 40% of what number?

9. 91 is 28% of what number?

10. 2084.58 is 37% of what number?

Ratio and Proportion

A **ratio** is the numerical relation between two quantities. If one man has two airplanes and another has three airplanes, the ratio of their airplane ownership is 2 to 3. This may also be expressed as 2/3 or $\frac{2}{3}$. Thus we see that a ratio is actually a fraction, and it may also be used mathematically as a fraction. Another method for expressing a ratio is 2:3.

A ratio may be reduced to lowest terms in the same manner as a fraction. For example, the ratio 24:36 may be reduced to 2:3 by dividing each term of the ratio by 12. If a certain store has 60 customers on Friday and 80 on Saturday, the ratio is 60:80 or 3:4.

Proportion expresses equality between two ratios. For example, 4:5::12:15. This may also be expressed 4:5 = 12:15 or $\frac{4}{5} = \frac{12}{15}$.

In a proportion problem the outer numbers, such as 4 and 15 above, are called the **extremes** and the two inside numbers (5 and 12) are called the **means. In any ratio, the product of the means is equal to the product of the extremes.** We may show this from the preceding problem.

$$5 \times 12 = 4 \times 15 = 60$$

We may use the above rule to find an unknown term in a proportion.

$$6:16 = 9:?$$

Using x to denote the unknown quantity, we can say

$$6 \times x = 16 \times 9 \quad \text{or} \quad 6x = 144$$

Then:

$$\frac{6x}{6} = \frac{144}{6} \quad \text{or} \quad x = 24$$

We can prove the foregoing answer by using it in the original proportion

$$6:16 = 9:24 \quad \text{or} \quad \tfrac{6}{16} = \tfrac{9}{24} = \tfrac{3}{8}$$

Practice Problems

Find the unknown quantities:

1. 2:4 = 6:?   *3.* 3:6 = 5:?   *5.* 10:6 = ?:18

2. 7:9 = 21:?   *4.* 2:? = 7:35   *6.* ?:12 = 5:30

Powers and Roots

A **power** of a number represents the number multiplied by itself a certain number of times. For example, 5 × 5 = 25; hence the **second power** of 5 is 25. If we multiply 5 × 5 × 5 to obtain 125, we have found the **third power** of 5. The third power of 5 is indicated thus: 5^3. It is read "5 cubed" or "5 to the third power."

The second power of a number is called the **square** of a number. This terminology is derived from the fact that the area of a square is equal to the length of one side multiplied by itself. The term **cube** is derived in a similar manner because the volume of a cube is equal to the length of one edge raised to the third power. Any power of any number may be found merely by continuing to multiply it by itself the indicated number of times. For example, 2^6 is equal to $2 \times 2 \times 2 \times 2 \times 2 \times 2 = 64$.

The small index number placed above and to the right of a number to indicate the power of the number is called an **exponent.** The number to be raised to a power is called the **base.** In the expression 25^4, the small number 4 is the exponent and the number 25 is the base. If we multiply 25 by itself the number of times indicated by the exponent 4, we find that $25^4 = 390\ 625$.

A **factor** of a number is another number which will divide evenly into the first number. For example, 3 is a factor of 12. Other factors of 12 are 2, 4, and 6, because each of these numbers will divide evenly into 12. A **root** of a number is a factor which when multiplied by itself a certain number of times will produce the number. For example, 2 is a root of 4 because it will give a product of 4 when multiplied by itself. It is also a root of 8 because $2 \times 2 \times 2 = 8$. A **square root** is the root of a number which when multiplied by itself once will produce the number. For example, 3 is the square root of 9 because $3 \times 3 = 9$. It is the **cube root** of 27 because $3 \times 3 \times 3 = 27$. A root which must be multiplied by itself three times to produce a certain number is the fourth root of that number. Hence 3 is the fourth root of 81 because $3 \times 3 \times 3 \times 3 = 81$.

When the square root of a number is indicated, we place the number under the **radical sign,** thus: $\sqrt{64}$. If a larger root is to be extracted, we place the index of the root in the radical sign, thus: $\sqrt[3]{27}$. This indicates that the value expressed is the cube root of 27, or 3.

Practice Problems

Find the powers and roots indicated:

1. $6^2 =$ *5.* $265^2 =$ *8.* $\sqrt{64} =$

2. $25^2 =$ *6.* $\sqrt{25} =$ *9.* $\sqrt[3]{125} =$

3. $12^3 =$ *7.* $\sqrt{100} =$ *10.* $\sqrt[3]{1000} =$

4. $5^4 =$

Many formulas in technical work require extraction of square roots. The development of the electronic calculator has greatly simplified this task. The speed and accuracy of the calculator has made learning the procedure for manually extracting a square root questionable. The procedure is shown in Fig. 1-2 for those who are interested. Many tables, such as Table 1.1, that contain various powers and roots are available. Tables of this type can be very useful when performing manual calculations.

Problem: *Extract the square root of 104 976.*

1. Place the number under the radical sign and separate it into periods of two digits each starting from the right of the number.

$$\sqrt{10'49'76}$$

2. Determine the nearest perfect square smaller than the first period on the left, and subtract this square from the first period. Place the root of the square above the first period. Bring down the next period to form the new dividend 149.

$$
\begin{array}{r}
3 \\
\sqrt{10'49'76} \\
9 \\
\hline
60)\overline{1\ 49}
\end{array}
$$

3. Multiply the root 3 by 20 and place the product to the left of the new dividend. The product 60 is the **trial divisor**. Determine how many times the trial divisor will go into the dividend 149. In this case 60 will go into 149 two times. Add 2 to the trial divisor to make 62, which is the complete divisor. Place 2 above the second period and then multiply the complete divisor by 2. Place the product 124 under the dividend 149 and subtract. Bring down the next period to make the new dividend.

$$
\begin{array}{r}
3\ 2 \\
\sqrt{10'49'76} \\
9 \\
\hline
60)\overline{1\ 49} \\
2 \\
\hline
62\quad 1\ 24 \\
25\ 76
\end{array}
$$

4. Multiply the partial answer 32 by 20 to obtain the new trial divisor 640. Determine how many times 640 will go into 2576. Inspection indicates that it will go into 640 four times. Add the 4 to 640 to obtain 644, which is the complete divisor. Place the 4 above the third period. Then multiply the complete divisor 644 by 4 to obtain 2576. This product is equal to the dividend; hence the computation is complete.

$$
\begin{array}{r}
3\ \ 2\ \ 4 \\
\sqrt{10'49'76} \\
9 \\
\hline
60)\overline{1\ 49} \\
2 \\
\hline
62\quad 1\ 24 \\
640)\overline{25\ 76} \\
4 \\
\hline
644 \\
25\ 76
\end{array}
$$

FIG. 1-2 Procedure for extracting a square root.

Practice Problems

Extract the square root of the following:

1. 15 129
2. 124 6641
3. 1314.0625
4. 760.89
5. 233.652

Powers and Roots of Fractions. When a fraction is to be raised to a certain power, the numerator is mul-

tiplied by the numerator and the denominator is multiplied by the denominator. For example, if we wish to find the third power (cube) of $\frac{3}{4}$, we multiply $3 \times 3 \times 3$ for the new numerator and $4 \times 4 \times 4$ for the new denominator. Since $3 \times 3 \times 3 = 27$ and $4 \times 4 \times 4 = 64$, the cube of $\frac{3}{4}$ is $\frac{27}{64}$.

To extract a particular root of a fraction, we must extract the roots of both the numerator and the denominator. For example, the square root of $\frac{4}{9} = \frac{2}{3}$. This is because the square root of 4 is 2 and the square root of 9 is 3.

Scientific Notation

Scientific notation is the process of using powers of 10 to simplify mathematical expressions and computations. Figure 1-3 shows the values of 10 for various powers. By using powers of 10 to express very large numbers or very long decimals, the amount of computation necessary for multiplication, division, and extracting roots will be reduced. Many calculators and computer programs use scientific notation to display large numbers or long decimals.

With scientific notation long numbers can be simplified. The number to be simplified is divided by a power of 10.

Example. Express 2 600 000 in scientific notation with one digit to the left of the decimal point. *Note:* $10^6 = 1\ 000\ 000$.

Solution

$$\frac{2\ 600\ 000}{1\ 000\ 000} = 2.6$$

Thus

$$2\ 600\ 000 = 2.6 \times 1\ 000\ 000$$

or

$$2\ 600\ 000 = 2.6 \times 10^6$$

Other examples are:

$$37\ 542\ 000 = 3.7542 \times 10^7$$
$$123\ 000 = 1.23 \times 10^5$$

The exponent of 10 can easily be determined for scientific notation by counting the number of places that the decimal point is moved to the left. In the first example (37 542 000) above the decimal has been moved seven places to the left. In the second example it has been moved five places, giving 10 the exponent of 5.

For numbers less than 1.0 the exponent will be neg-

$10^0 = 1$	$10^{-1} = 0.1$
$10^1 = 10$	$10^{-2} = 0.01$
$10^2 = 100$	$10^{-3} = 0.001$
$10^3 = 1000$	$10^{-4} = 0.0001$
$10^4 = 10000$	$10^{-5} = 0.00001$
$10^5 = 100000$	$10^{-6} = 0.000001$
$10^6 = 1000000$	$10^{-7} = 0.0000001$
$10^7 = 10000000$	

FIG. 1-3 Power of 10.

TABLE 1-1 Tables of Squares, Cubes, Square Roots, and Cube Roots

No. n	Sq. n^2	Cube n^3	Square root $\sqrt{n}$	Cube root $\sqrt[3]{n}$	No. n	n^2	n^3	$\sqrt{n}$	$\sqrt[3]{n}$
1	1	1	1.000	1.000	51	2 601	132 651	7.141	3.708
2	4	8	1.414	1.259	52	2 704	140 608	7.211	3.732
3	9	27	1.732	1.442	53	2 809	148 877	7.280	3.756
4	16	64	2.000	1.587	54	2 916	157 464	7.348	3.779
5	25	125	2.236	1.710	55	3 025	166 375	7.416	3.803
6	36	216	2.449	1.817	56	3 136	175 616	7.483	3.825
7	49	343	2.645	1.913	57	3 249	185 193	7.549	3.848
8	64	512	2.828	2.000	58	3 364	195 112	7.615	3.870
9	81	729	3.000	2.080	59	3 481	205 379	7.681	3.893
10	100	1000	3.162	2.154	60	3 600	216 000	7.746	3.914
11	121	1331	3.316	2.224	61	3 721	226 981	7.810	3.936
12	144	1728	3.464	2.289	62	3 844	238 328	7.874	3.957
13	169	2197	3.605	2.351	63	3 969	250 047	7.937	3.979
14	196	2744	3.741	2.410	64	4 096	262 144	8.000	4.000
15	225	3375	3.873	2.466	65	4 225	274 625	8.062	4.020
16	256	4096	4.000	2.519	66	4 356	287 496	8.124	4.041
17	289	4913	4.123	2.571	67	4 489	300 763	8.185	4.061
18	324	5832	4.242	2.620	68	4 624	314 432	8.246	4.081
19	361	6859	4.358	2.668	69	4 761	328 509	8.306	4.101
20	400	8000	4.472	2.714	70	4 900	343 000	8.366	4.121
21	441	9261	4.582	2.758	71	5 041	357 911	8.426	4.140
22	484	1 0648	4.690	2.802	72	5 184	373 248	8.485	4.160
23	529	1 2167	4.795	2.843	73	5 329	389 017	8.544	4.179
24	576	1 3824	4.899	2.884	74	5 476	405 224	8.602	4.198
25	625	1 5625	5.000	2.924	75	5 625	421 875	8.660	4.217
26	676	1 7576	5.099	2.962	76	5 776	438 976	8.717	4.235
27	729	1 9683	5.196	3.000	77	5 929	456 533	8.775	4.254
28	784	2 1952	5.291	3.036	78	6 084	474 552	8.831	4.272
29	841	2 4389	5.385	3.072	79	6 241	493 039	8.888	4.290
30	900	2 7000	5.477	3.107	80	6 400	512 000	8.944	4.308
31	961	2 9791	5.567	3.141	81	6 561	531 441	9.000	4.326
32	1024	3 2768	5.656	3.174	82	6 724	551 368	9.055	4.344
33	1089	3 5937	5.744	3.207	83	6 889	571 787	9.110	4.362
34	1156	3 9304	5.831	3.239	84	7 056	592 704	9.165	4.379
35	1225	4 2875	5.916	3.271	85	7 225	614 125	9.219	4.396
36	1296	4 6656	6.000	3.301	86	7 396	636 056	9.273	4.414
37	1369	5 0653	6.082	3.332	87	7 569	658 503	9.327	4.431
38	1444	5 4872	6.164	3.362	88	7 744	681 472	9.380	4.448
39	1521	5 9319	6.245	3.391	89	7 921	704 969	9.434	4.464
40	1600	6 4000	6.324	3.420	90	8 100	729 000	9.486	4.481
41	1681	6 8921	6.403	3.448	91	8 281	753 571	9.539	4.497
42	1764	7 4088	6.480	3.476	92	8 464	778 688	9.591	4.514
43	1849	7 9507	6.557	3.503	93	8 649	804 357	9.643	4.530
44	1936	8 5184	6.633	3.530	94	8 836	830 584	9.695	4.546
45	2025	9 1125	6.708	3.556	95	9 025	857 375	9.746	4.562
46	2116	9 7336	6.782	3.583	96	9 216	884 736	9.798	4.578
47	2209	10 3823	6.855	3.608	97	9 409	912 673	9.848	4.594
48	2304	11 0592	6.928	3.634	98	9 604	941 192	9.899	4.610
49	2401	11 7649	7.000	3.659	99	9 801	970 299	9.949	4.626
50	2500	12 5000	7.071	3.684	100	10 000	1 000 000	10.000	4.641

ative. It can be determined by counting the number of places the exponent has moved to the right.

Examples

$$0.0372 = 3.72 \times 10^{-2}$$
$$0.000\ 045\ 67 = 4.567 \times 10^{-5}$$

To change from scientific notation to the actual number the decimal point will be moved the number of places indicated by the exponent. It will be moved to the right if the exponent is positive and to the left if it is negative.

Multiplication and division are simplified with scientific notation. In multiplying two numbers, as shown in Fig. 1-4, the two base numbers are multiplied. The exponent for the answer is determined by adding the two exponents.

Two numbers stated in scientific notation may be divided as shown in Fig. 1-5. After dividing the base numbers, the exponent is assigned a value obtained by subtracting the exponent of the divisor from that of the dividend.

The square root of a number in scientific notation can be found by finding the square root of the base number and dividing the exponent by 2 (see Fig. 1-6).

Problem: *Multiply 12 000 000 by 173 000.*

1. Convert to scientific notation.

$$(1.2 \times 10^7) \times (1.73 \times 10^5)$$

2. Multiply both base numbers.

$$1.2 \times 1.73 = 2.076$$

3. Add the exponents.

$$7 + 5 = 12$$

4. Answer.

$$2.076 \times 10^{12} \text{ or } 2\ 076\ 000\ 000\ 000$$

FIG. 1-4 Multiplication by scientific notation.

Practice Problems. *Express the following in scientific notation with no more than one digit to the left of the decimal point.*

1. 123 450 000   *3.* 173   *5.* 0.312 508

2. 325 000   *4.* 0.000 314 67   *6.* 0.012 24

Write the number indicated.

7. 8.42×10^9 *9.* 3.674×10^{-4}

8. 1.226×10^2 *10.* 1.344×10^{-1}

● ALGEBRA

Introduction

Algebra may be defined as the branch of mathematics which uses positive and negative quantities, letters,

Problem: *Divide 12 000 000 by 173 000.*

1. Convert to scientific notation.

$$\frac{1.2 \times 10^7}{1.73 \times 10^5}$$

2. Divide both base numbers.

$$\frac{1.2}{1.73} = 0.694$$

3. Subtract the exponents.

$$7 - 5 = 2$$

4. Answer.

$$0.694 \times 10^2 \text{ or } 69.4$$

FIG. 1-5 Division with scientific notation.

Problem: *Find the square root of 4.0×10^6.*

$$\sqrt{4.0 \times 10^6}$$

1. Square root of $4 = 2$
2. Square root of $10^6 = 10^3$
3. $\sqrt{4.0 \times 10^6} = 2.0 \times 10^3$

FIG. 1-6 Finding a square root with scientific notation.

and other symbols to express and analyze relationships among units of quantitative data. The process of algebra enables us to make calculations and arrive at solutions which would be difficult or impossible through normal arithmetic methods. All mathematical systems beyond arithmetic employ the methods of algebra for computation.

Aviation maintenance technicians use many algebraic formulas and expressions daily. In many cases these operations have become so routine that many do not realize that algebra is being used. The formula for the area of a rectangle, $A = l \times w$, or $A = lw$ is an algebraic expression (A = area, l = length, and w = width). The formula for finding the force on a hydraulic piston may be expressed as $F = P/A$, where F is force in pounds, A is area of the piston in square inches, and P is the pressure of the fluid in pounds per square inch (psi). In computing weight and balance of an aircraft, the technician is not only working with algebraic formulas but with positive and negative quantities. Most computations involving these operations have been simplified to where the problem can be solved by placing the proper numbers into the formula. However, a knowledge of algebra is essential for the technician to understand what is happening in the procedure or system.

Equations

An equation is a mathematical expression of equality. For example, $2 + 6 = 8$ is a simple equation. In the general terms of algebra this equation would be $a + b = c$. If the value of any two of the symbols is known, the other one can be determined. If $a = 2$ and $b = 6$ in the equation, then we know that $c = 8$ because $2 + 6 = 8$.

If we ask the question "What number added to 6 will produce 10?" we can express the question in algebraic terms thus: $6 + x = 10$. To find the value of x, we must subtract 6 from 10, and so we rearrange the equation to $x = 10 - 6$. Note that we changed the sign of the 6 when we **transposed** it (moved it to the opposite side of the equals sign). We complete the solution and state the simplified equation as $x = 4$.

Positive and Negative Numbers

In algebra we use the same signs that are used in arithmetic; however, in algebra the signs sometimes have a greater significance than they do in arithmetic. All terms in algebra must have either a positive or a negative value. Terms having a positive value are preceded by a plus sign (+) or by no sign at all. Negative terms are preceded by a minus sign (−). A positive number or expression has a value greater than zero, and a negative number has a value less than zero. This may be understood by considering temperature. If we were told that the temperature was 10°, we would not know for sure what was meant unless we knew whether it was 10° above zero or 10° below zero. If the temperature were above zero, it could be shown as +10°, and if below zero, it could be shown as −10°.

Algebraic Addition

Algebraic addition is the process of combining terms to find the actual value of the terms. The sum $5 + 6 + 8 = 19$ is algebraic addition as well as arithmetic addition. The sum $-5 - 6 + 4 = -7$ is also algebraic addition, but we do not use this method in arithmetic. **To add the terms in an algebraic expression when there are both negative and positive quantities we combine the terms with the same sign and then subtract the smaller value from the larger and give the answer the sign of the larger.** To add $8 - 9 - 4 + 6 + 7 - 3$, we combine the 8, 6, and 7 to obtain +21 and then combine −9, −4, and −3 to obtain −16. We then subtract the 16 from 21 to obtain +5. If the negative quantity were greater than the positive quantity, the answer would be negative.

When we are combining numbers or terms containing letters or other symbols, we cannot add those terms having different letters or symbols. For example, we cannot add $3b$ and $5c$. The indicated addition of these terms would merely be $3b + 5c$. We can add $3a$ and $5a$ to obtain $8a$, in which case we would show the expression as $3a + 5a = 8a$.

In the term $3a$ the figure 3 is called the **coefficient** of a. Thus we see that a coefficient is a multiplier. Remember also that a number placed above and to the right of another number or symbol to show a power is called an **exponent**. For example, in the term x^2 the figure 2 is the exponent of x, and the term is read "x square."

In algebra the letters used in place of numbers are called **literal** numbers. Thus, the equation $x + y = z$ contains all literal numbers. When we wish to add terms containing different literal numbers, we combine those terms having the same letter or symbol. To solve or simplify the expression $4a + 5b - 2a - 6c + 9b - 3a + 8c - 3c - 4b + 3c$, we may proceed as follows:

$$
\begin{array}{rrr}
+4a & +5b & -6c \\
-2a & +9b & +8c \\
-3a & -4b & -3c \\
    &   & +3c \\
\hline
-a & +10b & +2c
\end{array}
$$

Then:

$$4a + 5b - 2a - 6c + 9b - 3a + 8c$$
$$-3c - 4b + 3c = -a + 10b + 2c$$

Algebraic Subtraction

The rule for subtraction in algebra is **change the sign of the subtrahend and add.**

If we wish to subtract 4 from 10, we change the sign of the 4 to minus and add. We then have $10 - 4 = 6$.

Subtract:

$$-2x + 3y + 4z \qquad \text{from} \qquad 6x + 5y - 8z.$$

$$
\begin{array}{l}
\begin{array}{r}
6x + 5y - 8z \\
-2x + 3y + 4z \\
\hline
\end{array}
\quad
\begin{array}{l}
\textit{Change signs of} \\
\textit{subtrahend:}
\end{array}
\quad
\begin{array}{r}
6x + 5y - 8z \\
2x - 3y - 4z \\
\hline
8x + 2y - 12z
\end{array}
\end{array}
$$

Use of Parentheses

Parentheses are used in algebra to indicate that two or more terms are to be considered as a single term. For example, $3 \times (5 + 2)$ means that the 5 and the 2 are both to be multiplied by 3. Also, the 5 and 2 can be added and then multiplied by 3, and the answer will be the same.

$$3 \times (5 + 2) = 3 \times 5 + 3 \times 2 = 15 + 6 = 21$$
$$3 \times (5 + 2) = 3 \times 7 = 21$$

If the parentheses were not used in the above example, the solution would be

$$3 \times 5 + 2 = 15 + 2 = 17$$

From this we observe that parentheses cannot be ignored in the solving of an algebraic problem.

In an expression where no multiplication or division is involved and parentheses are used, careful attention must be paid to the signs of the various terms. In the expression $(3a + 7b - 6c) + (4a - 3b - 2c)$, the parentheses actually have no effect and the expression could be written $3a + 7b - 6c + 4a - 3b - 2c$. If, however, a minus sign precedes the term enclosed by parentheses, then the signs of the quantities in the parentheses must be changed when the parentheses are removed:

$$(3a + 7b - 6c) - (4a - 3b - 2c)$$
$$= 3a + 7b - 6c - 4a + 3b + 2c$$

The foregoing expression means that the quantity $4a - 3b - 2c$ is to be subtracted from the quantity $3a + 7b - 6c$. From the rule for subtraction we know that the sign of the subtrahend must be changed and then the terms added.

$$
\begin{array}{r}
3a + 7b - 6c \\
-4a + 3b + 2c \\
\hline
-a + 10b - 4c
\end{array}
$$

Brackets [ ] are also used to group terms which are to be considered as one term. Usually the brackets are used only when parentheses have already been used inside the bracketed expression.

The following expression illustrates the use of brackets:

$$9 + [7a - (3b + 8x) - 2y + 4z] - 2c$$

The most common use of parentheses is to indicate multiplication of terms. For example, $(2x + 3y)(4x - 7y)$ indicates that the quantity $2x + 3y$ is to be multiplied by the quantity $4x - 7y$. The expression $5a(x + y)$ means that the quantity $x + y$ is to be multiplied by $5a$.

Multiplication

In order to explain multiplication clearly, certain arrangements of algebraic terms not previously defined must be discussed. These are **monomials, binomials, and polynomials.** A monomial is an expression containing only one term such as x, ab, $2z$, xy^2m, $2x3y$, and a^2b^3y. A binomial is an expression containing two terms connected by a minus $(-)$ or plus $(+)$ sign, as $a + b$, $2x + 3y$, $abc + xyz$, and $4y^2 - 3z$. *Polynomial* is a general term describing expressions containing two or more algebraic terms.

In the multiplication of algebraic terms, monomials, binomials, and other polynomials can be multiplied by any other expression regardless of whether it is a monomial, binomial, or other polynomial. Fractional terms and expressions can be multiplied by any other term or expression.

In the algebraic multiplication of terms and expressions, the signs of each term or expression must be carefully noted and properly handled. The following rules apply:

1. When two terms of *like signs* are multiplied, the sign of the product is positive.

2. When two terms of *unlike signs* are multiplied, the product is negative.

These rules may be demonstrated as follows:

1. $2 \times 3 = 6$
2. $-2 \times 3 = -6$
3. $-2 \times -3 = 6$
4. $2x \times -3y = -6xy$
5. $-4x \times -6y = 24xy$
6. $-5(a + b) = -5a - 5b$
7. $-3a(2x - 4y) = -6ax + 12ay$

To multiply purely literal terms which are unlike, the terms are merely gathered together as a unit.

$$a \times b = ab \quad ab \times cd = abcd \quad aby \times cdx = abcdxy$$

To multiply literal terms by like terms, the power of the term is raised.

$$a \times a = a^2 \quad ab \times ab = a^2b^2$$
$$abx \times aby = a^2b^2xy \quad bc \times bc \times bc = b^3c^3$$
$$abc \times bcx \times cxy = ab^2c^3x^2y$$

Multiplication in algebra can be indicated in three ways.

For example:

$$ab = a \times b = a \cdot b = (a)(b)$$
$$xyz = x \times y \times z = x \cdot y \cdot z = (x)(y)(z)$$

The technician performing multiplication in algebra must be alert to observe what the indications are and to perform the computation accordingly. It must be noted whether there is **no sign,** a regular **times sign,** or a **dot** placed between the terms to be multiplied.

To multiply a binomial by a monomial, multiply each term of the binomial separately by the monomial.

$$a(b + c) = ab + ac$$
$$a(ab + xy) = a^2b + axy$$
$$-4a(2b - 3c) = -8ab + 12ac$$
$$2b(a + 3c) = 2ab + 6bc$$
$$3x^2(4xy - 2z) = 12x^3y - 6x^2z$$

Since the purpose of this chapter is to serve as a review or refresher, multiplication and division of binomials and polynomials are not covered. Those needing more information on algebra topics should consult an algebra textbook.

Division

Division in algebra may be considered the reverse of multiplication, just as in arithmetic. The division sign $\div$ is not usually employed, and division is indicated by making the dividend the numerator of a fraction, and the divisor the denominator of the fraction.

For example:

$$a \div b \quad \text{is usually written} \quad \frac{a}{b}$$

and

$$(2a + 5b) \div (x + y) \quad \text{is written} \quad \frac{2a + 5b}{x + y}$$

A simple division may be performed as follows:

$$\frac{4a + 6ab}{2a} = 2 + 3b$$

Note that the monomial divisor $2a$ was divided into both terms of the binomial dividend (numerator). In this example the divisor divided evenly into both terms of the dividend. If, however, the divisor will not divide evenly into both terms, a part of the quotient will have to be fractional.

$$\frac{3x + 2y}{x} = 3 + \frac{2y}{x}$$

Order of Operations

In solving an algebraic expression or equation, certain operations must be performed in proper sequence. Indicated multiplications and divisions must be completed before additions are made. This is demonstrated in the following equation:

$$7x + 3(x - y) - \frac{8x - 4}{2} = 2(x + 5) - 3y$$

The terms enclosed in parentheses are to be multiplied by the coefficients 3 and 2, respectively. Then $3(x - y)$ becomes $3x - 3y$, and $2(x + 5)$ becomes $2x + 10$. As a result of the division indicated $-(8x - 4)/2$ becomes $-(4x - 2)$. The original equation is then

$$7x + 3x - 3y - 4x + 2 = 2x + 10 - 3y$$

To solve for x, all x terms are transposed to the left side of the equation. Note that when a term is moved from one side of the equation to the other, the sign must be changed to maintain the equality.

Transposing:

$$7x + 3x - 4x - 2x = 10 - 2 - 3y + 3y$$

Combining:

$$4x = 8 \qquad x = 2$$

In the foregoing equation, the term

$$-\frac{8x - 4}{2}$$

must be treated as a single quantity. Therefore, when the division is made, the term is placed in parentheses to indicate that the negative sign applies to the complete term: $-(4x - 2)$. When the parentheses are removed, the $4x$ takes a negative sign and the 2 becomes positive. When the value of 2 is substituted for x, $-(4x - 2)$ becomes $-(8 - 2)$ or -6. It must be remembered that whenever a mathematical expression is enclosed in parentheses or brackets, it is treated as a single quantity. If it is preceded by a minus sign, all the terms within the parentheses or brackets must have their signs changed when the parentheses are removed. Note the following examples:

$$-(a + b + c) = -a - b - c$$
$$-(x - y + z) = -x + y - z$$

Solution of Problems

When an algebraic expression contains only one unknown quantity, expressed by a letter, it is comparatively simple to find the value of the unknown quantity. In the equation $5x + 2 - 3x = 14 - 4x$, we can easily find the value of x by transposing and combining. A rule to be remembered at this point is that **when a term or quantity is moved from one side of an equation to the opposite side, the sign of the term or quantity must be changed.** The solution of the above equation is as follows:

$$5x + 2 - 3x = 14 - 4x$$
$$5x - 3x + 4x = 14 - 2$$
$$6x = 12$$
$$x = 2$$

Note that in the above operation the sign of $-4x$ was changed and the sign of the $+2$ was changed. This was done because the $-4x$ and the $+2$ were **transposed** or moved from one side of the equation to the other. When the quantities were combined, $6x$ was found equal to 12. It is quite apparent then that x is equal to 2. This is also shown by dividing both sides of the equation by 6.

Algebra is particularly useful in solving certain problems which are more difficult to solve by arithmetic or which may not be solved by arithmetic. The examples given below show how some of the less difficult types may be solved by algebraic methods.

Example. One number is 3 times another number. The sum of the numbers is 48. If this is true, what are the numbers?

Solution

Let x = the smaller number.
Then $3x$ = the larger number.

$$x + 3x = 48$$
$$4x = 48$$
$$x = 12, \text{ the smaller number}$$
$$3x = 36, \text{ the larger number}$$

Example. One number increased by 5 is equal to one-half another number. The sum of the numbers is 55. What are the numbers?

Solution

Let x = the smaller number.
Then $2(x + 5)$ = the larger number.

$$x + 2(x + 5) = 55$$
$$x + 2x + 10 = 55$$
$$3x = 55 - 10 = 45$$
$$x = 15$$

The larger number is $2(x + 5) = 2(15 + 5) = 40$.

$$15 + 40 = 55$$

Example. A piston-type airplane leaves Los Angeles for New York at 10:00 A.M., flying at 350 miles per hour (mph). One hour later a jet airplane flying at 550 mph leaves Los Angeles for New York. At what time will the jet airplane overtake the piston-type airplane?

Solution

Let x = the time in hours traveled by the slower plane. Then $x - 1$ = the time in hours traveled by the jet airplane.

Example. A man has five times as many dimes as he has quarters. The total value of his dimes and quarters is $5.25. What number of each does he have?

Solution

Let x = the number of quarters. Then $5x$ = the number of dimes.

$$25x + 10(5x) = 525$$
$$25x + 50x = 525$$
$$75x = 525$$
$$x = 7, \text{ the number of quarters}$$
$$5x = 35, \text{ the number of dimes}$$

Note: x must be multiplied by 25 to find the total number of cents represented by the quarters. Since $x = 7$, we find that the money represented by quarters was 175 cents, or $1.75. Also, we find that $5x \times 10 = 35 \times 10$, or 350 cents. Then $1.75 + $3.50 = $5.25.

There are many types of problems which may be solved with methods similar to those shown for the foregoing problems. Skill in solving such problems may be gained by practice. The ability to interpret word problems and reduce them to equation form is the most important requirement.

● GEOMETRY

The word **geometry** is derived from *geo,* a greek word meaning earth, and *metria,* meaning measurement. Geometry can be said to literally mean the measurement of earth or land. In actuality, geometry deals with the measurement of areas, volumes, and distances.

The proof of geometrical propositions by means of axioms, postulates, or corollaries constitutes the major portion of most geometry courses. It is expected that the student using this text will have previously had, or is taking, additional math courses, including geometry. In this section, definitions and applications of geometry will be emphasized. Theory will be introduced only to the extent necessary to enhance application.

Definitions

The following terms are essential to the understanding of the application of geometrical principles. Figure 1-7 provides graphic examples of the terms.

Point. A point is that which has no length, breadth, or thickness but has only **position.**

Line. A line has no breadth or thickness but has length.

Surface. A surface has no thickness but has length and breadth.

Plane or plane surface. A plane or plane surface may be defined in several ways, as follows:

1. A surface such that a straight line that joins any two of its points lies wholly in that surface.

2. A two-dimensional extent of zero curvature.

3. A surface any intersection of which by a like surface is a straight line.

Solid. A solid, in the geometric sense, is that which has three dimensions, that is, length, breadth, and thickness.

Lines

Straight line. A line having the same direction throughout its length. If a portion of a straight line is placed so that both ends fall within the ends of the other part, the portion must lie wholly within the line.

Equal lines. Two lines are equal if, when placed one upon the other, their ends can be made to coincide.

Curved line. A line which continuously changes direction.

Broken line. A line consisting of a number of different straight lines.

Parallel lines. Lines in the same plane which can never intersect no matter how far they are extended.

Angles. The following terms are used to define and describe angles. Figure 1-8 provides illustrations to assist with understanding the definitions.

Angle. An angle is the opening between two straight lines drawn in different directions from the same point.

Acute angle. An angle which is less than a right angle.

Right angle. An angle which is one-fourth of a circle, that is, 90°.

Obtuse angle. An angle of more than 90°.

Straight angle. An angle whose sides form a straight line, that is, an angle of 180°.

Bisector. A bisector is a point, line, or surface which divides a magnitude into two equal parts.

Adjacent angles. Two angles having a common side and the same vertex.

Vertex of an angle. The common point from which the two sides of an angle proceed.

Vertical angles. Two angles with the same vertex and with sides that are prolongations of the sides of each other.

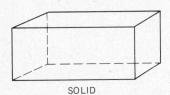
SOLID

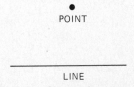
● POINT

LINE

SURFACE

FIG. 1-7 Geometric terms.

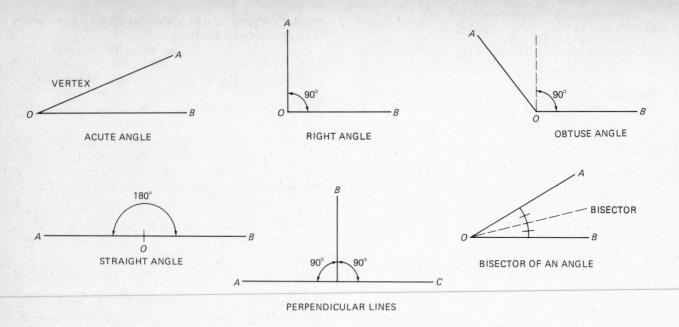

FIG. 1-8 Angles.

Perpendicular line. A straight line which makes a 90° angle with another straight line.

A common practice is to identify angles by upper-case letters. Most of the angles shown in Fig. 1-8 can be identified by a three-letter combination as either angle *AOB* or *BOA*. The center letter of the combination is the one located at the vertex. Three different angles can be identified in the illustration of the adjacent angles, *AOB*, *BOC*, and *AOC*. By using the three-letter combination, the specific angle will be clearly identified.

Shapes

Circles. A **circle** is a closed curve, all portions of which are in the same plane and equidistant from the same point (see Fig. 1-9.). The **diameter of** the circle is the length of a straight line passing through the center of a circle and limited at each end by the circle. The **radius** of the circle is a straight line from the center of the circle to the circle perimeter. The radius is equal to one-half the diameter. An **arc** is any portion of the circle. A major arc is one of more than 180° while a minor arc is less than 180°. A **semicircle** is an arc of 180°. A **sector** is the area within a circle bounded by two radii and the arc connecting the two radii. A **quadrant** is a sector with an arc of 90°. A **chord** is any straight line connecting two points on a circle. A **secant** is a straight line which intersects a circle. A **tan-**

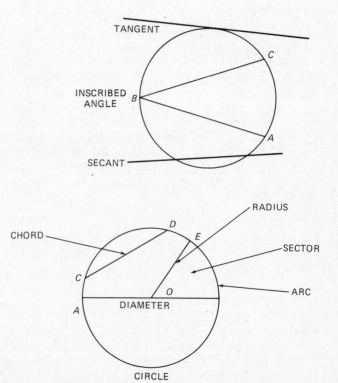

FIG. 1-9 Parts of a circle.

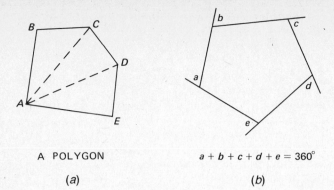

A POLYGON

$a + b + c + d + e = 360°$

(a)   (b)

FIG. 1-10   Angles of polygons.

gent is a straight line of unlimited length which only has one point in common with a circle. An **inscribed angle** in a circle is an angle whose vertex is on the circle. A **central angle** is an angle whose vertex is at the center of the angle.

The **circumference** of a circle is the length of the perimeter. The circumference can be computed by multiplying the diameter times **pi**. Pi (greek letter π) is a constant equal to 22 divided by 7. Pi is usually rounded off to 3.14 or 3.1416. The formula for circumference is

1. Circumference = pi $\times$ diameter
or $C = \pi D$

Since the diameter is 2 times the radius, an alternate formula is

2. Circumference = 2 $\times$ pi $\times$ radius
or $C = 2\pi r$

Polygons. A **polygon** is a plane, a closed figure bounded by straight lines joined end to end. Polygons may have any number of sides from three upward. A

regular polygon has all sides and angles equal. Some common polygons are

3 sides—triangle
4 sides—quadrilateral
5 sides—pentagon
6 sides—hexagon
8 sides—octagon

All polygons can be considered as being made of a number of triangles as shown by the pentagon in Fig. 1-10a. The number of triangles will be equal to the number of sides minus 2. The sum of the interior angles of a triangle is 180°. Thus the sum of the interior angles of any polygon is equal to the number of sides minus 2 times 180°. A triangle has a total of 180° [(3 − 2) × 180°]; a quadrilateral, 360° [(4 − 2) × 180°], a pentagon, 540° [(5 − 2) × 180°], etc. By definition a regular polygon has equal angles; therefore a regular hexagon would have included angles of 120°.

If the sides of any polygon, such as the pentagon in Fig. 1-10b are extended consecutively in the same direction, an angle will be formed that will be the supplement of the internal angle. The sum of the supplementary angles will always equal 360°.

The aviation maintenance technician can expect to encounter a number of different polygon shapes in his or her work. The majority of the shapes encountered will be in the categories of triangles, quadrilaterals, and hexagons.

Triangles. A **triangle** is a plane bounded by three sides, or a three-sided polygon with a total included angle of 180°. A number of various types of triangles exist within this definition. Variations include the sizes of the angles and the length of the legs.

An **acute triangle** is one in which all angles are less than 90° (see Fig. 1-11). An **obtuse triangle** has one angle greater than 90°. A **right triangle** has one 90°, or right, angle. Remember that the sum of the three angles must be 180°.

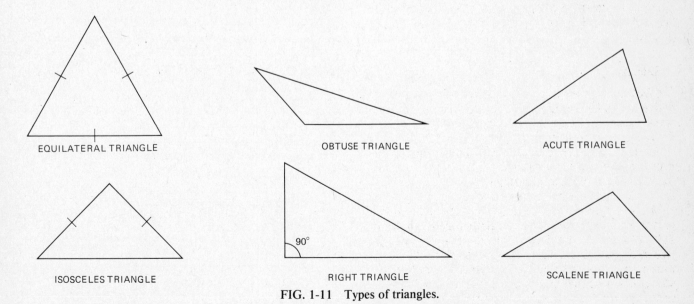

EQUILATERAL TRIANGLE   OBTUSE TRIANGLE   ACUTE TRIANGLE

ISOSCELES TRIANGLE   RIGHT TRIANGLE   SCALENE TRIANGLE

FIG. 1-11   Types of triangles.

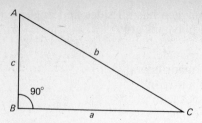

FIG. 1-12   Right triangle.

An **equilateral triangle** has all sides of equal length. An equilateral triangle is also a regular polygon. The length of the sides and the included angles are all equal. The angles are each 60°. An **isosceles triangle** has two sides equal in length and two equal angles. A **scalene triangle** has no equal sides or angles.

A drawing of a triangle will normally have the angles identified by an uppercase letter and the sides by a lowercase letter. Each side will have the same letter as the angle it is opposite. This allows each angle and/or side to be clearly identified.

The side opposite the 90° angle in a right triangle is called the **hypotenuse,** side b in Fig. 1-12. The **pythagorean theorem** states that in a right triangle the square of the hypotenuse is equal to the sum of the squares of the other two sides. This can be written algebraically as $a^2 + b^2 = c^2$. With this theorem, if we know the length of two sides of a right triangle, we can easily calculate the third.

In addition to angles and length of sides, triangles are dimensioned for computation purposes by **base** and **height.** Any one of the sides may be picked as the base. The height of the triangle is the length of a line perpendicular to and extending from the base to the vertex of the opposite angle. Figure 1-13 shows the correct way to measure height for different-shaped triangles.

Quadrilaterals. A **quadrilateral** is a four-sided polygon with the sum of the included angles equal to 360°. A regular quadrilateral with all sides of equal length and with equal angles is a **square.** A **rectangle** has four angles of 90° and two pair of parallel sides. One pair is longer than the other as shown in Fig. 1-14. A **parallelogram** is a four-sided figure whose opposite sides are equal and parallel. This definition would also include the rectangle. In common practice, a parallelogram does not have any right angles. A **trapezoid** is a four-sided plane that has two parallel sides and two that are not parallel.

Squares are dimensioned by the length of any two of the equal sides. Rectangles are dimensioned by the two different lengths referred to as length times width or length times height. In common practice the shorter side will be specified as the width or height. Parallelograms are dimensioned by base and height. The longest dimension is normally used as the base. The height is the length of a line perpendicular to, and extending from, the base to the opposite parallel line (see Fig. 1-14). Trapezoids are dimensioned for height in the same way as the parallelogram. Since the parallel sides are of unequal length, both are identified for dimensions, such as base 1 and base 2 ($b1$, $b2$).

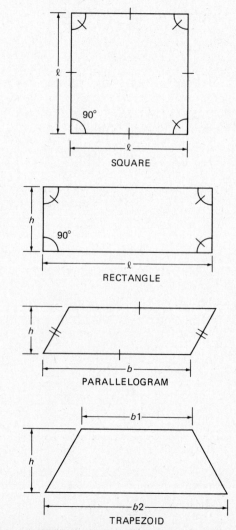

FIG. 1-14   Dimensions of quadrilaterals.

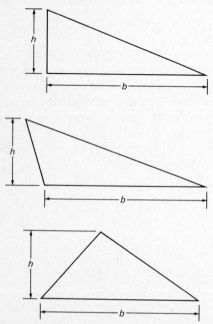

FIG. 1-13   Dimensions of triangles.

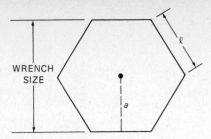

FIG. 1-15   Dimensions of a hexagon.

Hexagon. The six-sided regular hexagon is a familiar shape for the aircraft maintenance technician. This is a standard shape used for aircraft nuts and bolts. When used for aircraft hardware, a hexagon shape is dimensioned "across the flats" as shown in Fig. 1-15. For calculation purposes the length of each flat and the distance across corners is also used.

Formulas

The aviation maintenance technician needs to be able to calculate the amount of area in a plane, the volume of a solid, and the surface area of a solid. The need for this ability will be found when working in such areas as sheet-metal layout and repair, power plants, hydraulic systems, and fuel systems. While rote memorization of formulas may suffice for this task, it is recommended that the student learn how the formula is derived. This will enable better retention of the information and a wider use of its application.

Area. Area is measured in units of **square inches** (in^2) or square centimeters (cm^2). The **area of a circle** is a function of the radius. Three different formulas are in common use. They are given in the examples that follow.

Example. Find the area of a circle with a radius of 2 in (Fig. 1-16).
1. Area equals pi times the radius squared, $A = \pi r^2$.

Solution

$$A = 3.14 \times 2 \times 2 = 12.56 \text{ in}^2$$

2. Area equals pi times the diameter squared divided by four, $A = (\pi d^2)/4$ or $0.7854d^2$. *Note:* 3.1416/4 = 0.7854.

Solution

$$A = 0.7854 \times 4 \times 4 = 12.56 \text{ in}^2$$

3. Area equals one-half the radius times the circumference, $A = \frac{1}{2}rC$. *Note:* $C = \pi d$.

Solution

$$A = \frac{1}{2} \times 2 \times 3.14 \times 4 = 12.56 \text{ in}^2$$

All three formulas will give the same answer. The one to use depends on the data available and personal preference.

To find the **area of a sector,** first find the area of a full circle and divide the answer by $N/360$. N is the number of degrees included in the sector.

Example. Find the area of a 60° sector with a radius of 2 in (Fig. 1-16).

Solution

$$\text{Area of a circle with 2-in radius} = 12.56 \text{ in}^2$$
$$\text{Area of sector} = \frac{12.56}{360/60}$$
$$= \frac{12.56}{6} = 2.09 \text{ in}^2$$

The **area of any regular polygon** can be found by using the formula $A = \frac{1}{2}ap$ with a being the perpendicular distance from a side to the center (apothem) and p being the perimeter or sum of the sides. The similarity should be noted between this formula and one of the formulas for the area of a circle ($A = \frac{1}{2}rC$).

Example. Find the area of the hexagon shown in Fig. 1-15.

Solution

$$\text{Distance across flats } \tfrac{5}{8} \text{ in, } a = \tfrac{5}{16} = 0.3125$$
$$\text{Length of each side} = 0.360 \text{ in} \qquad p = 2.16 \text{ in}$$
$$A = \tfrac{1}{2}ap = \tfrac{1}{2} \times 0.3125 \times 2.16$$
$$A = 0.338 \text{ in}^2$$

The above formula will work with any regular polygon. However, in the case of squares and equilateral

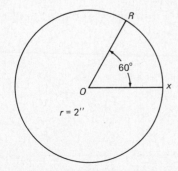

FIG. 1-16   Calculating the area of a circle or sector.

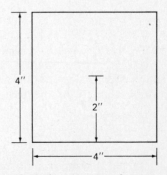

FIG. 1-17   Calculating the area of a square.

21

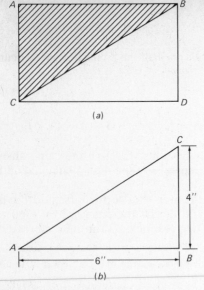

(a)

(b)

FIG. 1-18   Calculating the area of a triangle.

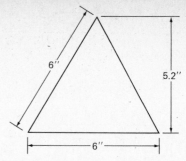

FIG. 1-19   Calculating the area of an equilateral triangle.

Example. Find the area of the triangle in Fig. 1-19 using both formulas.

Solution

1. $A = \frac{1}{2}bh$
            $= \frac{1}{2} \times 6 \times 5.2 = 15.6 \text{ in}^2$
2. $A = \frac{1}{4}b^2\sqrt{3}$
            $= \frac{1}{4} \times 6 \times 6 \times 1.732 = 15.6 \text{ in}^2$

Figure 1-20 illustrates the derivation of a formula for the **area of a parallelogram.** The parallelogram *ABCD* is divided with a diagonal from *B* to *C*. As with the rectangle in Fig. 1-18*a*, we have created two equal triangles. Each triangle has a base equal to the base dimension of the parallelogram and a height also equal to the height of the parallelogram. By visual inspection it can be seen that the area of parallelogram *ABCD* is equal to two times the area of triangle *BCD*. The formula for the area of the triangle is $A = \frac{1}{2}bh$. The area of a parallelogram can be found by multiplying the base times the height.

Example. Find the area of the parallelogram in Fig. 1-20.

Solution

$$A = bh$$
$$= 16 \times 10 = 160 \text{ in}^2$$

Care should be taken that the height is measured with a line that is perpendicular to the base and not by the length of an end.

Figure 1-21 illustrates a method for calculating the **area of a trapezoid.** Trapezoid *ABCD* has height *h*, base 1 equals line *AB*, and base 2 equals line *CD*. From point *B* extend line *AB* a length equal to line *CD* to point *B'*. From point *D* extend line *CD* a length

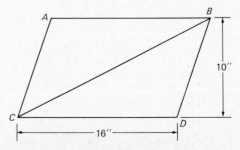

FIG. 1-20   Calculating the area of a parallelogram.

triangles other formulas are usually more convenient. In Fig. 1-17 is a square that is 4 in on each side. To use the polygon formula we determine that $a = 2$ and $p = 16$. Using the formula $A = \frac{1}{2}ap$ gives an area of $\frac{1}{2} \times 2 \times 16$, or 16 in². The same result can be found by simply multiplying the length by the height, $4 \times 4 = 16$. Since all four sides are equal in length, the area of a square may be stated as $A = l^2$. To find the area of a rectangle with two different lengths of sides, the formula will be $A = lh$ with *l* equal to length and *h* equal to height.

To find the **area of a triangle** look at the example in Fig. 1-18*a*. Rectangle *ABCD* is drawn with a diagonal from *B* to *C*. The diagonal creates two triangles, *ABC* and *BCD*. Each triangle has a height and a base equal to the height and base of the rectangle. The two triangles are equal and therefore the area of each triangle equals one-half the area of the rectangle. To find the **area of one triangle** the formula would be $A = \frac{1}{2}bh$, with *b* representing the length of the base and *h* representing the height.

Example. Find the area of a triangle with a base of 6 in and a height of 4 in (Fig. 1-18*b*).

Solution

Area = one-half times the base times the height
$A = \frac{1}{2}bh$
$= \frac{1}{2} \times 6 \times 4 = 12 \text{ in}^2$

This formula is valid for any type of a triangle.

The area of an equilateral triangle can be found by an alternate formula which states that the area is equal to one-quarter of the square of the base times the square root of 3.

$$A = \frac{1}{4}b^2\sqrt{3}$$

Figure 1-19 shows an equilateral triangle with 6 in on a side. The height of this triangle has been computed to be 5.2 in.

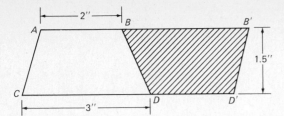

FIG. 1-21 Calculating the area of a trapezoid.

equal to line *AB* to point *D'*. Connect point *B'* and point *D'*. Line *AB'* = *AB* + *CD*. Line *CD"* = *CD* + *AB*. Thus line *AB'* = *CD'*. A parallelogram has been created. Further visual inspection of the figure will show that *ABCD* and *BB'CC'* are equal in area. The area of *ABCD* would then be half of the area encompassed by *AB'CD'*. The formula for the **area of a trapezoid** is one-half of the sum of base 1 plus base 2 times the height. This can be expressed in a formula as $A = \frac{1}{2}(b1 + b2)h$.

Example. Find the area of the trapezoid in Fig. 1-21.

Solution

$$b1 = 3 \qquad b2 = 2 \qquad h = 1.5$$
$$A = \tfrac{1}{2}(b1 + b2)h$$
$$= \tfrac{1}{2} \times (3 + 2) \times 1.5$$
$$= 3.75 \text{ in}^2$$

The area of an irregular polygon can be found by dividing it into shapes such as triangles or rectangles. Figure 1-10*a* shows a pentagon divided into three triangles. Using trigonometry it is possible to calculate the area of each triangle. The sum of the areas would be the area of the pentagon.

Volume. Volume requires that an object have length, breadth, and depth. Volume is expressed in units of cubic inches (in³) or cubic centimeters (cm³). The volume of a rectangular solid is equal to the product of the height, length, and width; $V = hwl$ (see Fig. 1-22). A solid cube has equal edge dimensions; thus, the volume of a cube will equal the cube of one dimension, or $V = l^3$.

The volume of a cylinder is equal to the product of its cross-sectional area and its height ($V = Ah$ or $V = \pi r^2 h$). The volume of a cone or pyramid is equal to one-third the product of the area of the base and the altitude ($V = a\pi r^2/3$). The correct way to measure altitude *a* is shown in Fig. 1-23*a*. The volume of a

$$V = hwl$$

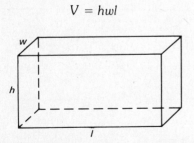

FIG. 1-22 Volume of a rectangular solid.

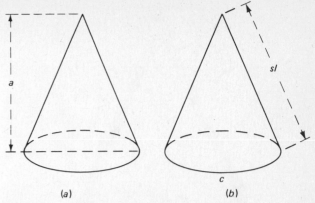

FIG. 1-23 Volume and surface area of a cone.

sphere is equal to the product of one-third times 4 pi and the cube of the radius [$V = (4\pi/3)r^3$].

Surface Area. It is occasionally necessary to calculate the surface area of an object. For an item which has surfaces made up of circles or polygons, it is a simple matter of finding the sum of the individual surfaces. The surface area of a cylinder may be found by multiplying the circumference by the height. This would give the surface area of a cylinder that had no ends. If the ends are to be included, they can be calculated as the area of a circle and added to the other answer. The surface area of a cone, called the **lateral area,** is equal to one-half the product of its **slant height** and the circumference of the base [$A = (\text{sl} \times C)/2$].

Figure 1-23*b* shows the correct way to measure slant height (sl). The lateral area does not include the base. If the surface area of the base is required, it can be calculated as the area of a circle. The surface area of a sphere is equal to the product of 4π and the square of the radius ($A = 4\pi r^2$).

Geometric Constructions

During maintenance of aircraft, it is often necessary to lay out or transfer geometric shapes to new material. It is possible with a compass, a ruler, and a protractor to construct many geometric figures which accurately fulfill their definitions or descriptions. In most cases the protractor is not needed. A compass and a protractor are shown in Fig. 1-24.

1. Bisect a straight line (Fig. 1-25*a*). Adjust the compass so that it spans a greater distance than one-half the length of the line. Place the point on *A* and strike an arc *CDE* as shown. Using *B* as a center, strike an arc *FGH*. Connect the points *J* and *K* with a straight line. The line *JK* bisects the line *AB*.

2. Draw a perpendicular from a point to a line (Fig. 1-25*b*). From point *P* use the compass to strike arcs at *A* and *B*, using the same radius in each case. Then from points *A* and *B* strike intersecting arcs at *C*. Connect the points *P* and *C* with a straight line. The line *PC* is perpendicular to *AB*.

3. Bisect an angle (Fig. 1-25*c*). Given the angle *AOB*, place the point of a compass at *O* and strike arcs at *A* and *B* so that *OA* = *OB*. From points *A* and *B* strike intersecting arcs at *C* using equal radii. Draw the line *OC*. *OC* is then the bisector of the angle.

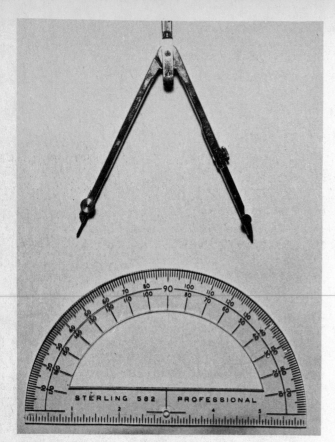

FIG. 1-24   A compass and protractor.

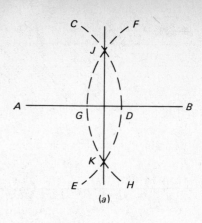

(a)

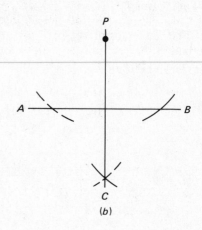

(b)

4. *Duplicate a given angle* (Fig. 1-25d). Given the angle *AOB*, draw the line *O'D*. Strike an arc *AB* such that *OA* = *OB*. Draw the arc *CD* using the radius *OA*. Using the distance *AB* as a radius, strike an arc at *C* with *D* as a center. Draw the line *OC*. The angle *CO'D* is equal to angle *AOB*.

5. *Duplicate a given triangle* (Fig. 1-25e). Given the triangle *ABC*, draw a horizontal line *DX*. Using *AB* as a radius and *D* as the center, draw an arc cutting *DX* at *E*. Using *AC* as a radius and *D* as a center, draw an arc in the vicinity of *F*. Using *CB* as a radius and *E* as a center, draw an arc to intersect the other arc at *F*. Draw the lines *DF* and *EF*. *DEF* is the duplicate of triangle *ABC*.

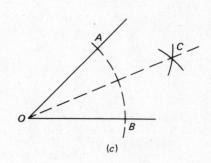

(c)

Practice Problems

1.   If the complement of an angle is 50°, what is the angle?

2.   What is the supplement of an angle of 60°?

3.   If one straight line meets another straight line and makes an angle of 80°, what other angle is also made?

4.   Define an isosceles triangle.

5.   If the sum of two angles in a triangle is 120°, what is the value of the other angle?

6.   What is the length of the hypotenuse of a right triangle when the lengths of the other two sides are 6 in and 8 in?

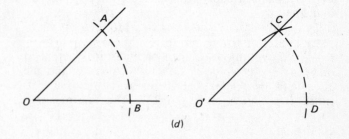

(d)

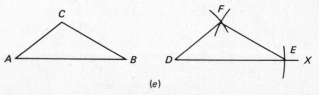

(e)

FIG. 1-25   Geometric constructions.

7. One side of a parallelogram is 8 in long and the perpendicular distance from this side to the opposite side is 5 in. What is the area of the parallelogram?

8. The base of a triangle is 10 in and its height is 7 in. What is the area of the triangle?

9. Give the area of a circle having a diameter of 9 in.

10. If a rectangular fuel tank is 2 feet (ft) long, 18 in wide, and 8 in deep, what is its volume in cubic inches? If there are 231 in^3 in a gallon, how many gallons can the tank hold?

11. If a 12-in sphere is submerged in a tank of water, how many gallons of water will it displace?

12. What is the volume of a cone having a base with a 10-in diameter and an altitude of 12 in?

● TRIGONOMETRY

Trigonometry is the branch of mathematics which makes possible the solution of unknown parts of a triangle. When the values of certain angles and sides of a triangle are known, it is possible to determine the values of all the parts through the use of trigonometric processes.

Trigonometric Functions

Trigonometric functions are based on the ratios of the sides of a right triangle to one another. In the diagram of Fig. 1-26, the right triangle $AB'C'$ is superimposed on right triangle ABC with the angles at A coinciding. The lines $B'C'$ and BC are parallel; hence the triangles are similar. In similar triangles the ratios of corresponding sides are equal and so $AB/AC = AB'/AC'$. In like manner, the ratios of the other sides are also equal. Furthermore, any right triangle which has an acute angle equal to A will have the same ratios as those for the triangles shown in Fig. 1-26.

In trigonometry the ratios of the sides of a right triangle to one another are given particular names. These are **sine, cosine, tangent, cotangent, secant,** and **cosecant.** These ratios are called **trigonometric functions** and may be explained by the use of the triangle in Fig. 1-27.

In the triangle ABC, side c is the hypotenuse, side b is the side adjacent to angle A, and side a is the side opposite angle A. The functions of angle A are then as follows:

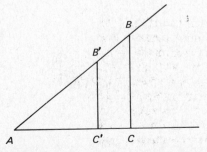

FIG. 1-26 Similar right triangles superimposed.

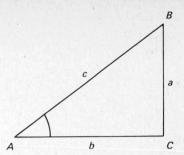

FIG. 1-27 Triangle to show functions of an angle.

The sine of angle A, called *sin A*, is

$$\frac{\text{side opposite}}{\text{hypotenuse}} \quad \text{or} \quad \frac{a}{c}$$

The cosine of A, called *cos A*, is

$$\frac{\text{side adjacent}}{\text{hypotenuse}} \quad \text{or} \quad \frac{b}{c}$$

The tangent of A, called *tan A*, is

$$\frac{\text{side opposite}}{\text{side adjacent}} \quad \text{or} \quad \frac{a}{b}$$

The cotangent of A, called *cot A*, is

$$\frac{\text{side adjacent}}{\text{side opposite}} \quad \text{or} \quad \frac{b}{a}$$

The secant of A, called *sec A*, is

$$\frac{\text{hypotenuse}}{\text{side adjacent}} \quad \text{or} \quad \frac{c}{b}$$

The cosecant of A, called *csc A*, is

$$\frac{\text{hypotenuse}}{\text{side opposite}} \quad \text{or} \quad \frac{c}{a}$$

The importance of the foregoing functions lies in the fact that a particular function always has the same value for the same angle. For example, sin 50° is always equal to 0.7660. This means that in a right triangle which has an acute angle of 50° the sine of 50° will always be 0.7660 regardless of the size of the triangle. The table "Trigonometric Functions" in the Appendix of this book may be used to determine the values of the functions of any angle.

In the triangle shown in Fig. 1-27, the functions of the angle B are the **cofunctions** of angle A. That is,

$$\begin{aligned} \sin B &= \cos A & \cos B &= \sin A \\ \tan B &= \cot A & \cot B &= \tan A \\ \sec B &= \csc A & \csc B &= \sec A \end{aligned}$$

These relationships can easily be shown by noting the sides adjacent to and opposite angle B.

25

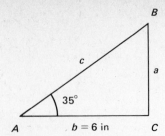

FIG. 1-28 Solution of a right triangle.

Solution of Right Triangles

If any side and one of the acute angles of a right triangle are known, all the other values of the triangle may be determined. For example, if an acute angle of the right triangle in Fig. 1-28 is 35° and the side adjacent to this angle is 6 in long, we may determine the other values as follows:

From the table of functions we find cos 35° = 0.8192.

Then:

$$\frac{b}{c} = 0.8192 \quad \text{or} \quad \frac{6}{c} = 0.8192 \quad c = \frac{6}{0.8192} = 7.32$$

From the table of functions we find tan 35° = 0.7002.

Then:

$$\frac{a}{b} = 0.7002 \quad \text{or} \quad \frac{a}{6} = 0.7002 \quad a = 4.2012$$

Since the sum of the angles of a triangle is 180°, the other acute angle of the triangle is 55°.

The sides of the triangle are 4.2012, 6, and 7.32. We can verify these answers by the formula $a^2 + b^2 = c^2$, which shows that the square of the hypotenuse of a right triangle is equal to the sum of the other two sides.

If the sides of a right triangle are known, the angles can also be determined. This is shown in the problem of Fig. 1-29. In the triangle ABC, side $a = 8$, $b = 15$, and $c = 17$.

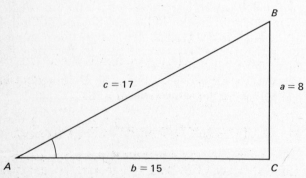

FIG. 1-29 Solving for an angle.

Then:

$$\sin A = \frac{a}{c} = \frac{8}{17} = 0.4706$$

From the table of functions:

$$0.4706 = \sin 28° \ 4' \quad \text{(approximately)}$$

Then:

$$\text{Angle } A = 28° \ 4' \quad \text{(approximately)}$$

Changes in Values of Functions

An examination of Fig. 1-30 reveals two right triangles with different angles but the hypotenuse of each being equal in length. The denominator of both the sine and the cosine of an angle is the value of the hypotenuse. Assigning a value of 1 to each hypotenuse in Fig. 1-30 results in the value of the sine being represented by the length of AB and $A'B'$. The cosine for each triangle will be represented by the length of OA and OA'. An examination of the diagram shows that as angle AOB increases, the value of the sine will also increase. As the angle decreases, the value of the sine will decrease. As the angle becomes smaller, this value will continue to decrease until at an angle of 0° the sine equals 0. At 90° the side opposite, AB, will equal the length of the hypotenuse and the value of the sine will be 1.0. The value of the sine will vary from 0.0 to 1.0 as the angle goes from 0° to 90°. The cosine will increase as the angle gets smaller. At 0° the length of the side adjacent, OA, is equal to the length of the hypotenuse and will equal 1.0. At 90° the side adjacent, OA, will have decreased to 0 and the value of the cosine will be 0.0.

Figure 1-31 shows a graph of the functions of a sine and cosine through 360°. The graph for the sine is called a sine wave. The sine wave shows that the value of the sine increases from 0 to 90° and decreases in value from 90 to 180°. From 180° the value increases,

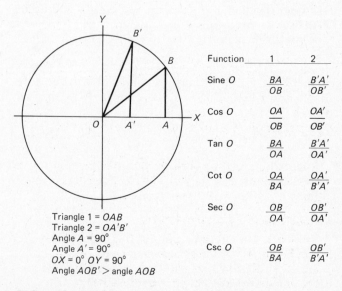

Function	1	2
Sine O	$\dfrac{BA}{OB}$	$\dfrac{B'A'}{OB'}$
Cos O	$\dfrac{OA}{OB}$	$\dfrac{OA'}{OB'}$
Tan O	$\dfrac{BA}{OA}$	$\dfrac{B'A'}{OA'}$
Cot O	$\dfrac{OA}{BA}$	$\dfrac{OA'}{B'A'}$
Sec O	$\dfrac{OB}{OA}$	$\dfrac{OB'}{OA'}$
Csc O	$\dfrac{OB}{BA}$	$\dfrac{OB'}{B'A'}$

Triangle 1 = OAB
Triangle 2 = $OA'B'$
Angle $A = 90°$
Angle $A' = 90°$
$OX = 0° \ OY = 90°$
Angle $AOB' >$ angle AOB

FIG. 1-30 Changes in values of functions.

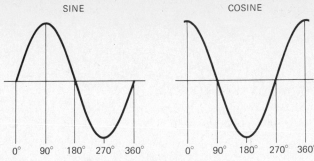

FIG. 1-31   Functions of the sine and cosine.

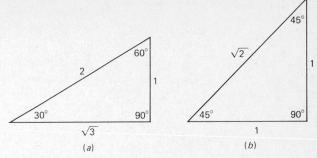

FIG. 1-32   Functions of 30, 45, and 60° angles.

but in a negative direction until it reaches 270°, at which time it begins decreasing to 0 at 360°. At this point the sine value has completed a full cycle of 360° and starts over again. Figure 1-31 also shows that the cosine has a similar waveform but is reaching maximum and minimum values 90° from the sine wave. The aviation technician student will find the sine wave used extensively in other subjects, such as electricity.

The tangent in Fig. 1-30 becomes larger as the angle increases. The side opposite becomes longer and the side adjacent becomes shorter. At 0° the value of AB would be 0 making the value of the tangent 0.0. As the angle approaches 90°, the value of the side adjacent approaches 0. As a result, the tangent value becomes very large and at 90° is said to be infinite (∞). Figure 1-30 can also be used to show the following ranges of values. The cotangent will have an infinite value at 0° and a value of 0.0 at 90°. The secant will have values from 1.0 to infinity in the 0° to 90° range. The value of the cosecant will vary from infinity to 1.0 over the same range.

The values, as the angle changes from 0 to 90°, can be summarized as follows:

sin a, 0 to 1
cos a, 1 to 0
tan a, 0 to ∞
cot a, ∞ to 0
sec a, 1 to ∞
csc a, ∞ to 1

It should also be noted that in a right triangle, such as Fig. 1-30, the sine of one angle will equal the cosine of the second. For example:

$$\text{sine of angle } AOB = \frac{AB}{OB} = \text{cosine of angle } OBA$$

Many other relationships exist among the trigonometric functions, but they are beyond the scope of this text.

Functions of Particular Angles

The angles of 30°, 45°, and 60° occur frequently. Triangles containing these angles have relationships that are easy to remember and that frequently will reduce the amount of calculation required for a problem. Figure 1-32a shows the relationship of length for the sides

of a 30-60-90° triangle. The shortest side, opposite the 30° angle, is assigned the value of 1. The hypotenuse of this triangle will be two times the length of the shorter side. The length of the third side, adjacent to the 30° angle, will be equal to the length of the shortest side times the square root of three. Thus:

$$\sin 30 = \tfrac{1}{2} = 0.5000 \text{ (also cosine 60)}$$
$$\text{cosine } 30 = \tfrac{\sqrt{3}}{2} = 0.8660 \text{ (also sine 60)}$$
$$\text{tangent } 30 = \tfrac{1}{\sqrt{3}} = 0.5773 \text{ (also cotangent 60)}$$
$$\text{cotangent } 30 = \tfrac{\sqrt{3}}{1} = 1.7320 \text{ (also tangent 60)}$$

The 45° triangle in Fig. 1-32b has two equal sides each assigned a value of 1. The hypotenuse has a value of the square root of 2, or 1.414.

$$\sin 45 = \tfrac{1}{\sqrt{2}} = 0.707$$
$$\cos 45 = \tfrac{1}{\sqrt{2}} = 0.707$$
$$\tan 45 = \tfrac{1}{1} = 1.000$$
$$\cot 45 = \tfrac{1}{1} = 1.000$$

The student should also remember the pythagorean theorem ($a^2 + b^2 = c^2$, where c is the hypotenuse of a right triangle). If any two sides of a right triangle are known, the third side can be calculated.

Practice Problems. In the following problems, the right angle will always be angle C. Angle A will be the smaller of the two acute angles if they are not equal. Uppercase letters refer to angles and lowercase letters refer to sides.

1. $A = 30$, $a = 1.8$. Find c.

2. $A = 45$, $b = 3.5$. Find c.

3. $a = 3$, $b = 4$. Find c.

4.   What is the value of angle A in Problem 3?

5.   Use the table of trigonomic functions in the Appendix to find the value of the sin of 39.

● ALTERNATIVE NUMBER SYSTEMS

All of the mathematics in this chapter have used a system with a base of 10. This system, using 10 digits and known as the **decimal system,** was discussed in the first part of the chapter. The development of digital electronics introduced the need for the use of other systems.

Every number system has three concepts in common: (1) a **base,** (2) **digit value,** and (3) **positional notation.** The base is the number of digits used in the system. Each digit of a specified system has a distinct value. Each number position carries a specific weight depending upon the base of the system.

Example. 546 (decimal)

Solution

$$6 \times 10^0 = 6 \times 1 = 6$$
$$4 \times 10^1 = 4 \times 10 = 40$$
$$5 \times 10^2 = 5 \times 100 = \underline{500}$$
$$546$$

Electronic devices work with a system of only two numbers, 0 and 1. This system is known as the **binary** system and uses powers of 2. A binary number of 1101 is equal to a decimal number of 13.

Example. 1101 (binary)

Solution

$$1 \times 2^0 = 1 \times 1 = 1$$
$$0 \times 2^1 = 0 \times 2 = 0$$
$$1 \times 2^2 = 1 \times 4 = 4$$
$$1 \times 2^3 = 1 \times 8 = \underline{8}$$
$$\text{(decimal) } 13$$

The **octal** system uses a base of 8 and digits from 0 through 7. The **hexadecimal** system uses 16 digits. These include the 10 digits from 0 to 9 and the first six letters of the alphabet, A through F. The hexadecimal positional value is based on powers of 16. Table 1-2 shows the relationship for some numbers in the four systems. An examination of Table 1-2 reveals that large quantities expressed in the binary system require a large number of positions and can become awkward to handle. The octal and hexadecimal systems can express large quantities in numbers with three or four positions. These two systems are easily converted to binary information.

● CHARTS AND GRAPHS

Charts and graphs are extensively used in aircraft maintenance and operation manuals to present mathematical data and to aid in its use. The time and effort required to make mathematical calculations or to understand an operation can often be greatly reduced by using such aids. Charts and graphs of many types are found in technical literature related to aircraft maintenance. The student should know the different types and when they would be used.

A chart may be used to present many types of information, and the information may be presented in a variety of ways. Table 1-1 is a chart that presents various functions of powers and roots. It is made up of

TABLE 1-2 Comparison Between Number Systems

Decimal	Binary	Octal	Hexadecimal
0	0	0	0
1	1	1	1
2	10	2	2
3	11	3	3
4	100	4	4
5	101	5	5
6	110	6	6
7	111	7	7
8	1000	10	8
9	1001	11	9
10	1010	12	A
11	1011	13	B
12	1100	14	C
13	1101	15	D
14	1110	16	E
15	1111	17	F
16	10000	20	10
17	10001	21	11
.	.	.	.
.	.	.	.
.	.	.	.
100	1100100	144	64
1000	1111101000	1750	3E8

numerical lists. Other charts may have combinations of numerical data and text, as in the case of a troubleshooting chart. The service section of a maintenance manual may have a lubrication chart that combines pictorial diagrams, text, numbers, and symbols.

Graphs

Graphs are charts which provide numerical or mathematical information in graphical form, that is, with lines, scales, bars, sectors, etc. The graph usually shows the changes in the value of one or more variables as another variable changes.

A **broken-line** or **bar graph** is used to show comparative quantitative data. The broken-line graph is useful to show trends in quantitative data over a period of time. The use is illustrated in Fig. 1-33. In the illustration, the broken-line and bar graphs provide the same information.

The **circular,** or **pie,** chart is used to graphically represent the division or distribution of a whole. Figure 1-34 is a circle graph indicating how and in what proportion a company's operating expenses were distributed. If this year's distribution was to be compared with last year's, a bar or broken-line graph would be the better choice.

Continuous-Line Graphs. The broken-line graph is made up of a number of finite points connected with a line. The space between the points has no significance. The **continuous-line graph** has a line connecting points which have been measured or calculated. The line provides continuous information in that a reading could be taken at any point on it. Figure 1-35 is an example of a continuous-line graph. The two variables are stress and strain of a metal. The strain, plotted on the horizontal axis, produced by varying amounts of stress, plotted on the vertical axis, is shown by this graph. If values were recorded on the graph, the

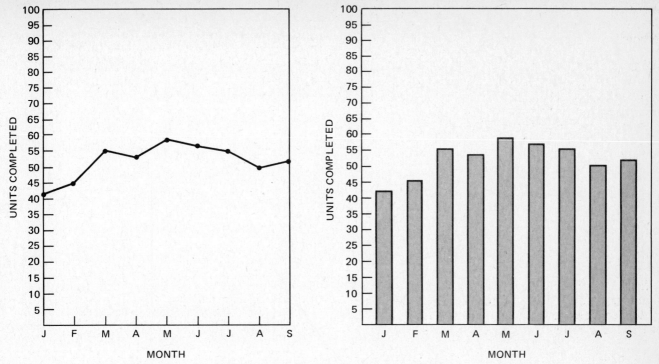

FIG. 1-33   Broken line graph and bar graph.

amount of strain produced by a given amount of stress could be determined. In addition to the values that can be read, the shape of the line will have significance to those working with it. Figure 1-36 shows three continuous lines on one graph. All three lines represent an independent variable that is being compared to a common variable. The common variable is the angle of attack. The relationship to the angle of attack is shown for each independent variable. By having several variables on one graph, it is possible to determine what values or ranges will give the best combined results.

Graphs may be used to show limits. In Fig. 1-37 the altitude and airspeed combinations safe for the auto-rotation of a helicopter have been plotted. The portions of the graph which represent unsafe combinations have been shaded. The safe combinations can be readily determined. For example, 30-knots (kn) airspeed is safe only above 350 ft. Figure 1-38 shows two graphs used for aircraft loading. The first graph converts weights at specific locations, i.e., front seats, into index units. The total index units are then plotted against the weight of the aircraft on the second chart. If the plotted point is within the limits shown on the chart, the aircraft is properly loaded.

Figure 1-39 shows a graph of three variables: distance, time, and speed. If any two of the three vari-

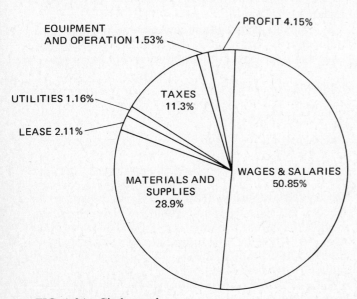

FIG. 1-34   Circle graph.

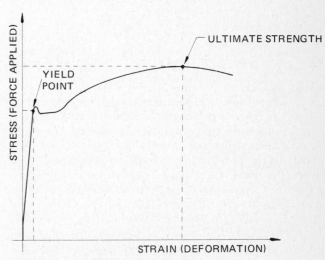

FIG. 1-35   A continuous line curve.

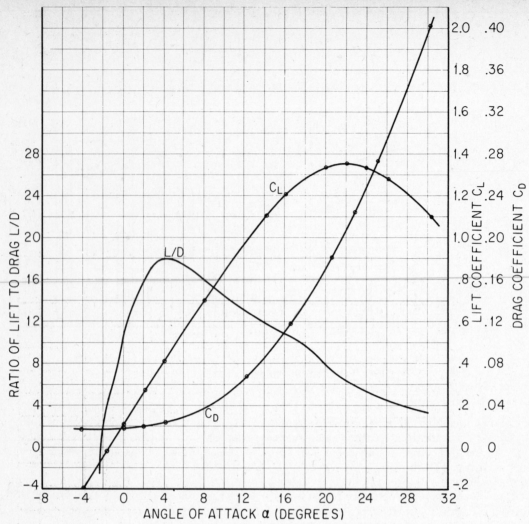

FIG. 1-36   Three continuous curves on a graph.

ables are known, the approximate value of the other can be quickly determined. The dotted line is an example of a known time and speed giving a distance. In the example a speed of 375 kn for 2.5 hours (h) would result in a distance of approximately 940 miles (mi). A **nomograph,** also called an alignment chart, is a chart used for calculations. It has scales showing the values of three or more variables. The distances between the scales and the values on each scale are placed in such a manner that the user may use a straightedge to line up two known values and obtain a third value.

Graphs of Mathematical Functions. Graphs can be used to help solve mathematical problems. Equations involving the values x and y can be plotted on a graph to provide a visual indication of the value of each variable as the other changes. Such a graph is shown in Fig. 1-40. The first equation plotted on this graph is $x + y = 8$. When x is given a value of 0, $y = 8$ and is plotted on the y coordinate (axis) at $+8$. In the same manner, when y is given the value of 0, $x = 8$ and is plotted on the x coordinate at $+8$. The line drawn between the plotted points provides all the values of x for any value of y and vice versa. When the equation

$x - y = 5$ is plotted on the same graph, the line for the equation intersects the first line at a point where $x = 6.5$ and $y = 1.5$. These values satisfy both equations.

It is apparent that the functions of x and y for two independent equations, also called simultaneous equations, can be solved graphically. For example,

Equation 1:

$$2x + 3y = 12$$

Equation 2:

$$x - 2y = -6$$

are solved graphically in Fig. 1-41. In Equation 1, when $x = 0$, $y = +4$, and when $y = 0$, $x = +6$. In Equation 2, when $x = 0$, $y = +3$, and when $y = 0$, $x = -6$. When the lines are plotted on the graph, they intersect at a point where $x = \frac{6}{7}$, and $y = 3\frac{3}{7}$. These values satisfy both equations.

Graphs of algebraic equations will produce other than straight lines as is shown in Fig. 1-42.

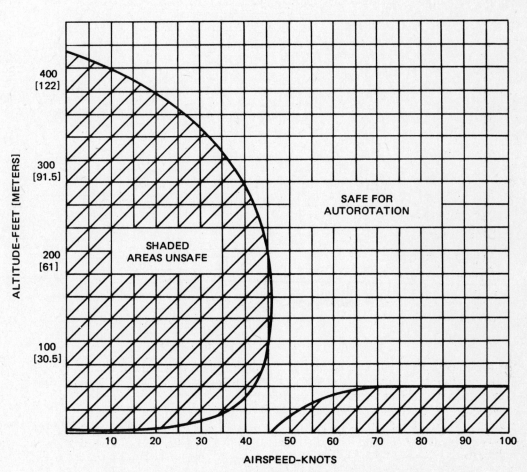

FIG. 1-37   A continuous-line graph showing operating limits.

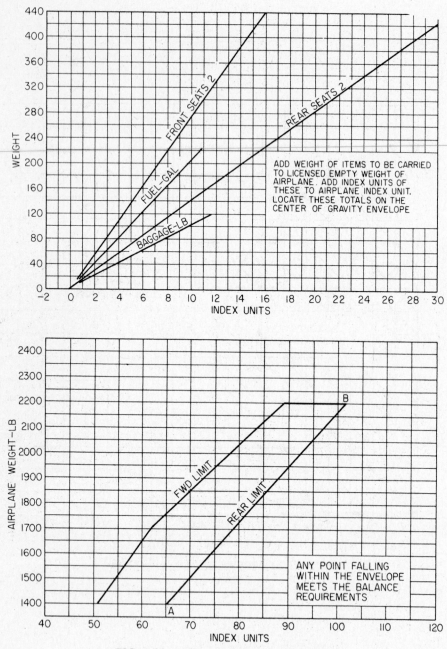

ADD WEIGHT OF ITEMS TO BE CARRIED
TO LICENSED EMPTY WEIGHT OF
AIRPLANE. ADD INDEX UNITS OF
THESE TO AIRPLANE INDEX UNIT.
LOCATE THESE TOTALS ON THE
CENTER OF GRAVITY ENVELOPE

ANY POINT FALLING
WITHIN THE ENVELOPE
MEETS THE BALANCE
REQUIREMENTS

FIG. 1-38 Using a graph for calculations.

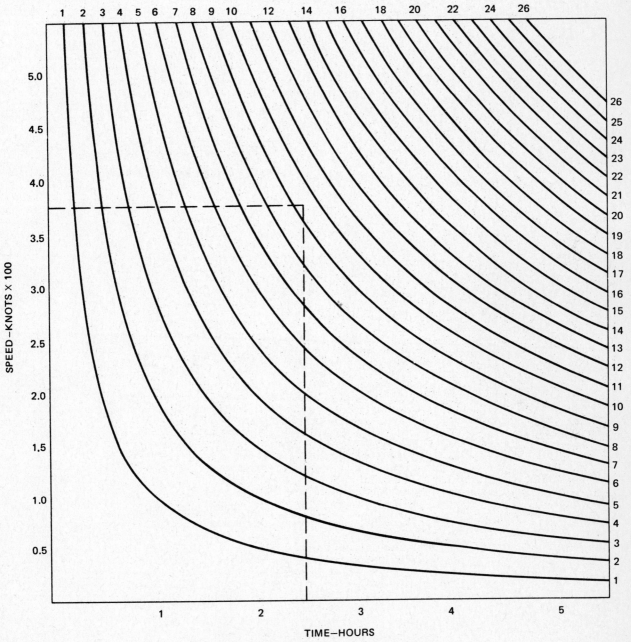

FIG. 1-39 Graph relating three variables.

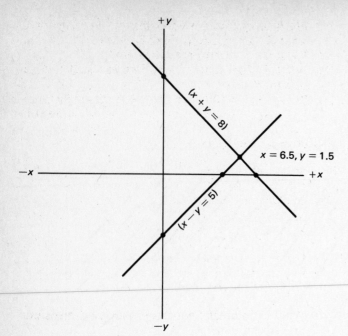

FIG. 1-40   Graphical solution of equations.

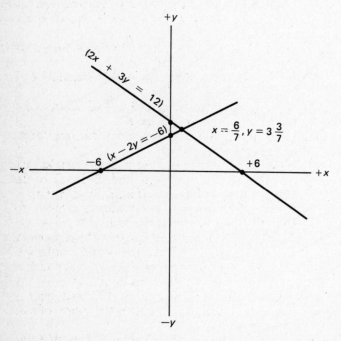

FIG. 1-41   Graphical solutions.

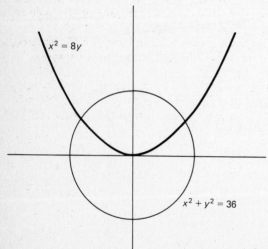

FIG. 1-42   Graphs of algebraic equations.

● REVIEW PROBLEMS

1. What do we mean by *decimal system?*
2. What is a *whole number?*
3. Explain *even* and *odd* numbers.
4. Add 324, 5623, 29, and 425.
5. Subtract 2465 from 3182.
6. What is the product of 785 and 56?
7. If a jet airplane holds 6400 gal of fuel and the fuel weighs 8 lb/gal, what is the total weight of the fuel load?
8. Divide 2888 by 76.
9. If a jet airplane flies 2500 mi in 4 hrs, what is its speed in miles per hour?
10. A rocket travels 300 mi in 4 minutes (min). What is its speed in feet per second? (5280 ft = 1 mi)
11. Add: $1\frac{5}{8} + 3\frac{3}{4} + 5\frac{5}{6}$.
12. Multiply $28\frac{4}{7}$ by $12\frac{3}{5}$.
13. Divide $78\frac{1}{8}$ by $6\frac{1}{4}$.
14. A strip of aluminum sheet is 36 in long. How many pieces $3\frac{3}{4}$ in long can you cut from the strip?
15. Add 5.25 + 4.036 + 235.4.
16. Convert 0.625 to a fraction.
17. Convert $\frac{9}{16}$ to a decimal.
18. Divide 3.7 by 4.9 to the nearest ten-thousandth.
19. What is 30 percent of 284?
20. A rocket is fueled with 6000 lb of oxygen. Before the rocket is fired, 7 percent of the oxygen is lost through "boil-off." How many pounds of oxygen remain?
21. If the fuel tank of an airplane is filled to a capacity of 155 liters (L) and if 93 L are consumed during a subsequent flight, what percentage has been consumed?
22. If a person borrows $8500.00 to purchase an airplane and if the interest charged for 1 year (yr) is $722.50, what is the rate of interest?
23. What is the ratio of 225 to 25?
24. If one airplane has a capacity of 75 passengers and another airplane has a capacity of 100 passengers, what is the ratio between their capacities?
25. Solve for *x*: 12:30 = *x*:45.
26. Simplify 14^2, 12^3, and 5^4.
27. Extract the square root of 106, 929, and 601.23.
28. Define *right angle, straight angle, acute angle,* and *obtuse angle.*
29. Define *diameter, radius, arc, chord,* and *sector.*
30. Define *right triangle, acute triangle, isosceles triangle,* and *equilateral triangle.*
31. Define *square, rectangle, parallelogram,* and *trapezoid.*
32. Define *complementary angles* and *supplementary angles.*
33. If two straight lines in the same plane can never intersect regardless of how far they are extended, what is their relationship to each other?
34. What is the pythagorean theorem?
35. What is the sum of the interior angles of a polygon?
36. Draw a circle and then using the radius distance strike arcs entirely around the circle. What do you find?
37. Give the following formulas:

    Area of a rectangle
    Area of a triangle
    Area of a trapezoid
    Area of a circle
    Volume of a rectangular solid
    Surface area of a sphere
    Volume of a sphere
    Lateral area of a cone
    Volume of a cone

38. What is the area of a rectangle having dimensions of 6 in and 9 in?
39. What is the area of a square having sides equal to 30 in each? How many square feet are in the square?
40. What is the area of a triangle having a base of 24 in and a height of 15 in?
41. What is the area of a circle with a diameter of 7 in?
42. If a trapezoid has bases of 9 in and 12 in and a height of 5 in, what is its area?
43. What is the volume of a cube that has one corner dimension of 35 cm?
44. What is the volume of a rectangular solid having dimensions of 13 in, 16 in, and 22 in? Give the volume in cubic feet as well as in cubic inches.
45. What is the volume of a cylinder that has a diameter of 16 in and a length of 60 in?
46. Draw a right triangle, designating the sides as *a, b,* and *c* and the angles as *A, B,* and *C,* making *C* the right angle. Now show the side ratios for sin *A,* cos *A,* tan *A,* cot *A,* sec *A,* csc *A.*
47. If side *a* of a right triangle is 15 ft and angle *A* (opposite side *a*) is 32°, what is the length of side *c* (the hypotenuse)?
48. What is the height of a power pole if a sight line from a point 100 ft from the base of the pole to the top of the pole makes an angle of 40°?
49. If an airplane flies 100 kilometers (km) north and 70 km west, how far will it be from the point at which it started?
50. An airplane is at 1000 ft altitude and approaching a landing field. How far does it travel horizontally to reach the field if the average glide angle is 5°?

2 SCIENCE FUNDAMENTALS

● INTRODUCTION

During the twentieth century, more progress has been made in technology and science than was made in all the previous centuries since the beginning of time. The space shuttle program has established the reality of civilians traveling in space (see Fig. 2-1). Interplan-

FIG. 2-1 Space Shuttle. *(NASA)*

etary communication and exploration has been accomplished, and it is possible that human beings will visit one or more of the other planets in our solar system before the end of this century. Although less spectacular than space travel, great advances have been made in air transportation and the automatic control of aircraft. The jumbo jets, air buses, and supersonic transport aircraft have brought air travel throughout the world to millions of people who could not have had this advantage a few years ago (see Fig. 2-2). All

FIG. 2-2 Boeing 767. *(Boeing Commercial Airplane Co.)*

these accomplishments are the result of the application of the laws of physics.

In our present age, the person concerned with the operation, maintenance, or design of any of the thousands of mechanical and electronic devices is constantly faced with the application of scientific law. A technician must have an understanding of the common laws of physics in such areas as motion, heat, light, and sound if he or she is to acquire the basic technical knowledge necessary to maintain and repair today's advanced aircraft and space vehicles.

● MEASUREMENTS

In order to arrive at values of distance, weight, speed, volume, pressure, etc., it is necessary to become familiar with the accepted methods for measuring these values and the units used to express them. Through the ages, human beings have devised many methods for measuring. In this text we shall concern ourselves principally with the **English system** and the **metric system,** both of which are used extensively throughout the world.

Length and Distance

The system commonly used in the United States is the **English system;** it involves such units as the inch, foot, yard, mile, pint, gallon, pound, ton, etc.

Originally, the units inch, foot, yard, and mile were not exact multiples or factors of one another, but, for the sake of convenience, the foot was made equal to 12 in, the yard was made 3 ft, and the mile was made 5280 ft, or 1760 yards (yd). The nautical mile (nmi), used internationally for navigation, is based on one-sixtieth of 1° of the earth's circumference at the equator. It is approximately 6080 ft. A speed of 1 nmi/h is called 1 kn. Most airplane airspeed indicators are calibrated in knots.

Most other countries, and scientists in this country, make use of a system called the **metric system,** which has many advantages over the English system. One aspect of the metric system which makes it extremely convenient is the existence of a set of prefixes which are applied uniformly to all the standard units. Kilo, for example, always means 1000: a kilometer is 1000 meters (m), and a kilogram (kg) is 1000 grams (g) (the gram is the metric unit of mass).

In the metric system all the measurements of length are either multiples or subdivisions of the meter based on multiples of 10. The following table shows how the units of length are related:

10 millimeters = 1 centimeter
10 centimeters = 1 decimeter
10 decimeters = 1 meter
10 meters = 1 decameter
10 decameters = 1 hectometer
10 hectometers = 1 kilometer

One meter is equal to 39.37 in, which is a little longer than the U.S. yard. Thus 1 decimeter (dm) is equal to 3.937 in, 1 cm is equal to 0.3937 in, and 1 millimeter (mm) equals 0.03937 in. In practice, the units of length most commonly used are the millimeter, the centimeter, the meter, and the kilometer.

Since today's aircraft are utilized all over the world, many manufacturers will provide both systems of measurement in their technical manuals. An example of this is illustrated in Fig. 2-3.

Area

Measurements of area are usually indicated in units that are the squares of the units of length. In the English system the units of area are the **square inch** (in^2), the **square foot** (ft^2), the **square yard** (yd^2), and the **square mile** (mi^2). Another unit of area commonly used for measuring land is the **acre**, which is equal to 43 560 ft^2.

Area in the metric system is indicated in square metric units. These are the **square centimeter** (cm^2), the **square meter** (m^2), and the **square kilometer** (km^2). Land measure in the metric system is indicated by means of the **hectare**. The hectare is equal to 10 000 m^2, or 2.47 acres.

Volume and Capacity

Volume and capacity are indicated in three-dimensional units, or cubes, of the basic linear units. The **cubic inch** (in^3), **cubic foot** (ft^3), and **cubic yard** (yd^3) are the common units of volume in the English system. Other units of volume and capacity are the **pint, quart,** and **gallon.** In general, when speaking of volume, we use the cubed units, and when speaking of capacity, we use the pint, quart, or gallon.

The metric system employs the cubed metric units as units of volume. The most commonly expressed units are the **cubic centimeter** (cm^3) and the **cubic meter** (m^3). For capacity the **liter** is generally employed. The liter (L) is equal to 1000 cm^3. It is also equal to 1.056 U.S. liquid quarts. One m^3 = 1000 L.

Weight

Units of weight in the English system are no more standardized than other units of measure. Among units of weight in common use are **grain, troy ounce, avoirdupois ounce, troy pound, avoirdupois (avdp) pound,** and **ton** (short ton). The following table shows the relationship among the units of weight mentioned:

7000 grains = 1 avdp pound
5760 grains = 1 troy pound

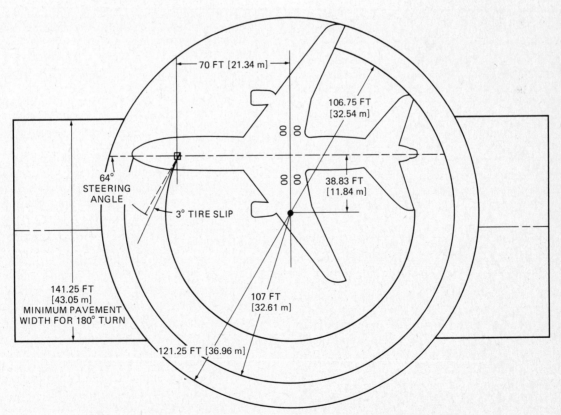

FIG. 2-3   Turning radii for the L-1011 given in both English and metric measurements.   *(Lockheed Corp.)*

12 troy ounces = 1 troy pound
16 avdp ounces = 1 avdp pound
2000 avdp pounds = 1 ton (short)

The **grain** is the smallest unit of weight in the English system and was derived from the weight of a grain of wheat. It is used principally for the measurement of medicinal components or drugs. The **troy** ounce and pound are used for weighing precious metals such as platinum, gold, and silver. **Avoirdupois** weights are used for almost all materials and objects in the English-speaking countries. These are the common ounce (oz), pound, hundredweight (cwt), and ton, with which we are most familiar.

The most uniform system of weights is the metric system. The following shows the relationship among the metric units of weight:

1000 grams = 1 kilogram
1000 kilograms = 1 metric ton

The **gram** (g) is the weight of 1 cm^3 of pure water at a temperature of 4°C or 39.2°F, which is the point of greatest density for water. The gram is equal to about 15.432 grains, and the **kilogram** is 2.2046 lb (avdp).

It is readily seen that the metric system of weights is much less complex than the English system because the weights can be converted from one unit to another merely by moving the decimal point to the right or left. For very small measurements of weight the **milligram** (mg) is used in the metric system. The milligram is 0.001 g.

Units of Measurement

It is readily understandable that units of measurement are required for many more purposes than those mentioned in the foregoing paragraphs. For example, we must be able to measure **force, density, electrical values, light intensity, sound intensity, velocity, energy,** and numerous other values. In this section we have explained only the most commonly known units of measurement; however, as we discuss other areas of physical laws and phenomena, we shall also define the units of measurement required in each area.

Conversion factors to show the relationships among various units are given in the Appendix of this text.

● GRAVITY, WEIGHT, AND MASS

Gravity, or **gravitation,** is the universal force that all bodies exert upon one another. It is defined by the **universal law of gravitation,** which states, **"The attraction between particles of matter is directly proportional to the product of their masses and inversely proportional to the square of the distance between them."** This can be expressed by the equation

$$F = G \frac{m_1 m_2}{r^2}$$

where
F = attractive force
r = distance between two bodies (particles)
m_1 and m_2 = masses of bodies
G = the universal gravitation constant
($G = 6.67 \times 10^{-11}$ newton·m^2/kg^2)

The terms weight and mass are, many times, used interchangeably; however, weight and mass have different definitions.

Weight is the pull exerted upon a body by the gravitation of the earth. The weight of a body may change depending upon its distance from the center of the earth. The farther away an object is from the center of the earth the less it will weigh.

Mass is the property of a body that is a measure of the amount of material it contains. Regardless of the location to the earth's center, the mass of a body will never change as long as no matter is added to or removed from it. It must be remembered, however, that 1 lb of mass will not be exactly 1 lb of weight if the mass is not at the proper distance from the center of the earth.

Both mass and weight are measured in units of pounds or kilograms. Another unit of measuring mass that may be encountered is that of the slug. A **slug** is defined as being a unit of mass having a value of approximately 32.175 lb under standard atmospheric conditions.

Density

An important physical property of a substance is its density (d) which is its mass per unit volume. Table 2-1 is a list of the densities of various common substances. In metric units, density is properly expressed in kilograms per cubic meter. The density of water is 1000 kg/m^3 since 1 m^3 of water has a mass of 1000 kg.

In the English system, density should be expressed in slugs per cubic foot since the slug is the unit of mass in this system. In these units the density of water is 1.94 slugs/ft^3.

Because weights rather than masses are usually specified in the English system, a quantity called weight density is commonly used. As the name suggests, weight density is weight per unit volume, and its units are pounds per cubic foot.

Changes in temperature will not change the mass of a substance but will change the volume of the substance by expansion or contraction, thus changing its density. To find the density of a substance, its mass and volume must be known. Its mass is then divided by its volume to find the mass per unit volume.

Specific Gravity

The **specific gravity** of a substance is the ratio of the density of the substance to the density of water. To determine the specific gravity of a substance when the density is known, we merely divide the weight of a given volume of the substance by the weight of an

TABLE 2-1 Weights of Materials

Gas

Gas	Density*	
	lb/ft³	kg/m³
Air (at 59°F)	0.076 510	1.225 60
Air	0.080 710	1.293 00
Carbon dioxide	0.123 410	1.977 00
Carbon monoxide	0.078 070	1.250 70
Helium	0.011 140	0.178 46
Hydrogen	0.005 611	0.089 89
Nitrogen	0.078 070	1.250 70
Oxygen	0.089 212	1.429 20

Liquids

Liquid	Specific gravity	Density*	
		lb/U.S. gal	kg/m³
Alcohol (methyl)	0.81	6.80	810
Benzine	0.69	5.80	690
Ethylene glycol	1.12	9.30	1 120
Gasoline	0.72	6.00	720
Glycerine	1.26	10.50	1 260
Jet fuel	0.80	6.70	800
Lubricating oil	0.89	7.40	890
Mercury	13.55	113.00	13 550
Sulfuric acid	1.84	15.40	1 840
Water	1.00	8.35	1 000

Metals

Metal	Specific gravity	Density*	
		lb/ft³	kg/m³
Aluminum	2.700	168.56	2 700
Brass	8.400	524.40	8 400
Copper	8.920	556.88	8 920
Gold	19.300	1 204.90	19 300
Iron	7.860	490.70	7 860
Lead	11.344	708.21	11 344
Mercury	13.550	845.93	13 550
Platinum	21.450	1 339.12	21 450
Silver	10.500	655.52	10 500
Titanium	4.500	280.95	4 500
Zinc	7.140	445.75	7 140

*At standard sea-level pressure and 0°C.

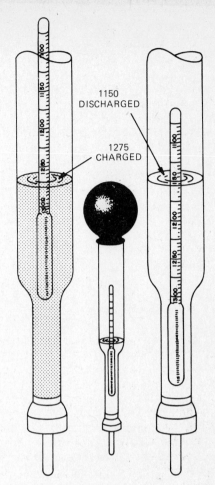

FIG. 2-4   A hydrometer.

equal volume of water. For example, if we wish to know the specific gravity of lead and we know that the density in pounds per cubic foot is 708.21 lb, we divide 708.21 by 62.4 (the density per cubic foot of water) and obtain 11.34, which is the specific gravity of lead. Table 2-1 lists the specific gravity for various substances.

A device called a hydrometer is used for measuring the specific gravity of liquids. This device consists of a tubular-shaped glass float contained in a larger glass tube (see Fig. 2-4). The float is weighted and has a vertically graduated scale. To determine specific gravity, the scale is read at the surface of the liquid in which the float is immersed. A specific gravity of 1000 is read when the float is immersed in pure water. When immersed in a liquid of greater density, the float rises, indicating a greater specific gravity. For liquids of lesser density the float sinks, indicating a lower specific gravity.

An example of the use of the hydrometer is its use in determining the specific gravity of the electrolyte (battery liquid) in an aircraft battery. When a battery is discharged, the calibrated float immersed in the electrolyte will indicate a specific gravity of approximately 1150. The indication of a charged battery is between 1275 and 1310.

The specific gravity and density of solids and liquids are expressed as explained in the foregoing paragraphs; however, gases require a different treatment. It is apparent that the weight of a given volume of a gas will depend upon the pressure and temperature. For this reason, the density of a gas is given according to standard pressure and temperature conditions, that is, a pressure of 76 cm of mercury and a temperature of 0°C. Under these conditions it is found that the density of dry air is 1.293 g/L or 0.081 lb/ft³. Since the density of air is used as a standard, the specific gravity of air is given as 1. The specific gravity of any other gas is the ratio of the mass of a given volume of the gas to an equal volume of dry air, with both the gas and the air being under standard conditions.

Speed

Speed is a measure of how fast something is moving. It tells us how fast an object has moved during a certain time without respect to direction. Speed involves only the length of the path traveled by a body and the time required to travel the path. In Fig. 2-5 the average speed of the automobile would be 35 mph because it required 1 h to travel a distance of 35 mi.

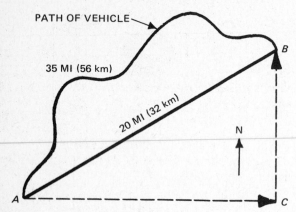

FIG. 2-5 Diagram to illustrate speed and velocity.

Speed is indicated as the ratio of the distance traveled to the time of travel. This definition is expressed in the equation

$$\text{Speed} = \frac{\text{distance traveled}}{\text{time required}}$$

Velocity

We often hear the words **velocity** and **speed** used in the same sense, that is, to indicate how fast something is moving. The two words are similar in some respects, but there is an important difference. When we use the word velocity, we include direction, distance in a straight line from point to point, and the **time** required to move from one point to another. If an automobile travels from point *A* to point *B* (in Fig. 2-5) in 1 h following the irregular path, the **velocity** of the automobile is 20 mph east-northeast.

● FORCE AND MOTION

Newton's First Law of Motion

Force is required to produce motion in a body that was previously at rest, and force is also required to stop the motion of a body. The concepts of force and motion are expressed by Newton's laws of motion. Newton's first law of motion states: **A body at rest tends to remain at rest and a body in motion tends to remain in motion in a straight line unless forced to change its state by an external force.**

The first law of motion concerns the property of matter called **inertia.** Inertia is defined as the tendency of matter to remain at rest if at rest or to continue in motion in a straight line if in motion. Hence it is seen that Newton's first law of motion defines inertia. The property of inertia is demonstrated by the fact that ar-

tificial satellites will continue to orbit around the earth for months or years even though they have no means of propulsion. This is because there is no appreciable outside force in space to retard the speed of the satellite. The only substantial force acting on a satellite orbiting the earth is the force of gravity. Gravity causes the satellite to curve around the earth instead of shooting off in space.

Inertia can also be demonstrated by a simple experiment with a glass of water. If a glass of water is placed on a piece of paper on a smooth surface, the paper can be jerked from under the glass without disturbing the glass or its contents. The inertia of the glass of water causes it to remain at rest when the paper is moved.

Units of Force

Force may be defined as a push or a pull upon an object. A unit of force in the metric system is the **newton** (N). The newton is the force required to accelerate a mass of 1 kilogram (kg) 1 meter per second per second (m/s^2).

The **dyne** (dyn) is also employed in the metric system as a unit of force. One dyne is the force required to accelerate a mass of 1 g 1 centimeter per second per second (cm/s^2). One newton is equal to 100 000 dynes [0.225 lb].

In the English system the **pound** (lb) is used to express the value of the force. For example, we say that a force of 30 lb is acting upon a hydraulic piston.

Newton's Second Law of Motion

Newton's second law of motion explains **acceleration.** Acceleration may be defined as **the change in velocity of a body.** If the body increases in velocity, it has **positive** acceleration; and if it decreases in velocity, the acceleration is **negative.** The second law of motion states: **The acceleration of a body is directly proportional to the force causing it and inversely proportional to the mass of the body.** This means that a given body will accelerate in proportion to the force applied to it. For example, if a 10-lb [4.536-kg] weight accelerates at 32.2 ft/s^2 [9.82 m/s^2 in the metric system] when a 10-lb [44.5-N] force is applied to it, it will accelerate at 64.4 ft/s^2 when a 20-lb [89-N] force is applied to it.

Airplanes must have rapid acceleration in order to take off from relatively short runways. In the case of the airplane pictured in Fig. 2-6, the acceleration is

FIG. 2-6 Gates Model 55 Learjet. *(Gates Learjet Corp.)*

provided by two jet engines each capable of producing 3700 lb [16 000 N] of thrust.

If we ignore the friction of air acting on a freely falling body, we find that it will accelerate at the rate of 32.2 ft/s² [9.82 m/s²]. This is called the **acceleration of gravity** and is indicated by the letter symbol g. From this we know that, when a force equal to the weight or mass of a body is applied to the body, the body will accelerate at 32.2 ft/s² [9.82 m/s²] if there is no friction or other force opposing the applied force. This knowledge is useful in developing an equation based upon Newton's second law.

$$F = \frac{Ma}{g}$$

where F = force, lb [N]
M = mass, lb [N]
a = acceleration, ft/s² [m/s²]
g = 32.2 ft/s² [9.82 m/s²]

With this formula, if we know the weight or mass, we can easily determine the force required to produce the acceleration.

If we wish to know how much force is required to accelerate a 4000-lb [1814-kg] automobile to a speed of 60 mph [26.8 m/s] in 10 s, we must first determine the rate of acceleration. Sixty miles per hour is 88 ft/s; hence if the automobile is accelerated from 0 to 88 ft/s in 10 s, the rate of acceleration is 8.8 ft/s² [2.683 m/s²]. Then, by the formula

$$F = \frac{4000 \times 8.8}{32.2} = 1093.1 \text{ lb [4862 N]}$$

From the above equation we know that it requires a force (push) of 1093.1 lb [4862 N] to accelerate the automobile from 0 to 60 mph in 10 s.

Newton's Third Law of Motion

Action and reaction are explained by Newton's third law of motion: **For every action there is an equal and opposite reaction.** This law indicates that no force can exist with only one body but that existence of a force requires the presence of two bodies. One body applies the force, and the other body receives the force. In other words, one body is **acting** and the other is **acted upon.** This is clear in the operation of an automobile. The wheels of an automobile exert a force against the road tending to force the road to the rear. Since the road cannot move to the rear, the automobile is forced to move forward. Other examples of action and reaction are the recoil of a gun, the backward force of a fire hose when the water is turned on, the thrust of a propeller, and thrust of a rocket, and the thrust of a jet engine as is shown in Fig. 2-7.

Thrust

Thrust is defined as being a reaction force which is measured in pounds.

The **thrust** of a propeller, rocket, or gas-turbine en-

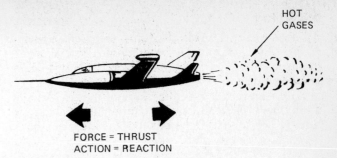

FORCE = THRUST
ACTION = REACTION

FIG. 2-7 Newton's third law of motion.

gine depends upon the acceleration of a mass (weight) in accordance with Newton's second law. A propeller accelerates a mass of air, a rocket accelerates the gases resulting from the burning of fuel, and a turbine engine accelerates both air and fuel gases. The quantity (mass) of air and gases accelerated and the amount of acceleration determine the thrust produced.

The basic formula for the thrust of a gas-turbine engine is

$$F = \frac{w}{g}(V_2 - V_1)$$

where F = force, lb
w = flow rate of air and fuel gases
g = acceleration of gravity (32.2 ft/s²) [9.82 m/s²]
V_2 = final velocity of gases
V_1 = initial velocity of gases

The Space Shuttle, illustrated in Fig. 2-8, is capable of producing in excess of 5 000 000 lb of thrust [25 000 000 N] on lift-off.

FIG. 2-8 The Space Shuttle on lift-off. *(NASA)*

Momentum

Another quantity that often must be taken into account in situations that involve moving bodies is momentum. There are two kinds of momentum: **linear** and **angular.**

Linear momentum is a measure of the tendency of a moving body to continue in motion along a straight line. It is defined as the product of the mass of a body times its velocity:

$$\text{Momentum} = \text{mass} \times \text{velocity}$$

Angular momentum is a measure of the tendency of a rotating body to continue to spin about an axis. It is the rotational quantity that is similar to linear momentum. A simple example of angular momentum is a top, which would spin indefinitely if not for the friction between its tip and the ground and between its body and the air.

The precise definition of angular momentum is more complicated than linear momentum because it depends not only upon the mass of the body and speed that it is turning but also upon how the mass is distributed in the body. The farther away from the axis of rotation the mass is distributed, the more angular momentum it will have.

● CENTRIFUGAL AND CENTRIPETAL FORCE

Everyone is familiar with the fact that a weight attached to the end of a cord and twirled around as shown in Fig. 2-9 will produce a force tending to cause

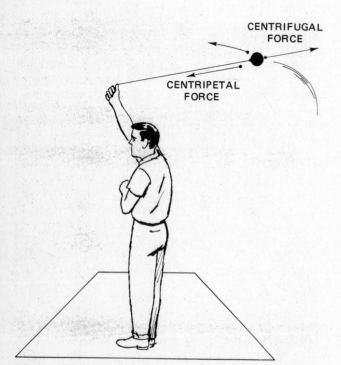

FIG. 2-9 Centrifugal and centripetal forces.

the weight to fly outward from the center of the circle. This outward pull is called **centrifugal force.** There is an equal and opposite force pulling the weight inward and preventing it from flying outward; this is called **centripetal force.**

From Newton's first law of motion we know that a body in motion tends to continue in motion in a straight line. Hence, when we cause a body to move in a circular path, a continuous force must be applied to keep the body in the circular path. This is centripetal force.

Centripetal force is always directly proportional to the mass of the object in circular motion. Thus, if the mass of the object in Fig. 2-9 is doubled, the pull on the string must be doubled to keep the object in its circular path, provided the speed of the object remains constant.

Centripetal force is inversely proportional to the radius of the circle in which an object travels. If the string in Fig. 2-9 is shortened and the speed remains constant, the pull on the string must be increased since the radius is decreased and the string must pull the object from its linear path more rapidly.

We can use either one of two formulas for determining the magnitude of centripetal force, depending upon whether we are employing the metric system or the English system of measurements. In the metric system we determine the centripetal force in **dynes** by the formula

$$F = \frac{mv^2}{r}$$

where F = force, dyn or N
  m = mass, g or kg
  v = speed, cm/s or m/s
  r = radius of circle, cm or m

The result in dynes if divided by 100 000 will give the force in newtons.

To find the force in pounds we employ the formula

$$F = \frac{Wv^2}{gr}$$

where F = force, lb [N]
  W = weight, lb [N]
  v = speed, ft [m/s]
  g = acceleration of gravity, 32.2 ft/s² [9.82 m/s²]
  r = radius, ft [m]

If we wish to determine the value of the centripetal force required to keep a 10-lb [4.5-kg] weight in a circle of 20-ft [7-m] radius while moving at a speed of 60 ft/s [1.83m/s], we proceed as follows:

$$F = \frac{10 \times 60^2}{32.2 \times 20} = \frac{36\ 000}{644} = 55.9 \text{ lb [248.7 N]}$$

COMPOSITION AND RESOLUTION OF FORCES

Vectors

A vector quantity is any quantity involving both magnitude and direction. A **vector** is represented by a straight arrow pointing in the direction in which the quantity is acting, and the length of the arrow represents the magnitude of the quantity. For example, the arrow **OY** in Fig. 2-10 represents a force of 40 lb [18.1 kg] acting upward.

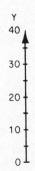

FIG. 2-10 Vector representing 40 lb directed upward.

In Fig. 2-11 a force of 10 lb [4.5 kg] is acting upward from a point O and a force of 5 lb [2.3 kg] is acting to the right from the point as shown by the vectors. We can determine the **resultant** force, or **resultant,** by drawing the parallelogram (rectangle in this case) $OXAY$ and then measuring the diagonal vector **OA.** **OA** is the resultant and by measurement is found to be approximately 11.2 lb [5 kg]. Since OAY is a right triangle, we can use the formula for solving right triangles as follows:

$$OA = \sqrt{10^2 + 5^2} = 11.18$$

To solve a problem involving a right triangle such as that shown in Fig. 2-11, it is convenient to use basic trigonometric formulas. These formulas are based upon trigonometric functions such as the **sine, cosine,** and **tangent** of a particular angle as explained in the

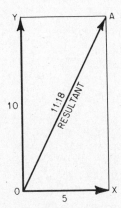

FIG. 2-11 Resultant of two vectors acting from the same point.

previous chapter of this text. That is, the values of the functions of a right triangle such as that shown in Fig. 2-12 may be expressed as follows:

$$Sine = \frac{side\ opposite}{hypotenuse} \quad or \quad \sin A = \frac{a}{c}$$

$$Cosine = \frac{side\ adjacent}{hypotenuse} \quad or \quad \cos A = \frac{b}{c}$$

$$Tangent = \frac{side\ opposite}{side\ adjacent} \quad or \quad \tan A = \frac{a}{b}$$

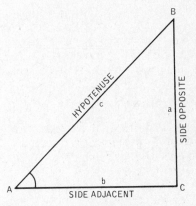

FIG. 2-12 Functions of a right triangle.

If we find the sine or tangent of the angle OAX in Fig. 2-11 on a table of trigonometric functions, we see that the angle is about 63°27'. The tangent of the angle is $\frac{10}{5} = 2$, which is the tangent of the angle 63°27' as shown in the tangent table. Hence we know that the direction of **OA** is 63°27' counterclockwise from **OX.**

The determination of the resultant force is particularly useful in calculating the effects of lift and drag forces that are imposed on airplanes.

The method of composition of forces can also be used to determine the path of flight of an airplane flying in a crosswind. In Fig. 2-13 an airplane is flying at 150 mph on a heading of 340° and there is a crosswind of 50 mph blowing from 45°. To determine the flight path and the speed made good, we draw a triangle or parallelogram of forces. OB is drawn to a convenient scale to represent 150 mph 340° clockwise from 0°. OA is drawn from 45° through the point O and scaled to represent 50 mph. Then the parallelogram $OACB$ is drawn, and the points O and C connected by the diagonal OC. This diagonal represents the direction and speed of the airplane along its actual flight path. By measurement we find that the length of the line OC represents a speed of about 137 mph and a flight path of approximately 320°.

Triangles such as those shown in Fig. 2-13 may also be solved by the law of sines or the law of cosines, as explained in Chap. 1 of this text.

From any one point we can combine any number of forces and obtain a resultant. In Fig. 2-14, if a force OA has a value of 20 lb [89 N], a force OB has a value of 15 lb [66.7 N], and a force OC has a value of 26 lb [115.7 N] we can easily determine the resultant force OE and find that it is about 56 lb [249 N]. First we

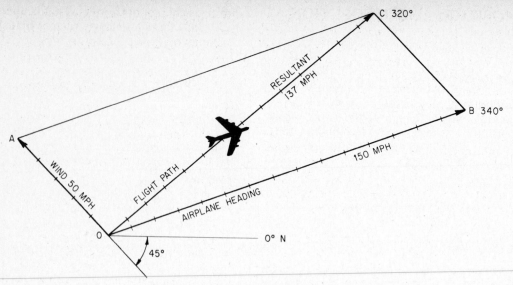

FIG. 2-13   Composition of forces in a flight problem.

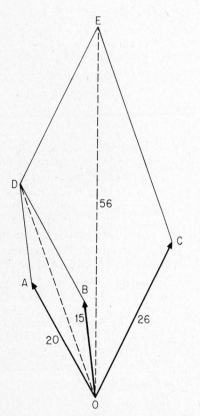

FIG. 2-14   Resultant of three forces exerted at one point.

draw the parallelogram *OADB* and then draw the diagonal *OD*, which is the resultant of *OA* and *OB*. Then using *OD* as a side, we draw the parallelogram *ODEC*. The diagonal *OE* is the resultant of the three forces and has a value of about 56 lb as previously stated.

A force can be **resolved** into its components if certain facts are known. For example, in Fig. 2-15 the force *OA* is composed for forces *OB* and *OC* whose directions are known but whose values are unknown. To determine the values of forces *OB* and *OC* it is merely necessary to draw *AC′* parallel to *OB* to the

point where it intersects with *OC* and *AB′* parallel to *OC* to the point where it intersects with *OB*. If *OA* is 30 lb [133.5 N], then we find that *OC′* (*OC*) is about 24 lb [107 N], and *OB′* (*OB*) is about 17 lb [76 N]. This determination of the components of a force is called **resolution of forces.**

It must be noted that the combination of vector forces, that is, forces having both magnitude and direction, requires something more than the mere addition of quantities. Vectorial addition is accomplished by means of parallelograms or triangles as demonstrated in the foregoing paragraphs. This is important in the solution of problems of alternating current, where we must combine electrical quantities having both magnitude and phase angle.

● WORK, ENERGY, AND POWER

Work

Work, in the scientific or engineering sense, refers to the application of force to a body and the displacement of the body in the direction of the force.

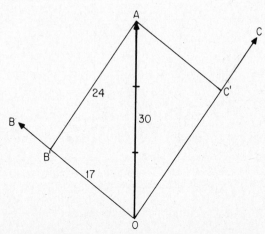

FIG. 2-15   Resolution of a force into its components.

The physical definition of work is not the same as the psychological; if we try unsuccessfully to lift a heavy weight, we may feel that we are doing a great deal of work. If the weight does not move, however, no physical work is done. Motion is a necessary part of the physical definition of work, but motion alone does not mean that work is done. To keep a ball rolling along a perfectly smooth, horizontal surface does not require work; no force is being overcome. Lifting a ball, however, does require work; the gravitational force of attraction of the earth for the ball must be overcome.

More precisely, work is defined as the product of the force overcome and the distance moved.

$$\text{Work } (W) = \text{force } (F) \times \text{distance } (D)$$

If the distance is zero, no work is done by the force no matter how great it is. Even if something moves through a distance, work is not done on it unless a force was acting.

In the English system of units, the unit in which work is measured is called the **foot pound** (ft·lb). One foot pound is the amount of work done by a force of 1 lb that acts through a distance of 1 ft.

In the metric system, work is measured in **joules**, where 1 joule (J) is the amount of work done by a force of 1 N that acts through a distance of 1 m. The joule is based on the erg, which is the work done when 1 g of mass is moved through a distance of 1 cm. The joule is equal to 10^7 ergs, or 10 000 000 ergs. Therefore,

$$1 \text{ J} = 1 \text{ N·m}$$

To convert from one system of units to another, we note that

$$1 \text{ J} = 0.738 \text{ ft·lb}$$
$$1 \text{ ft·lb} = 1.36 \text{ J}$$

It is safe to say that work is performed whenever a force is applied and a movement occurs as the result of the force. If we use a lever to raise a weight, we can easily conceive how a machine performs work. In the drawing of Fig. 2-16 a lever is used to raise a box weighing 500 lb. The distance from the **fulcrum** f to the center of gravity of the box B is 2 ft, and the distance from the fulcrum to the applied force A is 4 ft. By measurement we shall find that to lift the box a distance of 6 in ($\frac{1}{2}$ ft) we must move the opposite end A of the lever a distance of 1 ft. The work applied to the lever is then

$$250 \text{ (lb)} \times 1 \text{ (ft)} = 250 \text{ ft·lb } [\, 340 \text{ J}]$$

The work done on the box is expressed by the equation

$$W = FD = 500 \text{ (lb)} \times \tfrac{1}{2} \text{ (ft)} = 250 \text{ ft·lb}$$

Thus we see that the work applied to the lever is equal to the work done by the lever, or **output = input.** This is true for any machine that is 100 percent efficient. There is always some loss in a machine because of friction, so the output cannot be quite as great as the input.

We can find hundreds of examples of work all about us, and new types of work are being developed constantly. A modern example is the lifting of a large rocket vehicle by means of a rocket engine. If a rocket engine lifts a 10 000-lb payload to a height of 100 mi, we can see that 5 280 000 000, or 528×10^7, ft·lb of work has been performed on the payload.

Energy

Energy may be defined as the capacity for doing work. There are two forms of energy: **potential energy** and **kinetic energy. Potential energy** is the form of energy possessed by a body because of its position or configuration. For example, if a 10-lb weight was raised against the pull of gravity 2 ft higher than it was before, it is now capable of exerting a force of 10 lb (its weight) through a distance of 2 ft in returning to its original position. As long as the weight is elevated, the 20 ft·lb of work done on it is stored, ready to be released whenever the weight is lowered. It has, in other words, the potentiality of doing this amount of work, which is called its **gravitational potential energy.**

The notion of potential energy is not necessarily associated only with the force of gravity. A tightly wound spring or a gas compressed in a metal cylinder

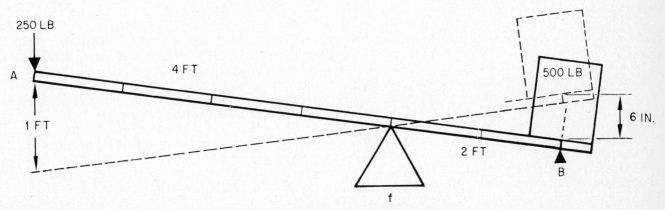

FIG. 2-16 Work done by means of a lever.

is also able to produce mechanical work that can be measured in the same units. Potential energy in chemical, rather than mechanical, form is stored in an automobile fuel tank that is filled with gasoline or in a charge of high explosive in an artillery shell. Potential energy in still another form lies in the nuclear energy of the fissionable fuel rods in a nuclear reactor.

Kinetic energy is the energy possessed by a body because of its motion. When a hammer is used to drive a nail, the kinetic energy of the hammer does the work of driving the nail. In the case of a water-driven turbine, when the water is stored, it possesses potential energy; but when it is released through the turbine, it has kinetic energy, and this energy is imparted to the turbine.

The **law of conservation of energy** states that energy can be neither destroyed nor created; it can be changed only in form. We can also say that the total amount of energy in the universe always remains constant. This means, of course, that the amount of energy imparted to a body is equal to the energy released by the body when the body is returned to its former state, that is, the condition of position, temperature, or configuration it was in before energy was imparted to it.

One of the principal facts to remember concerning energy is that any change in the state of a body requires either that energy be given to the body or that energy be given up by the body. If an automobile moving at a constant speed on a straight and level highway is to be accelerated, the engine must be given more fuel energy, which in turn is transmitted as mechanical energy to the wheels. If the automobile speed is to be decreased, the automobile must give up energy to the air (air friction) and to the road in the form of friction. For a more rapid decrease in speed the brakes must be applied, thus converting the kinetic energy of the automobile to heat energy at the brake drums.

Energy is expressed in the same units as those used for work. If a weight of 20 lb [9 kg] is raised 10 ft [3 m], the potential energy it acquires is 200 ft·lb [271 J] because

$$E = Fs \quad \text{or} \quad E = 20 \times 10 = 200$$

where E = energy
F = force
s = distance

In all cases the energy is the product of a force times a distance.

Kinetic energy can be expressed by the equation

$$E_k = \frac{Wv^2}{2g}$$

where W = weight
v = velocity
g = acceleration of gravity

Potential energy is expressed as

$$E_p = Fs \quad \text{or} \quad E_p = Wh$$

where W is the weight of a body and h is the height to which it has been raised.

Power

The rate of doing work is called **power,** and it is defined as the work done in unit time. As a formula, this would be

$$\text{Power} = \frac{\text{work done}}{\text{time taken to do the work}}$$

Power is expressed in several different units such as the **watt, ergs per second,** and **foot pounds per second.** The most common unit of power in general use in the United States is the **horsepower.** One horsepower (hp) is equal to 550 ft·lb/s or 33 000 ft·lb/min. In the metric system the unit of power is the **watt** or the kilowatt (kW). One watt is equal to $\frac{1}{746}$ hp; that is, 746 watts = 1 hp and 1 kW = 1.34 hp.

If we wish to compute the power necessary to raise an elevator containing 10 persons a distance of 100 ft in 5 s and the loaded elevator weighs 2500 lb, we proceed as follows:

$$\text{Power} = \frac{2500 \times 100}{5 \times 550} = 90.9 \text{ hp } [67.81 \text{ kW}]$$

If there were no friction, the power required would be 90.9 hp; however, there is always a substantial amount of friction to overcome, and so the actual power required would be considerably greater than that indicated.

In the foregoing problem the amount of work to be done is 2500×100 ft·lb. This work is to be done in 5 s or at a rate of 50 000 ft·lb/s. Since 1 hp = 550 ft·lb/s, we divide 50 000 by 550 to find the horsepower.

Let us suppose that we wish to find the horsepower required to fly a light airplane at 100 mph [86.84 kn] when we know that it requires 200 lb of thrust to overcome the drag at this speed. One hundred miles per hour is equal to 146.67 ft/s and 200 lb at 146.67 ft/s is equal to 29 334 ft·lb/s. We then divide 29 334 by 550 (ft·lb/s) to obtain the horsepower, which we find to be about 53.1 hp [39.6 kW].

● **MACHINES**

Nearly any mechanical device which aids man in doing work can be called a machine. Machines are devices which make use of **the law of conservation of energy** to change either the direction or the magnitude of a force.

No machine is capable of doing more work than the driving agent does on the machine. We know this to be true without examining the machine in detail since energy must be conserved in these systems. If more work could be done by the machine than was done upon the machine to make it run, the machine would be creating the ability to do work within itself. This, of course, it cannot do. Moreover, since any practical machine will have some friction, energy will be lost due to friction work within the machine. Hence more work has to be done on the machine than is required to do the work without the machine. An understanding of the principles of simple machines provides a necessary foundation for the study of compound machines, which are combinations of two or more simple machines.

The Lever

The **lever** is one of the simplest machines which enables a person to exert greater force than the person's direct effort can produce. Figure 2-17 shows a lever being used to raise one end of a heavy box. The distance from the fulcrum f to point A is 3 times the distance from f to B. Under these conditions a 100-lb force applied at A will lift a 300-lb weight at B.

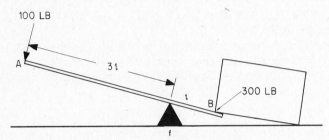

FIG. 2-17   Mechanical advantage of a lever.

Levers may provide mechanical advantages since they can be applied in such manner that they can magnify an applied force. The **mechanical advantage (MA)** of a machine may be determined by dividing the weight (W) lifted by the force (F) applied

$$ MA = \frac{W}{F} $$

Mechanical advantage is usually expressed in terms of a ratio: for the lever in Fig. 2-17, it would be 3:1.

Another principle of the lever is illustrated in Fig. 2-18. When the lever is balanced, it will be found that

$$ F_1 D_1 = F_2 D_2 $$

In the illustration $50 \times 40 = 200 \times 10$. The product of the force times the distance is called a **moment**. The distance from the reference point (in this case f) is called the **arm**. In a balance problem the force is usually a weight, and so we can say that the **moment is equal to the weight times the arm**. This principle is utilized in determining the weight and balance conditions of an airplane.

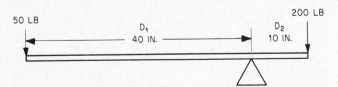

FIG. 2-18   Balancing a lever.

Pulleys

Pulleys are often used to provide a mechanical advantage. In Fig. 2-19 a single pulley is shown at a with a rope to support a weight of 50 lb [22.7 kg]. In order to

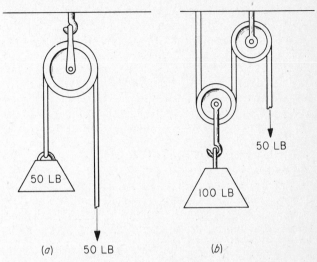

FIG. 2-19   Mechanical advantage of pulleys.

raise the weight, at least 50 lb must be applied downward on the end of the rope. At b two pulleys with a rope provide a 2:1 advantage. Observe here that the weight is being supported by two sections of rope; hence one section of rope only must support one-half the total weight. It is therefore possible to apply 50 lb at the end of the rope and raise 100 lb [45.4 kg] with the pulley. (At b we are not considering the weight of the lifting pulley.) In any arrangement of pulleys (block and tackle) the number of ropes actually supporting the weight determines the mechanical advantage of pulley combination.

In Fig. 2-20 a set of double pulleys is shown. It will be noted that there are four sections of rope supporting the weight of 80 lb [36.3 kg]. This means that each rope is required to support only 20 lb [9 kg], disregarding the weight of the pulleys. It is necessary to apply only slightly over 20 lb to the traction rope to raise the 80-lb weight. The mechanical advantage is therefore 4:1, since the weight is four times as great as the effort required to raise it.

Gears also are employed to provide a mechanical advantage. If one gear having a diameter of 2 in is meshed with a gear having a diameter of 4 in, as shown in Fig. 2-21, the mechanical advantage is 2:1. The small gear must turn two revolutions in order to rotate the larger gear one revolution. The turning force applied to the small gear need be only one-half the force applied to the large gear by the driven load. The turning force is called **torque,** or **twisting moment.** A train of many gears can be used to provide an extreme mechanical advantage. However, if the number of gears is too great, the friction will eventually become so large that the mechanical advantage is lost.

When a machine is driven by a belt, mechanical advantage can be obtained by driving a large pulley with a small pulley as shown in Fig. 2-22. The diameter of the large pulley is twice that of the smaller pulley;

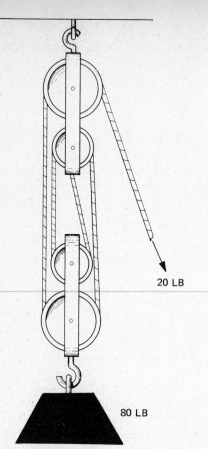

FIG. 2-20 Multiplication of forces by means of pulleys.

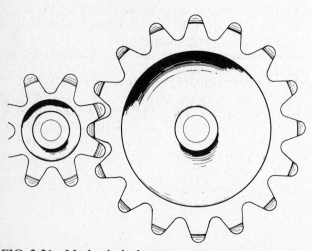

FIG. 2-21 Mechanical advantage produced by gears.

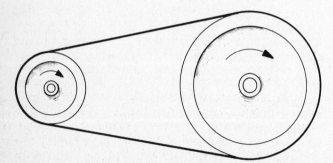

FIG. 2-22 Mechanical advantage with a pulley drive.

hence the mechanical advantage is 2:1. The large pulley will turn one-half the speed of the small pulley, but the torque required on the small pulley shaft will be only one-half the torque of the large pulley shaft.

Inclined Plane

The **inclined plane** offers a simple example of mechanical advantage that is used in many devices. Figure 2-23 illustrates the principle of the inclined plane. Assume that B is a 120-lb barrel of flour and it must be raised 2 ft. The work to be done is 240 ft·lb. If the barrel is moved a distance of 6 ft to do the work, then the force need be only 40 lb because $6 \times 40 = 240$.

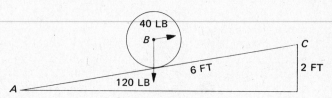

FIG. 2-23 Principle of the inclined plane.

Therefore, the ratio of the length of the inclined plane to the vertical distance is the mechanical advantage, disregarding friction.

The screw is actually an adaptation of the inclined-plane principle. A screw jack can be used to raise buildings through human power by providing a large multiplication of the human effort. There is considerable friction in a screw arrangement, but even with the friction the screw makes possible a great multiplication of force.

A combination of the screw and a gear, called a "worm-gear" arrangement, is often used in machines to provide a large mechanical advantage. A worm-gear drive is shown in Fig. 2-24. One revolution of the drive shaft will move the rim of the driven gear the distance of one tooth. The mechanical advantage is therefore equal to the number of teeth on the driven gear. If the gear has 20 teeth, the mechanical advantage is 20:1.

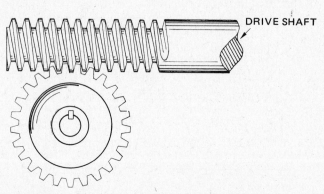

FIG. 2-24 Worm-gear drive to develop mechanical advantage.

HEAT

Heat is a form of energy, and it is manifested in matter by the motion of the molecules. As heat is increased, the motion of the molecules increases. This adds to the internal energy of the material to which the heat is applied. If heat is applied to one end of a metal rod, it will be found that the other end of the rod gradually becomes warmer. This is because the molecules in the heated end of the rod increase their motion and strike other molecules along the rod with greater force, which increases the motion of the molecules progressively all along the rod. When this occurs, we say that the rod is conducting heat.

When a heated object is in contact with a cold object, the heat transfers from the hot object to the cold object. This also is brought about by the motion of the molecules in the hot object striking the molecules of the cold object, thus increasing the motion of the molecules in the cold object.

Temperature is the degree of heat or cold (heat energy) measurable in a body. The measurement of temperature is accomplished with a thermometer, and the value is expressed in degrees Fahrenheit or Celsius.

Absolute zero, one of the fundamental constants of physics, is commonly used in the study of gases. It is usually expressed in terms of the centigrade scale. If the heat energy of a given gas sample could be progressively reduced, some temperature would be reached at which the motion of the molecules would cease entirely. If accurately determined, this temperature could then be taken as a natural reference, or as a true absolute zero value. Absolute temperatures are expressed in kelvins or degrees Rankine (°R). The various temperature scales are compared in Table 2-2.

TABLE 2-2 Comparison of Temperature Scales

	°Fahrenheit	°Celsius	°Rankine	Kelvins
Water boils	212	100	672	373
Ice melts	32	0	492	273
Absolute zero	−460	−273	0	0

It will be noted from Table 2-2 that there is a difference of 180°F or °R between the point where water boils and ice melts. For this same range of temperature there is a difference of 100°C or kelvins. From this we know that Fahrenheit and Rankine degrees have the same size and Celsius degrees and kelvins have the same size. Furthermore, we can see that 100° in the Celsius scale is equivalent to 180° in the Fahrenheit scale. A detailed conversion table is given in the Appendix.

To convert one type of scale to the other we can use the following formulas:

$$F = \tfrac{9}{5}C + 32 \qquad C = \tfrac{5}{9}(F - 32)$$

Since the kelvin is the same size as the Celsius degree and begins counting 273° higher on the scale, it is apparent that in order to convert a temperature given in degrees Celsius into kelvins, we need only add 273. Also, since the Fahrenheit and Rankine scales are the same, to convert degrees Fahrenheit to Rankine we need to add 460.

Effects of Heat

The effects of heat make possible many of the powerful machines that we use in our modern world. Various fuels such as gasoline may be burned to cause a great expansion of air and the gases of combustion. The expanded gases are used to move the pistons in gasoline engines, thus causing the crankshaft to rotate and develop power for turning a propeller. In jet engines the burning of fuel with oxygen causes a great expansion of gases, which drives the turbine of the engine to compress the air, and the exhausted gases cause the jet thrust. Similarly, the burning of either liquid or solid fuels in a rocket causes a great expansion of gases, which produces the thrust due to the acceleration of the gases as they are ejected from the rocket nozzle.

The energy available from a fuel is determined by the amount of heat it produces when burned. In order to measure heat energy it is necessary to employ heat units. A heat unit has been established on the basis of heating value. In the metric system the heat unit is called the **calorie (cal).** One calorie is the amount of heat required to raise the temperature of 1 g of water 1°C and is equal to approximately 4.186 J. In the English system the unit of heat measurement is that amount of heat necessary to raise the temperature of 1 lb of water through 1°F. This quantity of heat is called the **British thermal unit (Btu).**

The amount of work that can be performed by a certain amount of heat has been determined. For example, it has been found that 1 Btu can do 778 ft·lb of work. Also, 1 cal can produce 4.186 J of work, or about 3.09 ft·lb. From these values we can easily determine how much work can be obtained from a certain amount of fuel provided that we know the heat value of the fuel. The heat values of a few common fuels are given in Table 2-3.

TABLE 2-3 Heat Value

Fuel	Heat value	
	Btu/lb	Cal/g
Wood	7 000–8 000	4 000–4 500
Gasoline	20 000–20 500	11 000–11 400
Coal	13 500–15 000	7 600–8 400
Gas	9 900–11 500	5 500–6 400

From the foregoing we can determine how much power can be developed when a certain amount of gasoline is being burned in a given time. For example, if an engine is burning 40 lb/h of gasoline, how much power will the engine deliver if it is 35 percent efficient?

40 lb/h of gasoline would produce 800 000 Btu/h

800 000 Btu/h = 13 333.3 Btu/min

13 333.3 Btu/min = 10 373 307.4 ft·lb/min

Since 1 hp = 33 000 ft·lb/min,

10 373 307.4 ft·lb/min = 314 hp

314 × 0.35 = 110 hp, approximately

From the foregoing example, it is apparent that we can obtain approximately 110 hp [82.03 kW] from an engine burning 40 lb/h [18.4 kg/h] of fuel when the engine is 35 percent efficient.

Specific Heat

The **specific heat** of a substance is the number of calories required to raise the temperature of 1 g of the substance 1°C or the number of Btu's required to raise 1 lb of the substance 1°F. The value is the same for each. The specific heat of water is 1, and that of other substances is usually less than 1. Table 2-4 gives the specific heats of a variety of common substances.

TABLE 2-4 Specific Heat for Various Substances

Substance	Specific heat
Water	1.000
Ice	0.500
Alcohol	0.590
Aluminum	0.220
Copper	0.093
Iron	0.110
Silver	0.056
Lead	0.031
Mercury	0.033
Platinum	0.032

Table 2-4 shows that only 0.22 Btu is required to raise the temperature of 1 lb of aluminum 1°F or 0.11 Btu to raise the temperature of 1 lb of iron 1°F. It is therefore obvious that the specific heat of different substances varies substantially.

Another interesting heat phenomenon is noted when a substance melts or when it is converted to a vapor. For example, when 1 g of water changes to ice at 32°F [0°C], it gives up 80 cal. When 1 g of ice is melted, it absorbs 80 cal. This accounts for the fact that ice can be forced to melt by the application of salt, and this melting process absorbs heat and lowers the temperature of the water-salt mixture that we call **brine.** The home ice-cream freezer utilizes this principle in freezing the ice cream. The heat absorbed by a melting substance is the **heat of fusion.**

The effect of heat on metals is particularly important in the design and operation of heat engines. Metals usually expand with an increase in temperature, and this expansion must be accounted for in the design of an engine. The increase in length of a metal per unit length per degree of rise in temperature is called the **coefficient of linear expansion.** For iron the coefficient of linear expansion is 0.000 012 cm/°C. This means that 1 cm of iron will have a length of 1.000 012

cm after the temperature is increased 1°C. The coefficient of expansion for aluminum is twice that of iron. As illustrated in Table 2-5, the modern alloys used in jet engines often expand much more than ordinary iron or steel. For this reason a jet engine must be de-

TABLE 2-5 Coefficient of Thermal Expansion for Selected Materials

Material	Coefficient of thermal expansion 10^{-6}/°C
Steel	11.7
Iron	12.0
Aluminum	23.6
Brass	20.0
Magnesium	26.0
Titanium	9.5
Rubber	162.0

signed to "grow" as its temperature increases. A large engine may increase in length more than an inch at operating temperature. This is one of the reasons the technician must be careful to allow correct clearances when assembling a jet engine. This same precaution must be taken in assembling any device that is subject to large changes in temperature during operation.

Laws of Thermodynamics

There are two principal laws of thermodynamics which are of particular interest to the aviation technician. The term **thermodynamics** is defined as **the branch of the science of physics dealing with the mechanical actions and relations of heat.**

The first law of thermodynamics is similar to the law of the conservation of energy. That is, **heat energy cannot be destroyed; it can only be changed in form.**

The second law of thermodynamics states that **heat cannot flow from a body of a given temperature to a body of a higher temperature.** That is, heat will only flow from a warmer body to a cooler body. **Coldness** is the absence of heat.

● HEAT TRANSFER

The ability of a substance to either retain or transfer heat plays an important role in selecting the materials from which aircraft components and engines will be constructed. There are three basic methods by which heat is transferred between locations and substances; they are **conduction, convection,** and **radiation.**

Conduction

Conduction is the transfer of energy through a conductor by means of molecular activity and without any external motion.

Materials that are poor conductors are used to prevent the transfer of heat and are called heat insulators. Certain materials such as finely spun glass or asbestos are particularly poor heat conductors. The heat conductivities of some familiar materials are shown in Table 2-6. These figures were determined by utilizing

TABLE 2-6 Heat Conductivity of Different Materials*

Material	Heat conductivity (at 18°C)
Silver	0.970 00
Copper	0.920 00
Aluminum	0.480 00
Iron, cast	0.110 00
Lead	0.080 00
Mercury	0.016 00
Glass	0.002 50
Brick	0.001 50
Water	0.001 30
Wood	0.000 30
Asbestos	0.000 20
Cotton wool	0.000 04
Air	0.000 06

*Expressed in calories per second flowing through 1 cm² when the temperature gradient is 1°C/cm

a cube of the material 1 cm on a side, with one face of the cube kept just 1°C cooler than the opposite face. Heat will flow from the warmer face to the cooler, and the number of calories per second flowing through the cube is the heat conductivity or thermal conductibility of the material.

Convection

Convection is the process by which heat is transferred by movement. It has its basis in the fact that heated bodies increase their volume and therefore decrease their density. In a teakettle, the water near the bottom is heated by immediate contact with the hot metal. It becomes lighter than the rest of the water in the kettle and floats up, its place being taken by the cooler water from the upper layers. These convection currents carry the heat up "bodily," and mix the water in the kettle. A similar phenomenon takes place in the atmosphere when, on a hot summer day, air heated by contact with the ground streams up to be replaced by cooler air masses from above.

As the air rises to higher and cooler layers of the atmosphere, the water vapor in the air condenses into a multitude of tiny water droplets and forms the cumulus clouds so characteristic of hot summer days.

Radiation

A third way in which heat energy can be transferred from one body to another is by **radiation.** Standing outdoors at an open fire on a winter day, the heat received does not come to you by conduction through the air or the ground since both of these are cold. The heat is not transferred by convection since the hot air over the fire rises into the sky, taking its heat away with it. Just as the bright flames and the glowing coals radiate light, they also send out an even greater amount of radiant heat that travels unimpeded through the air, to be absorbed by skin and clothing.

The term *radiation* refers to the continual emission of energy from the surface of all bodies. All the energy we receive from the sun has been radiated in this way across 93 million miles of vacuum. Only a small part of this energy is in the form of light; most of the rest is radiant heat. Conduction and convection usually take place very slowly while radiation takes place at the speed of light.

Effects of Heat on Aircraft Materials

In maintaining, servicing, repairing, and operating aircraft and power plants, aviation technicians must be constantly aware of the effects of heat on the different materials and structures utilized in the design and construction of the part or parts upon which they are working. A seemingly simple operation, such as sharpening a drill or cutting a piece of steel can generate sufficient heat to damage the drill or soften the teeth of the metal-cutting saw. Special lubricants or cutting oils are often used in these cases.

Aluminum alloys used in aircraft structures are usually heat-treated to provide the maximum strength, hardness, and toughness possible. If such an alloy should be subjected to temperatures beyond a prescribed level, the alloy would lose its strength and might fail during operation. For this reason, some aluminum alloys cannot be welded.

Carbon steels and alloy steels are damaged by excessive temperatures. Some steels will maintain adequate strength after welding, while others must be reheat-treated after welding. The technician must, therefore, make sure to follow the prescribed procedure and use the correct materials.

Steels and other metals in an engine are often damaged when the engine is operated beyond safe temperature limits. Excessive heat in an engine causes such problems as feathered piston rings, scored cylinders, burned valves, burned piston heads, burned bearings and bushings, and a number of other damages. The operator of an aircraft engine must observe carefully all operating limitations specified by the manufacturer.

The very high temperatures experienced in the operation of gas-turbine engines make it necessary that certain parts of the engines be constructed of high-temperature alloys often referred to as "exotic metals." Such alloys often contain large percentages of cobalt, nickel, columbium, and other metals which, when alloyed, provide great strength at high temperatures as well as high-temperature corrosion resistance. During the operation of gas-turbine engines, great care must be taken to observe the temperature limitations.

In servicing aircraft heating and air-conditioning systems, the technician must ascertain that ducting for hot air be of a type and material adequate to withstand the temperatures involved. The temperature of air in hot-air ducts is often at a level which will burn or otherwise damage materials that are not designed for high temperatures. In replacing ducts, the technician must replace parts with the correct part number as specified by the manufacturer.

Many types of plastic materials are used throughout modern aircraft. Some plastics can withstand relatively high temperatures, while others will melt or burn. **Thermoplastic** plastics will melt or soften when exposed to excessive temperatures, while **thermosetting** plastics will burn, char, shrink, or crack when overheated. The technician must assure that the type of plastic material being installed is correct for the temperatures to which it will be subjected.

As explained previously, metals expand and contract with changes in temperature. An increase in temperature usually means an expansion of the metal. Since different metals have different coefficients of expansion, designers must allow clearances and other features which will permit the expansions and contractions to occur without warping or overstressing the materials involved. For example, an aluminum alloy fuselage expands more rapidly than the steel control cables when temperature increases. It is necessary, therefore, to install automatic tension adjusters with control cables for large aircraft.

● FLUIDS

Properties of Liquids

A **liquid** is defined as a substance that flows readily and assumes the shape of the container but does not tend to expand indefinitely. Liquids retain their total volume and a gallon of water will remain a gallon whether it is poured into a flat dish or into a tall, narrow container. The molecules in a liquid are free to move throughout the confining space, but they are bound with a force, one with another, so that they tend to remain together. Liquids are virtually incompressible. The highest pressures obtainable with modern laboratory equipment are able to squeeze water only into about three-fourths of its original volume.

Viscosity

When a force is applied to a liquid, the fluid deforms permanently under the force and we say that the liquid flows. Some liquids are more fluid, i.e., flow more readily, than others. It is convenient to have a quantity which measures the resistance of the liquid to flow. Such a quantity is the **viscosity** of the liquid. Its precise definition is the molecular property of a fluid which enables it to support stresses for a finite time and thus resist deformation.

The viscosity of several liquids is given in Table 2-7. The viscosity of air is also given. Notice the tremendous range of viscosities in this table. It should also be mentioned that the viscosity of liquids is very temperature sensitive. For most liquids, viscosity decreases with increasing temperature. This is a reflection of the fact that the molecules are less tightly bound together at the higher temperatures, and hence the friction between them is less. For water at 0, 50, and 100°C, the viscosity is 1.79, 0.55, and 0.28 centipoise (cP), respectively.

TABLE 2-7 Viscosities of Liquids at 30°C

Material	Viscosity, cP
Air	0.019
Acetone	0.295
Methanol	0.510
Water	0.801
Ethanol	1.000
SAE No. 10 oil	200.000

Archimedes' Principle

The **buoyancy** principle, first discovered by **Archimedes,** is as follows: **A body placed in a liquid is buoyed up by a force equal to the weight of the liquid displaced.** We can easily see from this that a floating body will displace its own weight in liquid. This can be demonstrated by placing a block of wood in a container of liquid that is filled to the overflow point, as shown in Fig. 2-25. When the block is placed in the liquid, an amount of the liquid will flow out that is equal in weight to the weight of the block of wood. This can be proved by weighing the liquid that has flowed out of the container.

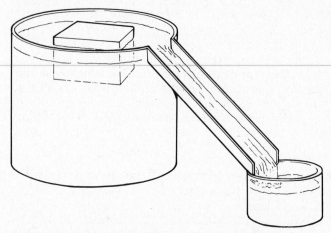

FIG. 2-25   Displacement of a liquid by a floating solid.

Buoyancy is the effect of liquid force on a body immersed or submerged in the liquid. For example, if a particular object which has a volume of 3 ft³ [0.08 m³] is submerged in water, the buoyant force (BF) on it will be equal to the weight of the displaced water, namely, the weight of 3 ft³ of water. Since the weight density of water is 62.4 lb/ft³ [1000 kg/m³], the buoyant force will be 3 times 62.4, or 187 lb [85 kg].

Fluid Pressure

Many types of liquids are known, but the most common is water. We shall, therefore, look at the characteristics of water as an example of a liquid. Consider water in a cylindrical glass. Because of its weight, the water exerts a force on the bottom of the glass, a force the same as that which would be produced by a cylindrical piece of ice if the water were frozen and the glass walls removed. This force is distributed over the entire bottom of the glass so that each square centimeter of the area of the bottom carries its own equal share of the load. This force per unit area is called the pressure P and is equal to the total force divided by the area over which it is exerted:

$$F = PA \quad \text{and} \quad P = \frac{F}{A}$$

Consider the bowl of water shown in Fig. 2-26. The arrows represent the direction of force acting on the sides and bottom of the bowl. The amount of force exerted at any particular point depends upon the vertical

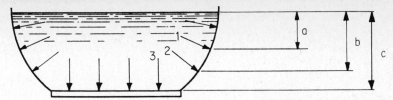

FIG. 2-26 Force exerted by a liquid.

distance from the surface of the water to the point where the force is to be measured. The force exerted at point 1 is determined by the distance *a*. Likewise, the force at 2 is determined by the distance *b*, and the force on the bottom at 3 or any other point on the bottom of the bowl is determined by the distance *c*. To compute total force we must employ force per unit area, or pressure. Pressure is expressed in pounds per square inch, in grams per square centimeter, or in kilopascals (kPa). One pascal is a pressure of $1 \text{ N} \cdot \text{m}^2$.

Pressure due to fluid height (*h*) also depends on the density (*d*) of the fluid. Water, for example, weighs 62.4 lb/ft^3 or 0.036 lb/in^3, but a certain oil might weigh 55 lb/ft^3, or 0.032 lb/in^3. In the illustration of Fig. 2-26, if the distance *c* is 4 in [10.16 cm], the pressure *P* at the bottom of the bowl will be

$$P = hd = 4 \text{ in} \times 0.036 \text{ lb/in}^3$$
$$= 0.14444 \text{ psi } [0.996 \text{ kPa}]$$

Water, or any other fluid, since it does not have a rigid shape, will exert equal pressure in all directions. Thus our formula $P = hd$ is equally useful in calculating the pressure against the wall of a container at any depth no matter at what angle the wall happens to be. For this reason, the shape of the container makes no difference. In the illustrations of Fig. 2-27, we see three containers of water. The areas of the bottoms of the containers are equal, and the depth of the water *h* is the same in each container. We know from these conditions that the total force on the bottom of each container is the same as that on the other two containers.

In Fig. 2-27, the containers are filled with a liquid of density *d*, the pressure at the bottom is *hd* for all, and at the points marked *A* the pressure in all three is *hd*.

Pressure and Force in Fluid Power Systems

In a hydraulic system the fluid is confined, and so pressure applied to the fluid at any point is immediately transmitted to every other point touched by the fluid. Figure 2-28 illustrates **Pascal's law,** which states that a liquid under pressure in a closed container transmits pressure undiminished to all parts of the enclosing wall. If the 10-lb weight is acting on an area of 1 in^2, then a pressure of 10 lb [68.95 kPa] is transmitted to every square inch of the enclosing wall. This principle is the key to the use of fluids to transmit power or force from one point to another.

The terms **area, pressure,** and **force** are mathematically related. This relationship establishes the foundation upon which hydraulic systems are based. It permits the engineer to determine the operating pressures required for certain units in a system, the size of pump required, and the requirements for the material strength in system units.

Consider the relationship of force, pressure, and area. If any two of these factors are known, it is possible to calculate the third. Force equals pressure times area ($F = P \times A$), pressure equals force divided by area ($P = F/A$), and area equals force divided by pressure ($A = F/P$). A simple aid for the solution of problems involving these factors is the diagram shown in Fig. 2-29. For example, suppose a force of 25 lb is exerted on a piston whose area is 5 in^2. That pressure is the amount of force per unit of area expressed in pounds per square inch; therefore, on each square inch of the piston there is 5 lb of force, or 5 psi.

Great force can be applied by means of a hydraulic system merely by selecting the size of piston to produce the desired force. The multiplication of force through the use of hydraulic pistons is illustrated in Fig. 2-30. If a force of 10 lb is applied on the small piston that has an area of 1 in^2, the pressure of the fluid becomes 10 psi. This pressure is transmitted undiminished to the large piston, which has an area of 50 in^2. Then $50 \times 10 = 500$ lb [2224 N], which is the force exerted by the large piston.

In the landing-gear actuating system for a large airplane we may find that the pressure of the system is 3000 psi [20 685 kPa]. If this pressure is applied to a piston with an area of 15 in^2, we can easily see that a

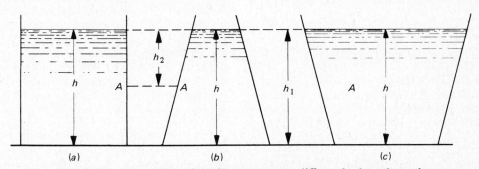

FIG. 2-27 Total effect of liquid pressures on differently shaped vessels.

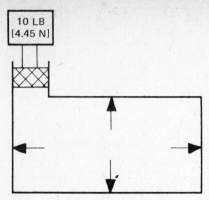

FIG. 2-28   Pressure transmitted by a liquid.

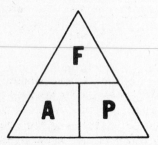

FIG. 2-29   Device for determining the arrangement of the force, pressure, and area formulas.

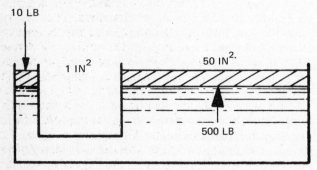

FIG. 2-30   Multiplication of force by means of hydraulic pistons.

force of 45 000 lb [200 160 N] is developed to raise the landing gear. In practice this amount of force is not actually required, although the system could develop it.

● THE NATURE AND LAWS OF GASES

There are three states of matter: **solid, liquid,** and **gas.** All three are collections of **molecules** bound together by forces of varying strengths. There is a difference in the freedom of movement which molecules have that depends on whether they are part of a solid, a liquid, or a gas; the three states of matter can be defined in terms of the freedom of motion of their molecules. If the molecules are restrained so that the distance between them is constant as is their relative position, then the material is called a solid. If the molecules remain more or less at a constant distance apart but are free to change their relative position, then the material is called a liquid. If neither the relative position nor the distance between the molecules is maintained constant, the material is called a gas.

The molecules in a gas are constantly in motion, and the velocity of the motion is dependent upon the temperature of the gas. Each molecule of a gas travels in a straight line until it strikes another molecule, and at this time both of the two colliding molecules continue their travels in different directions. The movement of molecules in a gas causes the gas to dissipate quickly when it is not confined. Because of the movement of the molecules in a gas, the gas will always completely fill any container in which it is placed; that is, the molecules will distribute themselves evenly throughout the space in the container.

A gas can be easily compressed, and as it is compressed, its pressure increases as its volume decreases, the temperature remaining constant. This is in accordance with **Boyle's law,** which states that **the volume of a confined body of gas varies inversely as its absolute pressure, the temperature remaining constant.** This can be expressed by the following equation:

$$\frac{V_1}{V_2} = \frac{P_2}{P_1} \quad \text{(temperature constant)}$$

In this equation V is the symbol for volume and P is the symbol for pressure. The subscript figures identify the first volume and pressure and the second volume and pressure.

Absolute pressure means the pressure above zero pressure, keeping in mind that the pressure of the atmosphere at sea level is approximately 14.7 psi [101.325 kPa] or 29.92 in [76 cm] of mercury (Hg). Therefore, if we have a confined gas in a cylinder with the gas at atmospheric pressure and then compress the gas to one-half its former volume, the pressure exerted by the gas will then be approximately 29.4 psi [202.65 kPa]. We must assume here that the temperature remains constant, although under normal conditions, when a gas is compressed, its temperature increases.

Just as changes in gas volume are related to pressure changes, so are they also related to temperature changes. This characteristic of a gas is expressed by the law attributed to the French physicist Jacques A. C. Charles (1746–1823). **Charles' law** states that **the volume of a gas varies in direct proportion to the absolute temperature.** Absolute temperature is temperature related to absolute zero, or the condition where there is a complete absence of heat. Absolute zero is −460°F [−273°C]. The application of Charles' law to a gas requires that the pressure remain constant. If we confine a gas so that the volume remains constant, then we find that the pressure varies in accordance with absolute temperature. Charles' law can be expressed by the following equation:

$$\frac{V_1}{V_2} = \frac{T_1}{T_2} \quad \text{(pressure constant)}$$

The equation relating to the pressure of a gas where the volume is constant is

$$\frac{P_1}{P_2} = \frac{T_1}{T_2} \quad \text{(volume constant)}$$

In each of the above equations we must remember that the temperature must be expressed in absolute units, that is, in degrees above absolute zero.

In order to determine the amount of expansion in a gas as the result of an increase in temperature, we must know the **coefficient of expansion.** This value is approximately the same for all gases and is found to be $\frac{1}{273}$, or 0.00366, for each degree Celsius with the gas at 0°C.

The **general gas law** is derived by combining Boyle's law and Charles' law. It is expressed by the following equation:

$$\frac{P_1 V_1}{T_1} = \frac{P_2 V_2}{T_2}$$

This equation can be used to determine a change in volume, pressure, or temperature of a gas when the other conditions are changed. It must be remembered that the pressures and temperatures expressed in this equation must be stated in absolute values.

● SOUND

The Nature of Sound

Sound may be defined in a number of ways, but the simplest definition is **that which can be heard.** Since sound is actually a vibration of a substance (solid, liquid, or gas), we can say that a sound can exist even though there may be no human ear in the vicinity to hear it. From this we can also understand that sound cannot travel in a vacuum. Our interest at this time is not the hearing of a sound but the nature of the vibration that causes the sensation of sound in the ear.

Vibration

Vibration is a rhythmic motion back and forth across a position of equilibrium of the particles of a fluid or of an elastic solid when its equilibrium has been disturbed. Such a state is clearly demonstrated in the plucking of a string on a musical instrument. The effects of vibration in mechanical devices create many of the problems that plague engineers in the design of such devices. It is therefore necessary in many instances to conduct vibration studies before the design of a particular machine can be approved.

To obtain a clear picture of a simple vibratory or harmonic motion, we can use a device such as that shown in Fig. 2-31. A T-shaped bar with a slotted head is mounted so that a pin on the rim of a wheel will fit into the slot in the bar. The bar is mounted in guides so that its motion is limited to two directions, up and down. When the wheel is rotated at a constant speed,

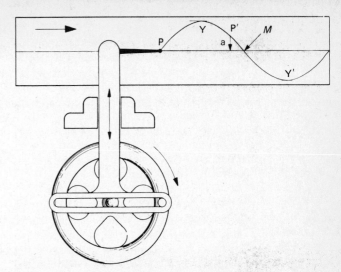

FIG. 2-31   Demonstration of harmonic motion.

the movement of the bar up and down will be a harmonic motion. It we attach a marking pen P to the end of the bar so that it will mark on a strip of paper moved at a uniform speed under the pen, then the pen will describe a sine curve on the paper. Thus we can see that the point P will move with a constantly changing speed, with the velocity being zero at points y and y' and maximum at the midpoint M.

Vibratory motion is **periodic** in nature and has characteristics by which it can be described. The time required for the motion to complete one cycle (one rotation of the wheel in Fig. 2-31) is called the **period.** The number of complete cycles occurring per second is the **frequency** of the vibration. The unit for frequency is the **hertz** (Hz). One hertz is equal to one cycle per second. The **amplitude** of the vibration is the distance from the midpoint of the swing to the point of maximum displacement. **Displacement** is the distance of the vibrating point from the midpoint of vibration at any particular time. In Fig. 2-31 the displacement of the point p' is a.

The wavelength is the distance measured along the direction of propagation between two corresponding points of equal intensity that are in phase on adjacent waves. This length can be represented by the distance between the adjacent maximum rarefaction (expansion) points in the traveling sound wave (see Fig. 2-32).

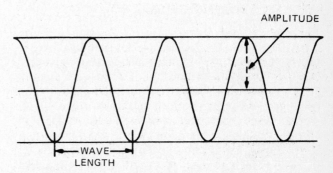

FIG. 2-32   Sound wavelength.

Wave Motion

To understand sound it is first necessary to examine **wave motion** because sound travels in waves. Sound is produced by initiating a series of compression waves in a medium capable of transmitting the vibrational disturbance. The particles of the medium acquire energy from the vibrating source and enter the vibrational mode themselves. As they do, they pass on the energy to adjacent particles. If the energy source continues to vibrate, a train of periodic waves travels through the medium, and a transfer of energy takes place.

Almost everyone has seen waves in water resulting from a disturbance in or on the water. The effects of sound in the atmosphere are similar to disturbances in water, but the difference in the compressibility of the two media also makes a difference in the nature of the waves. Figure 2-33 shows how waves emanate from a point in water where an object such as a small stone has been dropped. The illustration above the series of circles shows how a cross section of the water surface will appear when an object is dropped into the water.

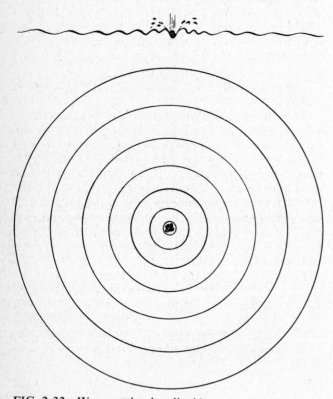

FIG. 2-33   Wave motion in a liquid.

Waves occur as two different kinds of motion. Wave movements are therefore described as **transverse** and **longitudinal.** If you tie one end of a rope to a stationary point and then, after stretching the rope to its full length, move the free end up and down rapidly with uniform motion, **transverse** waves will be produced in the rope (see Fig. 2-34). From this it will be seen that a transverse wave is one in which the material moves back and forth, sideways, or up and down from a zero

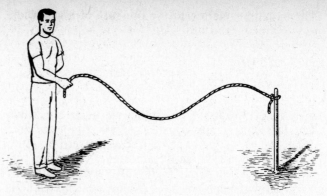

FIG. 2-34   Wave motion demonstrated with a rope.

reference line. The surface waves on water are of the transverse type.

One of the simplest methods for illustrating a **longitudinal** wave is to use a long coil spring stretched between two stationary points. After the spring is stretched, if one end is compressed and then released, a longitudinal wave will travel the length of the spring. The longitudinal wave is formed by a series of alternately compressed and expanded coils of the spring, as illustrated in Fig. 2-35. Sound travels through matter in the form of longitudinal wave motions.

FIG. 2-35   Wave motion in a coiled spring.

If we secure a piece of spring steel solidly at one end as shown in Fig. 2-36, we can observe harmonic motion when we pull the open end of the strip of steel to point A and release it. On the return swing the end of

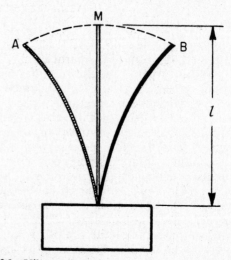

FIG. 2-36   Vibratory motion.

the strip will spring back to point B, which is almost as far from the center point M as was point A. Without added energy to maintain the vibration, the amplitude of the swing will decrease rapidly, but the frequency will remain constant. This requires that the speed of the motion also decrease. The frequency of the vibration of the steel spring in Fig. 2-36 depends upon the length of the strip l and the mass. If we mounted a weight on the end of the strip, we would find that frequency of vibration decreases.

As the spring steel moves from a to b, it does work on the gas molecules to the right by compression; the spring steel thus transfers energy to the molecules in the direction in which the compression occurs. At the same time, the gas molecules to the left expand into the space behind the spring steel as it moves and become rarefied. This motion also represents energy which is transferred to other molecules in the medium to the left of the spring steel. The combined effect of the simultaneous compression and rarefaction transfers energy to the molecules in both directions of the motion of the spring steel.

At the same time, of course, a corresponding series of rarefactions and compressions is produced to the right. The vibration of the spring steel thus generates longitudinal trains of waves in which vibrating gas molecules move back and forth along the path of the traveling waves, receiving energy from adjacent molecules nearer the source and passing it on to adjacent molecules farther from the source.

Sound Transmission

Most sounds come to us through the air which acts as the transmitting medium. Sound is a series of expansions and compressions in the molecules of the air. It is also of a similar nature when passing through a liquid or solid; however, the velocity will be different. The velocity of sound through a substance depends upon both the density and the elasticity of the substance that is conducting the sound.

Sound is transmitted better at low altitudes than at high altitudes because the air is less dense at higher altitudes. In a vacuum, sound will not be transmitted at all. Liquids are better transmitters of sound than gases because they have a higher elastic modulus and transmit the sound energy more readily. In general, because of their still higher elastic moduli, solids are better transmitters of sound than are liquids or gases.

The speed of sound in air is about 331.5 meters/second (m/s) at 0°C. As temperature increases, the speed of sound increases about 2 ft/s [60.96 cm/s] for each degree Celsius rise in temperature. From this, we know that at very high altitudes where the temperature is many degrees below zero, the speed of sound is much lower than it is at sea level. The speed of sound in water is about 4 times that in air; in water at 25°C sound travels about 1500 m/s. In some solids, the speed of sound is even greater. In a steel rod, for example, sound travels approximately 5000 m/s—about 15 times the speed in air. In general, the speed of sound varies with the temperature of the transmitting medium. Table 2-8 gives the speed of sound through

TABLE 2-8 Speed of Sound in Various Substances

Medium	Temperature, °C	Speed ft/s	Speed m/s
Air	0	1 087	331.4
Hydrogen	0	4 220	1 286.7
Oxygen	0	1 040	317.1
Aluminum	0	16 700	5 091.8
Copper	0	13 000	3 963.7
Glass	0	17 000	5 183.3
Iron, cast	0	14 200	4 329.6
Lead	0	4 040	1 231.8
Steel	0	16 000	4 878.4
Water	15	4 760	1 451.3

several common substances at the indicated temperatures.

Measurement of Sound Intensity

The intensity of sound is defined in terms of the energy being carried by the sound wave. It was pointed out previously that the vibrating source transmits motion or kinetic energy to the particles near it and that the wave is the flow of this energy away from the source. The vibrating source does work on the surrounding air, and this work appears as the energy of the sound wave. Hence, the energy carried away from the source by the sound wave in each second, the power, is actually a measure of the energy carried by the wave.

We define the intensity of the sound wave to be the energy per second transported through a unit area by the sound wave. The area is to be taken perpendicular to the direction of propagation of the wave. Sound intensity levels are measured by a unit called the **decibel** (dB). Originally the bel was used. It is 10 times larger than the decibel. It was named after Alexander Graham Bell. The decibel is defined according to a logarithmic scale. The intensity levels of a number of familiar sounds are listed in Table 2-9.

TABLE 2-9 Intensity Levels of Sounds

Type of sound	Intensity, dB
Threshold of hearing	0
Whisper	10–20
Very soft music	30
Average residence	40–50
Conversation	60–70
Heavy street traffic	70–80
Thunder	110
Threshold of pain	120
Jet engine	170

Resonance

Another interesting wave phenomenon is that of **resonance**. Resonance can be observed if two objects have the same natural vibrational frequency. Resonance can occur in an airplane by matching the vibration of the aircraft structure with the engine vibration. In the case of resonance, the air molecules will transmit their vibrations from one to another and eventually to the second object as well. This may allow vibration levels to build to dangerous and even destructive levels. For this reason aircraft undergo extensive vibration testing.

Doppler Effect

Today we often encounter the term **Doppler effect** in discussions of electronic nagivation and control systems as well as in discussions of sound. This is possible because both electromagnetic energy and sound travel in waves. The Doppler effect is observed in sound when the source of a sound wave changes its direction with respect to the hearer of the sound so that the number of sound waves per second reaching the ear is changed. Let us assume that a whistle is emitting a sound with a frequency of 1100 Hz and that the whistle is mounted on an automobile approaching the listener at a speed of 100 ft/s [30.5 m/s]. Assume also that the temperature is such that the speed of sound is 1100 ft/s [33.5 m/s]. Then with the frequency at 1100 Hz and the speed of sound 1100 ft/s, we know that there will be one sound wave (cycle) for each foot distance from the sound source. Since the sound source is approaching the listener at 100 ft/s, the listener will hear 1100 + 100 or 1200 Hz; that is, the sound will have a higher pitch than that at which it is emitted. When the sound source reaches and then goes away from the listener, the pitch will suddenly change, so that the listener will hear a pitch of 1000 Hz. This apparent change in pitch is called the Doppler effect. The formula for determining the change in frequency as a result of the Doppler effect is

$$p = f \frac{V}{V - S} \qquad \text{source moving toward listener}$$

$$p = f \frac{V}{V + S} \qquad \text{source moving away from listener}$$

where p = apparent frequency of sound heard by listener
f = frequency of sound at source
V = speed of sound
S = speed of source

The Doppler effect principle is used in electronic indicating systems because electronic signals are transmitted by means of waves and the apparent frequency of a signal from an approaching signal source will be higher than the frequency of a signal from a signal source that is moving away from the receiver. One of the principal applications of the Doppler effect in electronics is in navigation radar equipment.

● REVIEW QUESTIONS

1. What is the advantage of the metric system of measurements over the English system?
2. What is the basis for the *nautical mile?*
3. Compare the distances represented by the meter and the yard.
4. Compare 1 kg with 1 lb.
5. What is the basis for the gram weight?
6. State the universal law of gravitation.
7. Compare mass and weight.
8. Define *specific gravity.*
9. Define *density.*
10. What is the mass of 5 ft³ of a material whose density is 11.6?
11. Define the unit *newton.*
12. If the thrust of a gas-turbine engine is 11 500 lb, what is the thrust expressed in newtons?
13. What is the difference between velocity and speed?
14. State Newton's three laws of motion.
15. What is the acceleration of gravity?
16. What is the basic formula for the thrust of a gas-turbine engine?
17. What is the difference between linear and angular momentum?
18. Explain the difference between centripetal and centrifugal forces.
19. Describe a vector.
20. Define *work.*
21. Define *energy.*
22. Differentiate between potential energy and kinetic energy.
23. Give the law of conservation of energy.
24. Define *power.*
25. What is the unit of power in the metric system?
26. If an engine is delivering 140 kW of power, what is the horsepower being delivered?
27. Name two common devices by which mechanical advantage can be obtained.
28. What mechanical advantage can be obtained when a drive gear has 30 teeth and a driven gear has 120 teeth?
29. How is heat manifested in matter?
30. What causes the conduction of heat through a metal rod?
31. If a thermometer reads 30°C, what would be the Fahrenheit temperature?
32. What is the temperature reading for absolute zero Fahrenheit? Celsius?
33. What amount of work (in foot pounds) can be done by the energy of 1 Btu?

3 BASIC AERODYNAMICS

INTRODUCTION

An understanding of the basic principles of aerodynamics is as important to the aviation technician as it is to the pilot and the aerospace engineer. The technician is concerned with the strength of structural members of an aircraft because of the stresses applied through the forces of aerodynamics when the aircraft is in flight. Often responsible for the repair or restoration of aircraft structures, the technician must know that the repair work will restore the required strength to the parts that are being repaired. There are certain physical laws which describe the behavior of airflow and define the various aerodynamic forces acting on a surface. These principles of aerodynamics provide the foundations for a good understanding of what may be termed the "theory of flight."

Aerodynamics as studied by the engineer or scientist involves the use of advanced mathematics and physics; however, this chapter presents only the basic principles of the subject and their application to the flight of aircraft without the necessity of advanced mathematical analysis. The subject can therefore be more easily understood by the individual whose primary concern lies with the maintenance, operation, and repair of the aircraft.

PHYSICAL PROPERTIES OF THE AIR

The Atmosphere

The aerodynamic forces acting on a surface are due in great part to the properties of the air mass in which the surface is operating.

Air is a mixture of several gases. For practical purposes it is sufficient to say that air is a mixture of one-fifth oxygen and four-fifths nitrogen. **Pure, dry air contains about 78 percent (by volume) nitrogen, 21 percent oxygen, and 0.9 percent argon.** In addition, air contains about 0.03 percent carbon dioxide and traces of several other gases, such as hydrogen, helium, and neon. The distribution of gases in the air is shown in Fig. 3-1.

Static Pressure

The atmosphere is the whole mass of air extending upward hundreds of miles. It maybe compared to a pile of blankets. Air in the higher altitudes, like the top blanket of the pile, is under much less pressure than the air at the lower altitudes. The air at the earth's surface may be compared to the bottom blanket because it supports the weight of all the layers above it. The

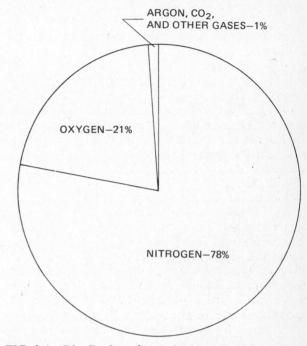

FIG. 3-1 Distribution of gases in the atmosphere.

static pressure of the air at any altitude results from the mass of air supported above that level.

Pressure may be defined as force acting upon a unit area. For example, if a force of 5 lb is acting against an area of 1 in, we say that there is a pressure of 5 psi; also, if a force of 20 lb is acting against an area of 2 in, the pressure is 10 psi. We have seen that air is always pressed down by the weight of the air above it. The atmospheric pressure at any place is equal to the weight of the column of air above it and may be represented by a column of water or mercury of equal weight. If the cube-shaped box shown in Fig. 3-2 has dimensions of 1 in^2 on all sides and is filled with mercury, the weight of the mercury will be 0.491 lb [222.72 g], and a force of 0.491 lb will be acting on the square inch at the bottom of the box. This means that there will be a pressure of 0.491 psi on the bottom of the box. If the height of the box were extended to 4 in with the cross-sectional area remaining at 1 in^2, the pressure at the bottom would be 4 × 0.491 psi, or 1.964 psi. The pressure exerted by a column of mercury does not change with the area of the cross section. If a 1-in column of mercury has a cross-sectional area of 10 in^2, the pressure will be 0.491 psi even though the total volume of mercury weighs 4.91 lb. Likewise,

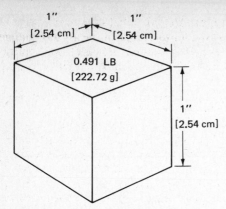

FIG. 3-2 Weight of a 1-in cube of mercury.

if the 1-in column of mercury has a cross-sectional area of $\frac{1}{4}$ in², the pressure will still be 0.491 psi.

Atmospheric pressure at sea level under standard conditions is 29.92 inches of mercury (inHg), or 14.69 psi. When we remember that 1 in of mercury produces a pressure of 0.491 psi, we can easily see that 29.92 inHg will produce a pressure of 14.69 psi (0.491 × 29.92 = 14.69).

Atmospheric pressure may be designated by a number of different units. Those more likely to be encountered are inches of mercury, millibars (mbar), pounds per square inch, kilopascals, and millimeters of mercury (mmHg). Standard atmosphere at 59°F [15°C] is approximately as follows in the units given above:

29.92 inHg
1013 mbar (0°C)
14.69 psi
101.04 kPa (60°F) [15.56°C]
760 mmHg

The effect of atmospheric pressure was demonstrated early in the seventeenth century by the Italian mathematician and scientist Evangelista Torricelli (1608–1647). Torricelli had worked with Galileo and had noted his theories regarding the "law" that **nature abhors a vacuum.** To explore the idea, Torricelli filled a long glass tube, having one end closed, with mercury. He then placed his thumb over the open end of the tube. Holding the tube in a vertical position with the closed end up, he placed the open end of the tube in a container of mercury and removed his finger from the end of the tube. Some of the mercury immediately flowed out of the tube into the container, leaving a vacuum in the upper end of the tube, as indicated in Fig. 3-3. The height of the column of mercury remaining in the tube was measured and found to be approximately 30 in [762 mm]. At sea-level standard conditions, the height of such a column of mercury is 29.92 in [760 mm]. Hence we say that standard atmospheric pressure at sea level is 29.92 in high. Barometers and sensitive altimeters are scaled to provide pressure information in inches of mercury.

As mentioned previously, in Fig. 3-3 the space above the mercury in the tube is a vacuum; this means that the pressure at this point is 0 psia. **Psia** indicates *pounds per square inch absolute.* Any gage marked for

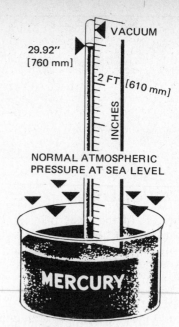

FIG. 3-3 Torricelli's experiment.

psia measures pressure from absolute zero rather than from ambient pressure zero.

Atmospheric pressure pressing down on the surface of any liquid will cause the liquid to rise in an evacuated tube in the same manner as mercury; however, the height to which a liquid will rise depends upon the density or specific gravity of the liquid. For example, water will rise to approximately 33.9 ft [10.34 m] in a completely evacuated tube. Sometimes pressure gages are scaled for inches of water (inH₂O) rather than for inches of mercury because such a gage is more sensitive and will measure lower pressure differences.

A mercury barometer is essentially a mercury-filled glass tube scaled to show the height of a mercury column. The upper end of the tube is sealed, and the lower end is exposed to the pressure being measured. The barometer can be scaled for pounds per square inch, inches of mercury, or other unit of pressure. On weather maps, the unit of pressure is the **millibar**, which is approximately one-thousandth of a **bar.** For standard purposes, the sea-level pressure is set at 1013 mbar (standard conditions). The bar is therefore the approximate atmospheric pressure at sea level. By computation, we can find that 1 inHg = 33.86 mbar.

For the convenience of engineers, a "standard" atmosphere was adopted by the National Advisory Committee for Aeronautics (now the National Aeronautics and Space Administration, or NASA). This standard atmosphere is entirely arbitrary, but it provides a reference and standard of comparison and should be known by all persons engaged in work involving atmospheric conditions. As previously explained, standard atmosphere at sea level is a pressure of 29.92 inHg [760 mmHg] with a temperature of 59°F [15°C] when the air is perfectly dry. This is supposed to be the average condition prevailing at latitude 40°N, although on different days or at different times during a day the temperature and pressure at latitude 40°N might be much different.

Since air has weight, it is easy to recognize that the pressure of the atmosphere will vary with altitude. This is illustrated in Fig. 3-4. Notice that at 20 000 ft [6097.56 m] the pressure is less than half sea-level pressure. This means, of course, that more than half the atmosphere lies below the altitude of 20 000 ft even though the "outer" half extends hundreds of miles above the earth. Table 3-1 shows the pressures and temperatures at various altitudes above the earth. This table is based upon standard conditions established by the International Civil Aviation Organization (ICAO). The table also shows the density of the air in **slugs** and the speed of sound at each altitude.

Air Temperature

Under standard conditions, temperature decreases at approximately 1.98°C for each increase of 1000 ft [304.88 m] of altitude until an altitude of 38 000 ft [11 585.44 m] is reached. Above this altitude the temperature remains at approximately −56.5°C.

Textbooks on meteorology often state that the temperature normally decreases with altitude at a rate of approximately 0.5°C per 100 m, or about 1°F per 300 ft. This amounts to a decrease of about 1.52°C for each increase of 1000 ft, which is different from the decrease under standard conditions. It must be remembered that the textbooks using the foregoing values are discussing **average** rather than standard conditions.

Adiabatic Lapse Rate

As shown in Table 3-1, the temperature of the air decreases as pressure decreases with an increase in altitude. This decrease of temperature with altitude is defined as **lapse rate.** An **adiabatic** temperature change means that the temperature of the air has changed, but the air has neither gained nor lost heat energy. The temperature of the air in such a case is due to a change in pressure.

Atmospheric pressure differences cause the air to flow from an area of higher pressure to an area of lower pressure. Thus air may flow up and over mountains or from higher elevations down into valleys. As air flows to higher altitudes, it becomes cooler, and as it flows to lower altitudes, it becomes warmer. This is in accordance with Charles' law (explained in Chap. 2). The **adiabatic lapse rate** is the increase or decrease in the temperature of the air for a given change in altitude. The adiabatic lapse rate varies from 3°F [1.67°C] per 1000 ft [304.88 m] for moist air to more than 5°F [2.78°C] per 1000 ft for very dry air. The standard rate shown in the ICAO chart is approximately 3.5°F per 1000 ft.

It must be remembered that the temperature of the air does not often conform to standards. For example, sometimes the air temperature 1000 ft or more above the surface of the earth is higher than it is at the surface. This condition is called an **inversion.** Mountains, clouds, surface winds, bodies of water, and sunshine all affect the temperature of the air.

Density

The **density** of the air is a property of great importance in the study of aerodynamics. Density has been defined previously; however, additional discussion is given here to relate density to the study of aerodynamics. Air is compressible, as illustrated in Fig. 3-5. As

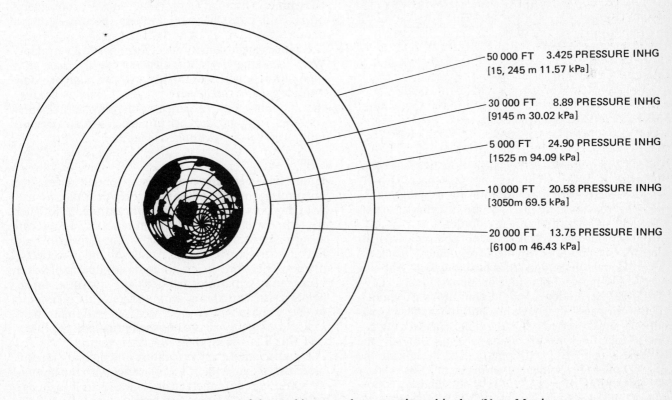

50 000 FT   3.425 PRESSURE INHG [15, 245 m 11.57 kPa]

30 000 FT   8.89 PRESSURE INHG [9145 m 30.02 kPa]

5 000 FT   24.90 PRESSURE INHG [1525 m 94.09 kPa]

10 000 FT   20.58 PRESSURE INHG [3050m 69.5 kPa]

20 000 FT   13.75 PRESSURE INHG [6100 m 46.43 kPa]

FIG. 3-4   Pressure of the earth's atmosphere at various altitudes. (Note: Metric equivalents given in round figures.)

TABLE 3-1 ICAO Standard Atmosphere

| Altitude | | t | | P | | | $\rho \times 10^3$, | |
ft	m	F	C	inHg	kPa	lb/ft²	slugs/ft³	c_s, ft/s
−2 000	−609.76	66.1	18.9	32.100	108.40	2273.70	2.520	1124.54
0	0.0	59.0	15.0	29.920	101.04	2116.20	2.380	1116.89
1 000	304.88	55.4	13.0	28.860	97.46	2040.80	2.310	1113.05
2 000	609.76	51.9	11.0	27.820	93.95	1967.70	2.240	1109.19
3 000	914.63	48.3	9.1	26.820	90.57	1896.60	2.180	1105.31
4 000	1 219.51	44.7	7.1	25.840	87.26	1827.70	2.110	1101.43
5 000	1 524.39	41.2	5.1	24.900	84.09	1760.80	2.050	1097.53
10 000	3 048.78	23.3	−4.8	20.580	69.50	1455.30	1.760	1077.81
15 000	4 573.17	5.5	−14.7	16.890	57.04	1194.30	1.500	1057.73
20 000	6 097.56	−12.3	−24.6	13.750	46.43	972.50	1.270	1037.26
25 000	7 621.95	−30.2	−34.5	11.100	37.46	785.30	1.070	1016.38
30 000	9 146.34	−48.0	−44.4	8.890	30.02	628.40	0.890	995.06
36 089	11 002.74	−69.7	−56.5	6.680	22.56	472.70	0.710	968.46
40 000	12 195.12	−69.7	−56.5	5.540	18.71	391.70	0.5850	968.46
50 000	15 243.90	−69.7	56.5	3.425	11.57	242.20	0.3620	968.46
60 000	18 292.68	−69.7	−56.5	2.118	7.15	149.80	0.2240	968.46
70 000	21 341.46	−69.7	−56.5	1.322	4.46	93.52	0.1388	968.46

t Standard temperature
P Pressure, lb/ft² or inHg
ρ Density
c_s Standard speed of sound

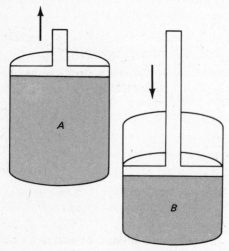

FIG. 3-5 Air expanded and compressed.

the air is compressed, it becomes more dense because the same quantity of air occupies less space. **Density varies directly with pressure with the temperature remaining constant.** In Fig. 3-5, air in cylinder B has twice the density of the air in cylinder A.

For the purposes of aerodynamic computations, air density is represented by the Greek letter ρ (rho), indicating mass density in **slugs** per cubic foot. The slug is a unit of mass with a value of approximately 32.175 lb [14.59 kg] under standard conditions of gravity. The word **mass** designates the pull in standard gravitational units exerted by the earth upon a piece of matter. Generally speaking, mass is approximately the same as weight. Since the slug is used to indicate the density of air, it is sufficient for the technician to know that the value of ρ can be found in standard atmospheric tables.

Air at standard sea-level conditions weighs 0.0765 lb/ft³ and has a density of 0.002378 slug/ft³. At an altitude of 40 000 ft [12 192 m] the air density is approximately 25 percent of the sea-level value.

A **general gas law** defines the relationship of pressure, temperature, and density when there is no change of state or heat transfer. Simply stated, this would be **density varies directly with pressure, inversely with temperature.** On a hot day, air expands, becoming "thinner," or less dense; conversely, on a cold day, the air contracts, becoming more dense.

Changes in air density affect the flight of an airplane. With the same thrust, an airplane can fly faster at a high altitude, where the density is low, than at a low altitude, where the density is greater. This is because the air offers less resistance to the airplane when it contains a smaller number of particles of air per unit volume. This is illustrated in Fig. 3-6.

Humidity

Humidity is a condition of moisture or dampness. The maximum amount of water vapor that the air can hold depends entirely on the temperature; the higher the temperature of the air, the more water vapor it can absorb. By itself, water vapor weighs approximately five-eighths as much as an equal volume of perfectly dry air. Therefore, when air contains 5 parts of water vapor and 95 parts of perfectly dry air, it is not as heavy as air containing no moisture. This is because water is composed of hydrogen (an extremely light gas) and oxygen. Air is composed principally of nitrogen, which is almost as heavy as oxygen.

Assuming that the temperature and pressure remain the same, the density of the air varies with the humidity. On damp days the density is less than it is on dry days.

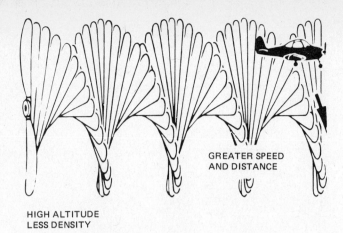

GREATER SPEED
AND DISTANCE

HIGH ALTITUDE
LESS DENSITY

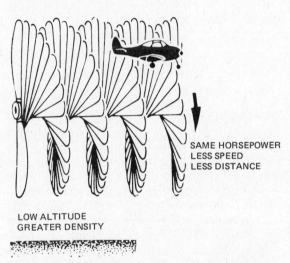

SAME HORSEPOWER
LESS SPEED
LESS DISTANCE

LOW ALTITUDE
GREATER DENSITY

FIG. 3-6   Effect of air density on aircraft in flight.

● BERNOULLI'S PRINCIPLE

The physical forces that support an aircraft in flight may be explained by two basic laws of physics, **Bernoulli's principle** and **Newton's third law of motion.**

Daniel Bernoulli, a Swiss scientist of the eighteenth century, discovered that **as the air velocity increases, the pressure decreases, and as the velocity decreases, the pressure increases.** Actually, in technical language, Bernoulli's principle states that the total energy of a particle in motion is constant at all points on its path in a steady flow. The most appropriate means of visualizing the effect of airflow and the resulting aerodynamic pressures is to study the fluid flow within a closed tube.

Suppose a stream of air is flowing through the venturi tube shown in Fig. 3-7. The airflow at station 1 in the tube has a certain velocity and static pressure. As the airstream approaches the constriction at station 2, certain changes must take place. Since the airflow is enclosed within the tube, the mass flow at any point along the tube must be the same and the velocity or pressure must change to accommodate this continuity of flow. As the flow approaches the constriction of sta-

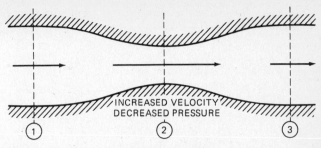

INCREASED VELOCITY
DECREASED PRESSURE

① ② ③

FIG. 3-7   Fluid flow within a closed tube.

tion 2, the velocity increases to maintain the same mass flow. As the velocity increases, the static pressure will decrease. The total energy of the airstream in the tube is unchanged. However, the airstream energy may be in two forms. The airstream may have a potential energy which is related by the static pressure and a kinetic energy represented by its dynamic pressure (velocity). As the total energy is unchanged, an increase in velocity (kinetic energy) will be accompanied by a decrease in static pressure (potential energy). Therefore, it can be said that **the sum of static and dynamic pressure in the flow tube remains constant.**

Figure 3-8 illustrates the variation of static, dy-

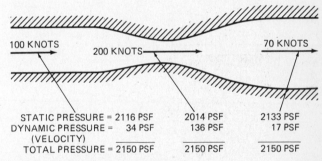

100 KNOTS   200 KNOTS   70 KNOTS

STATIC PRESSURE = 2116 PSF   2014 PSF   2133 PSF
DYNAMIC PRESSURE =   34 PSF   136 PSF   17 PSF
(VELOCITY)
TOTAL PRESSURE = 2150 PSF   2150 PSF   2150 PSF

FIG. 3-8   Bernoulli's principle.

namic, and total pressure of air flowing through a closed tube. Note that the total pressure is constant throughout the length and any change in dynamic pressure produces the same magnitude change in static pressure.

The effect produced by a wing moving through the air is illustrated in Fig. 3-9. When the air strikes the leading edge of the wing, the passage of the air is obstructed and its velocity is reduced. Some of the par-

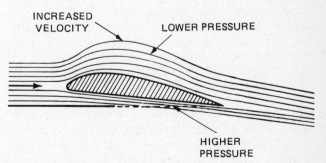

INCREASED
VELOCITY   LOWER PRESSURE

HIGHER
PRESSURE

FIG. 3-9   Pressure differential created by a wing in flight.

ticles of air flow over the upper surface and some flow under the lower surface, but all separating particles of air must reach the trailing edge of the wing at the same time. Those particles that pass over the upper surface have farther to go and therefore must move faster than those passing under the lower edge. In accordance with Bernoulli's principle, the increased velocity above the wing results in a lower static pressure than that existing below the wing. Since there is a difference of pressure, the greater must prevail, and there is an upward force exerted on the wing; this force is called lift.

● NEWTON'S THIRD LAW OF MOTION

Newton's third law of motion states: **For every action there is an equal and opposite reaction.** This also plays a role, along with Bernoulli's principle, in the development of lift. When there is an angle between the wing and the direction of the airstream, the air is forced to change direction. If the wing is tilted upward against the airstream, the air flowing under the wing is forced downward. The wing therefore applies a downward force to the air, and the air applies an equal and opposite upward force to the wing; this is lift. This is illustrated in Fig. 3-10. The angle through which an airstream is deflected by any lifting surface is called the downwash angle. It is especially important when control surfaces are studied, because they are normally placed to the rear of the wings where they are influenced by the downward deflected airstream known as the downwash.

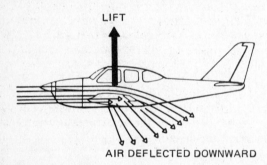

FIG. 3-10 Wing deflecting the air downward.

● AIRFOILS

An **airfoil** is technically defined as any surface, such as an airplane aileron, elevator, rudder, or wing, designed to obtain reaction from the air through which it moves. An airfoil section is a cross section of an airfoil, which can be drawn as a silhouette. If the wing of an airplane were sawed through from the leading edge to the trailing edge, the side view of the section through the wing at that point would be its airfoil section. An **airfoil profile** is merely the outline or shape of an airfoil section. It is a common practice among aircraft people to use the word *airfoil* when "airfoil section" or perhaps "airfoil profile" is meant.

Figure 3-11 illustrates five airfoil profiles of different shapes together with their chords. A **chord** may be de-

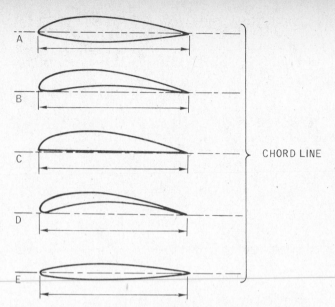

FIG. 3-11 Airfoil profiles of different shapes.

fined as the reference line from which the upper and lower contours of an airfoil are measured. A chord is also defined as a straight line directly across an airfoil from the leading edge to the trailing edge. In Fig. 3-11, profile A has a double convex shape. The chord is simply the straight line from the leading edge to the trailing edge. Profile B has a convex upper curvature and a concave lower curvature. The chord is the straight line connecting the imaginary perpendiculars erected at the leading and trailing edges. Profile C has a flat lower surface; hence the chord is the straight line connecting the leading and trailing edges. Profile D resembles profile B, and again the chord is the straight line connecting imaginary perpendiculars erected at the leading and trailing edges. Profile E is designed for supersonic flight and is almost symmetrical.

Skin Friction/Viscosity

Skin friction is illustrated in Fig. 3-12. A thin, flat plate is held edgewise to an airstream. The particles of air separate at the leading edge and flow smoothly over the upper surface and under the lower surface, reuniting behind the trailing edge. The resistance of skin friction is caused by a tendency of the particles of air to cling to the surface of the plate. There are two reasons why the air clings to the surfaces of the plate. First, the plate has a certain amount of roughness, relatively speaking. Regardless of how much we attempt to smooth a surface, it is impossible to make it perfectly smooth. This can be proved quickly by ob-

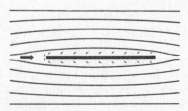

FIG. 3-12 Skin friction.

serving a polished surface through a high-powered microscope. To eliminate skin friction altogether, it would be necessary to have all the molecules of the material in perfect alignment on the surface of the material.

The second reason why air tends to cling to the surface is the **viscosity** of the air. Technically, viscosity is the resistance offered by a fluid (gas or liquid) to the relative motion of its particles, but in the common use of the term, it means the adhesive or sticky characteristics of a fluid. Even though it is not always apparent, air does have "thickness" as does oil. Viscosity may best be visualized by thinking of the difference between syrup and water; the syrup is considerably more viscous than water. The viscosity of gases is unusual in that the viscosity is generally a function of temperature alone and a decrease in temperature increases the viscosity. Therefore viscosity will generally increase with altitude.

Laminar Flow

Laminar flow employs the concept that air is flowing in thin sheets or layers close to the surface of a wing with no disturbance between the layers of air; that is, there is no cross-flow of air particles from one layer of air to another. Also, there is no sideways movement of air particles with respect to the direction of airflow.

Laminar flow is most likely to occur where the surface is extremely smooth and especially near the leading edge of an airfoil. Under these conditions the boundary layer will be very thin. The **boundary layer** is that layer of air adjacent to the airfoil surface. The air velocity in this layer varies from zero on the surface of the airfoil to the velocity of the airstream at the outer edge of the boundary layer. The cause of the boundary layer is the friction between the surface of the wing and the air.

Ordinarily, the airflow at the leading edge of a smooth-surfaced wing will be laminar, but as the air moves toward the trailing edge of the wing, the boundary layer becomes thicker and laminar flow diminishes. This is illustrated in Fig. 3-13.

FIG. 3-13 Development of the boundary layer as a result of skin friction.

Reynolds Number

Osborne Reynolds studied the flow of liquids in pipes and found that at a low speed the flow is smooth but at a high speed the flow is turbulent. By experimenting with pipes of various sizes and different liquids, he found a value which he called the **critical Reynolds number** (R). The flow was laminar (smooth) for values below the critical R and turbulent for the values above the critical R. This value worked well for the flow of liquids inside circular tubes, but it had to be handled differently when it was applied to the flow of air around objects which were unconfined, such as airfoils.

Whether a laminar or turbulent boundary layer exists around an airfoil depends on the combined effects of velocity, viscosity, density, and the size of the chord. It is the combined effects of these important parameters which produce the Reynolds number.

In finding the Reynolds number, the velocity V must be in feet per second and the linear dimensions L (sometimes written as lowercase l) of the object to be tested must be in feet. When wings are being tested, the length of the chord is usually chosen for L. If the test is conducted under standard atmospheric conditions (15°C and 760 mmHg or 29.92 inHg pressure), the density of the air ρ is 0.002 378 slug/ft^3 and the coefficient of viscosity μ is 0.000 000 373 slug/ft · s. If standard atmospheric conditions are not used, corrections must be made for density and viscosity. The formula for the Reynolds number is $R = \rho V(L/\mu)$, where ρ is air density, V is velocity, L is the dimension (usually the chord), and μ is the coefficient of viscosity.

If the Reynolds number for a model wing having a chord of 6 in is desired, and if the test is to be conducted at standard atmospheric conditions and at a velocity of 100 mph, the chord dimension is changed to feet, giving 0.5 ft, and the velocity of 100 mph is changed to its equivalent of 146.7 ft/s. We then have $R = (0.002\ 378 \times 146.7 \times 0.5)/0.000\ 000\ 373 = 467\ 630.9$.

While the actual magnitude of the Reynolds number has no physical significance, the quantity is used as an index to predict various types of airflow. High Reynolds numbers are obtained with large chord surfaces, high velocities, and low altitude; low Reynolds numbers result from small chord surfaces, low velocities, and high altitudes.

A term frequently used in discussing airplane design is **scale effect.** This is the change in any force coefficient, such as a drag coefficient, due to a change in the value of a Reynolds number.

Angle of Attack

The **angle of attack** is the acute angle between a reference line in a body and the line of the relative wind direction, as shown in Fig. 3-14. The angle of attack can be defined more simply as the acute angle between the chord of an airfoil and the relative wind. The arrowhead at the left in Fig. 3-14 represents the relative wind. Technically speaking, the **relative wind** is the velocity of the air with respect to a body in it. It is usu-

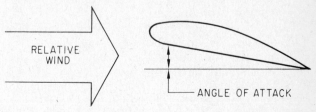

FIG. 3-14 Angle of attack.

ally determined from measurements made at such a distance from the body that the disturbing effect of the body upon the air is negligible. In other words, the relative wind refers to the velocity of the air before it strikes the leading edge of the airfoil and divides the flow around it. In calm, still air, the direction of the relative wind is opposite to the flight path of the airplane with reference to the ground. In the National Aeronautics and Space Administration's *Aeronautical Dictionary* the definition of **angle of attack** is: "The angle at which a body, such as an airfoil or fuselage, or a system of bodies, such as a helicopter rotor, meets a flow, ordinarily measured between a reference line in the body and a line in the direction of the flow or in the direction of movement of the body." The airfoil profile in Fig. 3-14 has a flat lower surface like C in Fig. 3-11; hence its chord is simply a straight line parallel to its lower surface. The angle of attack is the acute angle between the lower surface of the airfoil and the line parallel to the direction of the relative wind.

Drag

As explained previously, air has mass. When an airplane flies through air, the air is moved. When any mass is moved or accelerated, force is required, and the application of force produces an equal and opposite force. This is in keeping with Newton's law of motion. The impact of the air against the surfaces of the airplane applies force which tends to hold the airplane back. This is **drag**. Specifically, drag is a retarding force acting upon a body in motion. Total drag may be classified into two main types: **induced drag and parasite drag.**

Induced drag is the undesirable but unavoidable by-product of lift, and it increases in direct proportion to increases in the angle of attack. The greater the angle of attack up to the critical angle, the greater the amount of lift developed and the greater the induced drag. The airflow around the wing is deflected downward, producing a rearward component to lift, which is induced drag. The amount of air deflected downward increases greatly at higher angles of attack; therefore, the higher the angle of attack or the slower the airplane is flown, the greater the induced drag.

Parasite drag is the resistance of the air produced by any part of the airplane that does not produce lift. Several factors affect parasite drag. When each factor is considered independently, it must be assumed that other factors remain constant. These factors are (1) the more streamlined an object is, the less the parasite drag, (2) the more dense the air moving past the airplane, the greater the parasite drag, (3) the larger the size of the object in the airstream, the greater the parasite drag, and (4) as speed increases, the amount of parasite drag increases. If the speed is doubled, 4 times as much drag is produced.

Parasite drag can be further classified into **form drag, skin friction,** and **interference drag.** Form drag is caused by the frontal area of the airplane components being exposed to the airstream. A similar reaction is illustrated in Fig. 3-15, where the side of the airfoil is exposed to the airstream. This drag is caused by the form of the airfoil and is the reason streamlining is

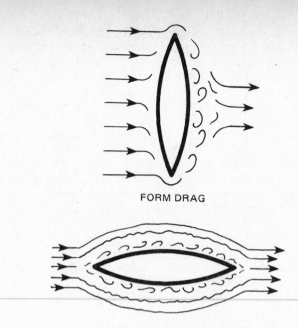

FORM DRAG

SKIN FRICTION DRAG

FIG. 3-15   Form drag and skin friction drag.

necessary to increase airplane efficiency and speed. Figure 3-15 also illustrates that when the leading edge of the airfoil is parallel to the airstream, the largest part of the drag is skin friction.

Skin friction drag is caused by air passing over the airplane's surfaces, and it increases considerably if the airplane surfaces are rough and dirty.

Interference drag is caused by interference of the airflow between adjacent parts of the airplane such as the intersection of wings and tail sections with the fuselage. Fairings are used to streamline these intersections and decrease interference drag.

In designing or repairing an airplane, one must consider the drag forces exerted on the structure of the airplane. As the airplane flies through the air, the effect is as if millions of tiny particles were striking against the forward parts of the airplane. If the total force of these particles becomes too great for the strength of the structure, damage will result. It is therefore necessary to make sure that the leading edges of the wings, the leading edges of the stabilizers, and the forward parts of the engine cowling are constructed of material with sufficient strength to withstand the maximum impact forces that will ever be imposed upon them. Furthermore, the devices by which the parts are attached to the main structure must have sufficient strength to withstand maximum forces.

Lift and Drag Components

The effects of **lift** and **drag** can be determined by employing **force vectors** as shown in Fig. 3-16. If the lift vector is represented by the line **AB** and the drag vector by the line **AD**, then the resultant of the two forces is **AC**. The resultant force is the combined net result of the two forces of lift which acts upward and perpendicular to the relative wind, and the drag which is parallel to the relative wind (see Fig. 3-17).

The lengths of the arrows **AB**, **AC**, and **AD** in Fig.

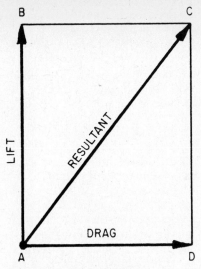

FIG. 3-16   Resultant of lift and drag.

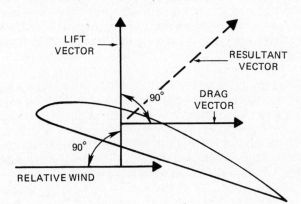

FIG. 3-17   Relationship between relative wind, lift, and drag.

The L/D ratio is a fraction with the lift for the numerator and the drag for the denominator. It is a perfectly simple principle of mathematics that if the numerator remains the same and the denominator increases, the value of the fraction decreases. For example, one-half is obviously greater than one-fourth. Likewise, if the denominator is small and the numerator large, the value of the fraction is greater. If the L/D ratio is 21:1 at a certain angle of attack, the value of the fraction and hence the effectiveness of the airfoil are greater than when the L/D ratio is 14:1.

Center of Pressure

The **center of pressure** (CP) is shown at A in Fig. 3-18. The CP is the point at which the chord of an airfoil section intersects the line of action of the resultant aerodynamic forces and about which the pressures balance.

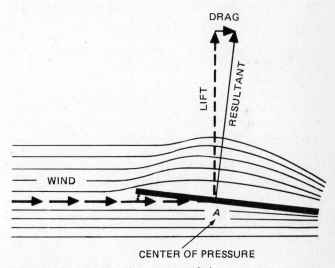

FIG. 3-18   Flat plate in a stream of air.

The location and direction in which the resultant will point depends upon the shape of the airfoil section and the angle at which it is set to the airstream. Throughout most of the flight range—that is, at the usual angles of attack—the CP moves forward as the angle of attack increases and backward as the angle of attack decreases. The most forward position is usually about three-tenths of the chord back from the leading edge for most airfoil sections, although this is by no means an infallible statement. If the airplane wing has a chord of 63 in [160 cm], for example, the most forward position of the CP may be 0.3×63 or 18.9 in [48 cm] behind the leading edge.

The farthest position to the rear that the CP may reach on some airfoil sections is four-tenths of the chord back from the leading edge. If the chord is 63 in long, for example, the CP in this case may be 0.4×63 or 25.2 in [64 cm] behind the leading edge. Therefore, in the example mentioned, the CP, at the usual angles of attack, will be somewhere between 18.9 and 25.2 in behind the leading edge on the chord. Notice carefully that we have said that these may be the positions at the *usual* angles of attack; the CP can still travel for-

3-16 are in proportion to the magnitudes of the forces they represent. For example, if lift is 4 times as great as drag, the arrow representing lift must be drawn 4 times as long as the arrow representing drag. These lift and drag arrows (vectors) are called components because the forces that they represent are component parts of the force represented by the resultant.

The **resultant** of lift and drag shows the direction and magnitude of the force created by the difference in pressure between the top surface and the bottom surface of an airfoil.

Lift/Drag Ratio

As the angle of attack of an airfoil increases, so do the components of lift and drag; however, the drag will increase in proportion more than the lift will. The **ratio of lift to drag** at any angle of attack is a measure of the "effectiveness" because lift is a beneficial force and drag is a detrimental force. However, drag must be accepted as a necessary evil to produce lift.

The ratio of lift to drag is called the **lift/drag ratio** and is commonly referred to as the **L/D ratio.** However, when any L/D ratio is given, the corresponding angle of attack should also be given because the L/D ratio has no meaning by itself.

ward or backward from these usual positions. For example, at a low angle of attack, the CP may run off the trailing edge and disappear because there is no more lift.

Figure 3-19 shows the travel of the CP along the chord at various angles of attack, as well as changes in the magnitude, direction, and location of the resultant. In actual flight, there is a different airspeed for different angles of attack; but in a wind-tunnel test, the velocity of the airstream can be constant while the angle of attack changes, as is the case in Fig. 3-19. At small angles of attack (shown at top of Fig. 3-19), the CP is quite a distance from the leading edge, the resultant is comparatively small, and it points upward and to the rear of the vertical. At medium angles of attack (center, Fig. 3-19), the CP has moved toward the leading edge, the resultant is greater, and its direction more nearly approaches right angles to the chord. At high angles of attack (bottom, Fig. 3-19), the CP has moved to its farthest forward position and the resultant is still greater.

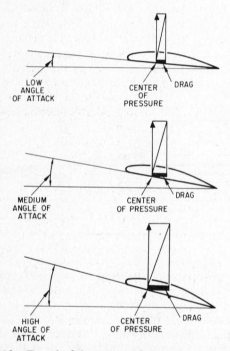

FIG. 3-19   Travel of the center of pressure.

Before proceeding further in the discussion of Fig. 3-19, it is necessary to introduce some new terms. The **critical angle of attack** is the angle of attack at which the flow about an airfoil changes abruptly accompanied by corresponding abrupt changes in the lift and drag. **Burble** is a breakdown of streamline airflow about a body. The **burble point** is the angle of attack at which the streamline flow about a body begins to break down. In Fig. 3-19, the high angles of attack, represented by C, approach the burble point, the CP either remains in the same position or moves back slightly, the resultant decreases in magnitude, and its departure from the vertical toward the rear becomes greater.

The CP behaves differently on a flat plate. There the

CP moves backward as the angle of attack increases and forward as it decreases. It is possible to design an airfoil section that will behave in this manner, and it is well worth considering. For example, if a rising column of air strikes the leading edge of an airplane wing with an airfoil section of this type, the angle of attack is increased. This causes the CP to move to the rear. Therefore, the lift is at the rear of the wing, where it raises the wing and decreases the angle of attack, which was unexpectedly raised by the column of air. This type of airfoil section is described as **stable** because forces are developed that tend to restore the airfoil to its original state. The question naturally arises as to the reason for not using airfoil sections having the CP travel characteristics of a flat plate. A full explanation will be given later. For the present, it is sufficient to say that the characteristic of stability produced by the direction of CP travel is less important than certain other features; hence most airfoil sections are designed so that the CP travels forward as the angle of attack increases and backward as it decreases. These usual airfoil sections are described as unstable because, after a disturbance, the forces tend to move the airfoil farther away from its original state.

One of the reasons for studying CP travel is that the CP is the point at which the aerodynamic forces can be considered to be concentrated; hence the airplane designer must make provisions for the CP travel by preparing a wing structure that will meet any stress imposed upon it. Technicians and inspectors cannot change the design, but they can perform their duties better if they know the characteristics and limitations of an airplane.

Area and Lift

One of the factors that determine the total lift of an airfoil is the area of the surface exposed to the airstream. If the area is small, the region of pressure differential is small and there is little lift. On the other hand, if the area is great, the region of pressure differential is great and there is a large amount of lift. Later in this text, we shall explain why more lift can be obtained from a long, narrow wing than from one in which the width more closely approaches the length (span), but for the present it is sufficient to remember one simply statement: **Lift varies directly with the area,** other factors being equal. In practice, the area of wings are measured in square feet.

Velocity and Lift

A positive angle of attack causes increased velocity and decreased pressure on the upper surface of a wing and decreased velocity and increased pressure on the lower surface. This is in accordance with Bernoulli's principle, which states that **as the velocity of the air increases, the pressure decreases.**

If the air flows slowly around the airfoil, a certain amount of lift is generated. If the velocity of the airstream increases, the pressure differential increases and the lift increases. When we use the word velocity in this sense, we mean the airspeed with respect to the airfoil.

Air Density and Lift

We have seen that lift depends upon the shape of the airfoil, the angle of attack, the area of the surface exposed to the airstream, and the airspeed. One more factor of lift remains: the air density. On hot days the density of the air is less than on cold days; on wet days, the density is less than on dry days. Also, density decreases with altitude. When the density is low, the lift will also be comparatively lower.

If an airplane flies at a certain angle of attack at sea level and then flies at the same angle of attack at a higher altitude, where density is less, the airplane must be flown faster. On hot days, when the density is less, the airplane must be flown faster for the same angle of attack than on cold days, when the density is greater. On wet days, the density is less; hence the airplane must be flown faster for the same angle of attack than on dry days. Therefore, greater airspeed is required for a particular airplane when the density is lower. To express the same idea in different words, the airspeed must increase as the density decreases in order to maintain the airplane at the same angle of attack in level flight.

Angle of Maximum Lift and Minimum Speed

Beginning with small angles of attack, the lift increases as the angle of attack increases until an angle of attack is reached where the lift has a maximum value. This angle corresponds to the burble point and is the angle of attack at which the streamline flow begins to break down over the upper surface of the wing and burbling begins at the trailing edge. This angle of attack is called the **stalling angle.** At angles greater than the angle of maximum lift, the lift decreases rapidly and the drag increases rapidly.

For each angle of attack there is a corresponding airspeed, assuming that other conditions, such as wing area and air density, remain constant. As the angle of attack increases, this airspeed decreases; hence the least possible airspeed exists at the angle of maximum lift (stalling angle). Therefore, we can say that another name for the angle of maximum lift is the **angle of minimum speed.** We also point out that the **stalling speed** of an airplane is the minimum speed at which the wing will maintain lift under a certain set of conditions.

● HIGH-SPEED FLIGHT

Thus far we have discussed airflow and aerodynamic principles with respect to subsonic airspeeds only. Developments in aircraft and power plants have produced high-performance airplanes with capabilities for very high-speed flight. The study of aerodynamics at these very high flight speeds has many significant differences from the study of low-speed aerodynamics. The behavior of an airfoil under subsonic conditions is easily predictable. However, when we consider operations at transonic and supersonic speeds, we find the reaction of an airfoil altogether different from that which is found at subsonic speeds. The reason for this is the reaction of the air itself.

Compressibility

At low flight speeds the study of aerodynamics is greatly simplified by the fact that air may experience relatively small changes in pressure with only negligible changes in density. This airflow is termed **incompressible** since the air may undergo changes in pressure without apparent changes in density. Such a condition of airflow is similar to the flow of water, hydraulic fluid, or any other incompressible fluid. However, at high flight speeds, the pressure changes that take place are quite large, and significant changes in air density occur. The study of airflow at high speeds must account for these changes in air density and consider that the air is **compressible.**

Speed of Sound

A factor of great importance in the study of high-speed airflow is the **speed of sound.** The speed of sound is the rate at which small pressure disturbances will be spread through the air.

The speed at which sound travels in air under standard sea-level conditions is 1116 ft/s [340.24 m/s], or 761 mph, or 661 kn. The speed of sound is not affected by a change in atmospheric pressure because the density also changes. However, a change in the temperature of the atmosphere changes the density without appreciably affecting the pressure; hence, the speed of sound changes with a change in temperature. The speed of sound can be calculated with the equation

$$a = 49.022\sqrt{T}$$

where a = speed of sound, ft/s
T = absolute temperature, °F

The temperature of the air decreases with an increase in altitude up to an altitude of about 37 000 ft [11 280 m], and it is then constant to an altitude of more than 100 000 ft [30 488 m]. Therefore, the speed of sound decreases with altitude to about 37 000 ft and then remains constant to more than 100 000 ft. For example, at 30 000 ft [9146.34 m] the temperature of standard air is −48°F [−44.44°C], and the speed of sound is 995 ft/s [303.35 m], or 589 kn. Table 3-2 il-

TABLE 3-2 Variation of Speed of Sound with Temperature

Altitude, ft	Temperature		Speed of sound, kn
	°F	°C	
Sea level	59.0	15.0	661.7
5 000	41.2	5.1	650.3
10 000	23.3	−4.8	638.6
15 000	5.5	−14.7	626.7
20 000	−12.3	−24.6	614.6
25 000	−30.2	−34.5	602.2
30 000	−48.0	−44.4	589.6
35 000	−65.8	−54.3	576.6
40 000	−69.7	−56.5	573.8
50 000	−69.7	−56.5	573.8
60 000	−69.7	−56.6	573.8

lustrates the variation of the speed of sound in the standard atmosphere.

Mach Number

Because of the relationship between the effect of the air forces at high speed and the speed of sound, and because the speed of sound varies with altitude, it is the ratio of the speed of the aircraft to the speed of sound that is important rather than the speed of the aircraft with respect to the air. This ratio, called the **Mach number,** is the true airspeed of the aircraft divided by the speed of sound in the air through which the aircraft is flying at the time. Thus, Mach 0.5 at sea level under standard condition is 558 ft/s [170 m/s]; however, Mach 0.5 at 30 000 ft is only 497 ft/s [151.52 m/s].

Types of High-Speed Flight

It is important to note that compressibility effects are not limited to flight speeds at and above the speed of sound.

The speed of the air flowing over a particular part of the aircraft is called the **local speed.** The local speed may be higher than the speed of the aircraft, for example, the speed of the air across the upper portion of the wing. Thus, an aircraft can experience compressibility effects at flight speeds well below the speed of sound. Since there is the possibility of having both subsonic and supersonic flows existing on the aircraft, it is convenient to define certain regimes of flight. These regimes are defined approximately as follows:

Subsonic. Mach numbers below 0.75
Transonic. Mach numbers from 0.75 to 1.20
Supersonic. Mach numbers from 1.20 to 5.00
Hypersonic. Mach numbers above 5.00

While the flight Mach numbers used to define these areas of flight are quite approximate, it is important to appreciate the types of flow existing in each area.

Subsonic Flight

At speeds less than 300 mph [483 km], the airflow around an aircraft behaves as though the air were incompressible. Pressure disturbances, or pressure pulses, are formed ahead of the parts of the aircraft such as the leading edge of the wing. These pressure pulses travel through the air at the speed of sound and, in effect, serve as a warning to the air of the approach of the wing. As a result of this warning, the air begins to move out of the way. Evidence of this "pressure warning" is seen in the typical subsonic flow pattern of Fig. 3-20 where there is upwash and flow direction

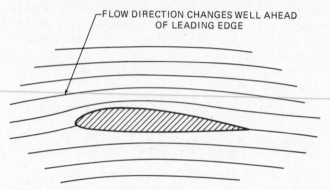

FIG. 3-20   Typical subsonic flow pattern.

change well ahead of the leading edge. If the object is traveling at some speed above the speed of sound, the airflow ahead of the object will not be influenced by the pressure field on the object since pressure disturbances cannot be propagated ahead of the object. Thus, as the flight speed nears the speed of sound, a **compression wave** will form at the leading edge and all changes in velocity and pressure will take place quite sharply and suddenly. The airflow ahead of the object is not influenced until the air particles are suddenly forced out of the way by the concentrated pressure wave set up by the object. Evidence of this phenomenon is seen in the typical supersonic flow pattern of Fig. 3-21.

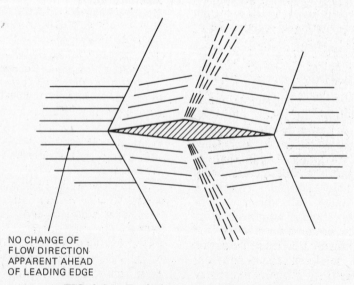

FIG. 3-21   Typical supersonic flow pattern.

Transonic Flight

During transonic flight a **shock wave** forms on both the top and the bottom surfaces of the wing. The magnitude and location of these shock waves are constantly changing. Airflow separation will occur with the formation of shock waves resulting in the loss of lift. Other phenomena that may be associated with transonic flight are aircraft buffeting, trim and stability changes, and a decrease in control surface effectiveness. These forces and the turbulence that accompanies transonic flight may cause the pilot to lose control, especially if the airplane is not designed to operate under transonic conditions. When the speed becomes supersonic, the shock waves move back and become attached to the trailing edge of the wing. When this takes place, control conditions become predictable and orderly again.

Supersonic Flight

When a wing is moving at a speed greater than the speed of sound, there can be no warning pressure changes because the wing is traveling faster than the pressure changes can travel.

When supersonic flow is clearly established, all changes in velocity, pressure, density, and flow direction take place quite suddenly and in relatively confined areas. The areas of flow change are very distinct and these areas are referred to as **wave formations.**

The wave formations formed at supersonic speeds are not imaginary or theoretical, and in a suitably arranged high-speed wind tunnel they can be photographed. Figure 3-22 is an unretouched photograph showing shock waves formed on a wind-tunnel model at low supersonic speed.

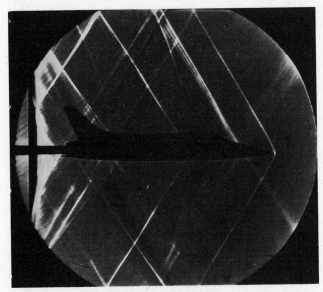

FIG. 3-22 Photograph of shock waves.

Various types of waves can occur in supersonic flow and the nature of the wave formed depends upon the airstream and the shape of the object causing the flow change. Essentially, there are three fundamental types of waves formed in supersonic flow: (1) **the oblique shock wave, (2) the normal shock wave,** and (3) **the expansion wave.**

Oblique Shock Wave

A typical case of **oblique shock wave** formation is that of a wedge pointed into a supersonic airstream. The oblique shock wave will form on each surface of the wedge as shown in Fig. 3-23. A supersonic airstream passing through the oblique shock wave will experience these changes:

1. The airstream is slowed down; the velocity and Mach number behind the wave are reduced, but the flow is still supersonic.
2. The flow direction is changed to flow along the surface of the airfoil.
3. The static pressure of the airstream behind the wave is increased.
4. The density of the airstream behind the wave is increased.
5. Some of the available energy of the airstream (indicated by the sum of dynamic and static pressure) is dissipated and turned into unavailable heat energy. Hence, the shock wave is wasteful of energy.

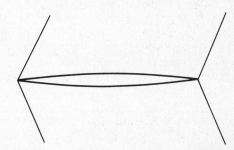

FIG. 3-23 Shock waves at the edges of an airfoil.

Normal Shock Wave

If a blunt-nosed object is placed in a supersonic airstream, the shock wave which is formed will be detached from the leading edge. Whenever the shock wave forms perpendicular to the upstream flow, the wave is termed a **normal shock wave,** and the flow immediately behind the wave is subsonic. Any relatively blunt object, such as is shown in Fig. 3-24, placed in a supersonic airstream will form a normal shock wave immediately ahead of the leading edge, slowing the airstream to subsonic so that the airstream may feel the presence of the blunt nose and flow around it. Once past the blunt nose, the airstream may remain subsonic or accelerate back to supersonic depending on the shape of the nose and the Mach number of the free stream.

In addition to the formation of normal shock waves described above, this same type of wave may be formed in an entirely different manner. Figure 3-25 illustrates the way in whch an airfoil at high subsonic speeds has local flow velocities which are supersonic. As the local supersonic flow moves aft, a normal shock wave forms, slowing the flow to subsonic. The transition of flow from subsonic to supersonic is smooth

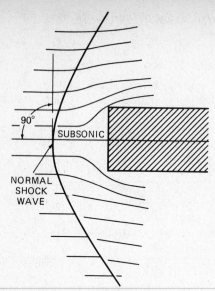

FIG. 3-24   Normal shock wave formation.

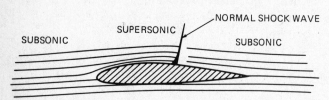

FIG. 3-25   Normal shock wave formation.

and is not accompanied by shock waves if the transition is made gradually with a smooth surface. The transition of flow from supersonic to subsonic without direction change always forms a normal shock wave.

A supersonic airstream passing through a normal shock wave will experience these changes:

1. The airstream is slowed to subsonic.
2. The airflow direction immediately behind the wave is unchanged.
3. The static pressure of the airstream behind the wave is increased greatly.
4. The density of the airstream behind the wave is increased greatly.
5. The energy of the airstream (indicated by total pressure—dynamic plus static) is greatly reduced. The normal shock wave is very wasteful of energy.

If a supersonic airstream were to flow "around a corner" as shown in Fig. 3-26, an **expansion wave** would form. An expansion wave will not cause sharp, sudden changes in the airflow except at the corner itself and thus is not actually a shock wave. A supersonic airstream passing through an expansion wave will experience these changes:

1. The airstream is accelerated; the velocity and Mach number behind the wave are greater.
2. The flow direction is changed to flow along the surface provided separation does not occur.
3. The static pressure of the airstream behind the wave is decreased.

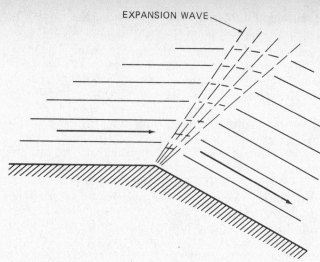

FIG. 3-26   Expansion wave formation.

4. The density of the airstream behind the wave is decreased.
5. Since the flow changes in a rather gradual manner, there is no shock and no loss of energy in the airstream. The expansion wave does not dissipate airstream energy.

Sonic Booms

When an airplane is in level supersonic flight, a pattern of shock waves is developed. Although there are many shock waves coming from an aircraft flying supersonically, these waves tend to combine into two main shocks, one originating from the nose of the aircraft and one from the tail.

If these waves extend to the ground or water surface, as shown in Fig. 3-27, they will be reflected, causing a **sonic boom.** An observer would actually hear two booms. The time between the two booms and their intensity is primarily a function of the distance the airplane is from the ground. The lower the aircraft is the closer together and louder the two booms would be.

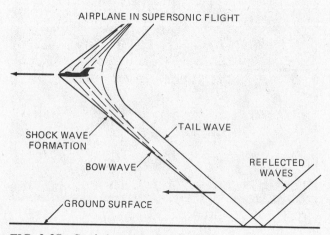

FIG. 3-27   Sonic boom wave formation.

Hypersonic Flight

When the speed of an aircraft or spacecraft is 5 times the speed of sound or greater, the speed is said to be

hypersonic. Practical experience has been gained by engineers working in the space program under the direction of the National Aeronautics and Space Administration (NASA), with the result that certain hypersonic vehicles are now practical and others will be developed. The principal hindrance to hypersonic flight is the extreme temperature generated by air friction at hypersonic speeds. New materials and cooling methods are being developed to overcome this problem.

FIG. 3-28   Hypersonic wind-tunnel test.   *(NASA)*

The high-velocity test of a Space-Shuttle model is shown in Fig. 3-28. The shuttle enters the earth's atmosphere at hypersonic speed. The model shown in Fig. 3-28 was tested in a Mach 20 helium tunnel at NASA's Langley Research Center, Hampton, Virginia. The configuration developed by the NASA Manned Spacecraft Center is being tested at a 20° angle of attack at simulated Mach 20 reentry speed. Bow shock patterns from the nose and fixed straight wing are made visible by exciting the helium flow with a high-energy electron beam.

● REVIEW QUESTIONS

1. Why should the aviation maintenance technician have a good understanding of the basic principles of aerodynamics?
2. What are the approximate percentages of the principal gases in our atmosphere?
3. What is the atmospheric pressure at sea-level standard conditions in pounds per square inch and inches of mercury?
4. Describe *Torricelli's experiment.*
5. Explain *adiabatic lapse rate.*
6. What is the value of a *slug?*
7. What effect does temperature have on the density of the air?
8. In simple terms, explain Bernoulli's principle.
9. Describe a venturi tube, and explain how it affects the pressure and rate of flow of a moving fluid.
10. Why is lift generated when an airfoil is moved through the air with the leading edge slightly higher than the trailing edge?
11. What is the cause of skin friction on a surface moving through the air?
12. What does the term *viscosity* mean?
13. Describe *laminar flow.*
14. What is the *boundary layer?*
15. Define *airfoil.*
16. Define *chord* for an airfoil.
17. Explain *angle of attack.*
18. What are the two main types of *drag?*
19. Define *resultant force.*
20. Explain what is meant by the term *center of pressure.*
21. How does angle of attack affect the CP on a wing?
22. How does wing area affect lift?
23. Discuss the effect of air density with respect to lift.
24. Explain *angle of maximum lift.*
25. What is meant by *compressibility?*
26. How is the speed of sound affected by temperature?
27. What is a *Mach number?*
28. Explain transonic flight as compared with supersonic flight.
29. What are three types of shock waves?
30. What causes a *sonic boom?*
31. What is *hypersonic speed?*
32. What is a main hindrance to the flight of aircraft in the atmosphere at hypersonic speeds?

4 AIRFOILS AND THEIR APPLICATIONS

● INTRODUCTION

A general knowledge and understanding of the nature of airfoils and the factors affecting their performance are of value to many technicians, particularly those involved in the structural repair of airfoils and those who may be interested in building their own airplanes. The previous chapter discussed basic aerodynamic principles and their application to airfoils in producing lift.

Since the early days of aircraft research when the Wright brothers tested airfoil shapes in a small wind tunnel, literally thousands of different airfoil shapes have been developed and tested. These range from the types that operate at low subsonic speeds to those designed for supersonic and hypersonic speeds. In this chapter the basic elements of airfoil design and the characteristics that determine airfoil selection for different aircraft applications will be discussed.

● AIRFOIL TERMINOLOGY

Since the shape of an **airfoil** and its angle to the airstream are so important in determining its characteristics and pressure distribution, it is necessary to properly define the airfoil terminology. Figure 4-1 shows a typical airfoil and illustrates the various items of airfoil terminology.

The **chord line** is a straight line connecting the leading and trailing edges of the airfoil. The chord is used as a reference line from which the upper and lower contours of an airfoil are measured.

Camber is defined as the curvature of an airfoil surface or an airfoil section from the leading edge to the trailing edge. The degree or amount of camber is expressed as the ratio of the maximum departure of the curve from the chord to the chord length. Figure 4-1 shows an airfoil that has a double convex curvature, which means that it has camber above and below the chord line. **Upper camber** refers to the curve of the upper surface of an airfoil, and **lower camber** refers to the curve of the lower surface.

The **mean-camber line** is a line drawn halfway between the upper and lower surfaces. The chord line connects the ends of the mean-camber line. Any point on this mean line should be the same distance from the upper and lower surfaces. Mean camber is the curvature of the mean line of an airfoil profile from the chord. Camber is positive when the departure from the straight line is upward and negative when it is downward. When the upper and lower camber of an airfoil are the same, the airfoil is said to be symmetrical.

The shape of the mean-camber line is very important in determining the aerodynamic characteristics of an airfoil section. The **maximum camber** (displacement of the mean line from the chord line) and the location of the maximum camber help to define the shape of the mean-camber line. These quantities are expressed as fractions or percentages of the basic chord dimension. A typical low-speed airfoil may have a maximum camber of 4 percent located 40 percent aft of the leading edge.

The **thickness** and **thickness distribution** of the profile are important properties of an airfoil. The maximum thickness and the locaton of the maximum thickness are expressed as fractions of the percentage of the chord. A typical low-speed airfoil may have a maximum thickness of 12 percent located 30 percent aft of the leading edge.

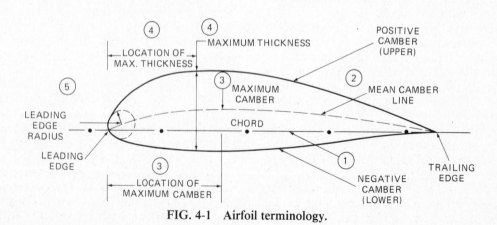

FIG. 4-1 Airfoil terminology.

The **leading-edge radius** of the airfoil is the radius of curvature given the leading-edge shape. It is the radius of the circle centered on a line tangent to the leading-edge camber connecting tangency points of upper and lower surfaces with the leading edge. Typical leading-edge radii are 0 (knife edge) to 4 or 5 percent.

● AIRFOIL PROFILES

An **airfoil** is any surface, such as an airplane wing, aileron, or rudder, designed to obtain reaction from the air through which it moves.

An **airfoil profile** is the outline of an airfoil section. An **airfoil section** is a cross section of an airfoil parallel to the plane of symmetry or to a specified reference plane. If we imagine that Fig. 4-2 is a cross section of

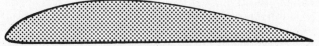

FIG. 4-2   Clark Y airfoil profile.

an actual airplane wing, it is correct to call it an airfoil section; but if it is merely the outline, it should be called an airfoil profile. The three terms are used interchangeably in conversation, even by people who know the technical distinctions. Airfoil profiles can be considered to be made up of certain profile thickness forms disposed about certain mean lines. The major shape variables then become two: the thickness form and the mean-line form. The thickness form is of particular importance from a structural standpoint. On the other hand, the form of the mean line determines almost independently some of the most important aerodynamic properties of the airfoil section.

Early textbooks on the theory of flight use the **Clark Y airfoil,** Fig. 4-2, to illustrate various statements, and this practice has been continued by authors who simply rewrote the existing material. However, the airfoil profiles developed by **NACA** (National Advisory Committee for Aeronautics) were described in detail in NACA Report 460, published November 1933, entitled *The Characteristics of Seventy-Eight Related Airfoil Sections from Tests in the Variable Density Wind Tunnel.* NACA was a government agency which performed thousands of tests on airfoil shapes to develop information regarding which were most efficient for various flight conditions. NACA is now **NASA,** the National Aeronautic and Space Administration, which has continued airfoil development and testing. The NACA and NASA airfoil reports do not contain information that a technician would normally utilize, but they are of interest to those who wish to widen their knowledge of aerodynamics.

NACA airfoil profiles are designated by a number consisting of four digits and also by numbers with five or more digits. In the four-digit numbers, the first digit indicates the camber of the mean line in percentage of the chord, the second digit shows the position of the maximum camber of the mean line in tenths of the chord from the leading edge, and the last two digits

indicate the maximum thickness in percentage of the chord. Thus the NACA2315 profile has a maximum mean camber of 2 percent of the chord at a position three-tenths of the chord from the leading edge and a maximum thickness of 15 percent of the chord. Likewise the NACA0012 airfoil is a symmetrical airfoil having a maximum thickness of 12 percent of the chord. Airfoil numbers with five or more digits can presently be found in the reports of NACA and NASA along with an explanation of the significance of the digits.

Figure 4-3 shows how to draw the NACA2421 profile. First, draw a base line and call it the chord. Sec-

STA	UP' R	L'W'R.
0	—	0
1.25	3.87	− 2.82
2.5	5.21	− 4.02
5.0	7.00	− 5.51
7.5	8.29	− 6.48
10	9.28	− 7.18
15	10.70	− 8.05
20	11.59	− 8.52
25	12.15	− 8.67
30	12.38	− 8.62
40	12.16	− 8.16
50	11.22	− 7.31
60	9.79	− 6.17
70	7.94	− 4.87
80	5.74	− 3.44
90	3.18	− 1.88
95	1.76	− 1.06
100	(.22)	(−.22)
100	—	0

L. E. RAD.: 4.85
SLOPE OF RADIUS
THROUGH END OF
CHORD: 2/20

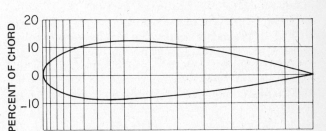

FIG. 4-3   Method for drawing an airfoil profile.

ond, divide the chord into 20 equal divisions; each division will represent 5 percent of the chord. Draw vertical lines of indefinite length at each of these divisions, and number them, starting with 0 at the leading edge and ending with 100 at the trailing edge. These lines are called **stations.** Then draw vertical lines at the other stations shown on the table. For this particular airfoil the additional stations are 1.25, 2.5, and 7.5 percent of the chord behind the leading edge.

Third, lay out the points for the upper and lower contour lines from the ordinates given in the second and third columns of the table shown in the figure. Notice that those in the second column are positive ordinates and those in the third column are negative ordinates. Positive ordinates are measured upward from the chord at their stations, and negative ordinates are measured downward from the chord at their stations in percentage of chord. Fourth, using a spline or French curve, connect all these points with a smooth line. This provides the airfoil section except that the nose requires more work.

Notice that the bottom of the table in Fig. 4-3 says: "L.E. rad.: 4.85. Slope of radius through end of chord: 2/20." Refer to Fig. 4-4, which shows how to draw the nose. Through the 0 percent station of the chord, that is, the leading edge, construct the line OC at an angle that will have a slope of 2:20. This is the same as a

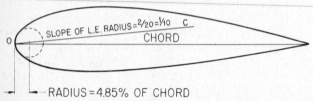

FIG. 4-4   How to draw the nose of an airfoil profile.

slope of 1:10, which means that for every 10 units of horizontal distance there will be a rise of 1 unit of vertical distance. You can establish this slope by measuring 10 in along the chord, erecting a perpendicular 1 in above the chord, and connecting the top of that perpendicular with the leading edge.

On the line OC, locate the center of the leading-edge radius by measuring 4.85 percent of the chord length from zero, the leading edge. Set a compass for 4.85 percent of the chord length, place the point of the compass at the point found to be the center of the leading-edge radius, and inscribe an arc that passes through the leading edge. This provides the nose. Now, with a spline or French curve, connect the leading-edge curve so that it blends smoothly into the curve previously constructed. (In practice, the nose is usually drawn before the ends of the ordinates are connected, but this part of the instruction was postponed so that the principal work could be explained before discussing the details.)

● PERFORMANCE OF AIRFOILS

Airfoil Characteristics

A particular airfoil, that is, one having certain definite dimensions, has specific lift, drag, and center of pressure (CP) position characteristics during flight. These features are collectively known as **airfoil characteristics** and they are classified as follows:

1. Lift coefficient
2. Drag coefficient
3. Lift/drag ratio
4. Center-of-pressure position

In place of the CP position, highly technical publications may use some equivalent characteristics, such as the moment of aerodynamic force about the leading edge or the aerodynamic center; but it is sufficient for the technician to consider only the CP position.

Symbols Relating to Airfoils

Aerodynamicists employ certain letter symbols to represent various quantitative factors in performing mathematical calculations on airfoil performance. The following are those applicable to the material in this chapter.

A Aspect ratio
C_D Coefficient of drag
C_L Coefficient of lift
C_M Coefficient of pitching moment
D Drag
L Lift
M Pitching moment
R Reynolds number
S Wing area
V Speed or velocity
a Wing lift-curve slope
b Wing span
c Wing chord
$\bar{c}$ Mean geometric chord, S/b
c Mean aerodynamic chord
c_s Root chord
c_t Tip chord
x Longitudinal coordinate (fore and aft)
y Lateral coordinate
z Vertical coordinate
α (alpha) Angle of attack
ρ (rho) Air density

Fundamental Equation for Lift

Lift has been defined as the net force developed perpendicular to the relative wind. The aerodynamic force of lift on an airplane results from the generation of a pressure distribution on the wing.

The lift of an airfoil can be determined mathematically by using given known values in the fundamental equation for lift. This formula is as follows:

$$L = C_L \frac{\rho}{2} V^2 S$$

where L = lift, lb
  C_L = lift coefficient
  ρ = mass density, slugs/ft^3
  V = velocity of wind relative to the body, ft/s
  S = airfoil area, ft^2

Coefficient of Lift

A coefficient, according to a dictionary, is a number or symbol which acts as a multiplier to a variable or unknown quantity. For example, 3 is a coefficient in $3ab$; x is a coefficient in $x(y + z)$. Thus, a coefficient is a number used as a multiplier.

The fundamental equation of lift can be used to determine the coefficient of lift by rewriting it as follows:

$$C_L = \frac{L}{\frac{1}{2}\rho V^2 S}$$

The coefficient for lift is a function of the airfoil shape and the angle of attack. For a given shape, the coefficient of lift varies with the angle of attack; hence, when the fundamental equation for lift is used, the angle of attack must be specified to make the computation meaningful. A certain airfoil has a C_L of 0.4 at a 4° angle of attack and a C_L of 1.2 at a 16° angle of attack. It is clear, then, that the angle of attack must be known before the answer to the equation has a usable value.

Let us assume that we now wish to know the amount of lift that can be obtained from an airplane wing having an area of 180 ft^2 [16.7 m^2] at a velocity of 120 mph [53.6 m/s] and an altitude of 1000 ft [304.8 m] with a C_L of 0.4 at a 4° angle of attack. First the velocity is converted into feet per second. A velocity of 120 mph is equal to 176 ft/s. Next, by reference to the NASA standard atmospheric tables, it is found that the value of ρ (density) is 0.002 309. Substituting values in the fundamental equation for lift, we have

$$L = 0.4 \times \frac{0.002\ 309 \times (176)^2 \times 180}{2}$$

or

$$L = 2574.85 \text{ lb } [1167.95 \text{ kg}]$$

Fundamental Equation for Drag

The fundamental equation for drag is found to be almost identical with the equation for lift except that we use drag and the coefficient of drag in place of the values for lift. The equation for drag is then

$$D = C_D \frac{\rho}{2} V^2 S$$

The symbol meanings are the same as those stated previously in the equation for lift except for D (drag) and C_D (coefficient of drag).

The equation for drag can be rewritten to obtain the coefficient of drag as follows:

$$C_D = \frac{D}{\frac{1}{2}\rho V^2 S}$$

Like the coefficient of lift, the coefficient of drag is a function of the airfoil shape and the angle of attack. For a given shape the coefficient of drag varies with the angle of attack. For example, a certain airfoil has a C_D of 0.11 at a 4° angle of attack, but at a 16° angle of attack it has a C_D of 0.24.

The fundamental equation for drag is used in the same manner as the equation for lift by substituting known values for the symbols.

The three basic types of drag were discussed in the previous chapter. **Induced drag** is the drag produced by an airfoil in the process of developing lift. **Parasite drag** is the drag produced by elements not involved in producing lift—for example, landing gear, struts, rivet heads, skin roughness, etc. **Profile drag** is the drag produced by the skin friction of the airfoil, and it varies according to the airfoil profile. Profile drag is generally considered as a part of parasite drag.

Lift/Drag Ratio

The **lift/drag ratio** is the ratio of the lift to the drag of any body in flight and is a measure of the effectiveness of an airfoil because the lift is the force required to support the weight while the drag is a necessary nuisance that must be accepted to obtain lift. For any angle of attack, the C_L divided by the C_D will give the L/D ratio.

The maximum value of the L/D ratio for the wing is always more than the maximum value of the L/D ratio for the complete airplane because the drag for the complete airplane includes not only the drag of the wing but also the drag contributed by the rest of the airplane. This is based upon the assumption that all the lift of a conventional airplane is obtained from the wing. It can be represented in a formula, thus:

$$\text{Airplane } L/D = \frac{\text{lift from wing}}{\text{wing drag} + \text{other drag}}$$

Center-of-Pressure Coefficient

It has been explained that the CP of an airfoil is the point in the chord of the airfoil that is at the intersection of the chord and the line of action of the resultant air force. The CP coefficient is the ratio of the distance of the CP from the leading edge to the chord length. In other words, the CP is given by stating that it is a certain percentage of the chord length behind the leading edge. For example, if the chord is 6 ft [1.82 m] long and the CP is 30 percent of the chord length behind the leading edge, then it is 30 percent of 6 ft, or 1.8 ft [0.546 m] behind the leading edge at the particular angle of attack. Since the pressure distribution varies along the chord with changes in the angle of attack, the CP, which is the point of application of the resultant, moves accordingly. We have previously explained that the CP generally moves forward as the angle of attack increases and backward as it decreases, although there are exceptions to all rules, including this one.

Characteristic Curves

At the beginning of this section we explained that the lift coefficient, drag coefficient, lift/drag ratio, and CP position are collectively known as airfoil characteristics. **Characteristic curves** are graphical representations of airfoil characteristics for various angles of attack. It is important to understand that the values of all the airfoil characteristics vary with the angle of attack and that they are different for different airfoil sections.

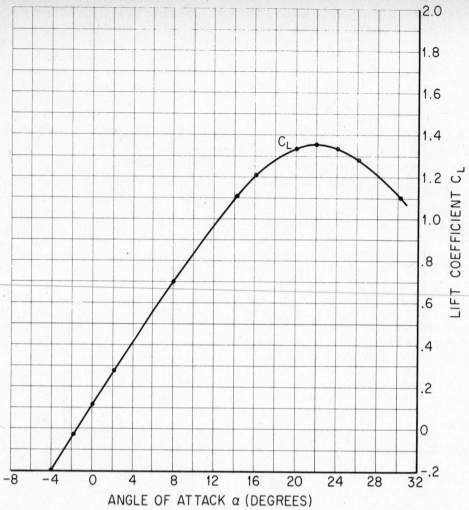

FIG. 4-5 Coefficient-of-lift curve.

Figure 4-5 is a **coefficient-of-lift** diagram for the NACA 2421 airfoil. In this diagram the angle of attack is plotted horizontally, and the value of the lift coefficient is plotted vertically. Notice that a horizontal line is drawn and along this line are located points which correspond to the various angles of attack from −8 to +32°. These points are marked with the angles represented. At each of these points the person preparing the illustration measures vertically upward a distance that represents to a suitable scale the coefficient of lift for that particular angle of attack and makes a dot or draws a tiny circle. A smooth curve is then drawn through the points located in this manner. This curve is technically called the **lift-coefficient curve,** or simply the **lift curve,** for the particular airfoil being represented.

Figure 4-6 is a **coefficient-of-drag** diagram for the NACA2421 airfoil. The angle of attack is again plotted horizontally, and the drag coefficient is plotted vertically for each angle of attack. The resulting curve is technically called the **coefficient-of-drag curve,** or simply the **drag curve,** for the airfoil.

Figure 4-7 shows the lift and drag curves for the NACA2421 airfoil drawn on the same diagram. This is common practice because it presents two types of information on one drawing. The coefficient-of-lift

and coefficient-of-drag curves were shown here separately to make it easier to understand their construction.

Figure 4-8 shows the lift/drag ratio curve drawn on the same diagram with the lift and drag curves. For each angle of attack the corresponding coefficient of lift is divided by the corresponding coefficient of drag to obtain the L/D ratio for that particular angle of attack. A distance is measured vertically upward to represent the L/D ratio value, and a dot or circle is drawn. The dots or circles found in this manner are connected by a smooth line to represent the L/D ratio curve on the diagram.

So that we can see how the L/D ratio curve is obtained from the C_L and C_D curves on the diagram, let us consider a specific instance. If the airfoil is at an angle of attack of 12°, the coefficient of lift C_L is approximately 0.95 and the drag coefficient C_D is approximately 0.07. Then the L/D ratio is $C_L/C_D =$ 0.95/0.07 = 13. Thus we see that the lift/drag or L/D ratio is equal to 13. If we observe the L/D ratio curve on the diagram of Fig. 4-8, we note that at a 12° angle of attack this curve has a value of 13. Figure 4-9 shows the curve representing the position of CP drawn on the same diagram with the lift, drag, and L/D ratio curves. For each of the angles of attack represented on

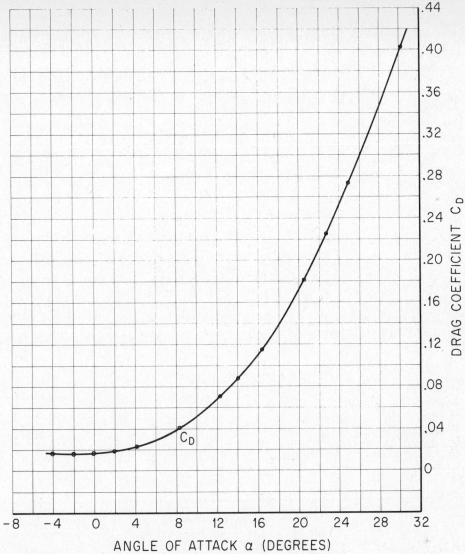

FIG. 4-6 Coefficient-of-drag curve.

the horizontal line, the location of the corresponding CP is plotted vertically upward in terms of the percentage of chord from the leading edge, and the points so found are connected by a smooth line to provide the curve.

Airfoil Profiles for General-Aviation Aircraft

Although general-aviation aircraft have been operating satisfactorily for many years with a variety of proven airfoil designs, research and development have been conducted to improve the efficiency of airfoils used for light aircraft. The research has been conducted at the NASA Langley Research Center utilizing information and technology derived from the development of supercritical airfoils designed for aircraft which operate at speeds just below the speed of sound.

The first of the new airfoils, now being used on some aircraft such as the Beechcraft Skipper shown in Fig. 4-10, is the **GAW-1** (general-aircraft wing-1). Its profile is approximately as shown in Fig. 4-11. This airfoil has a thickness ratio of 17 percent, and the de-

sign has improved the lift/drag ratio during climb by 50 percent. The maximum lift without flaps is improved by 30 percent, and the stall characteristics are good. The above data are in comparison with the wing which the GAW-1 replaced.

A newer design, the GAW-2 airfoil, has a thickness ratio of 13 percent and has a lower drag and higher C_L than the GAW-1. These features provide an increase in the rate of climb and a decrease in stall speed. The two airfoils can be used to advantage in a tapered wing with the GAW-1 at the root, where the extra thickness provides for greater structural strength, and the GAW-2 at the outer portion of the wing to give increased aerodynamic efficiency. The approximate profile of the GAW-2 airfoil is shown in Fig. 4-12.

It is not unusual for an airplane to utilize more than one airfoil profile in a wing. Oftentimes a manufacturer will use one airfoil profile for the inboard wing area, a second profile for the center section, and a third for the wing tip. While utilizing more than one profile may increase aerodynamic performance, it also adds greatly to the construction costs.

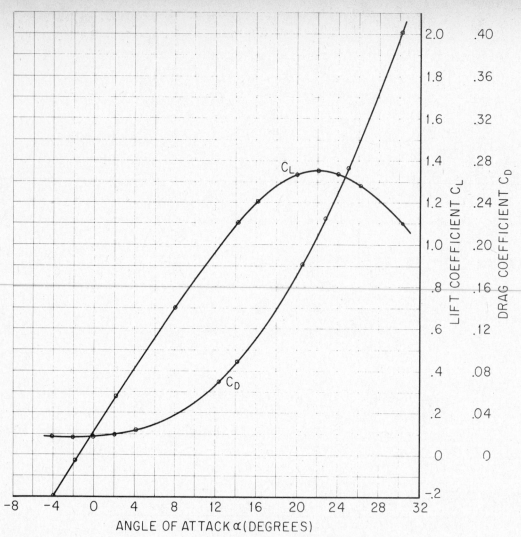

FIG. 4-7   Lift and drag curves on the same graph.

High-Lift Devices

There are many different types of high-lift devices such as flaps, slots, and slats that are used to increase the maximum lift coefficient for low-speed flight.

Wing Flaps

A **wing flap** is defined by NASA as a hinged, pivoted, or sliding airfoil, usually near the trailing edge of the wing. It is designed to increase the lift, drag, or both when deflected and is used principally for landing, although large airplanes use partial flap deflection for takeoff. Most flaps are usually 15 to 25 percent of the airfoil's chord. The deflection of a flap produces the effect of a large amount of camber added well aft on the chord. This makes it possible for the airplane to have a steeper angle of descent for the landing without increasing the airspeed. Flaps are normally installed inboard of the ailerons on a monoplane, although, in some cases, they may be placed both inboard and outboard of the ailerons.

Some of the basic types of flap design are illustrated in Fig. 4-13. A basic airfoil without a flap is shown at the top.

The next drawing is a **plain flap.** The plain flap is a simple hinged portion of the trailing edge. The effect of the camber added well aft on the chord causes a significant increase in both the coefficients of lift and drag.

The **split-edge flap** is usually housed flush with the lower surface of the wing immediately forward of the trailing edge. This flap is illustrated in the middle of Fig. 4-13. The split-edge flap is usually nothing more than a flat metal plate hinged along its forward edge. The split-edge flap produces a slightly greater change in lift than the plain flap. However, a much larger change in drag results from the great turbulent wake produced by this type of flap.

The **Zap flap,** as shown in the drawing next to the bottom in Fig. 4-13, is roughly similar to the plain split flap when it is fully retracted; but as the flap is opened, the hinged axis moves rearward to keep the trailing edges of the wing and the flap on a line perpendicular to the chord of the wing.

The **Fowler flap,** shown at the bottom of Fig. 4-13, is constructed so that the lower part of the trailing edge of the wing rolls back on a track, thus increasing the effective area of the wing and at the same time lowering the trailing edge. The flap itself is a small airfoil

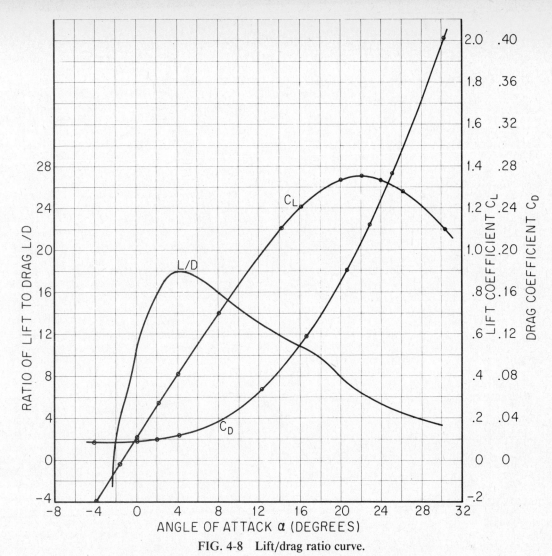

FIG. 4-8   Lift/drag ratio curve.

that fits neatly into the trailing edge of the main wing when closed. As shown in Fig. 4-14, when the flap opens, the small airfoil slides downward and backward on tracks until it reaches the position desired by the pilot, thus providing a wing with a variable coefficient of lift and a variable area. With the Fowler flap, the wing area can be increased causing large increases in lift with minimum increases in drag, the exact amount of increase of each depending upon the angle to which the flap is lowered. The Fowler flap is one of the designs which are particularly well adapted for use at takeoff as well as landing.

Slotted flaps have been developed to provide even more lift than the flaps described previously. When such flaps are extended, either partially or completely, one or more slots are formed near the trailing edge of the wing. The slots allow air from the bottom of the wing to flow to the upper portion of the flaps and downward at the trailing edge of the wing. This aids in delaying airflow separation and creates a downward flow of air which produces an upthrust to the wing. Triple-slotted flaps are shown in Fig. 4-15. The slotted flap can provide much greater increases in lift than the

plain or split flap and corresponding drag changes are much lower.

Effects of Flaps

At normal flying speeds, when flaps are fully retracted, that is, when they are all the way up, they have no effect on the lift characteristics of the wing. On the other hand, when they are lowered for landing, there is increased lift for similar angles of attack of the basic airfoil, and the maximum lift coefficient is greatly increased, often as much as 70 percent with the exact amount of increase depending upon the type of flap installed.

The effectiveness of flaps on a wing configuration depend on many different factors. One important factor is the amount of the wing area affected by the flaps. Since a certain amount of the span is reserved for ailerons, the actual wing maximum lift properties will be less than that of the flapped two-dimensional section. If the basic wing has a low thickness, any type of flap will be less effective than on a wing of greater thickness. The curves of Fig.4-16 illustrate the lift charac-

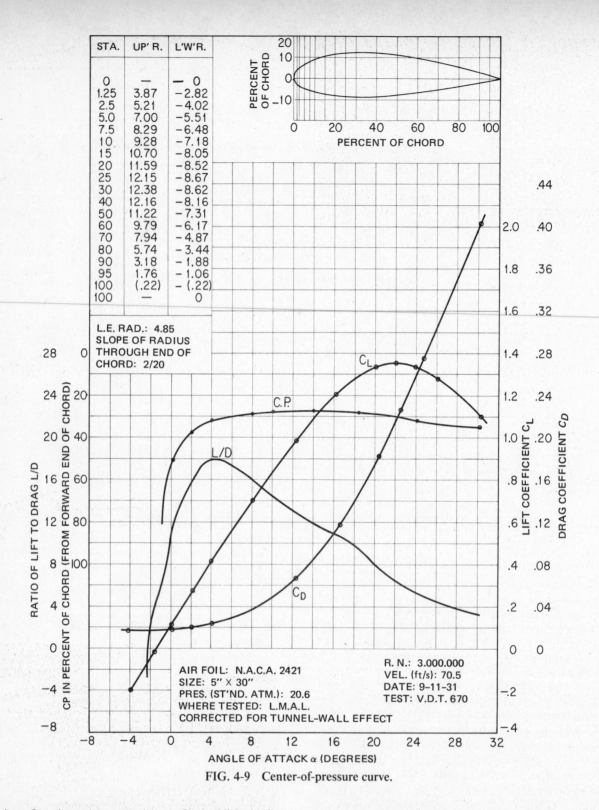

FIG. 4-9 Center-of-pressure curve.

teristics of a wing with and without flaps. With the increase of lift comes a decrease in landing speed; but there is also an increase of drag when the flap is down, and this requires a steeper glide to maintain the approach speed. The increase of drag also acts as a brake when the airplane is rolling to a stop on the landing strip.

Slots

A **slot** is also a high-lift device because it improves lift. It is a nozzle-shaped passage through a wing designed to improve the airflow conditions at high angles of attack and slow speeds. It is normally placed very near the leading edge and is formed by a main and an auxiliary airfoil, or slat. A **slat** is a movable auxiliary airfoil attached to the leading edge of the wing, which, when closed, falls within the original contour of the wing and which, when opened, forms a slot. Slots are illustrated in Figs. 4-17 and 4-18.

There are two general types of slots: the fixed and the automatic. When the fixed type is used, the airflow depends on the angle of attack. As the angle of attack

FIG. 4-10 Beechcraft Skipper utilizing the GAW-1 airfoil profile. *(Beech Aircraft Co.)*

FIG. 4-11 GAW-1 airfoil profile.

FIG. 4-12 GAW-2 airfoil profile.

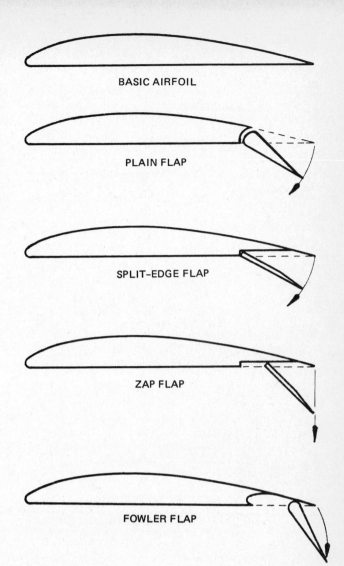

FIG. 4-13 Basic types of flaps.

of the wing increases, air from the high-pressure region below the wing flows to the low-pressure area above the wing as shown in the bottom drawing of Fig. 4-17. This flow of air postpones the breakdown of streamline flow that accompanies an increase in the angle of attack. If the slot is well designed, the burble point will not be reached until the angle of attack is much greater than that of a normal stall.

The automatic slot consists of an auxiliary airfoil that is nested into the leading edge of the wing while the wing is at low angles of attack but is free to move forward a definite distance from the leading edge at high angles of attack. This forms a slot through which a portion of the airstream flows and is deflected along the upper surface of the wing, thus maintaining a streamline flow around the wing. Figure 4-17 shows the effect of the airstream diverted by a slot and the advantage gained by its use. The top picture shows the airfoil with a slot closed at a high angle of attack. The airfoil is shown in a stalling position because the burbling of the air reaches almost the leading edge of the wing.

The automatic slot has disadvantages as well as advantages. The number of moving parts and the weight of the wing are increased. In the past, weight was also added because it was necessary, in the case of some airplanes, to have longer landing-gear struts to eliminate tail skid landings brought about by the increased angle of attack at the stall. The slots must be installed properly and operate equally well on both wings or they are useless. If a slot on one wing should open be-

FIG. 4-14 Operation of the Fowler flap.

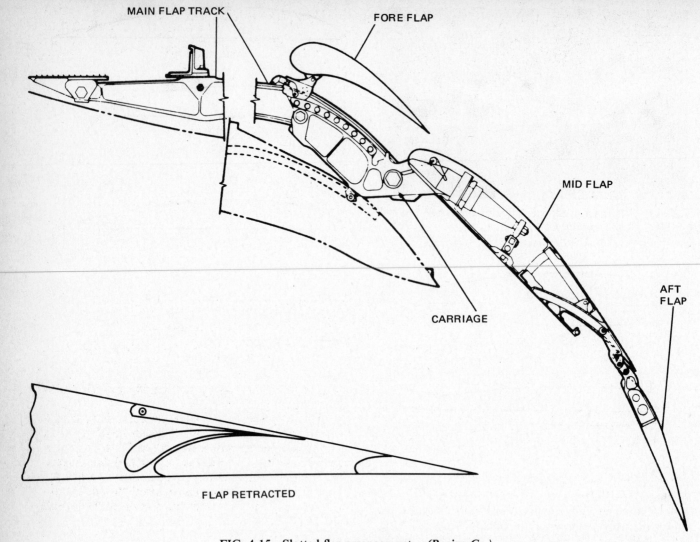

MAIN FLAP TRACK

FORE FLAP

MID FLAP

AFT FLAP

CARRIAGE

FLAP RETRACTED

FIG. 4-15   Slotted-flap arrangement.   *(Boeing Co.)*

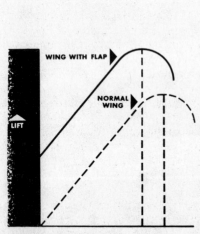

WING WITH FLAP

NORMAL WING

LIFT

FIG. 4-16   Effect of flaps on lift.

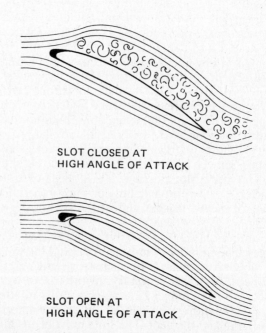

SLOT CLOSED AT
HIGH ANGLE OF ATTACK

SLOT OPEN AT
HIGH ANGLE OF ATTACK

FIG. 4-17   Effect of a slot on wing performance.

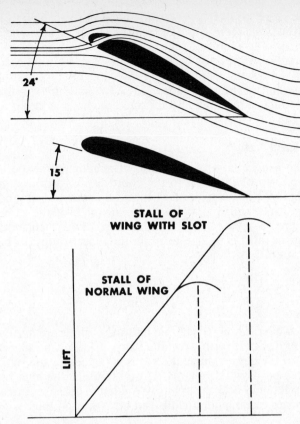

FIG. 4-18   Effect of the slot on stalling speed.

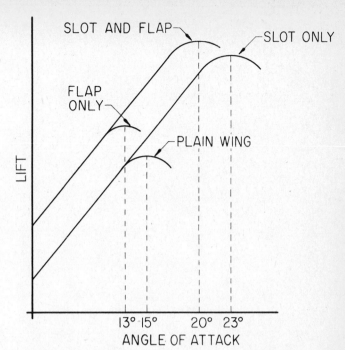

FIG. 4-19   Effect of the combination of slots and flaps.

fore the slot on the opposite wing opened, disastrous results could occur. The usual location of slots is such that they are subjected to ice formation, and in spite of any anti-icing or deicing equipment, they may fail to function. If any of these factors causes a lack of balance, lateral control may be impaired. For these reasons, a device is usually provided for locking slots in a closed position if they do not function properly.

Figure 4-18 illustrates the effect of the slot on the lift coefficient. Notice that at angles where the slot is opened, the lift is greater and the maximum C_L occurs at a much higher angle of attack. This indicates that an airplane with a slotted wing has a lower stalling speed than one without slots, other things being equal.

Figure 4-19 illustrates the effect of having a combination of slots and flaps. With this arrangement, it is possible to have a much lower landing speed, better control of the flight path, and at least a partial elimination of the nose heaviness that may result from the use of flaps alone. It should be understood that Fig. 4-19 is based upon a particular set of conditions and does not illustrate the effect produced by various airfoils and combinations of different flaps and slots. Other types of flaps and combinations with slots will produce values differing from those shown in this figure.

Leading-Edge Flaps

A leading-edge flap is a high-lift device which reduces the severity of the pressure peak above the wing at high angles of attack. This enables the wing to operate

at higher angles of attack than would be possible without the flap.

One method for providing a wing flap is to design the wing with a leading edge that can be drooped as shown in Fig. 4-20. Another method for providing a leading-edge flap is to design an extendable surface that ordinarily fits smoothly into the lower part of the leading edge. When the flap is required, the surface extends forward and downward as shown in the second drawing of Fig. 4-20.

● SHAPES AND DIMENSIONS OF AIRFOILS

Effects of Wing Planform

Previous discussions of aerodynamic forces have been limited to how they affect airfoil profiles. Since an airfoil section has no span, we have been concerned about airflow in two dimensions only. When the effects of wing **planform** (shape) are introduced (see Fig. 4-21), attention must be directed to the existence of airflow in the spanwise direction. In other words, airfoil section properties deal with flow in two dimensions while actual wings have flow in three dimensions.

The aspect ratio, taper ratio, and sweepback of a planform are the principal factors which determine the aerodynamic characteristics of a wing. These same quantities also have a definite influence on the structural weight and stiffness of a wing.

Wing Area

The **wing area** S is simply a measure of the total surface of the wing. Although a portion of this area may be covered by the fuselage or nacelles, pressure is still

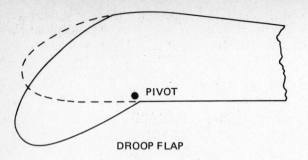

DROOP FLAP

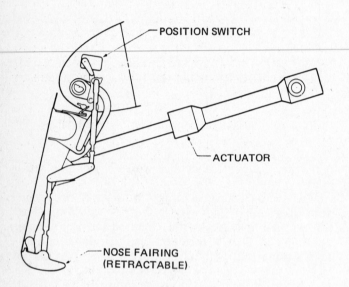

POSITION SWITCH

ACTUATOR

NOSE FAIRING
(RETRACTABLE)

KREUGER FLAP

FIG. 4-20 Types of leading-edge flaps.

acting on it; therefore it is included in the calculation of the total wing area. The wingspan b is measured tip to tip. The average wing chord c is simply a geometric average of the wing chords. As an example, a pointed tip delta wing would have an average chord equal to one-half of the root chord. As shown is Fig. 4-22, the product of the span and the average chord is the wing area ($b \times c = S$).

Aspect Ratio

The **aspect ratio** of an airfoil of rectangular shape is the ratio of the span to the chord. Thus, airfoil A in Fig. 4-23 has a span of 24 ft and a chord of 6 ft, and the aspect ratio is 4. Likewise, airfoil B of the same figure has a span of 36 ft and a chord of 4 ft; hence the aspect ratio is 9. The formula for aspect ratio can be written thus:

$$A = \frac{\text{span}}{\text{chord}}$$

However, for aerodynamic reasons, airfoils are hardly ever designed with a rectangular planform. The aspect ratio for nonrectangular airfoils is defined as the span squared divided by area. If we represent the area by the letter S and the span by the letter b, the formula for nonrectangular airfoils can be expressed thus:

$$A = \frac{b^2}{S}$$

The formula for the aspect ratio of a nonrectangular airfoil could be applied to a rectangular airfoil if any such object were in existence since S, the area, is a product of the width c and the length b. Typical aspect ratios vary from 35 for a high-performance sailplane to 3.5 for a jet fighter.

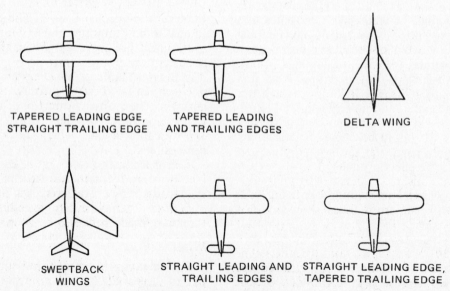

TAPERED LEADING EDGE,
STRAIGHT TRAILING EDGE

TAPERED LEADING
AND TRAILING EDGES

DELTA WING

SWEPTBACK
WINGS

STRAIGHT LEADING AND
TRAILING EDGES

STRAIGHT LEADING EDGE,
TAPERED TRAILING EDGE

FIG. 4-21 Wing planforms.

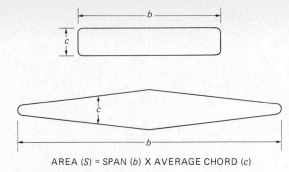

AREA (S) = SPAN (b) X AVERAGE CHORD (c)

FIG. 4-22   Area of a wing.

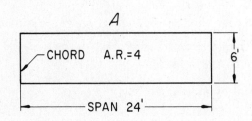

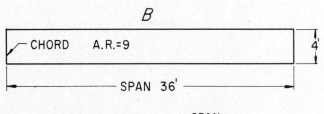

$$\text{ASPECT RATIO} = \frac{\text{SPAN}}{\text{CHORD}}$$

FIG. 4-23   Aspect ratio.

Wing-Tip Vortices

Air flows over and under the wing of an airplane, as shown in Fig. 4-24. This produces a region of relatively low pressure above the wing and a region of relatively high pressure under the wing. If the airfoil were of infinite span, that is, if it had no ends, the air-

FIG. 4-24   Airflow over and under the wing.

flow would be direct from the leading edge to the trailing edge, as shown in Fig. 4-24, because there would be no way for the air in the region of high pressure below the wing to flow into the region of low pressure above the wing. However, it is obvious that the airfoil is of finite area and span; that is, it has ends. Hence the air under the wing will seek the region of low pressure above the wing by "spilling over" the tips, as

shown in Fig. 4-25. Eddies, or regions of turbulence, are formed in this manner at the tips, causing the streamlines to form vortices, which are like tiny tor-

FIG. 4-25   Air spilling over wing tips.

nados such as those illustrated in Fig. 4-26. The turbulence produced absorbs energy and increases the induced drag. Induced drag is that part of the drag caused by lift. That is, drag caused by the change in direction of airflow. As induced drag increases, the lift is reduced by destroying the force of the airstream near the tips.

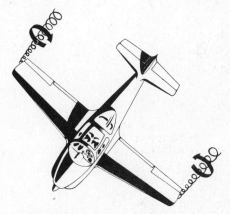

FIG. 4-26   Wing-tip vortices.

Effects of Aspect Ratio

The effect of aspect ratio on the lift and drag characteristics is shown in Fig. 4-27. The effect of increasing aspect ratio is principally to reduce induced drag for any given coefficient of lift. This, of course, improves the L/D ratio.

Since wing-tip vortices exert their influence for a distance inboard from the tips in any given airfoil, it is apparent that the percentage of area so affected is less for a long, narrow airfoil than it is for a short, wide airfoil. This is shown in Fig. 4-28, which illustrates the area affected by wing-tip vortices. The shape in the upper-left-hand corner of Fig. 4-28 is short and wide; the center shape is long and relatively narrow; and the lower-right shape is still longer and narrower. Although the area affected by wing-tip vortices remains the same for all these shapes, a smaller proportion of the total area is affected when the airfoil is long and narrow.

The relationship for induced drag coefficient emphasizes the need of a high aspect ratio for an airplane which is continually operated at high lift coefficients.

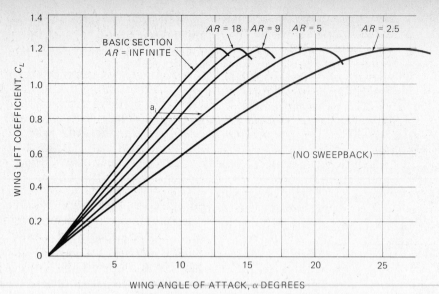

FIG. 4-27   Effect of aspect ratio on wing characteristics.

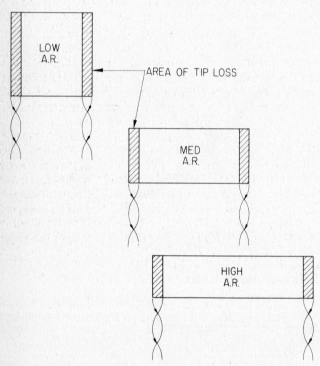

FIG. 4-28   How increased aspect ratio reduces the effect of wing-tip vortices.

In other words, airplane configurations designed to operate at high lift coefficients during the major portion of their flight, such as sailplanes, demand a high aspect ratio wing to minimize the induced drag. While the high aspect ratio wing will minimize induced drag, long, thin wings increase structural weight and have relatively poor stiffness characteristics. This fact will temper the preference for a very high aspect ratio.

Winglets

The utilization of **winglets,** which look like small jib sails mounted at and above the wingtips (see Fig.

4-29), is another method of reducing wing-tip vortices. A winglet reduces the cross-flow on the wing, which in turn reduces the trailing vortex in strength. This reduction of wing-tip vortices will cause a corresponding reduction in drag, thereby increasing the wings' lift/drag ratio.

FIG. 4-29   Winglets on a Learjet Model 55.   *(Gates Learjet Corporation)*

Taper

The root chord c_r is the chord at the wing center line and the tip chord c_t is the chord measured at the tip. An airfoil is tapered when one or more of its dimensions gradually decreases from the root to the tip. When the airfoil decreases from the root to the tip in both thickness and chord, the airfoil is said to have **taper in plan and thickness.** This is shown in the bottom drawing of Fig. 4-30. If the thickness and chord remain the same from the root to the tip, there is no taper.

When there is a gradual change (usually a decrease) in the chord length along the wing span from the root to the tip, with the wing sections remaining geometrically similar, the airfoil is said to have **taper in plan only,** as shown in the upper-right-hand drawing of Fig. 4-30. When there is a gradual change in the thickness

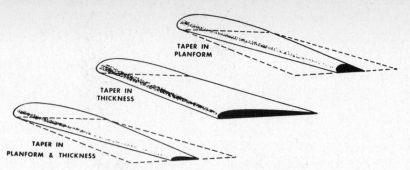

FIG. 4-30 Taper in aircraft wings.

ratio along the wing span, with the chord remaining constant, the airfoil is said to have **taper in thickness ratio only,** as shown in the middle drawing of Fig. 4-30.

The **taper ratio** λ (lambda) is the ratio of the tip chord to the root chord. Thus the airfoil in Fig. 4-31

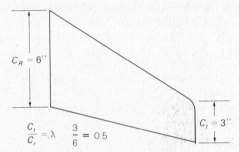

FIG. 4-31 Wing-taper ratio.

has a root chord of 6 ft [1.8 m], a tip chord of 3 ft [0.9 m], and a taper ratio of 0.5. A rectangular wing has a taper ratio of 1.0, and the pointed tip delta wing has a taper ratio of 0.0.

The taper ratio affects the lift distribution and the structural weight of the wing. When a wing is tapered in thickness in such a manner that the thickness near the tip is 60 percent of the thickness at the root and it is compared with an airfoil of constant section (not tapered) equal to the mean (average) section of the tapered wing, the following characteristics are observed on certain airfoils: (1) The CP moves less for changes in angle of attack. (2) The maximum C_L is greater and the peak of the characteristic curve is flatter because all of the wing does not attain the maximum C_L at the same time; that is, each section reaches its maximum C_L at a different angle of attack from any other section. (3) The C_D values are lower, the most noticeable decrease being at the low angle of attack from any other section. (4) The maximum L/D ratio are larger at small angles of attack. It is interesting to note that a tapered wing may also have a constantly changing airfoil section from the root of the wing to the tip.

When a wing is tapered in planform and is compared with a rectangular airfoil that has an equivalent aspect ratio, the following characteristics are observed on certain airfoils: (1) The CP moves more for changes in angle of attack. (2) There is a greater maximum C_L. (3) There are lower values of C_D, especially at low an-

gles of attack. (4) The L/D ratio is greater throughout the flight range, especially at the higher angles of attack. When a wing or any airfoil is tapered in both thickness and planform, it is possible to take advantage of the best aerodynamic features of the airfoil tapered in thickness only and the airfoil tapered in planform only.

When the distribution of the area of a tapered wing places the resultant force near the center line, it may be possible to build a wing of relatively light weight, having the thicker, heavier, and stronger portions near the root, where the greatest stresses normally occur. On the other hand, in a tapered airfoil, the spars must be tapered and different jigs must be used for building the ribs. For this reason the construction of the wing tapered in both planform and thickness becomes considerably more costly than the construction of other types of wings.

Sweep Angle

The **sweep angle** Λ (cap lambda) is usually measured as the angle between the line of 25 percent chords and a perpendicular to the root chord as shown in Fig. 4-32. The sweep of a wing causes definite changes in compressibility, maximum lift, and stall characteristics.

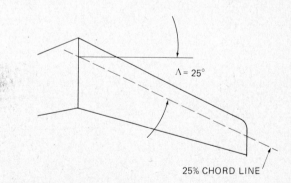

FIG. 4-32 Sweep angle.

Mean Aerodynamic Chord

The **mean aerodynamic chord** (MAC) is the chord drawn through the center of area. As an example, the pointed-tip delta wing with a taper ratio of zero would have an average chord equal to one-half the root chord but a MAC equal to two-thirds of the root chord. The MAC is located on the reference axis of the airplane and is a primary reference for longitudinal stability considerations.

Laminar Flow Control

Today's civil transports typically cruise at about 500 mph (800 km/h). One problem of traveling at such a speed occurs in the **boundary layer,** a thin sheet of flowing air that moves along the surfaces of the wing, fuselage, and tail of an airplane.

At low speeds, this layer follows the aircraft contours and is smooth, a condition referred to as **laminar.** At high speeds, the boundary layer changes from laminar to turbulent, creating friction and drag that waste fuel. This is illustrated in Fig. 4-33. Many experiments have been carried out in an effort to control the boundary layer and increase laminar flow.

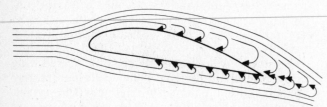

FIG. 4-33   Development of the boundary layer as a result of skin friction.

The **Laminar Flow Control System** calls for removing the turbulent boundary layer by suction, thus maintaining laminar flow as is shown is Fig. 4-34. Basically, this system includes the suction surface through which a portion of the boundary layer air is taken into the airplane, a system for metering the level and distribution of the ingested flow, a ducting system for collecting the flow, and pumping units which provide sufficient compression to discharge the suction flow at a velocity at least as high as the airplane velocity. The effect of this system is to keep the boundary layer thin and permit laminar flow to continue.

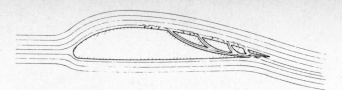

FIG. 4-34   One method for boundary-layer control.

Vortex Generators

Even though most modern jet airliners do not fly at the speed of sound (Mach 1), there are certain areas on the airplane where the airflow velocity will be greater than Mach 1. This is particularly true at the upper surface of parts of the wing where, because of the curvature of the wing, the air velocity must increase substantially above the airspeed of the airplane. This is illustrated in Fig. 4-35, which shows an airfoil profile moving through the air at high subsonic speed. A short distance back from the leading edge of the wing above the top surface, the air reaches supersonic speed. At the rear part of the supersonic area where the airflow returns to subsonic speed, a shock wave is formed. To the rear of this shock wave the air is very turbulent, and this area of the wing is, in effect, partially stalled. This, of course, causes a substantial increase in drag, which increases as airspeed increases.

In order to reduce the drag caused by supersonic flow over portions of the wing, small airfoils called **vortex generators** are installed perpendicular to the surface of the wing. On the Boeing 720 airplane a total of 96 vortex generators are mounted in two rows as shown in Fig. 4-36. There are 23 generators in the forward row and 25 in the aft row. The two rows are parallel and centered spanwise aft of the inboard nacelles, with the aft row extending slightly more inboard than the forward row. The vortex generators are mounted in complementary pairs, as shown in the drawing of Fig. 4-37. This arrangement causes the vortices being

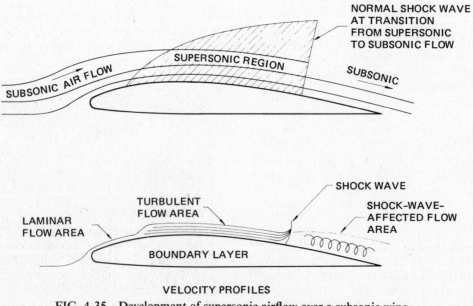

FIG. 4-35   Development of supersonic airflow over a subsonic wing.

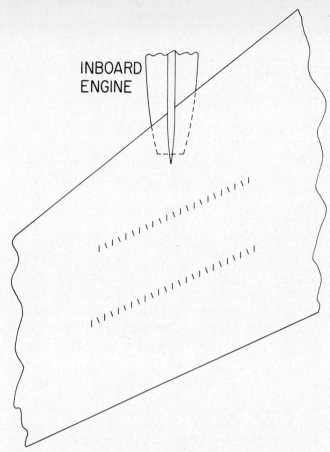

FIG. 4-36   Vortex generators mounted on an airplane wing.

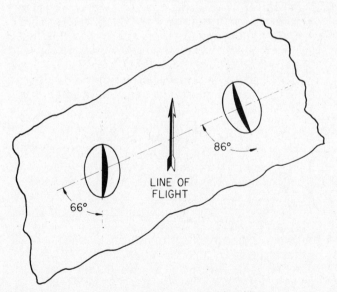

LINE OF FLIGHT

66°

86°

FIG. 4-37   Vortex generators arranged in pairs.

developed to add to one another, thus increasing the effect.

Because of the low aspect ratio of the vortex generators, they develop a strong tip vortex. The tip vortex causes air to flow upward and inward in a circular path around the ends of the airfoil. The vortex generated has the effect of drawing high-energy air from outside the boundary layer into the slower-moving air close to the skin. The strength of the vortex is proportional to the lift developed by the generator. To operate effectively, the generators are mounted forward of the point where separation begins.

Drag reduction achieved by the addition of vortex generators can be seen in the drag-rise curve. Since the generators effectively reduce the shock-induced drag associated with the sharp rise in the curve at speeds approaching Mach 1.0, the curve is pushed to the right as shown in Fig. 4-38.

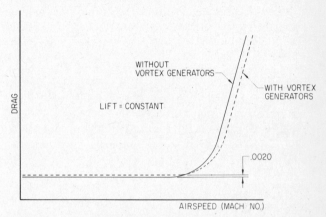

FIG. 4-38   Drag reduction achieved by vortex generators.

The addition of the vortex generators actually increases overall drag very slightly at lower speeds. However, the gains at cruise speeds more than balance out the losses at lower speeds. Since the airplane spends most of its flight time at cruise speeds, the net gain is significant.

Wing Fences

Ideally, air would always flow chordwise over a wing; however, as we have seen, air will tend to flow spanwise toward the tip. Spanwise flow is particularly a problem on swept wings. This spanwise flow of air may be partially controlled by the use of a **wing fence** such as is illustrated in Fig. 4-39. A wing fence is a stationary vane, projecting from the upper surface of an airfoil, which is used to prevent the spanwise flow of air.

FIG. 4-39   Wing fences.   *(Gates Learjet Corporation)*

Planforms for High-Speed Flight

Because of the complex nature of supersonic phenomena, aircraft and spacecraft designers have been required to resort to designs which at first seem rather unconventional. Many of these designs are decidedly inferior to the more familiar aircraft forms at subsonic flight speeds. When we consider the airflow across a wing, it becomes obvious that as a wing becomes thicker, the increase in local speed across the curved areas will become greater. If we wish to obtain an increase in critical Mach number, we must utilize wings that are as thin as possible. This presents difficulties, however, because a very thin wing does not have great strength and also because its lift is poor at low speeds.

Another method for increasing the critical Mach number is to sweep back the wing. This will improve the critical Mach number, but it will also present a problem in stability. If an aircraft with a pronounced sweepback changes heading as a result of rudder movement or a disturbance in the air, the wing that moves forward will have much greater lift than the other wing and the airplane will have a tendency to roll. At supersonic speed and higher, the advantage of the sweptback wing begins to decrease, and at Mach 2, the straight wing is superior.

Another method for increasing the critical Mach number is to reduce the aspect ratio. This is not a completely satisfactory solution, because a low aspect ratio adversely affects flight at low speeds.

The Concorde, pictured in Fig. 4-40, cruises at speeds of over Mach 2; yet its design which incorporates a delta planform wing with an aspect ratio of 1.7 is still capable of adequate performance at subsonic speeds.

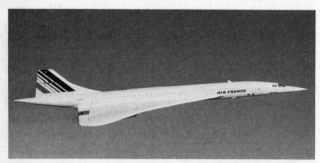

FIG. 4-40   The Concorde, designed for flight speeds of over Mach 2.   *(Air France)*

Supercritical Wing

Supercritical wings enhance a high-performance jet aircraft's maneuvering capabilities in transonic flight. Developed by NASA, this design delays and softens the onset of shock waves on the upper surface of a wing. The shock wave is far less severe than on the conventional wing, and fuel efficiency is substantially improved.

CONVENTIONAL WING DESIGN

SUPERCRITICAL WING DESIGN

THIN SUPERCRITICAL WING DESIGN

FIG. 4-41   Supercritical wing.

The supercritical airfoil, illustrated in Fig. 4-41, is slightly flatter on top, curving downward at the rear.

Airfoil Selection

The selection of the best airfoil for an airplane requires a careful consideration of the many factors that may conflict with one another, with the result that the final decision is usually a compromise. Some important factors to be considered in airfoil selection are (1) airfoil characteristics, (2) airfoil dimensions, (3) facility of airflow about the airfoil, (4) the speed at which the aircraft is designed to operate, and (5) flight operating limitations.

● REVIEW QUESTIONS

1. In what terms is *airfoil camber* expressed?
2. What is a *symmetrical airfoil*?
3. Define *profile thickness*.
4. What is the difference between airfoil profile and airfoil section?
5. Explain the meaning of the digits in the NACA airfoil number.
6. What do the following symbols indicate: A, C_D, C_L, D, L, S, V, c, c'.
7. Name four airfoil characteristics.
8. What is the fundamental equation for lift?
9. What is meant by *lift/drag ratio*?
10. What is the coefficient of lift?
11. Define *induced drag, parasite drag,* and *profile drag.*
12. In general, how does the center of pressure of an airfoil behave as the angle of attack changes?
13. What are characteristic curves for an airfoil?
14. What is the principal difference in design between a GAW-1 and GAW-2 airfoil?
15. What is the advantage of using flaps when landing an airplane?
16. What is the advantage of slotted flaps over plain flaps?
17. Describe slots and slats at the leading edge of a wing.
18. Explain why flaps, slots, and slats are called high-lift devices.
19. What is the purpose of leading-edge flaps?
20. What is meant by the *span of an airfoil?*
21. Define *aspect ratio* and explain how it may be determined.

22. What is the formula for determining the aspect ratio of a nonrectangular wing?
23. What are *wing-tip vortices?*
24. What is the effect of increasing the aspect ratio of a wing?
25. Give some of the limitations for increase of aspect ratio.
26. Describe three types of tapered wings.
27. What is the purpose of a winglet?
28. List design factors which must be considered in selecting an airfoil design.
29. Why are vortex generators employed on some airfoils?
30. What is the purpose of a wing fence?

5 AIRCRAFT IN FLIGHT

● INTRODUCTION

An aircraft must have satisfactory handling properties in addition to adequate performance. The aircraft must have adequate stability to maintain a uniform flight condition and recover from various disturbing influences.

Any person associated with aviation should have some understanding of the actual processes of flight. An individual who is responsible for the design, construction, and operation or maintenance of aircraft must be thoroughly familiar with the forces acting on an airplane in flight, the components of the airplane that control the flight forces, and the reactions of the airplane to its control.

This chapter includes the information concerning the stability and control of aircraft that today's technician needs to possess in order to make intelligent decisions affecting the flight safety of both airplanes and helicopters.

Understanding why the aircraft is designed with a particular type of primary and secondary control system is essential in maintaining today's complex aircraft.

Proper maintenance of an airplane requires that the person responsible for such maintenance have a knowledge of the forces acting on the airplane in flight in order to determine what repairs and adjustments will be needed. Any repair to an airplane should be just as strong as the original part. A maintenance technician who knows what stresses will be applied to a given part of the aircraft in flight will be in a better position to judge the suitability of a particular repair.

● FORCES ON THE AIRPLANE IN FLIGHT

Lift and Weight

The forces acting on an airplane in flight are lift, weight, drag, and thrust. The weight acts vertically downward from the **center of gravity** (CG) of the airplane. The lift acts in a direction perpendicular to the direction of the relative wind from the center of pressure (CP). In straight and level flight the lift and weight must be equal. When the airplane is flying at a constant speed, the thrust must equal the drag. Thrust, of course, is provided by the engine and propeller or by high-velocity gases ejected from the tail pipe of a jet engine.

Figure 5-1 illustrates the forces acting on an airplane in straight and level flight at a constant speed. In this case all the forces are in balance.

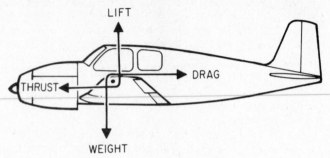

FIG. 5-1   Forces on an airplane in flight.

Drag and Thrust

The **thrust line** is an imaginary line passing through the center of the propeller hub perpendicular to the plane of the propeller rotation, as illustrated in Fig. 5-1. In the case of a jet-propelled airplane, the thrust line is parallel to the path of the ejected gases from the jet engine.

Propeller thrust is technically defined as the component of the total air force on the propeller that is parallel to the direction of advance. In simple terms, thrust is merely the force that drives the airplane forward. In a propeller-driven airplane, thrust acts from the hub of the propeller along the thrust line. **Thrust** is defined as the forward-directed pushing or pulling force developed by an aircraft engine. This includes reciprocating engines, turbojet engines, turboprop engines, or rocket engines.

Drag, in general terms, is the force which opposes the forward motion of the airplane. Previously, when we discussed drag, we referred to induced drag and parasite drag. The total drag of the airplane includes the wing drag and a number of other forms of drag, including that caused by the fuselage, the landing gear, and other parts of the airplane. The total drag of the airplane is in oppositon to thrust, as illustrated in Fig. 5-1.

Assuming that the airplane is flying straight and level in calm air, the drag acts parallel to the direction of the relative wind. As long as thrust and drag are equal, the airplane flies at a constant speed. If the engine power is reduced, the thrust is decreased and the speed of the airplane is reduced. If the thrust is lower than the drag, the speed of the airplane becomes less and less until it finally is lower than the speed required to maintain level flight and the airplane will descend. On the other hand, if the power of the engine is increased, the thrust is increased and the airplane gains speed, or, in technical terms, **accelerates.** While the airplane accelerates, the drag increases until it eventually equals the thrust. Then the airplane flies at a constant speed.

These facts can be summarized by saying that in straight and level flight at a constant speed, the sum of the components of the forces acting on an airplane equals zero.

Loads and Load Factors

During level flight the forces exerted on an airplane are at a minimum; however, they still exist. The drawings of Fig. 5-2 show the forces acting on an airplane in level flight. It will be noted that impact pressures exist at all points where the air strikes the surfaces from a forward direction. Negative pressures exist above and to the rear of the cabin, above the wings, and below the horizontal stabilizer. The total upward force (lift) exerted on the airplane is equal to the weight of the airplane; hence the airplane flies in a level flight path.

The internal structure of the airplane wing shown in Fig. 5-2 must be such that it can withstand the severe bending moments imposed by the combination of weight and lift. During flight, the wings of an airplane will support its maximum allowable gross weight. As long as the airplane is moving at a steady rate of speed and in a straight line, the load imposed upon the wings will remain constant.

A change in speed during straight flight will not produce any appreciable change in load, but when a change is made in the airplane's flight path, an additional load is imposed upon the airplane structure. This is particularly true if a change in direction is made at high speeds with rapid forceful control movements.

According to certain laws of physics, a mass (airplane, in this case) will continue to move in a straight line unless some force intervenes, causing the mass (airplane) to assume a curved path. During the time the airplane is in a curved flight path, it still attempts, because of inertia, to force itself to follow straight flight. This tendency to follow straight flight, rather than curved flight, generates a force known as centrifugal force which acts toward the outside of the curve.

Any time the airplane is flying in a curved flight path with a positive load, the load the wings must support will be equal to the weight of the airplane plus the load imposed by centrifugal force. A **positive load** occurs when back pressure is applied to the elevator, causing centrifugal force to act in the same direction as the force of weight. A **negative load** occurs when forward pressure is applied to the elevator control, causing centrifugal force to act in a direction opposite to that of the force of weight.

Curved flight producing a positive load is a result of increasing the angle of attack and, consequently, the lift. Increased lift always increases the positive load imposed upon the wings. However, the load is increased only at the time the angle of attack is being increased. Once the angle of attack is established, the load remains constant. The loads imposed on the wings in flight are stated in terms of **load factor**.

Load factor is the ratio of the total load supported by the airplane's wing to the actual weight of the airplane and its contents; i.e., the actual load supported by the wings divided by the total weight of the airplane. For example, if an airplane has a gross weight of 2000 lb [907 kg] and during flight is subjected to aerodynamic forces which increase the total load the wing must support to 4000 lb [1814 kg], the load factor would be 2.0 (4000/2000 = 2). In this example the airplane wing is producing lift that is equal to twice the gross weight of the airplane.

Another way of expressing load factor is the ratio of

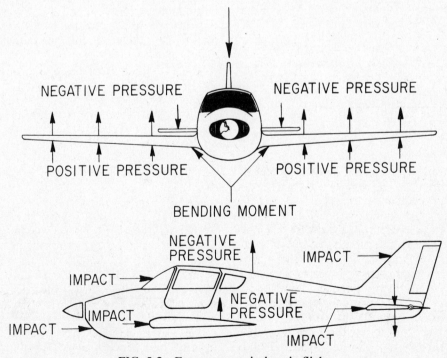

FIG. 5-2 Forces on an airplane in flight.

a given load to the pull of gravity; i.e., to refer to a load factor of 3 as "three g's," where g refers to the pull of gravity. In this case the weight of the airplane is equal to 1 g, and if a load of 3 times the actual weight of the airplane were imposed upon the wing due to curved flight, the load factor would be equal to 3 g's.

Load Factors and Airplane Design

To be certified by the Federal Aviation Administration, the structural strength (load factor) on airplanes must conform with prescribed standards set forth by Federal Aviation Regulations.

All airplanes are designed to meet certain strength requirements depending upon the intended use of the airplane. Classification of airplanes as to strength and operational use is known as the category system. Aircraft may be type-certificated into **normal, utility,** or **acrobatic categories.**

The **normal category** is limited to airplanes intended for nonacrobatic operation and has a load factor limit of 3.8. The **utility category** applies to airplanes intended for limited acrobatic operations and has a load factor limit of 4.4. **Acrobatic category** aircraft may have a load factor limit of 6.0 and are free to operate without many of the restrictions that apply to normal and utility category aircraft. Small airplanes may be certificated in more than one category if the requirements of each category are met.

The category in which each airplane is certificated may be readily found in the aircraft's Type Certificate Data Sheet or by checking the Airworthiness Certificate found in the cockpit.

An airplane is designed and certificated for a certain maximum weight during flight. This weight is referred to as the maximim certificated gross weight. It is important that the airplane be loaded within the specified weight limits because certain flight maneuvers will impose an extra load on the airplane structure which may, particularly if the airplane is overloaded, impose stresses which will exceed the design capabilities of the airplane. If during flight severe turbulence or any other condition causes excessive loads to be imposed on the airplane, a very thorough inspection must be given to all critical structural parts before the airplane is flown again. Damage to the structure is often recognized by bulges or bends in the skin, "popped" rivets, or deformed structural members.

Effects of Turns on Load Factor

In turns, the load factors on an airplane increase similarly to those experienced in pulling out of a dive. The reason for **increased wing loading** in a turn is illustrated in Fig. 5-3. In this diagram the airplane is in a turn with a 45° bank. Under these conditions, the gravity will still be pulling the airplane down with a force of 1 g and the centrifigal force will be pulling the airplane horizontally with a force of 1 g. When these forces are resolved, a resultant of 1.41 g is found. This means, of course, that the wing is required to carry 1.41 times the weight of the airplane. In a 60° bank, the load factor is 2; in a 70° bank, the load factor is nearly 3; and in an 80° bank, it is almost 6.

It is obvious that an airplane should not be turned with a bank so steep that the safe load factor of the airplane is exceeded.

Wing Loading

Load factors should not be confused with **wing loading.** Wing loading is the ratio of the total gross weight of the aircraft divided by the total wing area. Wing loading is expressed in pounds per square foot (lb/ft^2). Wing loading will range from approximately 10 lb/ft^2 for a small single engine aircraft to around 75 lb/ft^2 for a business jet.

● AXES OF THE AIRPLANE

While being supported in flight by lift and propelled through the air by thrust, an airplane is free to revolve or move around **three axes,** namely the **longitudinal axis,** the **lateral axis,** and the **vertical axis.** These are illustrated in Fig. 5-4.

The axis which extends lengthwise through the fuselage from the nose to the tail is the **longitudinal axis.**

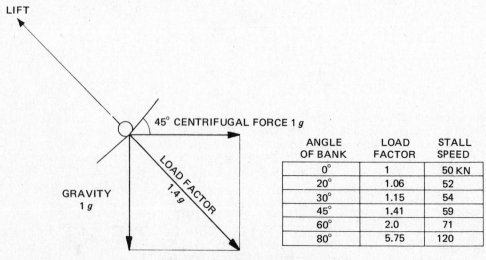

ANGLE OF BANK	LOAD FACTOR	STALL SPEED
0°	1	50 KN
20°	1.06	52
30°	1.15	54
45°	1.41	59
60°	2.0	71
80°	5.75	120

FIG. 5-3   Load factors in a turn.

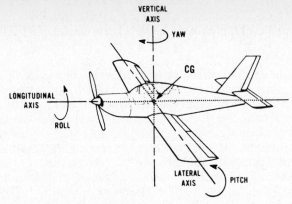

FIG. 5-4   Axes of the airplane.

The axis extending through the fuselage from wing tip to wing tip is the **lateral axis.** The axis which passes vertically through the fuselage at the center of gravity is the **vertical axis.**

During flight, an airplane is rotated about the three axes by means of the three primary flight controls. The **ailerons control roll** about the longitudinal axis, the **elevators control pitch** about the lateral axis, and the **rudder controls yaw** about the vertical axis.

● AIRCRAFT STABILITY

Definition

Because of their ability to revolve about these axes, all airplanes must possess **stability** in varying degrees for safety and ease of operation. Stability is the inherent ability of a body, after its equilibrium is disturbed, to develop forces or moments that tend to return the body to its original position. In other words, a stable airplane will tend to return to the original condition of flight if disturbed by a force such as turbulent air.

Static Stability

An aircraft is in a **state of equilibrium** when the sum of all forces and all moments is equal to zero. When an aircraft is in equilibrium, there are no accelerations and the aircraft continues in a steady condition of flight. If the equilibrium is disturbed by a gust or deflection of the controls, the aircraft will experience acceleration because of an unbalance of moment or force.

The **static stability** of a system is defined by the initial tendency to return to equilibrium conditions following some disturbance from equilibrium. If an object is disturbed from equilibrium and has the tendency to return to equilibrium, **positive static stability** exists. If the object has a tendency to continue in the direction of disturbance, **negative static stability,** or static instability, exists. If the object subject to a disturbance has neither the tendency to return nor the tendency to continue in the displacement direction, **neutral static stability** exists. This is an intermediate condition which could occur when an object displaced from equilibrium remains in equilibrium in the displaced position.

These three categories of static stability are illustrated in Fig. 5-5. The ball in a trough illustrates the condition of positive static stability. If the ball is displaced from equilibrium at the bottom of the trough, the initial tendency of the ball is to return to the equilibrium condition. The ball may roll back and forth through the point of equilibrium, but displacement to either side creates the initial tendency to return. The ball on a hill illustrates the condition of static instability. Displacement from equilibrium at the hilltop brings about the tendency for greater displacement. The ball on a flat, level surface illustrates the condition of neutral static stability. The ball encounters a new equilibrium at any point of displacement and has neither stable nor unstable tendencies.

The term "static" is applied to this form of stability since the resulting motion is not considered. Only the tendency to return to equilibrium conditions is considered in static stability.

Dynamic Stability

While static stability is concerned with the tendency of a displaced body to return to equilibrium, **dynamic stability** is defined by the resulting motion with time. If an object is disturbed from equilibrium, the time history of the resulting motion indicates the dynamic stability of the system.

Dynamic stability is the property which dampens

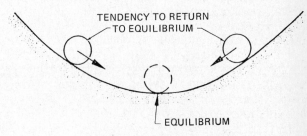

POSITIVE STATIC STABILITY

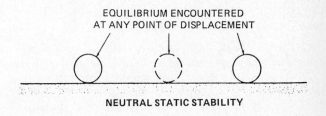

NEUTRAL STATIC STABILITY

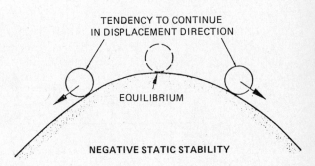

NEGATIVE STATIC STABILITY

FIG. 5-5   Static stability.

the oscillations set up by a statically stable airplane, enabling the oscillations to become smaller and smaller in magnitude until the airplane eventually settles down to its original condition of flight. Figure 5-6 illustrates the relationship of dynamic and static stability.

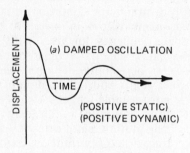

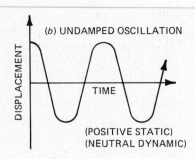

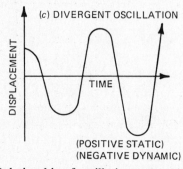

FIG. 5-6   Relationship of oscillation and stability.

The existence of static stability does not necessarily guarantee the existence of dynamic stability. However, the existence of dynamic stability implies the existence of static stability. An aircraft must demonstrate the required degrees of static and dynamic stability if it is to be operated safely.

● FACTORS AFFECTING AIRCRAFT PERFORMANCE AND STABILITY

Angle of Incidence

We have discussed and defined **angle of attack** in a previous chapter of this text. A related but different term is **angle of incidence.** The angle of incidence of a wing is the angle formed by the intersection of the wing chord line and the horizontal plane passing through the longitudinal axis of the aircraft. Angle of incidence is illustrated in Fig. 5-7. Airplanes are usually designed with a *positive* angle of incidence in which the leading edge of the wing is slightly higher than the trailing edge. The correct angle of incidence is essential for low drag and longitudinal stability.

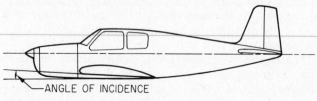

FIG. 5-7   Angle of incidence.

Many airplanes are designed with a greater angle of incidence at the root of the wing than at the tip; this characteristic of a wing is called **washout.** The purpose of washout is to improve the stability of the aircraft as it approaches a stall condition. The section of the wing near the fuselage will stall before the outer section, thus enabling the pilot to maintain good control and reducing the tendency of the aircraft to "fall off" on one wing. If a wing is designed so that the angle of incidence is greater at the tip than at the root, the characteristic is called **washin.**

A difference in the washout and washin of the right and left wings of an aircraft is used to compensate for **propeller torque.** Propeller torque causes the aircraft to roll in a direction opposite that of the propeller rotation. To compensate for this, the right wing is rigged or designed with a smaller angle of incidence at the tip than that of the left wing. Thus, the right wing is washed out more than the left.

Dihedral

A **dihedral angle** is an angle formed by the intersection of two planes. In aircraft terminology, *dihedral* means the lateral angle of the wing with respect to a horizontal plane; this is illustrated in Fig. 5-8. **Positive dihe-**

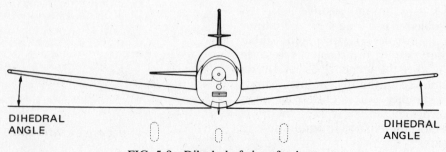

FIG. 5-8   Dihedral of aircraft wings.

dral exists when the tip of a wing is above the horizontal plane passing through the root of the wing. **Negative dihedral** exists when the tip of the wing is below the horizontal plane passing through the root of the wing. Negative dihedral is also called **cathedral.**

The purpose of positive dihedral is to provide lateral **stability** for the aircraft.

Lateral Stability

Lateral stability is the stability of an airplane about the longitudinal or roll axis. An airplane that tends to return to a wings-level attitude after being displaced from a level attitude by some force such as turbulent air is considered to be laterally stable.

The factors that primarily affect lateral stability are dihedral and sweepback. The stabilizing effect of dihedral occurs when the airplane sideslips slightly as one wing is forced down in turbulent air. Usually a wing will provide the greatest amount of lift if it is in a perfectly horizontal position laterally. When an airplane is designed with dihedral, both wings will form angles with the horizontal plane. If the aircraft rolls slightly to the right, the tip of the right wing moves downward and the lift of the wing increases. At the same time, the left wing lift is decreased because the angle with the horizontal plane is increasing. The airplane, therefore, is subjected to more lift from the right wing and less lift from the left wing. This causes the airplane to roll back to the left to resume a level position. These effects are illustrated in Figs. 5-9 and 5-10.

Sweepback is the angle at which the wings are slanted rearward from the root to the tip. The effect of sweepback in producing lateral stability is similar to that of dihedral but not as pronounced. If one wing lowers in a slip, the angle of attack on the low wing increases, producing greater lift. This results in a tendency for the lower wing to rise and return the airplane to level flight. Sweepback augments dihedral to achieve lateral stability. Another reason for sweepback is to place the center of lift farther rearward, which affects longitudinal stability more than it does lateral stability.

Horizontal Stability

The stability of an airplane about the lateral axis is **horizontal stability.** If the airplane is put into a dive or climb and then the control is released, the airplane should return to level flight automatically. If the airplane does not have horizontal stability, it may increase the angle of dive after being placed in a dive, or it may "porpoise," that is, oscillate through a series of

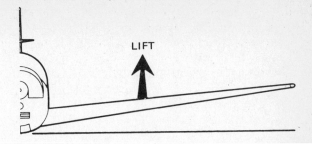

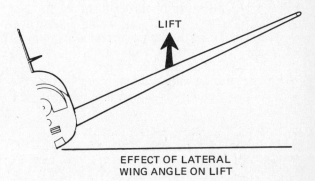

EFFECT OF LATERAL
WING ANGLE ON LIFT

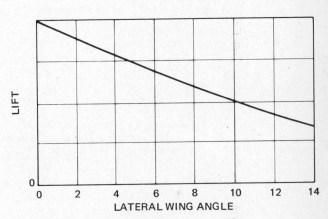

FIG. 5-9   Effect of lateral wing angle on lift.

dives and climbs (pitch up and down) unless controlled by the pilot.

The location of the center of gravity with respect to the center of lift determines to a great extent the longitudinal stability of the airplane.

Figure 5-11 illustrates neutral longitudinal stability. Note that the center of lift is directly over the center of gravity or weight. An airplane with neutral stability will produce no inherent pitch moments around the center of gravity.

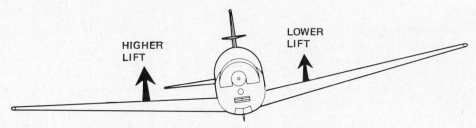

FIG. 5-10   Effect of dihedral on aircraft in flight.

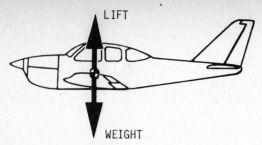

FIG. 5-11   Neutral stability.

Figure 5-12 illustrates the center of lift in front of the center of gravity. This airplane would display negative stability and an undesirable pitch-up moment during flight. If disturbed, the up and down pitching moment will tend to increase in magnitude. This condition can occur especially if the airplane is loaded so that the center of gravity is rearward of the airplane's aft loading limits.

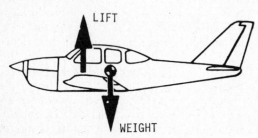

FIG. 5-12   Negative stability.

Figure 5-13 shows an airplane with the center of lift behind the center of gravity. Again, this produces negative stability. Some force must balance the down force of the weight. This is accomplished by designing the airplane in such a manner that the air flowing downward behind the trailing edge of the wing strikes the upper surface of the horizontal stabilizer. This creates a downward tail force to counteract the tendency to pitch down and provides positive stability.

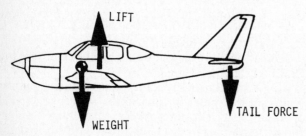

FIG. 5-13   Positive stability.

Directional Stability

The stability of an airplane about the vertical axis is called **directional stability.** This means that the airplane will return to a straight flight path after having been turned (yawed) one way or the other.

Directional stability is accomplished by placing a vertical stabilizer or fin to the rear of the center of gravity on the upper portion of the tail section. The surface of this fin acts similarly to a weathervane and

causes the airplane to weathercock into the relative wind. If the airplane is yawed out of its flight path, either by pilot action or turbulence, during straight flight or turn, the relative wind would exert a force on one side of the vertical stabilizer and return the airplane to its original direction of flight.

A problem encountered on single-engine airplanes is that as the propeller turns clockwise, a rotating flow of air is moved rearward, striking the left side of the fin and rudder, which results in a left yawing moment. To counteract this effect, many airplanes have the leading edge of the vertical fin offset slightly to the left, thereby allowing the slipstream to pass evenly around it (see Fig. 5-14).

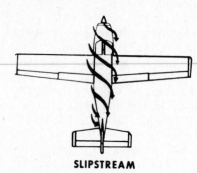

SLIPSTREAM

FIG. 5-14   Slipstream.

Sweptback wings aid in directional stability. If the aircraft yaws from its direction of flight, the wing which is farther ahead offers more drag than the wing which is aft. The effect of this drag is to hold back the wing which is farther ahead and to let the other wing catch up. This is illustrated in Fig. 5-15.

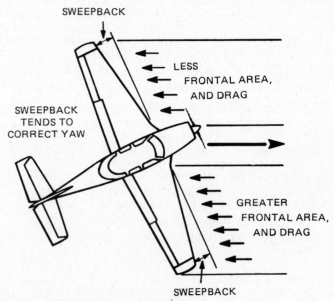

FIG. 5-15   Effect of sweepback.

Excessive Stability

It is possible to build into an airplane a degree of stability that reduces or makes difficult the control of the aircraft. The smooth and easy response of the airplane to its controls is called **controllability.** If the airplane does not have the proper degree of this quality, it will be difficult and tiring to fly, particularly through maneuvers.

Maneuverability is also an important characteristic of an airplane. This is the ability of an airplane to be directed along a selected flight path.

● AIRCRAFT CONTROL

Nomenclature

An airplane is equipped with certain fixed and movable surfaces, or airfoils, which provide for stability and control during flight. These are illustrated in Fig. 5-16.

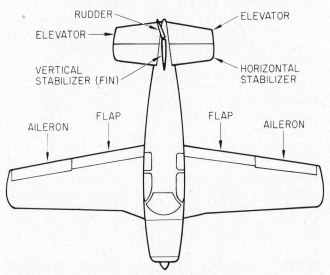

FIG. 5-16 Control surfaces of an airplane.

Each of the named airfoils is designed to perform a specific function in the flight of the airplane. The fixed airfoils are the **wings,** the **vertical stabilizer (fin),** and the **horizontal stabilizer.** The movable airfoils, called **control surfaces,** are the **ailerons, elevators, rudder,** and **flaps.** The ailerons, elevators, and rudder are used to "steer" the airplane in flight to make it go where the pilot wishes it to go and to cause it to execute certain maneuvers. The flaps are normally used only during landings and sometimes for takeoff.

Large jet aircraft, gliders, and some other types of aircraft are equipped with lift-control devises called **spoilers** (see Fig. 5-17). These are rectangular surfaces mounted on the top of the wing. When it is desired to reduce the lift of the wing, the spoilers are raised into the airstream.

Ailerons

An **aileron** may be defined as a movable control surface attached to the trailing edge of a wing to control an airplane in **roll,** that is, rotation about the longitudinal axis. The conventional monoplane has two ailerons, one attached to each wing. They are rigged so that when one is applying an upward force to one wing, the other is applying a downward force to the opposite wing.

The ailerons are moved by means of a control stick or a wheel in the cockpit. If it is desired to roll the airplane to the right, the wheel is turned to the right. After the desired degree of bank is obtained, the wheel is returned to neutral to stop the roll. During normal turns of an airplane, the movement of the ailerons is coordinated with movements of the rudder and elevators to provide a banked horizontal turn without "slip" or "skid." A slip, or sideslip, is a movement of

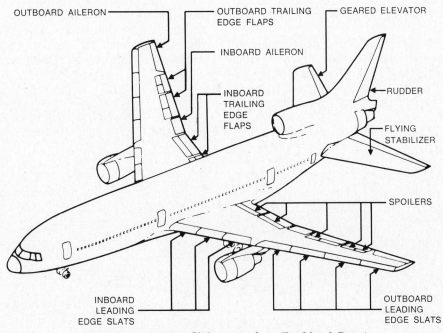

FIG. 5-17 L-1011 flight controls. *(Lockheed Corp.)*

an airplane partially sideways. In a turn the slip is downward and inward toward the turn. A skid in a turn is a movement of the airplane sideways and outward from the turn.

The correct rigging of the ailerons is of primary importance. After an airplane has been overhauled and during preflight inspections, the direction of aileron movement with respect to control-stick movement must be carefully noted. If the wheel is moved to the right, the right aileron must move up and the left aileron must move down. Reverse movement of the control should then cause a reverse of the position of the ailerons.

During flight of the airplane, a down movement of the aileron causes an increase in lift for the wing and the wing rises. At the same time the opposite aileron moves up and causes a decrease in lift for the opposite wing and this wing moves down. This action is illustrated in Fig. 5-18.

Aileron control in an airplane is complicated somewhat by an effect called **adverse yaw.** An aileron that moves down at the trailing edge of a wing creates considerably more drag than the aileron on the opposite wing that moves upward the same amount. Therefore, if the ailerons were rigged to move the same distance in response to the movement of the cockpit control, the drag of the downward-moving aileron would cause the airplane to turn toward the side on which the downward-moving aileron is located. Thus, a pilot wishing to make a left turn would move the control to the left, causing the right aileron to move downward, but the drag caused by the aileron would cause the airplane to turn to the right except for strong rudder control. To overcome adverse yaw, the ailerons of an airplane are rigged for differential movement. The differential control causes the up-moving aileron to move a greater distance than the down-moving aileron. The amount of differential is sufficient to balance the drag between the ailerons, thus eliminating the yaw effect. The design for differential control is explained in the associated text *Aircraft Maintenance and Repair.*

It was mentioned that conventional monoplanes are equipped with two ailerons, one being attached to the trailing edge of each wing. This does not apply to modern jet airliners. These high-speed airplanes usually employ at least two sets of ailerons. This is illustrated in Fig. 5-17.

At low speeds all the ailerons are used, but at high speeds only one small pair of ailerons is required because the effectiveness of aileron control increases with speed and because **spoilers** are used with the ailerons for effective control at high speeds.

Elevators

An **elevator** is defined as a horizontal, hinged control surface, usually attached to the trailing edge of the horizontal stabilizer of an airplane, designed to apply a **pitching moment** to the airplane. A pitching moment is a force tending to rotate the airplane about the lateral axis, that is, "nose up" or "nose down." When the control stick or wheel in the airplane is pulled back, the elevators are raised. The force of the relative wind on the elevator surfaces tends to press the tail down, thus causing the nose to pitch up and the angle of attack of the wings to increase. The reverse action takes place when the control stick or wheel is pushed forward. The action of the elevators is illustrated in Fig. 5-19.

During flight of an airplane the operation of the elevators is quite critical, especially at low speeds. When power is off and the airplane is gliding, the position of the elevators will determine whether the airplane dives, glides at the correct angle, or stalls. It is necessary to remember that an airplane will not necessarily climb when the control is pulled back. It is the power developed by the engine that determines the rate of climb of an airplane rather than the position of the elevators. As a matter of fact, if the elevators are held in a fixed position, the throttle alone can be used to make the airplane climb, dive, or maintain level flight. The position of the elevator is important, however, to establish the most efficient rate of climb and a good gliding angle when power is off. It is also most essential for proper control when "breaking the glide" and holding the airplane in landing position.

A special type of elevator that combines the functions of the elevator and the horizontal stabilizer is called a **stabilator.** When this type of control airfoil is installed on an airplane, there is no fixed horizontal stabilizer. The stabilator is an airfoil that responds to the normal elevator control and serves as an elevator as well as a stabilizer. A stabilator is illustrated in Fig. 5-20.

Rudder

A **rudder** is a vertical control surface that is usually hinged to the tail post aft of the vertical stabilizer and

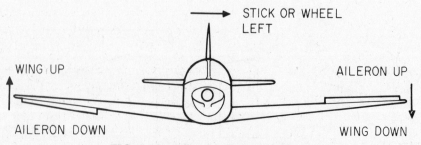

FIG. 5-18 Effect of ailerons in flight.

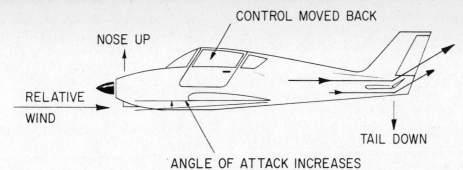

FIG. 5-19 Action of elevators.

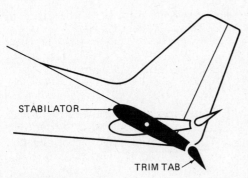

FIG. 5-20 Stabilator.

designed to apply **yawing** moments to the airplane, that is, to make it turn to the right or left about the vertical axis. The movement of the rudder is controlled by pedals or a "rudder bar" operated by the feet of the pilot. When the right pedal is pressed, the rudder swings to the right, thus bringing an increase of dynamic air pressure on its right side. This increased pressure causes the tail of the airplane to swing to the left and the nose to turn to the right. The operation of a rudder is shown in Fig. 5-21.

Although it appears that the rudder causes the airplane to turn, it must be pointed out that the rudder itself cannot cause the airplane to make a good turn.

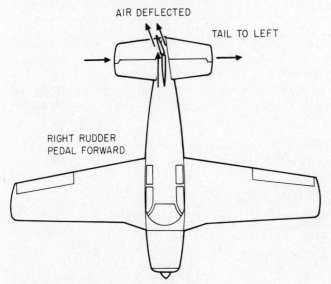

FIG. 5-21 Action of the rudder.

We remember from Newton's first law of motion that a moving body tends to continue moving in a straight line unless some outside force changes its direction. When rudder is applied to an airplane in flight, the airplane will turn but it will continue to travel in the same direction as before unless a correcting force is applied. Thus, with rudder only, we find that the airplane **skids.** In order to prevent this skid in a turn, we use the ailerons to **bank** the airplane. Anyone who drives a car will know that a banked turn is much easier to negotiate at comparatively high speeds in a car than a flat turn. It is the same with an airplane. To prevent skidding in a turn, the airplane must be banked.

Too much of a bank without sufficient rudder in a turn will cause **slipping;** that is, the airplane will slide down toward the inside of the turn. It is therefore necessary that the proper amount of rudder and aileron be applied when entering a turn in order to produce what is termed a **coordinated** turn. Usually, after the airplane is placed in a turn, the rudder pressure is almost neutralized to hold the turn. Likewise, it is necessary to reduce the amount of aileron used to place the airplane in the turn.

Another factor to note concerning turns is that the steeper the turn, the more the **elevator** will have to be used. Thus we see that a properly executed turn requires the use of all three of the primary controls.

Trim Tabs

In addition to the primary flight controls, there is, on most modern airplanes, a group termed **secondary controls.** These include **trim devices** of various types, **spoilers,** and **wing flaps.**

Trim tabs are essentially small auxiliary control surfaces that are usually hinged to the trailing edges of the main surface. They are commonly used to relieve the pilot of maintaining continuous pressure on the primary controls when correcting for an unbalanced flight condition resulting from changes in aerodynamic forces or weight. Some tabs also help to actuate the main control surfaces by exerting force on the main surface, thus reducing the amount of force on the controls to maneuver the airplane.

Most of the trim tabs installed on aircraft are mechanically or electrically operated from the cockpit through an individual cable system. However, some aircraft have trim tabs that are adjustable only when the airplane is on the ground. It is possible to adjust tabs of this type so that the airplane will only fly satisfactorily under one set of conditions. These conditions are usually those established for normal cruising power and speed. Trim tabs may be installed on elevators, rudders, and ailerons.

Controllable Trim Tabs

A **controllable trim tab** is illustrated in Fig. 5-22. The elevator trim tab control in the cockpit is usually a small wheel or knob arranged so that its plane of rotation is vertical and longitudinal with respect to the airplane. To raise the nose of the airplane, the top of the wheel is moved rearward; to lower the nose, the top of the wheel is moved forward. When the elevator trim tab is moved down, as shown in Fig. 5-23, by means of the cockpit control, the airstream develops a force that tends to push the tab up. This force is transmitted to the elevator and moves the elevator up, and this change, in turn, causes the tail of the airplane to move down and the nose to go up.

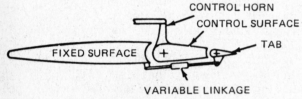

FIG. 5-22   Controllable trim tab.

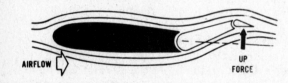

FIG. 5-23   Effect of trim tabs.

Balance Tab

A **balance tab** is linked to the airplane in such a manner that a movement of the main control surface will give an opposite movement to the tab. Thus, the balance tab will assist in moving the main control surface. Balance tabs are particularly useful in reducing the effort required to move the control surfaces of a large airplane. A balance tab is illustrated in Fig. 5-24.

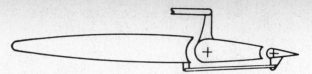

FIG. 5-24   Balance tab.

Servo Tab

Servo tabs, sometimes referred to as flight tabs, are used primarily on the large main control surfaces, A servo tab is one that is directly operated by the primary controls of the airplane. Only the servo tab moves in response to movement of the cockpit control. The force of the airflow on the servo tab then moves the primary control surface. The servo tab, illustrated in Fig. 5-25, is used to reduce the effort required to move the controls on a large airplane.

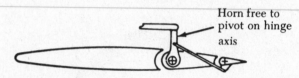

FIG. 5-25   Servo tab.

Spring Tabs

Spring tabs, like some servo tabs, are usually found on large aircraft that require considerable force to move a control surface. The purpose of the spring tab is to provide a boost, thereby aiding in the movement of a control surface. On the spring tab, illustrated in Fig. 5-26, the control horn is connected to the control surface by springs.

Balance of Control Surfaces

Aircraft control surfaces will function properly on an airplane only if they are properly balanced. **Static balance** is accomplished by installing weights forward of the hinge line of the control. Usually, static balance requires that the sum of the weights forward of the hinge line is approximately equal to the weight aft of the hinge line. The methods for checking the balance and adjusting the weights are provided in the manufacturer's maintenance manual.

Dynamic or **aerodynamic balance** is accomplished by designing the control airfoil such that aerodynamic forces during flight will tend to balance moments forward of the hinge line with moments aft of the hinge line. This is accomplished by placing the hinge line substantially aft of the leading edge of the control sur-

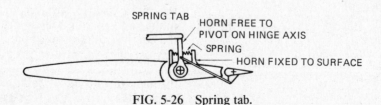

FIG. 5-26   Spring tab.

face, thereby extending a portion of the leading-edge surface a considerable distance forward of the hinge line.

New airplanes direct from the factory have the control surfaces statically and dynamically balanced. However, if a repair is made to a control surface, or if the surface is painted, it is necessary to check the balance against the manufacturer's specifications. Balancing of control surfaces is discussed in the associated text *Aircraft Maintenance and Repair*.

The need for proper balance of control surfaces cannot be overstressed. If a control surface is out of balance, either flutter or vibration will occur to the extent that the aircraft structure may be damaged.

● STALLS AND THEIR EFFECTS

Conditions Leading to a Stall

In Chap. 3, it was pointed out that a stall occurs when the angle of attack becomes so great that the laminar airflow separates from the surface of an airfoil, leaving an area of burbling that destroys the low-pressure area normally existing at the upper surface of a wing in flight. Figure 5-27 illustrates this condition, which also represents the maximum coefficient of lift.

When an airplane is in flight, there are a number of flight conditions that may lead to a stall. First, if an airplane is pulled up sharply until its forward speed diminishes to a point where lift is less than gravity, the airplane will begin to lose altitude. The angle of attack increases, and when it reaches the stalling value (about 20°), the wing stalls and the airplane stops flying. If the stall is balanced on both sides of the airplane, it will pitch forward and may soon regain flying speed.

Stalls may also occur at high speeds. Stalls occurring under these conditions are called **high-speed stalls,** and they occur when an airplane is pulled up so abruptly that the angle of attack exceeds the stall angle. This stall is not often encountered, because under ordinary conditions it is not necessary to pull an airplane up sharply enough to cause a stall.

Stalls are more likely to occur during turns than in level flight. This is because greater lift is required to maintain level flight in a turn.

Stall Warning

The experienced pilot can usually sense when a stall is about to happen because of the "feel" of the airplane controls and the reactions of the airplane. Often the airplane will start to shake or buffet because of the flow separation on the wing and the turbulent air buffeting the tail surfaces. The controls become "sloppy" and do not have the solid feel of normal flight.

Most airplanes are equipped with stall-warning devices. Typical of such devices is a small vane mounted near the leading edge of the wing and arranged so that it will actuate a switch when it rises as a result of an excessive angle of attack. The switch causes a warning horn to sound when the angle of attack approaches maximum, usually about 5 to 10 kn [2.57 to 5.14 m/s] above stalling speed.

Effect of Wing Design on Stall

The type of wing design for a particular airplane depends almost entirely on the purpose for which that airplane is to be used.

To achieve good stall characteristics, the root of the wing should stall first, with the stall pattern progressing outward to the tip. This type of stall pattern decreases undesirable rolling tendencies and increases lateral control when approaching a stall. It is undesirable for the wing tips to stall first, particularly if the tip of one wing stalls before the tip of the other wing, which usually happens.

A desirable stall pattern can be accomplished by (1) designing the wing with a twist so that the tip has a lower angle of incidence (*washout*) and, therefore, a lower angle of attack when the root of the wing approaches the critical angle of attack (see Fig. 5-28), (2) designing **slots** near the leading edge of the wing tip to allow air to flow smoothly over that part of the wing at higher angles of attack, therefore stalling the root of the wing first (see Fig. 5-29), and (3) attaching **stall strips** on the leading edge near the wing root (see Fig. 5-30).

The stall strip is a triangular strip mounted on the leading edge of the wing at the inboard end. At high angles of attack where stalling would be likely to occur, the strip causes the inboard portion of the wing

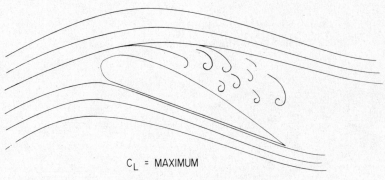

C_L = MAXIMUM

FIG. 5-27   Airfoil in a stall.

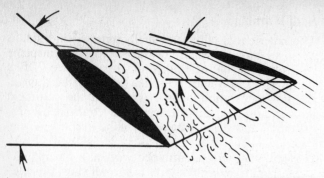

FIG. 5-28 View of wingtip twist. Ailerons are still effective even though wing root is in the stalled condition.

to stall before the outer portion. This enables the pilot to maintain control of the aircraft with the ailerons, and the airplane does not "fall off" on one wing.

● AIRCRAFT DESIGN VARIATIONS

There are many shapes and sizes of airplanes, most of which are similar in appearance. In its traditional form the airplane is marked by an arrangement of clearly distinguishable parts. The traditional design of the fuselage supported by wing lift, stabilized by the tail surfaces, and propelled by the engine in the nose has worked well over the years. Many variations of the standard design have been appearing lately which appear to work equally as well.

Canard Aircraft

The earliest powered aircraft, such as the *Wright Flyer* (see Fig. 5-31), had horizontal surfaces located ahead of the wings. This configuration, also utilized on the *Beech Starship,* shown in Fig. 5-32, which has two lifting surfaces, with the forward airfoil being called a **canard,** is an appealing way to assist in carrying some

of the airplane weight to reduce drag and increase cruising speed.

Conventional airplane designs that have tail surfaces located behind the wing use the horizontal tail to balance thrust and center-of-gravity loads. This usually means a down load on the tail as previously discussed in this chapter. This requires an increase in lift coefficient to support the added wing load. Since wing drag increases with wing lift, a climb and cruise penalty is paid for the stability offered by an aft-located horizontal tail.

The canard airplane has no stabilizing down loads because the canards share the lifting loads with the wing, thus making the aircraft somewhat unstable. This instability is referred to as "relaxed static stability," and it is used to reduce drag. The forward wing (foreplane) lifts a greater share of the total weight per square foot of wing area (i.e., it has a heavier wing loading) than the aft wing. This is achieved by having the center of gravity well ahead of the aft wing. The aft wing pitching moment also adds to the foreplane load.

In a well-designed canard, the forward wing will always stall at a lower angle of attack than the aft wing. The aft wing cannot be accidentally stalled; therefore the ailerons remain effective. The canard could be flown with the foreplane alternately stalling and unstalling, the nose bobbing up and down gently in a porpoising mode.

Forward Swept Wing

A revolutionary concept in aircraft design is that of the **forward swept wing** as illustrated in Fig. 5-33. The 30° forward swept wing of the X-29 provides drag reductions of up to 20 percent in the transonic maneuvering range, giving it performance equivalent to an aircraft with a more powerful engine. As illustrated in Fig. 5-34, air moving over the forward swept wing tends to flow inward rather than outward, allowing the

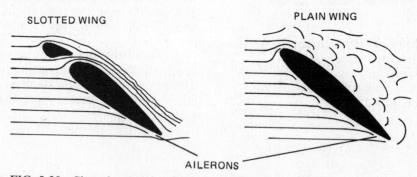

SLOTTED WING   PLAIN WING

AILERONS

FIG. 5-29 Slotted and plain wings at equal angles of attack.

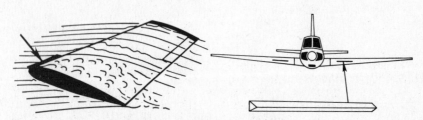

FIG. 5-30 The stall strip ensures that the root section stalls first.

FIG. 5-31 *Wright Flyer.*

FIG. 5-32 Beech Starship. *(Beech Aircraft Corp.)*

FIG. 5-33 X-29 Forward Swept Wing. *(Grumman Aircraft Corp.)*

wing tips to remain unstalled at high angles of attack and therefore easier to control in extreme maneuvers. Forward swept wings provide less drag, more lift, better maneuverability, and more efficient cruise speed. These improvements in performance are gained at the expense of reduced lateral and longitudinal stability.

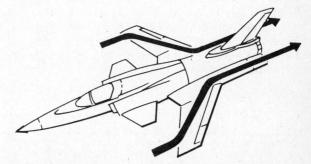

FIG. 5-34 Airflow over forward swept wing.

To control an aircraft designed with relaxed stability, the flight control system must provide an artificial stability. This is accomplished with a digital fly-by-wire flight control system. A fly-by-wire system enables the control surfaces of an airplane to be operated electronically through a computer system. The pilot moves the aircraft's stick, sending a command to the flight control computer. The computer calculates the control surface movements necessary and sends a command to the actuator to move the control surfaces.

T Tails

The **T-tail** arrangement positions the stabilizer and elevator at the top of the vertical fin. A T tail is illustrated in Fig. 5-35. The use of a T-tail configuration not only makes the fin and rudder more effective due to the end plate action of the stabilizer location, but it also positions the horizontal tail above wing turbulence.

A T-tail structure will be somewhat heavier than a conventional tail arrangement due to combined horizontal tail and fin bending loads which must be carried by the fin and fuselage.

FIG. 5-35 Beechcraft Duchess. *(Beech Aircraft Corp.)*

The T tail possesses excellent spin recovery characteristics since the vertical tail is positioned up and out of the turbulent airflow.

Unusual Controls

Some airplanes have been designed with special types of control surfaces that do not fit into the descriptions of the conventional controls. One such control is called **ruddervator.** The ruddervator is used on airplanes with "butterfly" tails, and the surfaces serve both as rudders and as elevators. When it is desired to increase the angle of attack, the control wheel is pulled back and both ruddervators move upward and inward as shown in Fig. 5-36. When the wheel is pushed forward, the ruddervators move down and outward as illustrated.

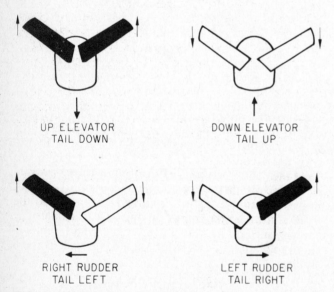

UP ELEVATOR
TAIL DOWN

DOWN ELEVATOR
TAIL UP

RIGHT RUDDER
TAIL LEFT

LEFT RUDDER
TAIL RIGHT

FIG. 5-36 Operation of ruddervators.

If it is desired to turn an airplane with ruddervators and the right rudder is applied, the right ruddervator will move down and outward while the left ruddervator will move up and inward. These movements will be in response to the movement of the rudder pedals and will provide the forces necessary to rotate the airplane about the vertical axis. The turning action of the ruddervators is also illustrated in Fig. 5-36.

A disadvantage of the V tail is that the heavier tail structure necessary to support combined horizontal and vertical surface loading along with a somewhat heavier control system make the V tail generally as heavy as the conventional design it would replace, and the stability characteristics are somewhat less desirable—particularly in rough air.

Another somewhat unconventional control is the **elevon.** Elevons are combination elevators and ailerons used on the outer tips of some delta wings. When used as elevators, they both move in the same direction; and when used as ailerons, they move in opposite directions. Elevons are especially needed for all-wing airplanes, or "flying wings."

Ailerons that are rigged to serve as ailerons or flaps are called **flaperons.** When employed as flaps, flaperons on opposite wings move either upward or downward together. When employed as ailerons, the flaperons move in opposite directions. The X-29, illustrated in Fig. 5-33, utilizes flaperons which allow the wing to vary its camber or curvature. By varying the wings' camber, the aircraft will possess better performance capabilities over a wider operating range.

● AIRFOILS ON BIPLANES

Biplane Pressure Interference

Even though it is recognized that the biplane type of airplane is not used as extensively today as it has been in the past, still there are many such airplanes in operation, and it is important for the technician to understand some of the details of construction and the aerodynamic characteristics of such aircraft. For this reason we shall discuss the aerodynamic characteristics of biplanes and point out some of their special features.

When the air flows over and under the wing of a monoplane, there is decreased pressure above the wing and increased pressure below. The region of high pressure seeks the region of low pressure, but these two regions can merge only at the tips and at the trailing edge. If the upper and lower wings of a biplane were far enough apart, the airflow would be the same for each of the two wings of the biplane as it would be for the wings of the monoplane, disregarding the effect of the struts connecting the wings of the biplane.

In actual practice, however, the wings of the biplane are so close together that the interference of the streamlines reduces the comparatively low pressure on the upper surface of the lower wing, because air always attempts to flow from a region of high pressure to a region of low pressure. The lift of both wings is therefore reduced; but the lower wing loses more lift than the upper wing, and the loss is so great that a biplane is usually less efficient than a monoplane that has an equivalent wing area.

In general, the greater the distance between the wings of a biplane, the smaller is the loss of lift due to **interplane interference.** The inverse is also true; that is, the closer the wings are together, the greater will be the interplane interference. The gap/chord ratio, stagger, and decalage are all involved in interplane interference.

Gap/Chord Ratio

Gap is the distance between the leading edges of the upper and lower wings of a biplane as shown in Fig. 5-37 and is measured perpendicular to the longitudinal axis of the airplane. Gap is sometimes defined as the distance separating two adjacent wings of a multiplane. These definitions mean the same thing.

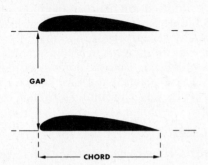

FIG. 5-37   Gap and chord.

The **chord** has been defined and explained before in this text. In addition to other definitions, it may be defined as the straight line tangent to the lower surface of the airfoil at two points or as the straight line between the trailing edge and the imaginary perpendicular line at the leading edge. The chord is indicated in Fig. 5-37. Instead of the gap of a biplane, it is customary to give the **gap/chord ratio.** If the gap/chord ratio is 1, it means that the gap and the chord have the same length. In practice the gap/chord ratio is usually close to 1. The principal determining factor for the gap/chord ratio is the interplane interference. The upper and lower wing should be as near to each other as possible and yet still be far enough apart so that interplane interference is kept at a minimum.

Gap/Span Ratio

Some textbooks on aerodynamics mention the **gap/span ratio.** This is the ratio of the gap to the span, but the gap is always much less than the span; hence the ratio is always less than 1. Gap/span ratio is also defined as the ratio of the gap between two superimposed surfaces to the span of the surfaces.

Stagger

Stagger is technically defined as the difference in the longitudinal position of the axes of two wings of an airplane. In simple words, stagger is the amount which the leading edge of one wing of a biplane is ahead of the leading edge of the other wing. Stagger is also used to define the distance of one compressor stator blade ahead of another and also the distance of one rotor of a tandem rotor helicopter ahead of the other.

The upper biplane of Fig. 5-38 has **positive stagger,** because the leading edge of the upper wing is ahead of the leading edge of the lower wing. The lower biplane of Fig. 5-38 has a **negative stagger,** because the leading edge of the upper wing is **behind** the leading edge of the lower wing. Stagger is expressed in inches, percentage of chord length, or degrees.

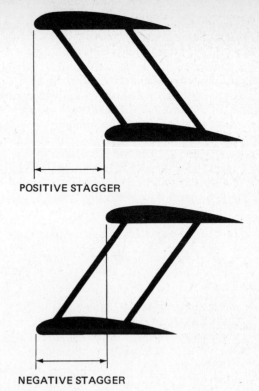

POSITIVE STAGGER

NEGATIVE STAGGER

FIG. 5-38   Stagger.

The aerodynamic advantages of stagger are small. A biplane may have stagger to improve the vision of the pilot and provide better access to the cockpit.

When stagger is expressed in degrees, a line is drawn between the leading edges, and the angle that this line makes with a line drawn perpendicular to the chord of the upper wing is the **angle of stagger.** For example, in Fig. 5-39 the angle formed by a line connecting the leading edges of the wings and a line drawn perpendicular to the chord of the upper wing is 23°. Since the leading edge of the upper wing is ahead of the leading edge of the lower wing, we can say that there is 23° positive stagger.

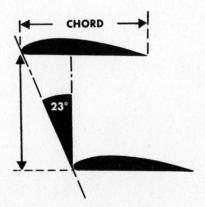

CHORD

23°

FIG. 5-39   Stagger expressed in degrees.

Decalage

Decalage is the angular difference between the mean aerodynamic chords of the wings of a biplane. The

angle of wing setting, or chord angle, is the same thing as the **angle of incidence.** Therefore, we can say that decalage is the difference between the angles of incidence of the wings of a biplane.

The decalage is measured by the angle (less than a right angle) between the chords in a plane parallel to the plane of symmetry. The decalage is considered positive if the upper wing of a biplane is set at the larger angle of incidence. In Fig. 5-40 there is an angle of incidence for the upper wing but none for the lower wing; hence there is a positive angle of decalage, which, in this particular case, happens to be the same as the angle of incidence of the upper wing.

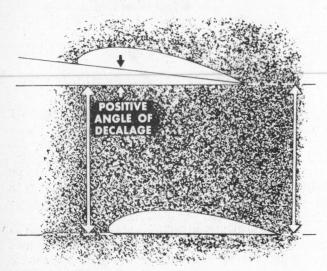

FIG. 5-40   Positive decalage.

In Fig. 5-41, the lower wing has an angle of incidence but the upper wing has none; hence there is a negative angle of decalage. In this particular case, the angle of incidence of the lower wing is the angle of decalage. If the upper wing were set at an angle of incidence of 3° and the lower wing were set at an angle of incidence of 2°, there would be a positive angle of decalage of 3° − 2° = 1°.

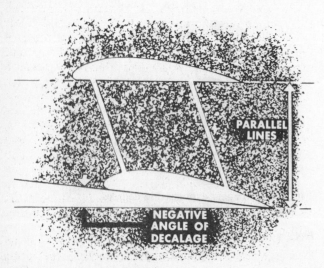

FIG. 5-41   Negative decalage.

If the chords of the upper and lower wings of a biplane are parallel, the downwash of the upper wing has the effect of decreasing the angle of attack of the lower wing. Setting the lower wing at a greater angle of incidence will more properly distribute the lift between the two wings. Since the upper and lower wings then have different angles of incidence, they have different angles of attack in flight and there will be a difference of pressure distribution between them. Positive decalage gives the upper wing an increase in load percentage, especially at high speed. Negative decalage gives the lower wing an increase in load percentage. Each of the wings will reach its burble point at a different angle of attack, with a result that a stall will be less abrupt. On lift curves, this condition is shown by a flatter peak.

Biplane Loads

A biplane is constructed with external bracing between the wings to support a large part of the loads that occur during flight and landing. These external supporting members are shown in Fig. 5-42. During flight the **flying wires** are under a high-tension stress and the wing struts are subjected to compression stress. Upon landing, the **landing wires** are under tension and the wing struts are still under compression. The cabane struts are always under compression, and the cabane wires are under tension.

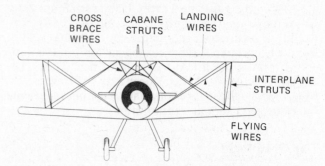

FIG. 5-42   Load-bearing members in a biplane.

Monoplanes Compared with Biplanes

The early airplane designers and builders reasoned that a biplane would provide more lifting surface than a monoplane of the same span, but they soon found that drag was caused by interplane interference and also by the struts and wires. Many designers were aware that a cantilever (unbraced) monoplane with tapered wings was superior on a basis of aerodynamic performance, but they had not yet developed a deep airfoil section that would eliminate the need for external bracing. Furthermore, they lacked materials that would provide the necessary structural strength and rigidity for an unbraced wing.

The early biplane designers also argued that even if they had a deep airfoil section and the proper materials for building cantilever wings, the biplane was structurally more efficient and was more maneuverable. There were other arguments that were extremely important, although they were seldom mentioned. One was that those making biplanes had a vast

110

amount of money invested in their equipment. Another was that the engineers and mechanics were reluctant to change from familiar procedures to untried methods. For all these reasons, the conversion from biplanes to monoplanes was a slow process. However, today the monoplane is the accepted type of aircraft, although a few new biplanes are still being produced for specialty uses such as aerobatic or agricultural use.

● THE HELICOPTER

One of the most versatile and useful aircraft for a wide variety of applications is the helicopter. The main difference between a helicopter and an airplane is the main source of lift. The airplane derives its lift from a fixed airfoil surface while the helicopter derives lift from a rotating airfoil called the rotor.

The word **helicopter** is derived from the Greek words meaning helical wing or rotating wing.

The **rotating wing** (main rotor) of a helicopter has two or more blades, depending upon the design and size of the helicopter. Each blade is an airfoil having a profile similar to that of an airplane wing. The same laws of aerodynamics that apply to other airfoils apply to the helicopter rotor blades.

The great usefulness of the helicopter is due to its ability to fly straight up, sideways, forward, or backward, or to remain still in a hovering position. Because of its ability to land in almost any small clear area, the helicopter is used for air taxi service, police work, intercity mail and passenger service, power-line patrolling, construction work, fire fighting, agricultural work, air-sea rescue, and a variety of other services.

Helicopter Flight

The helicopter flies in accordance with the same laws of aerodynamics that govern the flight of a conventional airplane. In the helicopter, however, these laws are applied differently. The helicopter is subject to the same four forces that affect other aircraft, that is, lift, weight, thrust, and drag. **Lift** supports the **weight** of the aircraft, and **thrust** overcomes **drag** and moves the aircraft in the direction desired.

As mentioned previously, a helicopter can fly in any direction with reference to the heading. For example, the helicopter can be headed north and flying south, east, west or any other direction. Furthermore, it can move straight up or straight down, or it can remain stationary. When the helicopter is in stationary flight, it is said to be **hovering.** If the helicopter is hovering in a no-wind condition, the plane of rotation of the rotor, or **tip-path plane,** is horizontal or parallel with the level ground. During hovering, the sum of the lift and thrust of the helicopter is equal to the sum of the weight and drag. This is illustrated in Fig. 5-43. During vertical flight, when the sum of the lift and thrust is greater than the sum of the weight and drag, the helicopter will rise. If the sum of the weight and drag is greater than the sum of the lift and thrust, the helicopter will descend.

For forward, rearward, or sideward flight, the tip-path plane of the rotor must be tilted in the direction in which it is desired to fly. The conditions for various

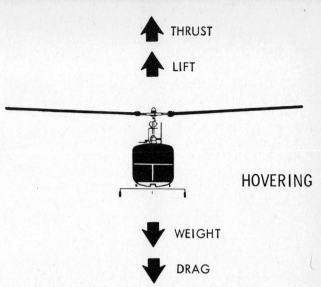

FIG. 5-43   Forces acting on a helicopter while hovering.

types of horizontal flight are illustrated in Fig. 5-44. When flying forward, the tip-path plane of the rotor is tilted forward as shown in the illustration. The forces of flight are resolved into vertical and horizontal components. If the helicopter is flying forward in a straight and level path, the thrust is equal to or greater than the drag in the horizontal direction and the lift is equal to the weight in the vertical direction.

When the helicopter is flying rearward, sideward, or in any other horizontal direction, the thrust is acting in the direction in which the vehicle is moving and the drag is acting in the opposite direction. Lift and weight always act in the vertical direction.

Conditions Affecting Rotor Operation

The spinning main rotor, often called a **rotary wing,** is subject to a number of special conditions that must be given consideration in the design of a helicopter. Among these are **dissymmetry of lift, gyroscopic precession, centrifugal force, torque,** and **Coriolis effect.**

Dissymmetry of lift occurs when the helicopter is in horizontal flight and is caused by the difference in airspeed over the advancing rotor blades and the retreating rotor blades. Assuming that the tip speed of the rotor blades is 350 kn [180 m/s] and the helicopter is flying in a forward direction at 100 kn [51.44 m/s] true airspeed, the tip of the rotor blade that is at a position perpendicular to the flight path on the right side of the helicopter will be moving through the air at 450 kn [213.48 m/s] and the tip of the rotor blade on the opposite side will be moving through the air at 250 kn [128.6 m/s]. It is obvious that the blade on the right side will produce more lift than the blade on the left side if both have the same angle of attack (pitch). Figure 5-45 illustrates the effects of horizontal flight on the lift of the individual rotor blades. It will be noted that the condition of hovering flight in still air requires equal lift on both sides of the rotor disk.

Dissymmetry of lift is compensated for by the design of the rotor, which permits **blade flapping,** and by the design of the **cyclic-pitch-control system.** An artic-

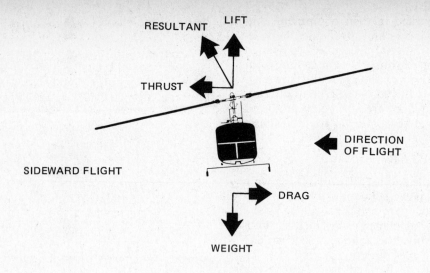

SIDEWARD FLIGHT

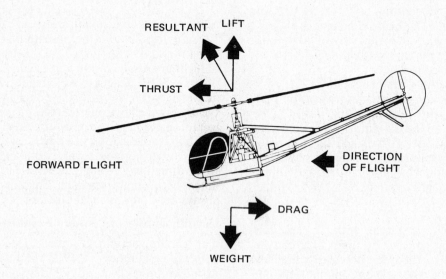

FORWARD FLIGHT

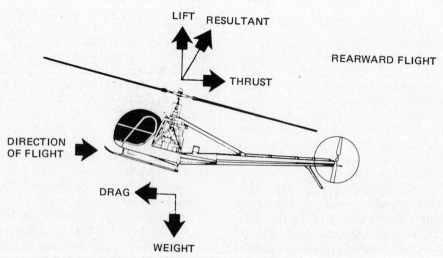

FIG. 5-44 Conditions for horizontal flight in a helicopter.

112

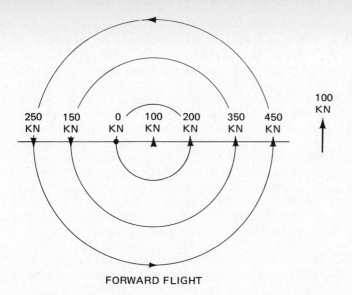

FORWARD FLIGHT

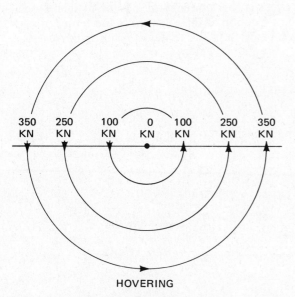

HOVERING

FIG. 5-45 Dissymmetry of lift caused by horizontal flight.

ulated rotor is designed with hinges at the root of each blade, which permits the blade to move up and down. When the helicopter is flying horizontally, the increased lift on an advancing blade causes it to flap up. The upward movement of the blade reduces the angle of attack because the relative wind direction with respect to the blade is more downward than before. This, of course, decreases the lift of the blade. At the same time, the retreating blade of the rotor flaps downward, thus producing a greater angle of attack and increased lift. The combined effect of decreased lift on the advancing blade and increased lift on the retreating blade tends to equalize the lift on the two sides of the rotor disk. A **semirigid** rotor is not hinged at the hub, but the hub itself can tilt as required to produce the flapping action. Thus, as the advancing blade flaps up, the retreating blade flaps downward. The design of the cyclic-pitch control also is such that it decreases the angle of attack of the advancing blade

and increases the angle of attack of the retreating blade. The total effect is to make the lift of one side of the rotor equal to the lift of the other side.

In studying the effect of forward aircraft speed on the helicopter rotor, it becomes obvious that there is a limit of speed beyond which the helicopter cannot fly. As the forward speed of the helicopter increases, the relative wind velocity on the retreating rotor blade decreases. At a certain relative wind velocity, the retreating blade will stall owing to the high angle of attack the blade requires to maintain lift equal to that of the advancing blade. The stall begins at the tip of the blade and works inward as forward speed increases. When this occurs, the helicopter will roll or fall off toward the stalled retreating rotor blade. Helicopters are placarded for maximum never-exceed (V_{ne}) speed.

Gyroscopic precession is a dynamic force that affects any rotating unit depending upon the mass, diameter, and speed of rotation of the unit. Since the rotor of a helicopter has a relatively large diameter and turns at several hundred revolutions per minute, precession is a prime factor in controlling the rotor operation.

The **cyclic-pitch** control causes a variation in the pitch of the rotor blades as they rotate about the circle of the tip-path plane. The purpose of this pitch change is, in part, to cause the rotor disk to tilt in the direction in which it is desired to make the helicopter move. When we consider only the aerodynamic effects of the blades, it would seem that when the pitch of the blades is high, the lift would be high and the blade would rise. Thus, if the blades had high pitch as they passed through one side of the rotor disk and low pitch as they passed through the other side of the disk, the side of the disk having the high pitch should rise and the side having the low pitch should fall. This would be true except for gyroscopic precession.

Gyroscopic precession is caused by a combination of a spinning force and an applied acceleration force perpendicular to the spinning force. Figure 5-46 is a drawing of a spinning disk that represents the main rotor of a helicopter. If the disk is spinning in the direction indicated by the arrow and a force is applied upward at F, the disk will precess (move) in the direction shown at P. Thus, if a force is applied perpendicular to the plane of rotation at one point near the rim, the precession will cause an upward force 90° from the applied force in the direction of rotation.

As a result of the foregoing principle, if it is desired

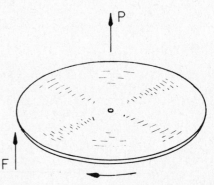

FIG. 5-46 Spinning showing the effect of precession.

to cause the main rotor of a helicopter to tilt in a particular direction, the applied force must be at an angular displacement of 90° ahead of the desired direction of tilt. The required force is applied aerodynamically by changing the pitch of the blades through the cyclic-pitch control. When the cyclic control is pushed forward, the blade at the left increases in pitch as the blade on the right decreases in pitch. This applies an "up" force to the left-hand side of the rotor disk, but the up movement takes place 90° ahead of this in the direction of rotation. The up movement is therefore at the rear of the rotor plane, and the rotor tilts forward. This, of course, applies a forward thrust and causes the helicopter to move forward. The action is illustrated in Fig. 5-47.

The helicopter can be caused to move in any direction desired merely by moving the cyclic-pitch control in that direction. The main rotor tilts in the direction called for by the control, and the helicopter moves as directed. If the control is in neutral, the helicopter will hover, that is, remain stationary in the air. In a wind the helicopter will drift in the direction in which the wind is blowing unless sufficient rotor thrust is applied to cancel the effect of the wind.

Before a helicopter takes off and when the rotor is turning, **centrifugal force** is the main force acting on the rotor blades. At this time the rotor blades are in a horizontal position. As the blade pitch is increased and power is applied to the rotor, the lift of the blades increases and the helicopter rises. The lift force of the blades causes the blade tips to rise above the horizontal plane and rotate through a conical path. This effect is called **coning** of the rotor blades and is illustrated in Fig. 5-48. When a helicopter is equipped with an **articulated** rotor having flapping hinges, the blades remain straight as they assume the coning position. If the rotor does not have flapping hinges, as in a rigid or semirigid rotor, the blades bend a limited amount when the helicopter is in flight.

The coning of rotor blades produces what is known as the **Coriolis effect,** which occurs as the center of mass of the rotor moves closer to the center of rotation when the blades rise. An examination of Fig. 5-49 shows why the center of mass (c_1, c_2, c_3) moves toward the center of the rotor disk as the blades rise. The Coriolis effect is the tendency of a rotor to increase the speed of rotation as the distance of the center of mass of each blade from the center of rotation decreases.

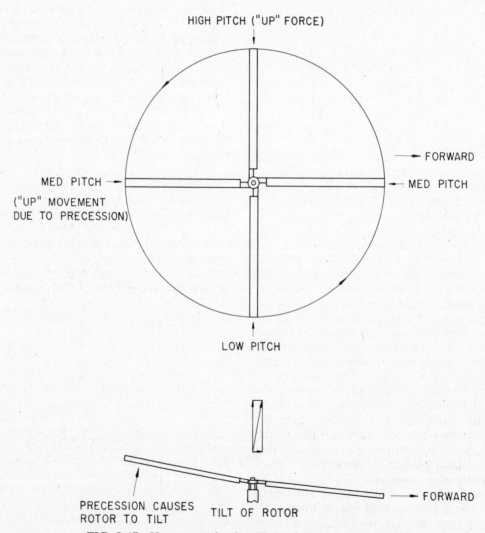

FIG. 5-47   How precession is utilized to tilt the main rotor.

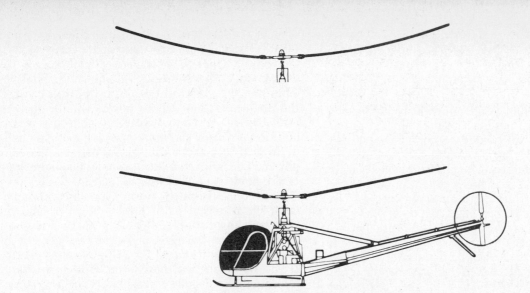

FIG. 5-48   Coning of the rotor blades.

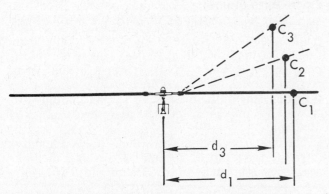

FIG. 5-49   Movement of center of mass as blades rise.

The result of Coriolis effect on a rotor is an acceleration of the blades as they flap up and a deceleration as they move downward. The direction of this acceleration and deceleration is parallel to the plane of the rotor disk. The movement is absorbed by the vertical drag hinges or by dampers and the structure of the blades. Two-blade rotors that are underslung and semirigid are not greatly influenced by Coriolis effect.

The Tail Rotor

According to Newton's third law of motion, for every force there is an equal and opposite force or reaction. The torque force applied to the rotor shaft of a helicopter to turn the rotor causes an equal and opposite **torque force** which would turn the fuselage of the helicopter in the opposite direction unless measures were taken to prevent it. This is the function of the antitorque rotor or tail rotor. Since the main rotor turns to the left (counterclockwise) as viewed from the top, the torque force causes the helicopter to turn to the right. Tail-rotor force must therefore be applied to the right to keep the heading steady. The combined effect of the tail-rotor thrust and the main-rotor torque is to apply a net force that causes the helicopter to drift to the right. This drift is corrected by rigging the main-rotor mast so that the main rotor will apply a force slightly to the left or by designing the cyclic-pitch con-

trol to provide a slight tilt of the tip-path plane to the left.

Ground Effect

When a helicopter is hovering near the ground, the downward stream of air strikes the ground and does not escape from beneath the helicopter as rapidly as it is being driven toward the ground. This causes a buildup of air pressure below the helicopter, which acts as a cushion to help support the machine in the hovering position. The ground cushion is usually effective to a height of approximately one-half the diameter of the main rotor while the helicopter is hovering. When the helicopter moves horizontally at 3 to 5 kn [1.54 to 2.57 m/s], the ground cushion is left behind.

Translational Lift

When a helicopter is moving horizontally in flight at more than 15 kn [7.7 m/s], the performance of the main rotor improves, owing to the increased volume of air passing through it. This effect is called **translational lift** because the lift of the rotor increases. It is apparent, therefore, that less engine power is required to maintain flight when the helicopter is flying horizontally than when it is hovering.

Transverse Flow Effect

A condition that must be compensated for in the design and rigging of a helicopter is caused by the change in the velocity of the air as it flows across the main-rotor disk, or tip-path plane. The air at the rear portion of the disk has a greater velocity than the air at the forward portion, owing to acceleration of the air as it flows across and through the disk. The effect is to produce a greater volume of air through the rear portion of the disk than through the forward portion, and this produces a greater force upward at the rear of the disk. The effect, because of precession, is to cause the rotor disk to tilt toward the left. This effect, called **transverse flow effect,** is corrected through the cyclic-pitch system.

● HELICOPTER CONTROLS

Heading Control

The **heading control** for a helicopter is similar to the rudder control for a conventional aircraft. The "rudder" pedals in a helicopter are not really rudder pedals because they control the pitch of the tail rotor rather than the deflection of a rudder. Some manufacturers refer to the tail rotor as a **rotary rudder,** and in this case the control pedals could logically be called rudder pedals. The tail rotor is often called an **antitorque rotor,** and the pedals are called **antitorque pedals.**

During straight and level flight, the pitch of the tail rotor is such that the rotor provides a thrust that exactly counterbalances the torque of the main rotor. If it is desired to turn the helicopter to the right, the pitch of the tail rotor is decreased by pressing the right pedal. The torque of the main rotor then causes the helicopter to turn to the right. If it is desired to turn to the left, the left pedal is pressed and the tail-rotor pitch is increased. This causes the additional thrust needed to push the tail to the right, and the helicopter then turns left.

It must be emphasized that the tail-rotor pedals in a helicopter do not control the direction in which the helicopter is flying. They control only the direction in which the fuselage is headed. Direction of flight is controlled through the cyclic-pitch system.

Collective-Pitch Control

The **collective-pitch control** increases or decreases the pitch of all the main-rotor blades simultaneously.

Hence, when it is desired to cause the helicopter to rise from the ground, collective pitch is increased and engine power is increased. The collective-pitch control is a lever (stick) that is usually situated at the pilot's left. The control is coordinated with the throttle control of the engine so that engine power will be increased as the lever is raised to increase the collective pitch. In addition, the lever has a motorcycle-type grip that can be rotated to make additional adjustments to engine power. The grip is rotated counterclockwise to decrease power and clockwise to increase power.

The collective-pitch control operates through a complex arrangement of levers, bearings, and linkages, as shown in Fig. 5-50. The drive and control mechanisms in the figure show the arrangement for a helicopter manufactured by Bell Helicopter Textron. It will be noted that the collective-pitch mechanism raises and lowers the swashplate as a unit, thereby causing both blades to change pitch simultaneously.

Cyclic-Pitch Control

The **cyclic-pitch control** causes a variation of the blade pitch as each blade rotates through the tip-path plane. This is accomplished through the tilting of the swashplate shown in Fig. 5-50. The purpose of the cyclic-pitch control, as mentioned previously, is to cause the tip-path plane of the main rotor to tilt as required to provide for movement of the helicopter in a desired direction.

The inner ring of the swashplate is stationary and is linked to the levers of the control. Through the linkage, the swashplate is tilted forward and rearward or

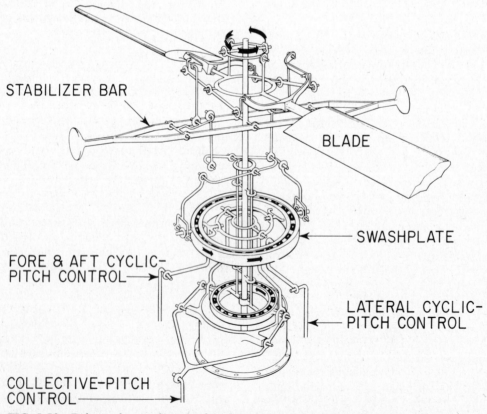

STABILIZER BAR

BLADE

SWASHPLATE

FORE & AFT CYCLIC-PITCH CONTROL →

LATERAL CYCLIC-PITCH CONTROL

COLLECTIVE-PITCH CONTROL →

FIG. 5-50 Drive and control mechanism for the main rotor of a helicopter. *(Bell Helicopter Textron)*

to either side. The outer ring of the swashplate turns with the main rotor and is linked through a stabilizer bar to the pitch-changing mechanism. When the swashplate is parallel to the plane of rotation, the pitch of the rotor blades is equal and uniform throughout rotation. When the swashplate is tilted, the pitch of the blades changes throughout the circle of rotation. On one side of the rotor disk the pitch will be decreasing, and on the other side of the disk the pitch will be increasing. If the cyclic control stick is moved to the right, the pitch of the blades in the forward half of the rotor disk will be higher than the pitch in the rearward half. This will cause an upward force at the forward side of the rotor disk, which, because of precession, will cause an upward movement of the left side of the rotor disk. The disk therefore will tilt to the right, and the thrust will pull the helicopter to the right.

The Power Train

The **power train** in a helicopter is the system of gears or belts and pulleys, clutch, and free-wheeling mechanism through which the rotors are driven by the engine. In large helicopters, a geared transmission is used to provide the correct ratio of speeds for the rotors. The ratio for the main rotor may vary from 6:1, engine-to-rotor speed, to 9:1 for helicopters with reciprocating engines and to as much as 100:1 for helicopters with gas-turbine engines. One model of the Sikorsky S61, for example, has an engine speed of 18 966 rpm [1985.7 radians per second (rad/s)] and a main-rotor speed of 203 rpm [21.25 rad /s]. The ratio in this case is 93.4289:1. The same helicopter has a tail-rotor speed of 3030 rpm [317.2 rad/s], which requires a reduction ratio of 6.26:1.

In some turbine engines, part of the gear reduction is built into the engine and the output shaft speed is much less than engine speed. Other engines do not include integral reduction-gear systems and the reduction in speed is accomplished entirely in the helicopter power-train gearbox.

The tail rotors for helicopters equipped with reciprocating engines may turn at the same speed as the engine or they may turn faster or slower, depending upon the design of the helicopter and rotor. The gearing necessary to attain the desired rotor speed is usually contained in the transmission gearbox. A typical tail-rotor speed is 3000 rpm [314.1 rad/s], more or less. For helicopters equipped with turbine engines, there must be substantial speed reduction, because the engine speed may be as high as 35 000 rpm [3664.5 rad/s] for some engines.

Autorotation

The helicopter must incorporate a safety feature to provide for the condition that exists in the event of power failure. This feature is called **autorotation** and is required before a helicopter can be certificated by the Federal Aviation Administration. If power failure occurs, the engine is automatically disengaged from the rotor system through a free-wheeling device associated with the transmission.

During autorotation, the airflow is upward through the main rotor rather than downward, and this upward flow causes the rotor to turn in the same direction as in normal operation. The turning of the rotor generates lift, which makes it possible to continue controlled flight while descending to a safe landing.

If engine failure occurs, the pilot immediately lowers the collective-pitch control, thus reducing the pitch of all rotor blades simultaneously. He also moves the cyclic-pitch control forward to establish the best forward speed for autorotation. Each helicopter has a characteristc forward speed, which produces maximum lift and lowest rate of descent.

Once the collective pitch is at the low-pitch limit, the rotor revolutions per minute can be increased only by a sacrifice in altitude or airspeed. If insufficient altitude is available to exchange for rotor speed, a hard landing is inevitable. Sufficient rotor rotational energy must be available to permit adding collective pitch to reduce the helicopter's rate of descent before final ground contact.

At low altitudes and low forward velocities, power failure in a helicopter is hazardous because of the difficulty in establishing sufficient autorotational lift to make a safe landing. Manufacturers provide **airspeed-vs-altitude limitations charts** to inform the pilot regarding the combinations of safe altitudes and speeds. A typical chart is shown in Fig. 5-51. It will be noted that it is comparatively safe to hover and fly at low speeds at very low altitudes. After attaining an indicated airspeed of 50 mph, or 44 kn [22 m/s], it is comparatively safe to fly at any altitude above 50 ft [16 m], because there is sufficient time to make the transition to the autorotation mode.

During autorotation, the outer 25 percent of the blades produces the lift, the section between 25 and 70 percent of the distance from the tip of the blades produces the driving force that keeps the rotor turning, and the inner 25 to 30 percent produces neither lift nor drive in any measurable degree.

● HELICOPTER CONFIGURATIONS

As in the case of fixed-wing aircraft, helicopters have many different configurations. The most popular helicopter arrangement is that of the **single rotor** utilizing a tail rotor such as is illustrated in Fig. 5-52. The single-rotor helicopter is relatively lightweight when compared to other configurations. This is due to its fairly simple design, with one rotor, one main transmission, and one set of controls.

The disadvantages of the single-rotor machine are that it has limited lifting and speed capabilities, as well as having a severe safety hazard during ground operation with the tail rotor positioned several feet behind the pilot and out of his or her line of vision.

Tandem-Rotor Helicopters

A **tandem-rotor** helicopter manufactured by Boeing Vertol, Division of Boeing Company, is shown in Fig. 5-53. This helicopter utilizes two synchronized rotors turning in opposite directions. The opposite rotation of the rotors causes one rotor to cancel the torque of the other, thus eliminating the need for an antitorque rotor. Each rotor is fully articulated and has three blades. Climb or descent is accomplished by means of the collective-pitch control. When the collective-pitch

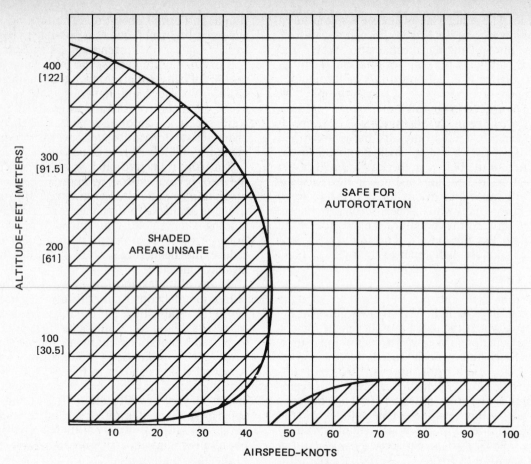

FIG. 5-51   Airspeed-versus-altitude limitations chart.

FIG. 5-52   Sikorsky's-76 Mark II Helicopter.   *(Sikorsky Aircraft)*

FIG. 5-53   A tandem-rotor helicopter.   *(Boeing Vertol, Division of Boeing Company)*

lever is raised, the pitch of all six rotor blades is increased simultaneously, causing the helicopter to ascend. This control, therefore, operates similarly to the collective-pitch control for a single-rotor helicopter. Descent is accomplished by lowering the collective-pitch control.

Directional control is achieved by tilting the plane of rotation of the rotors. Control motions to accomplish a turn are imparted by the control stick (cyclic pitch), directional pedals, or both. These controls tilt the swashplates in the rotor controls which, in turn, raise or lower the pitch links. The pitch links vary the pitch of the rotor blades during the rotation cycle.

Since the lift of the rotor blades is increased through part of the cycle and decreased through another part of the cycle, the plane of rotation is tilted. When the pilot applies directional-pedal-control movement in one direction, the plane of rotation of the forward rotor is tilted downward in that direction and the plane of rotation of the aft rotor is tilted downward in the opposite direction. This causes the helicopter to make a hovering turn around the vertical axis as illustrated in Fig. 5-54. The tandem helicopter is capable of lifting large loads since these loads may be distributed between the two rotors. A disadvantage of the tandem-rotor helicoptor is that it is not efficient in for-

118

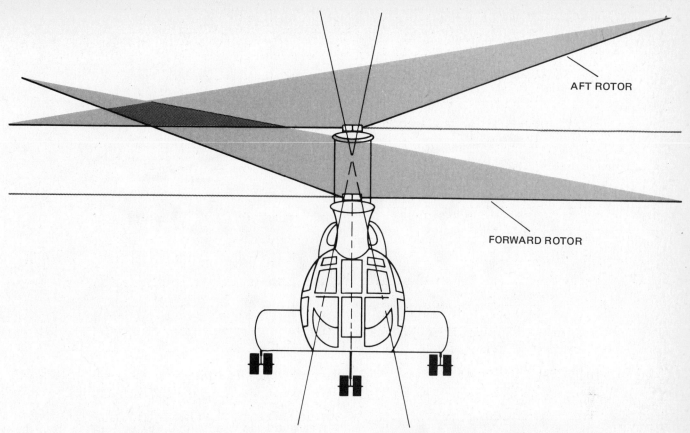

AFT ROTOR

FORWARD ROTOR

FIG. 5-54 Rotor plane movement during a turn. *(Boeing Vertol, Division of Boeing Company)*

ward flight because one rotor is working in the wake of the other. This loss of lift may be minimized by placing the rear rotor above the main rotor.

Side-by-Side Rotors

The **side-by-side** helicopter has two main rotors mounted on pylons or wings postitioned out from the sides of the fuselage, as illustrated in Fig. 5-55. The side-by-side configuration has the rotors turning in opposite directions, which eliminates the need for a tail rotor.

The advantages of the side-by-side configuration are that it has excellent stability, and the rotors are more efficient in forward flight than the tandem arrangement due to the fact that one rotor is not in the wake of the other. The side-by-side helicopter has the disadvantages of having high parasitic drag and high structural weight, both resulting from the structure necessary to support the main rotors.

Coaxial Rotors

In the **coaxial helicopter,** illustrated in Fig. 5-56, fuselage torque is eliminated by utilizing two counter-rotating rigid main rotors mounted one above the other on a common shaft. This type of configuration is also referred to as the **advancing blade concept** because the lift load at high forward speed is carried primarily by the advancing blades. It combines the advantages of a low-speed helicopter with those of a

high-speed aircraft without the need for a wing and without the need for a tail rotor.

By avoiding retreating blade stall, which is a key limiting factor in the speed and maneuverability of a pure helicopter, the coaxial helicopter has been made faster and more maneuverable. This design has successfully demonstrated forward speeds of more than 250 kn [129 m/s].

Tilt Rotor

After many years of research, the technology in the areas of composites and electronic flight control systems has finally been developed to make a functional **tilt-rotor** aircraft a reality. The tilt rotor has the ability to combine the vertical takeoff, low-speed capabilities of the helicopter with the high-speed performance of a turboprop airplane.

The XV-15 tilt rotor, illustrated in Fig. 5-57, resembles a twin turboprop with large-diameter propellers, or rotors, mounted on wing-tip nacelles. The engines rotate from a vertical position for helicopter flight to horizontal position for cruise flight.

Lift in the vertical or hover position is provided totally from the 38-ft [11.6-m] diameter **prop rotors.** When in the horizontal or cruise position, lift is provided by the wings. The tilt rotor operates effectively with the engines between the vertical and horizontal positions, which provides for a wide range of lift and speed combinations. When operated in this conversion configuration, lift is provided from both the

FIG. 5-55   Russian Mi-12 helicopter.   *(General Electric)*

FIG. 5-56   Sikorsky S-69 Advancing Blade Concept Demonstrator.   *(Sikorsky Aircraft)*

wings and the prop rotors. Full conversion from the hover to the fixed wing mode can be accomplished in approximately 12 s.

The X Wing

The **X wing** illustrated in Fig. 5-58 is a concept for a vertical takeoff and landing aircraft which uses a four-bladed helicopter-like rotor system that rotates for hover and low-speed flight and stops at approximately 200 kn [103 m/s] to become a fixed-wing aircraft for high-speed flight. The vehicle then flies as a fixed-wing aircraft while accelerating to speeds of approximately 450 to 500 kn [232 to 257 m/s] utilizing an auxiliary propulsion system.

In addition to employing computerized flight controls and the lastest in composite technology, the X

FIG. 5-57   XV-15 Tilt-Rotor Aircraft.   *(NASA)*

FIG. 5-58   Sikorsky X-Wing Aircraft.   *(Sikorsky Aircraft)*

wing relies on an air circulation system, which is essential to the X-wing concept.

The air circulation system is comprised of a compressor, control valves, and ducts to the leading and trailing edges of the rotor/wing. The compressor feeds pressurized air through hollow leading and trailing edges to slots in the rotor/wing. The air is blown out the slots over the rounded leading and trailing edges of the symmetrically shaped rotor/wing, providing circulation control lift. Lift is controlled by altering the airflow through the slots, thus allowing the air circulation system to substitute for collective control and cyclic control as well as for flaps and ailerons. During transition from a rotor/wing to a fixed-wing aircraft, all lift must be generated by the air circulation system.

● REVIEW QUESTIONS

1. Why is it important for the maintenance technician to understand the stresses that may be imposed on an aircraft in flight?
2. What are the four forces that are applied to an airplane in flight?
3. How is an airplane's load factor determined?
4. In what categories may airplanes be certified?
5. What is the load factor in a 60° turn?
6. What is *wing loading?*
7. Describe the three axes of an airplane.
8. Discuss static versus dynamic stability.
9. Describe *washin* and *washout.*
10. Describe *dihedral* and explain how it affects lateral stability.
11. Define *stability* as applied to an airplane in flight. How are the various types of stability attained in the design of an airplane?
12. What are the three principal controls of an airplane?
13. Which primary control affects aircraft movement around each of the three axes?
14. What is a *stabilator?*
15. Describe the use of the rudder in a coordinated turn.
16. Explain the use of trim tabs.
17. What is the difference between a balance tab and a controllable trim tab?
18. Explain the importance of properly balanced control surfaces.
19. What flight conditions may lead to a stall?
20. What may be done to a wing design so that it will have good stall characteristics?
21. How do the loads that are imposed on a canard vary from those that are imposed on a conventional tail?
22. Describe the advantages and disadvantages of a forward swept wing.
23. What are some advantages of a T tail?
24. Explain the operation of a ruddervator.
25. What type of aircraft require elevons?
26. What are *flaperons?*
27. What is meant by interplane interference in a biplane?
28. Define *gap/chord ratio* and *gap/span ratio.*
29. Explain the two types of stagger which may be designed into a biplane.
30. Define *angle of incidence.*
31. Describe the difference between positive and negative decalage.
32. What method is used to make a helicopter move in any particular direction?
33. What airfoil provides lift in a helicopter?
34. What is the cause of dissymmetry of lift?
35. What is meant by *blade flapping?*
36. What is an *articulated rotor?*
37. Describe a *semirigid rotor.*
38. How does dissymmetry of lift limit the forward speed of a helicopter?
39. How does gyroscopic precession affect the control of the main rotor of a helicopter?
40. What is meant by *coning of the rotor blades?*
41. Describe the *Coriolis effect* and explain how it affects the rotor blades.
42. How is the torque of the main rotor of a helicopter compensated for?
43. Why does the tail rotor cause the helicopter to drift sideways?
44. What causes translational lift and how is it of value in the operation of a helicopter?
45. How does a helicopter compensate for transverse flow effect?
46. Describe the *collective-pitch-control system.*
47. How does *cyclic pitch* operate?
48. Explain the function of the swashplate in contolling cyclic pitch.
49. What is the *power train?*
50. Compare tail-rotor revolutions per minute with main-rotor revolutions per minute.
51. Why is a tail rotor not required on a tandem-rotor helicopter?
52. What rotor action takes place when making a turn with a tandem-rotor helicopter?
53. What advantages does a tilt-rotor system have?
54. Describe *autorotation.*
55. What is the purpose of an airspeed-vs-altitude chart?

6 AIRCRAFT DRAWINGS

● INTRODUCTION

Engineering drawings and prints (copies of drawings) are essential tools in the design and manufacture of aircraft. An engineering drawing is used to describe an object by means of lines and symbols. With the use of a print or drawing, an engineer can convey to those who build, inspect, operate, and maintain aircraft and spacecraft the necessary instructions for ordering the materials, making the parts, assembling the units, and finishing the surfaces. Drawings constitute the abbreviated written language of the aerospace industry, a shorthand method for presenting information that would take many pages of manuscript to transmit. The aviation maintenance technician must be able to correctly interpret the information on many types of drawings in technical reference manuals. In addition to engineering drawings the technician will encounter schematic diagrams, installation and location drawings, and wiring charts. Reproduction of drawings was originally done by a process that printed white lines on a blue background. The term *blueprint* was used for these prints. Today, the term **blueprint** is commonly used to refer to many types of drawings and prints without regard to the production or copying process.

● TYPES OF DRAWINGS

Production Drawings

A major use of engineering drawings is for the fabrication or assembly of components. Drawings used for this purpose are also called **production drawings**, or **working drawings.** Production drawings can be categorized as detail drawings, assembly drawings, or installation drawings.

A **detail drawing** can consist of one part or several parts making up an assembly. The detail drawing will provide, by the use of lines, notes, and symbols, all specifications (size, shape, and material) needed to make the part. An example of a detail drawing is shown in Fig. 6-1.

An **assembly drawing** is used to show how parts produced from detail drawings fit together to form a component. The assembly drawing does not show the dimensions of the detail parts except as necessary for location purposes. Figure 6-2 shows how the parts of an electrical connector are assembled. Figure 6-3 is an assembly drawing of a structural fastener. The upper part of the drawing is a **pictorial drawing** of the three detail parts. A pictorial drawing is similar to a photograph and shows the parts as they would appear to the eye. The pictorial portion of the drawing is often re-

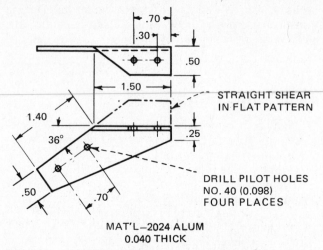

FIG. 6-1   A detail drawing for an angle bracket.

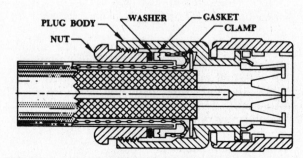

FIG. 6-2   An assembly drawing of a cable connector.

ferred to as an "exploded view." The detail parts are shown individually but arranged so as to indicate how they would be assembled. The lower part of the drawing shows the panel fastener correctly installed into the structure. Illustrated parts catalogs utilize exploded-view assembly drawings for identification of parts and part numbers.

An **installation drawing** shows how a part or component is installed in the aircraft. Figure 6-4 shows a bracket installed at a specific point in an aircraft fuselage. Dimensions and directional information are provided in this drawing. Figure 6-5 shows the location and directional orientation of a gear door latch. Dimensional information is not provided, or necessary, on this drawing. The individual parts of the latch are to be attached to the existing structural assembly. The dimensional information for the installation and assembly of the structure would be provided on other drawings.

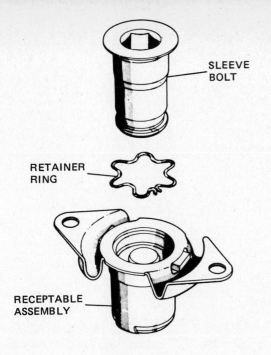

SLEEVE
BOLT

RETAINER
RING

RECEPTABLE
ASSEMBLY

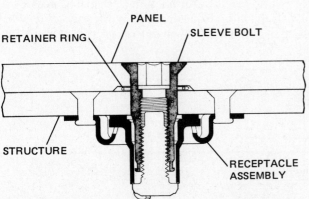

FIG. 6-3   A structural panel fastener.   *(VOI-SHAN Division of VSI Corp.)*

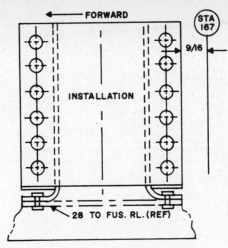

FIG. 6-4   Fuselage bracket installation.

Block Diagrams

A **block diagram** is a special drawing used to simplify the explanation of complex circuits. Block diagrams are widely used for electronic circuits but can be used for any type of aircraft system. A block diagram allows nonspecialized personnel to understand the function and relationship of various systems within a component. Various shapes may be used within the diagram to help explain function. Figure 6-6 is a block diagram of an autopilot system. Block diagrams are very useful for troubleshooting. By knowing the input and output for each component, a technician can identify and isolate those contributing to a malfunction.

Schematic Diagrams

A **schematic diagram** is also used to explain a system. A simple schematic would show the functional location of components within a system. This is done without regard to the physical location of the compo-

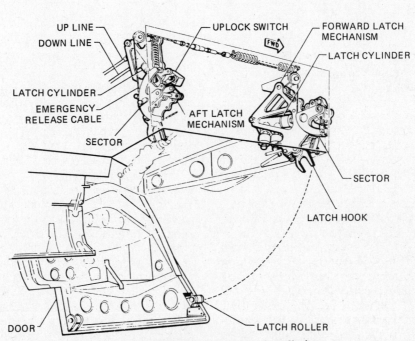

FIG. 6-5   Door latch assembly installation.

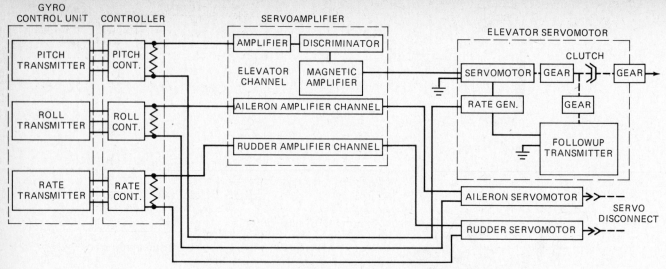

FIG. 6-6　Block diagram of an autopilot system.

nents in the aircraft. The flow of fluid in a lubrication system is indicated by the use of various types of shading as shown in Fig. 6-7. While some schematics are colored, black and white diagrams provide better copies and are usable for microfilm and microfiche.

Figure 6-8 is an illustration of the same lubrication system as in Fig. 6-7. In this drawing the components and lines are shown in relation to their physical location in the aircraft installation. While the same items are shown in each drawing, it is apparent that tracing

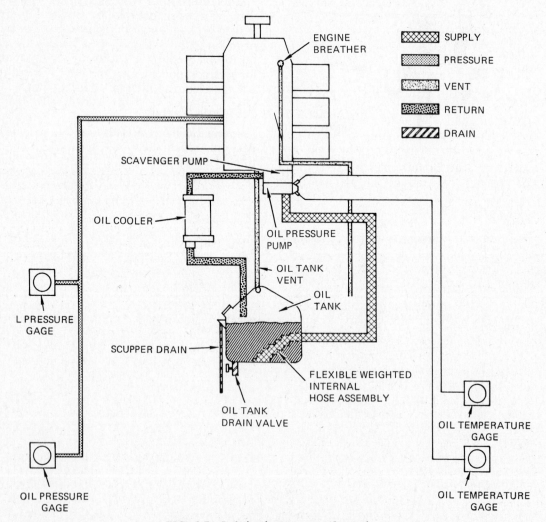

FIG. 6-7　Lubrication system schematic.

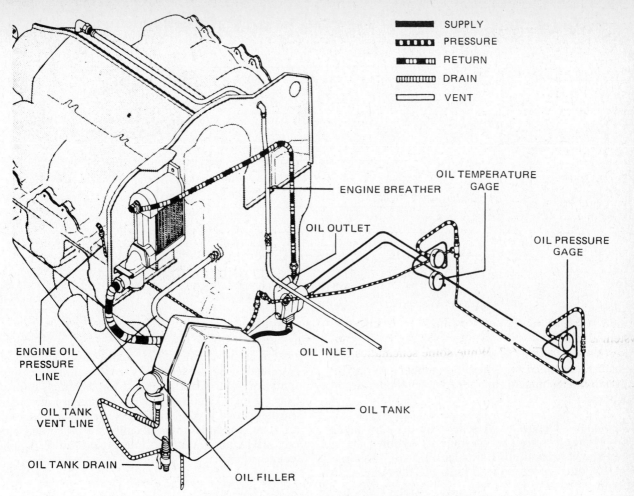

SUPPLY
PRESSURE
RETURN
DRAIN
VENT

ENGINE BREATHER

OIL TEMPERATURE GAGE

OIL OUTLET

OIL PRESSURE GAGE

ENGINE OIL PRESSURE LINE

OIL INLET

OIL TANK VENT LINE

OIL TANK DRAIN

OIL TANK

OIL FILLER

FIG. 6-8   Lubrication system perspective.

flow would be easier with the schematic. The perspective view would be useful for locating the component in the aircraft once it has been identified from the schematic.

Shop Sketches

A **shop sketch** may be anything from a simple line drawing such as that shown in Fig. 6-9 to a rather complex and detailed drawing such as a standard engineering drawing. The purpose of a shop sketch is to convey information concerning the repair of a part or structure, to illustrate a proposed modification, to provide information for engineering drafters from which they can make standard engineering drawings, and for various other uses where an illustration is necessary to convey technical information. Due care should be exercised in the preparation of a shop sketch so that it will present a good appearance, the information it contains is complete and accurate, and it conveys the information that it is intended to convey.

The aviation maintenance technician should develop as much skill as possible in preparing shop sketches. Such sketches are often necessary in preparing a repair proposal for approval by the FAA. The sketches become a part of the maintenance and repair record for repairs that are not documented by manufacturers or FAA publications.

Drawings for Electrical and Electronic Systems

Aircraft today are required to be equipped with extensive electrical circuitry and electronic units. As a result, the preparation of engineering drawings, wiring drawings, schematics, etc., has become quite complex. It was once a simple matter for a technician to take a system blueprint and extract a schematic circuit for the purpose of troubleshooting. This can still be done; however, maintenance manuals usually include schematics of all the circuits in the aircraft.

It is the purpose of this section to show some of the types of electrical and electronic system and circuit drawings that will be encountered by the maintenance technician. The use of wiring diagrams and schematic circuits in troubleshooting is covered in the associated text on aircraft electricity and electronics.

Wiring Diagrams. The purpose of a wiring diagram is to show all the wires and wire segments and all the connections in an electrical system or circuit. Figure 6-10 is a sample wiring diagram arranged in accordance with Air Transport Association (ATA) specifications. It will be noted that every wire segment is identified by an alpha-numerical code. The letters and numbers shown in the diagram are stamped on the wires in the aircraft at intervals of 15 in [38.1 cm] or

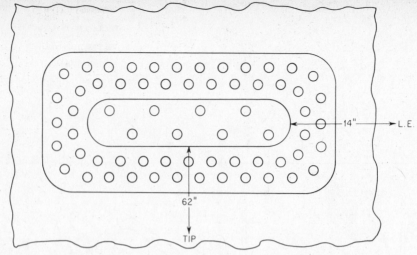

REPAIR OF CUT – BOTTOM, LEFT WING – 14" REAR OF L.E.–62" INBOARD OF WING TIP –
CUT 3 7/8" – DAMAGED MATERIAL CUT OUT–SKIN MATERIAL – 2024-T3 –0.032" ALCLAD –
PATCH SAME MATERIAL–RIVETS–AN 426-AD-4-3 –PATTERN EXCEEDS ORIGINAL IN STRENGTH
PATCH INSIDE WING –SKIN MADE FLUSH WITH PLUG

FIG. 6-9 A shop sketch.

less. The connections to connector plugs and electrical units are identified by letters or numbers. With this diagram the technician can prepare a schematic circuit diagram.

Schematic Diagrams. A schematic diagram taken from a manufacturer's service manual is shown in Fig. 6-11. The wires and units are identified so the technician can quickly see how all units are related electrically in the system.

The wire identification in this schematic utilizes letters and numbers as shown in the chart of Fig. 6-12. It will be noted that this system makes it possible to identify every segment of every wire in a circuit. If a particular unit is malfunctioning, the technician can quickly determine which wires are involved and test them for continuity.

Logic Circuitry for Electronic Systems. Because of the complexity of electronic systems and the use of logic systems involving solid-state electronic units, it has been necessary to simplify system drawings using symbols that have meaning to the electronic technician. Logic systems involve the use of **binary** mathematics. The binary system of mathematics utilizes only two digits, 1 and 0. If a circuit is conducting, the signal is 1, and if it is not conducting, the signal is 0. Thus a switch, transistor, or some other unit or combination of units can be used as a "gate" to provide the correct signal for the function involved.

As shown in Fig. 6-13, if an OR gate is in the circuit, the first input to become 1 produces a 1 in the output. For the symbols shown in the illustration, the input is on the left and the output is on the right. A NOR gate produces a 1 in the output if there are no 1s in the input. With the AND gate, all the inputs must be 1s to produce a 1 in the output. The NAND gate will produce a 1 signal in the output if any of the inputs is 0.

A diagram showing the use of logic symbols for a system is provided in Fig. 6-14. This illustration is provided so the student will recognize this type of presentation and is not intended for technical information on logic circuitry. Logic systems are covered in the associated text on aircraft electricity and electronics. The diagram shown is called an integrated block test (IBT) and is easily interpreted by a qualified technician.

● DRAFTING TECHNIQUES

An aircraft maintenance technician does not have to be a skilled draftsperson. However, in order to correctly read and interpret drawings, a knowledge of the techniques used to graphically communicate technical information is essential. These techniques include the use of different types of views, lines, symbols, standard dimensioning practices, and title block information.

Projections

The Perspective Drawing. Figure 6-15 is a photograph of an aircraft. Compare the photograph with the perspective drawing in Fig. 6-16 of the same aircraft. There is not any difference in the shape or arrangement of parts, but there is a difference in shading. Both illustrations show the aircraft as it would look to the eye, but they often do not present the detailed information needed by the aircraft maintenance technician. In a perspective drawing parallel lines will be shown as converging for the same reason that the rails of a railroad seem to meet in the distance. This produces an effect known as foreshortening. As a result of foreshortening, the various parts of the aircraft are not drawn in a true scale with each other. The perspective drawing is best used for overall views and providing location information such as the lubrication system shown in Fig. 6-8.

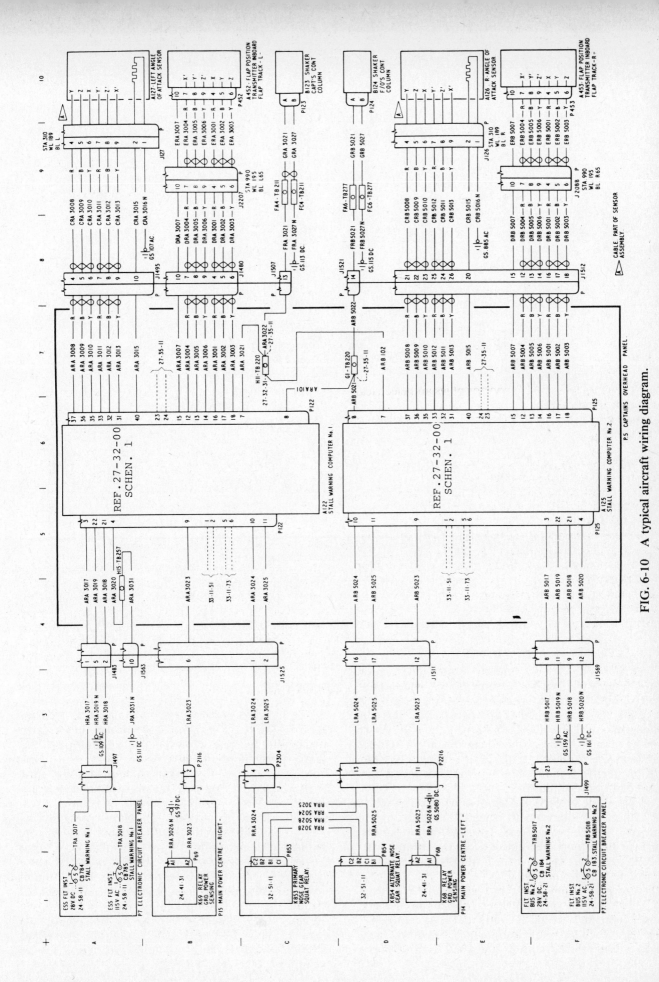

FIG. 6-10 A typical aircraft wiring diagram.

127

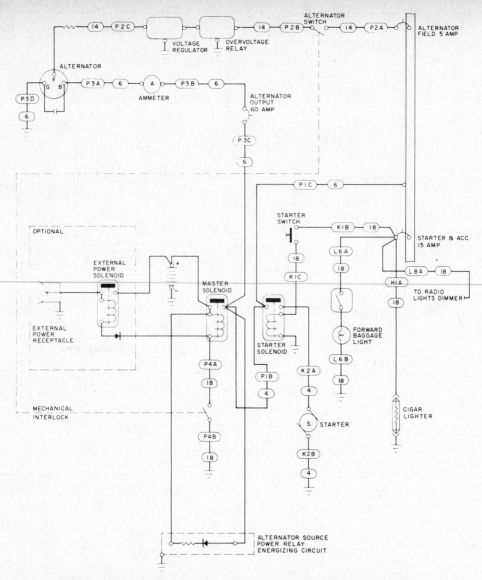

FIG. 6-11   A schematic for an electrical system.   *(Piper Aircraft Co.)*

The Oblique View—Isometric Projection. Figure 6-17 shows two views of the same object, one in **perspective** and the other an **oblique view.** The oblique view is somewhat similar to the perspective, but the line representing the edge farthest from the observer is drawn the same length as the line nearest the observer. The lines are parallel and of the same length; yet the farthest one from the observer seems to be longer because of an optical illusion.

An **isometric projection** is an oblique drawing without the optical illusion of perspective. In this type of drawing, the object appears distorted, but it shows equal distances on the subject as equal distances on the drawing, thus enabling the drafter to make the dimensions clear to the reader.

Figure 6-18 is an isometric drawing. All vertical lines are drawn as verticals and all horizontal lines are drawn at an angle of 30°. It shows that certain lines are parallel to each other and at right angles to other lines. There is no foreshortening, and there is no conver-

gence of lines as in the true perspective drawing. The isometric drawing provides a bird's-eye view, it gives proportions, and it is often very useful in explaining the design and construction of complicated assemblies. Its faults are that the shape of the object is distorted and the angles do not appear in their true size.

The term *isometric* is derived from two Greek words *isos* and *metron,* meaning **equal** and **measure.** Thus we see that isometric means **of equal measure.**

Orthographic Projections.   The three types of drawings we have already discussed (perspective, oblique, and isometric) give a bird's eye view of an object, but the person who reads a print must have more than one view, especially if the object shown is three-dimensional; and the reader must have each view presented without distortion in most cases.

Figure 6-19 is an **orthographic projection** of a bracket. The word *orthographic* means **the projections of points on a plane by straight lines at right angles to**

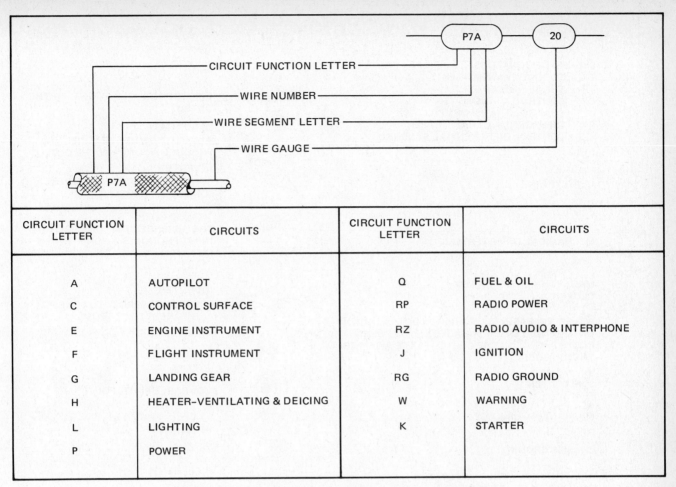

CIRCUIT FUNCTION LETTER	CIRCUITS	CIRCUIT FUNCTION LETTER	CIRCUITS
A	AUTOPILOT	Q	FUEL & OIL
C	CONTROL SURFACE	RP	RADIO POWER
E	ENGINE INSTRUMENT	RZ	RADIO AUDIO & INTERPHONE
F	FLIGHT INSTRUMENT	J	IGNITION
G	LANDING GEAR	RG	RADIO GROUND
H	HEATER–VENTILATING & DEICING	W	WARNING
L	LIGHTING	K	STARTER
P	POWER		

FIG. 6-12   Wire identification system.   *(Piper Aircraft Co.)*

the plane and is derived from the Greek word *orthos,* meaning **straight.** There is a **front view,** a **top view,** and a **right-side view.** The **front view** is what you would see if you were directly in front of the bracket; it would be impossible to see any of the top, the bottom, or the sides. The **top view** is what you would see if you were looking directly down on the bracket; you would not see the front, the bottom, or any of the other surfaces. The **right-side view** is what you would see if you looked directly at that side; you would not see the top, the bottom, or any other surface except the right side.

Figure 6-20 demonstrates three views for an orthographic projection. There are actually six views that can be used. Not shown are the back, left-side, and bottom views. For most objects, the three views shown are adequate to describe the object. By studying the three different views, it is possible to determine the configuration of the object drawn. In order to interpret the views one must understand the use of various types of lines.

The Meaning of Lines

Standards have been set for lines to be used in drafting. Figure 6-21 shows the types and forms of lines that are used to prepare drawings.

Most drawings use three widths, or intensities, of lines: **wide, medium,** and **narrow.** These lines may vary somewhat on different drawings; but on any one drawing there is a noticeable contrast between a wide line and a narrow line, and the medium line is somewhere between.

The **visible outline** (object line) is a medium-to-wide line which should be the outstanding feature of the drawing. The thickness may vary to suit the drawing, but it should be at least 0.015 in [0.038 cm] wide. This line represents edges and surfaces that can be seen when the object is viewed directly.

The **invisible outline,** more correctly called a **hidden line,** is a medium-width line made up of short dashes and is sometimes called a **dotted** or **broken line.** It represents edges and surfaces behind the surface being viewed and therefore is not visible to the observer.

The **center line** is drawn narrow or thin and consists of alternate long and short dashes. It shows the location of the center of a hole, rod, symmetrical part, or symmetrical section of a part. Center lines are the first lines drawn, and they provide the basic reference for the rest of the drawing.

Phantom lines are not used by all drafters; however, they serve a good purpose and should be understood. The phantom line is a medium-width line and is made up of a series of long dashes with two short dashes be-

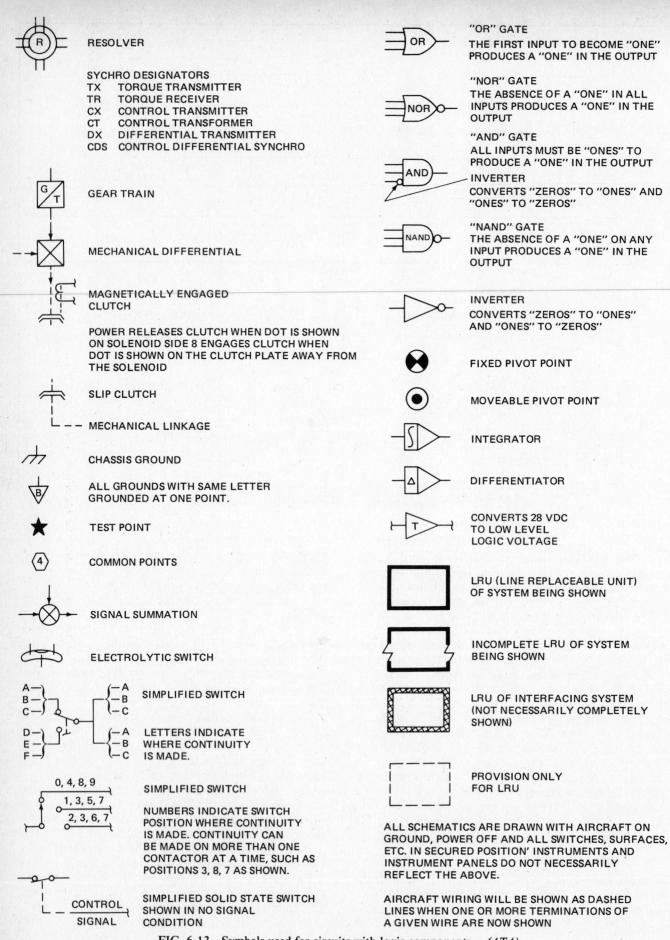

RESOLVER

SYCHRO DESIGNATORS
TX TORQUE TRANSMITTER
TR TORQUE RECEIVER
CX CONTROL TRANSMITTER
CT CONTROL TRANSFORMER
DX DIFFERENTIAL TRANSMITTER
CDS CONTROL DIFFERENTIAL SYNCHRO

GEAR TRAIN

MECHANICAL DIFFERENTIAL

MAGNETICALLY ENGAGED CLUTCH

POWER RELEASES CLUTCH WHEN DOT IS SHOWN ON SOLENOID SIDE 8 ENGAGES CLUTCH WHEN DOT IS SHOWN ON THE CLUTCH PLATE AWAY FROM THE SOLENOID

SLIP CLUTCH

MECHANICAL LINKAGE

CHASSIS GROUND

ALL GROUNDS WITH SAME LETTER GROUNDED AT ONE POINT.

TEST POINT

COMMON POINTS

SIGNAL SUMMATION

ELECTROLYTIC SWITCH

SIMPLIFIED SWITCH

LETTERS INDICATE WHERE CONTINUITY IS MADE.

0, 4, 8, 9
1, 3, 5, 7
2, 3, 6, 7

SIMPLIFIED SWITCH

NUMBERS INDICATE SWITCH POSITION WHERE CONTINUITY IS MADE. CONTINUITY CAN BE MADE ON MORE THAN ONE CONTACTOR AT A TIME, SUCH AS POSITIONS 3, 8, 7 AS SHOWN.

CONTROL
SIGNAL

SIMPLIFIED SOLID STATE SWITCH SHOWN IN NO SIGNAL CONDITION

"OR" GATE
THE FIRST INPUT TO BECOME "ONE" PRODUCES A "ONE" IN THE OUTPUT

"NOR" GATE
THE ABSENCE OF A "ONE" IN ALL INPUTS PRODUCES A "ONE" IN THE OUTPUT

"AND" GATE
ALL INPUTS MUST BE "ONES" TO PRODUCE A "ONE" IN THE OUTPUT

INVERTER
CONVERTS "ZEROS" TO "ONES" AND "ONES" TO "ZEROS"

"NAND" GATE
THE ABSENCE OF A "ONE" ON ANY INPUT PRODUCES A "ONE" IN THE OUTPUT

INVERTER
CONVERTS "ZEROS" TO "ONES" AND "ONES" TO "ZEROS"

FIXED PIVOT POINT

MOVEABLE PIVOT POINT

INTEGRATOR

DIFFERENTIATOR

CONVERTS 28 VDC TO LOW LEVEL LOGIC VOLTAGE

LRU (LINE REPLACEABLE UNIT) OF SYSTEM BEING SHOWN

INCOMPLETE LRU OF SYSTEM BEING SHOWN

LRU OF INTERFACING SYSTEM (NOT NECESSARILY COMPLETELY SHOWN)

PROVISION ONLY FOR LRU

ALL SCHEMATICS ARE DRAWN WITH AIRCRAFT ON GROUND, POWER OFF AND ALL SWITCHES, SURFACES, ETC. IN SECURED POSITION' INSTRUMENTS AND INSTRUMENT PANELS DO NOT NECESSARILY REFLECT THE ABOVE.

AIRCRAFT WIRING WILL BE SHOWN AS DASHED LINES WHEN ONE OR MORE TERMINATIONS OF A GIVEN WIRE ARE NOW SHOWN

FIG. 6-13 Symbols used for circuits with logic components. *(ATA)*

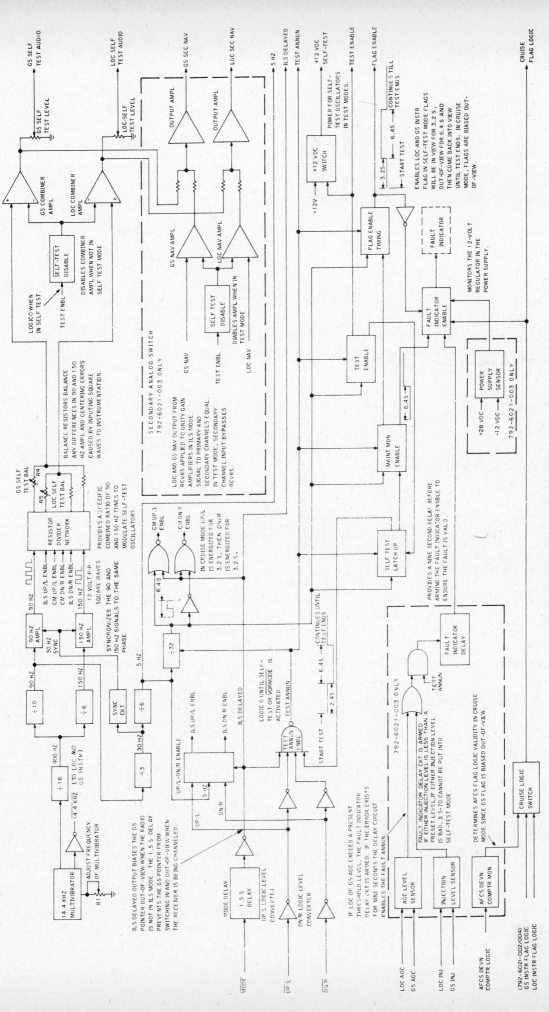

FIG. 6-14 An integrated-block-text diagram. (ATA)

FIG. 6-15   A photograph of an aircraft.   *(Beech Aircraft Co.)*

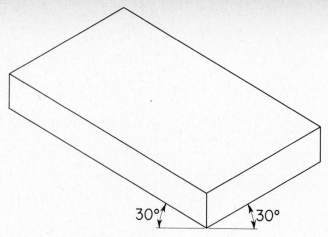

FIG. 6-18   Isometric drawing.

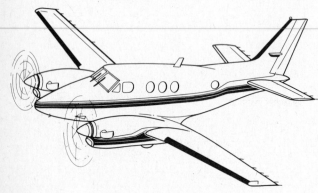

FIG. 6-16   A drawing of an aircraft.

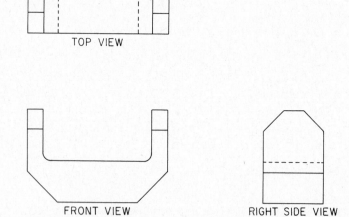

FIG. 6-19   Orthrographic projection of a bracket.

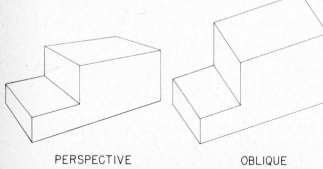

PERSPECTIVE   OBLIQUE

FIG. 6-17   Perspective and oblique views of the same object.

tween the long dashes. The purpose of the line is to indicate the position of an adjacent part for local reference.

The **dimension line** is a narrow line and is unbroken except where a dimension is written in. Having determined the general shape of an object, the person reading the blueprint wants to know the size. The length, width, or height of a dimensioned part is customarily shown by a fraction or a number placed in a break in the line.

An **extension line,** or **witness line,** is used to extend the line indicating an edge of the object for the pur-

pose of dimensioning. The extension line is drawn very narrow or thin.

The **cutting-plane line** is a heavy, wide, broken line made up of one long and two short dashes, alternately spaced. It is used where the drafter wants to refer the reader to another view and direct attention to a section, or "slice," that reveals the interior.

The **broken material line,** often simply called a **break line,** can be used where the drafter is cramped for space on the drawing. Where the drawing would run off the paper if complete, the drafter uses a medium line of the form illustrated to show that the drawing has been reduced, although the length of the actual object is not reduced. The narrow ruled line with zigzags is for long breaks. The wider freehand wavy line is used for short breaks, and it is especially useful where the drafter has removed an outer surface to reveal an interior part of the object.

The break line is also used simply to avoid the necessity of repeating the second half of a symmetrical part. This saves drafting time and has been done in the case of the pulley in Fig. 6-22, section A-A. There is

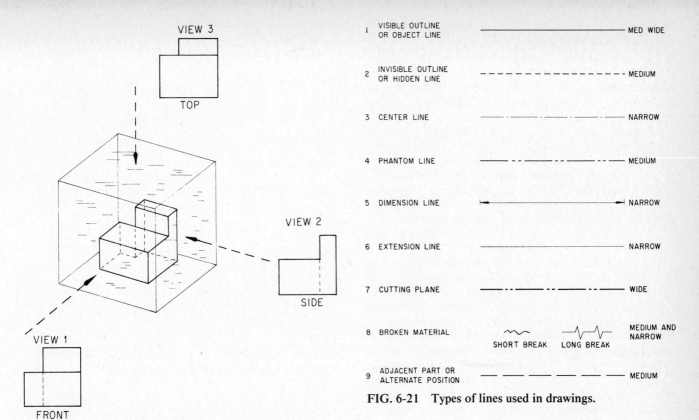

VIEW 3

TOP

VIEW 2

SIDE

VIEW 1

FRONT

FIG. 6-20   Demonstration of an orthographic projection.

1	VISIBLE OUTLINE OR OBJECT LINE		MED WIDE
2	INVISIBLE OUTLINE OR HIDDEN LINE		MEDIUM
3	CENTER LINE		NARROW
4	PHANTOM LINE		MEDIUM
5	DIMENSION LINE		NARROW
6	EXTENSION LINE		NARROW
7	CUTTING PLANE		WIDE
8	BROKEN MATERIAL	SHORT BREAK LONG BREAK	MEDIUM AND NARROW
9	ADJACENT PART OR ALTERNATE POSITION		MEDIUM

FIG. 6-21   Types of lines used in drawings.

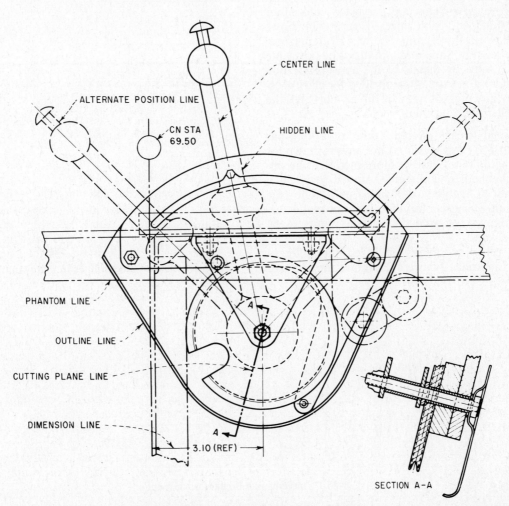

CENTER LINE

ALTERNATE POSITION LINE

CN STA 69.50

HIDDEN LINE

PHANTOM LINE

OUTLINE LINE

CUTTING PLANE LINE

DIMENSION LINE

A

A

3.10 (REF)

SECTION A-A

FIG. 6-22   Drawing showing the use of lines.

133

obviously enough room to complete the drawing, but it is quicker and simpler to do it as shown.

The **adjacent part,** or **alternate-position line,** is a medium, broken line made up of long dashes. It can be used to show the relationship between a part and an assembly or to show the alternate positions of a moving part.

Figure 6-22, an installation of a flap-control unit, illustrates the use of many of the lines described in the foregoing. Notice the visible outline line, the invisible outline line labeled **hidden line,** the center line, the dimension line, the phantom line, the alternate-position line, and the cutting-plane line. In this particular drawing, three alternate positions are shown for the lever since it is a moving part. The break lines are not labeled, but they are easily found.

When a surface has been cut away to reveal a hidden, inner feature of an object, **section lines** are used. These are narrow, solid lines spaced evenly to present a shaded effect. Notice that the large letter A appears twice in Fig. 6-22, connected by a cutting-plane line. This shows where a section was taken. In the lower-right-hand corner of the drawing, the **section** itself is shown. It must be pointed out here that the shading or line pattern used to indicate a cutaway section usually varies according to the material from which the object is constructed. Some of the patterns used for various materials are shown later in this chapter.

Visible and Hidden Lines. Figure 6-23 shows that all edges or sharp corners are projected into other views as outlines or visible edges. Figure 6-24 shows that any line or edge that cannot be seen from one particular view must be shown as a **hidden line** (invisible edge) in that view by means of the appropriate line symbol. Hidden lines are often omitted for clarity, especially on drawings of very complex parts, if the drawing is clear without them.

Conventional Breaks. A pipe, tube, or a long bar having a uniform cross section is not always drawn for its entire length. When one or more pieces of the object are broken out and the ends moved together, a larger and more legible scale can be used. The true length will not be shown, but that does not matter because the dimensions give the measurements to be taken on the work.

Different types of breaks are used for different shapes and materials. Figure 6-25 shows the conven-

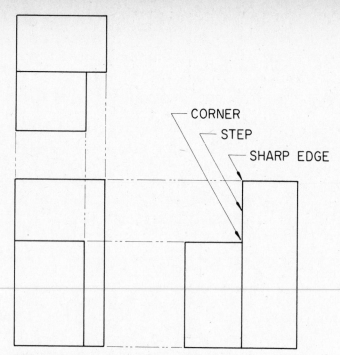

FIG. 6-23   Projection of edges and corners.

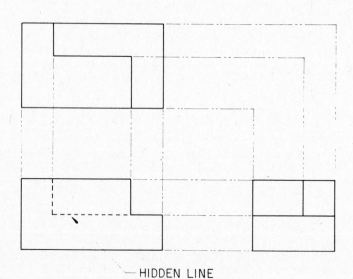

FIG. 6-24   Use of hidden lines in a drawing.

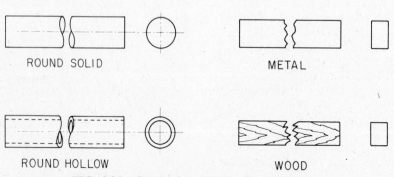

FIG. 6-25   Conventional breaks for drawings.

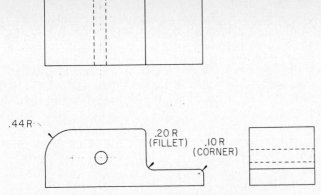

FIG. 6-26 Showing curved surfaces in a three-view drawing.

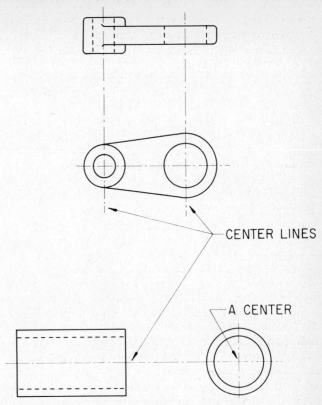

FIG. 6-27 Two-view drawings.

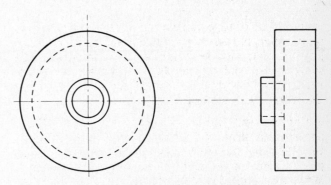

INVISIBLE CIRCLES

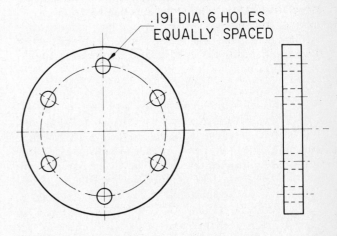

.191 DIA. 6 HOLES
EQUALLY SPACED

DRILLED HOLES

FIG. 6-28 Two-view drawings showing drilled holes.

tional breaks used in drawing a round, solid object; a round, hollow object; a metal object; and a wood object.

Curved Surfaces. Figure 6-26 illustrates the general rule that a curved surface or a circle appears in one view only in that form. In other views, the curve or the circle is shown as a straight line or by a pair of straight lines, according to circumstances. For example, there is a hole through the object in Fig. 6-26 which appears as a circle in the front view, as two parallel lines in the top view, and as two parallel lines in the right-side view. In the top and right-side views, the lines representing the hole are dashed lines because the hole cannot be seen from the top or the side of the object.

The fillet in Fig. 6-26 is shown as a curved line in the front view and as a straight line in each of the other views. Since it is visible, it is represented by a solid, straight line in the top and right-side views and by a solid curved line in the front view.

Selection of Views

Two-View Drawings. It is a general practice to present three views of an object, but it is also a common practice to illustrate such simple parts as bolts, collars, studs, and simple castings by means of only two views.

Figure 6-27 includes drawings of two entirely different objects, with two views of each object. The one at the top is a simple casting, illustrated by what we may call a front view and a top view, although the naming of the views in a case like this may vary according to the whims of the drafter. The second drawing shows a tubular object such as a bushing. In both drawings, a circle is shown in only one view as a circle in accordance with the general rule previously explained.

Figure 6-28 includes drawings of two different objects. The object at the top has invisible circles. It is customary to use two-view drawings such as these to illustrate objects having either drilled holes or invisible circles. The object at the bottom shows the drilled holes as small circles in the front view and as dotted lines in the side view.

Single-View Drawings. Some objects are so simple that their shape can be shown by only one view. Figure 6-29 shows two views of this type. The object at the top of the drawing is a cylindrical part with a groove that must be made according to the directions shown near the root of the arrow. The letter D indicates that the long cylindrical section has a uniform diameter throughout its length. The object at the bottom of the illustration is a piece of sheet metal of regular shape; hence only one view is needed. Since the kind of material to be used and its thickness are important, these are given in coded language in a note at the bottom of the drawing.

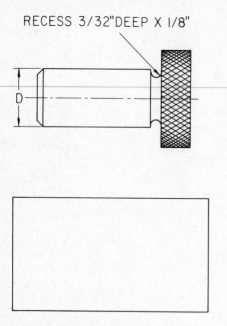

FIG. 6-29 Single view drawings.

Sectional Views. A **sectional view** is obtained by cutting away part of an object to show the shape and/or construction at the cutting plane. The surfaces that have been cut away are indicated with section lines. Figure 6-2 shows a sectional view of an electrical connector. When the whole view has been sectioned, it is called a **full section.** If the cutting plane extends only halfway across the object, leaving the other half of the object in exterior view, it is called a **half section.**

Detail Views. A **detail view** shows only a portion of the object but in greater detail than the principal view. The detail view may consist of a larger scale or a sectional view. The portion involved in the detailed view will be indicated on the principal view. This indication may take the form of a letter or cutting-plane line in the case of a sectioned detail view. The flap-control-assembly drawing in Fig. 6-22 shows a detailed sectional view of the shaft and pulley area.

Dimensions

Dimensions are required on any drawing used to fabricate or repair parts. Dimensions can be broken down into two principal classes according to their purpose.

When locating holes for drilling or positions for slots, the technician uses **location dimensions.** When cutting a piece of stock to the size and shape for a part, **size dimensions** are used.

Placement of Dimensions. Figure 6-30 shows how dimensions are located for a simple drawing. A **dimension line** has an arrowhead at each end, which shows

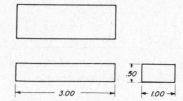

FIG. 6-30 Dimension lines for a simple object.

where the dimension ends. The **dimension** is the distance shown by the fraction or number at the break in the dimension line. The short lines at right angles to the arrowheads are **extension lines** made by the drafter before drawing the dimension lines. Dimensions common to two views are usually placed **between** the views. On drawings of simple parts, the length is given under the front view, while the width and thickness are given in the right-side view.

Figure 6-31 shows three views of an object that is solid but can be regarded for purposes of discussion as two blocks joined together. The front view shows the length of the small block as 0.875 in [2.22 cm] and the length of the large block as 1.0625 in [2.700 cm]. The front view also shows the height of the small block as 1.125 in [2.857 cm] and the height of the large block as 1.875 in [4.762 cm]. The right-side view gives the

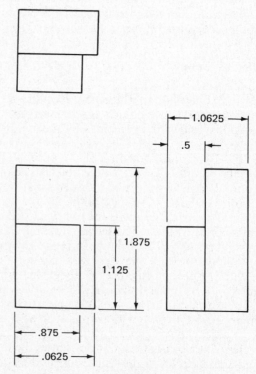

FIG. 6-31 Three-view drawing with dimensions.

width of the small block as 0.50 in [1.27 cm] and the width of the whole object as 1.0625 in [2.700 cm].

A slightly more complex object is shown in Fig. 6-32. The general rule that dimensions are given to visible and not to hidden lines is also illustrated in the drawing. The three views show that there is a cutaway section and a hole in the object. The hole is dimen-

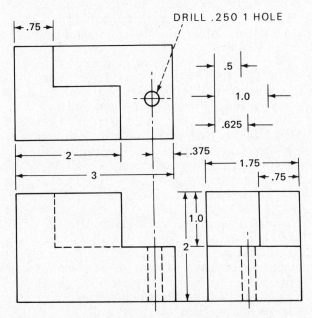

FIG. 6-32 Dimensioning an object with hidden lines.

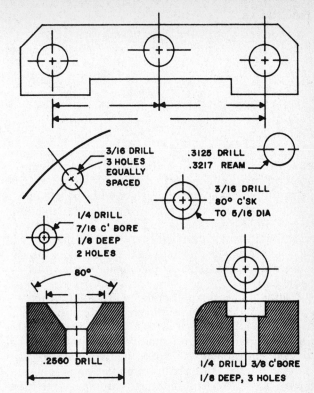

FIG. 6-33 Dimensioning drilled holes.

sioned in the top view, where it is indicated that the hole is 0.375 in [0.952 cm] from the right side. The right-side view shows that the top of the hole is 1 in [2.54 cm] below the upper surface of the object.

In the upper-right-hand corner of Fig. 6-32 are shown several approved methods for dimensioning parts or holes where the space on the drawing is limited. The one at the top is the method used for locating the center of the hole in the top view of the main drawing. The lower of the three methods shown is followed for giving the other dimensions on the drawing.

Dimensions are always given in inches unless some other unit is indicated, but the inch symbol (″) is omitted from aircraft drawings to save space and time. When dimension lines and extension lines cross each other, they are often broken at the intersection.

Holes to be drilled in a part are located by dimensions to the centers as shown in Fig. 6-33. The hole will be dimensioned by the size of the drill to be used, i.e., No. 7 drill, drill F, or $\frac{1}{4}$-in drill. For accuracy some hole dimensions are given in decimals. Notes are used to indicate the depth of hole and counterbores. Notes may also be used to indicate if the part has a number of equally sized or equally spaced holes.

Limits, Tolerance, and Allowance. The dimensions on a blueprint that represent the perfect size are sometimes called the **basic dimensions.** For example, the basic dimension for the length of an object might be 4 in. All basic dimensions have **limits.** From these limits, the person reading the blueprint can tell the ex-

treme permissible dimensions that can safely be allowed. For example, the limits for the 4-in dimension might be 4.005 and 3.995 in. The part could be made 0.005 in longer or 0.005 in shorter than the basic dimension and still pass inspection. In other words, the **limits** are the extremes of size allowable. The limits are usually written with one over the other. For example, a drawing may read DRILL $\frac{0293}{0290}$.

The **tolerance** is the difference between the extreme permissible dimensions. It is the range of error between the limits that will be accepted, or tolerated. For example, if the limits are ± 0.005, then the plus limit is the dimension plus 0.005, the minus limit is the dimension minus 0.005, and the tolerance is 0.010.

Allowance is generally defined as the difference between the nominal dimensions and the upper or lower limit. It represents the condition of the tightest permissible fit for the correct construction and operation of the mating parts. For example, when a permanent assembly is being put together, it may be desirable to have a driving or heavy force fit, which can be described as a **tight fit.** On the other hand, if clearance is required between moving parts, a **loose fit** is desired.

The American National Standards Institute (ANSI) has established eight classes of fits, ranging from the large allowance, or loose fit, to the considerable negative allowance, or heavy force and shrink fit. These classes of fits are used by the engineers to determine the limits or tolerances to be used, but a class is not ordinarily used as a designation on a drawing. Actual sizes will usually be "called out." The term *called out* means **indicated** on the drawing.

Title Blocks

The **title block** is the index to the drawing. It provides all necessary information that is not shown in or near the actual drawing. It is usually located in the lower-right-hand corner of the drawing, although it may occupy the whole bottom portion of a drawing made on a small sheet of paper.

Figure 6-34 illustrates a typical title block used in the aerospace industry. It is exceptionally complete and yet is flexible enough in its design to cover many unforeseen details. Since the one title block can be used on many different blueprints, it is standardized and printed on the drawing paper with the spaces left blank. When executing a drawing, the drafter fills in the required information as far as possible, and then additional blanks are filled in by various members of the engineering department as the drawing is submitted through channels to the higher authorities for approval.

Figure 6-35 is a small drawing prepared by the engineering department of an aerospace company. The title block is somewhat similar to the one shown in Fig. 6-34, but there are some variations. All companies do not prepare their drawings exactly alike, and even within the same organization it is impossible and often undesirable to standardize completely the manner of preparing drawings. Most manufacturers will have a **drafting-room manual** (DRM), or other standardized specifications, that sets forth their standards and procedures for producing drawings. The person reading the print must learn the practices used by that manufactuer. In order to explain the different parts of the title block, each feature will be discussed separately.

Drawing, or Print, Number. The **drawing, or print, number** is customarily printed in large numerals in the lower-right-hand corner of the title block and repeated elsewhere on the same drawing. This keeps the number available when the drawing is folded and even if one of the numbers should be torn off or otherwise obliterated. In some companies, the drawing number is repeated in the upper-left-hand corner or in some other location where it will not interfere with other information.

When a drawing consists of more than one sheet, each sheet is identified by the basic drawing number and the sheet number. The sheet number is entered in the number block.

In recent years a variety of number systems have been used by different companies. To understand any particular system, it is necessary to consult the **drafting-room manual** of the company concerned.

Part Number. Each part of an aerospace vehicle always has a number of its own. On a print, the part number and the print number may be the same; but if they are not, then the part number is a dash number shown in the title block.

Right-Hand and Left-Hand Parts. A profile or silhouette drawing of a person's right shoe is the same as the outline of the left shoe viewed from the side, even though the shoes have different shapes as viewed from the top. In like manner, many parts of large assemblies are right-hand and left-hand mirror images of each other and not two identical parts.

An aircraft has many such "mirror image" parts. To reduce the amount of paper work it is common practice to show the part for one side in the drawing and indicate that the other side is the opposite. A common practice is to show the left-hand part and add −1 to the part number. The right-hand part would carry the same number with −2 as a dash number. This is not

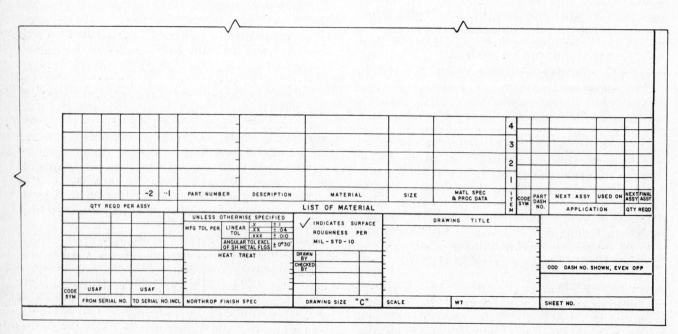

FIG. 6-34   A title block.

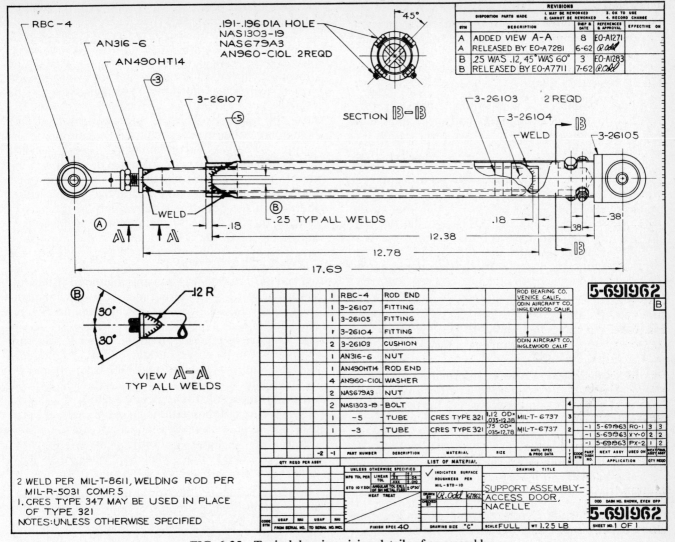

FIG. 6-35 Typical drawing giving details of an assembly.

a universal system as some manufacturers will apply separate part numbers to opposite parts. The system used, if any, can usually be determined by examining the drawing.

Scale. The word *scale* has many meanings. One of them is the proportion in dimensions between a drawing, map, or plan and the object that is represented on paper. For example, a map may be drawn on a scale of 1 in to 1 mi, which means that 1 in on the map represents 1 mi on the earth. In a like manner, a drawing from which a blueprint is to be made may be prepared to any desired scale, and the scale is shown in the title block on the drawing. Most aerospace drawings are made "full scale," "$\frac{1}{2}$ size," "$\frac{1}{4}$ size," or "$\frac{1}{10}$ size." Occasionally, other scales may be used.

Name of Part, Unit, or Assembly. The name of the part, unit, or assembly is given first followed by descriptive terms, just as the family name is given in a telephone directory followed by the given or Christian name, such as "Jones, John J." In like manner, in the

aerospace industry, the name of a carburetor air intake flange would read: "Flange, Carburetor Air Intake." Likewise, the assembly shown in Fig. 6-35 is "Support Assembly, Access Door, Nacelle." In other words, the noun is given first and is followed by the descriptive terms.

The location may be a part of the name, such as "Elevator and Tab Assembly-Left," showing that it belongs on the left side of the airplane. Another example is "Installation-C.N. sta. 77 flap control cable pul. brkt. assemb." When analyzed, it is apparent that this is an installation, it is located at crew nacelle station 77, and it is the flap-control pulley bracket assembly.

Revisions. On the drawing of Fig. 6-35 there is a space in the upper-right-hand corner labeled **"Revisions,"** and it will be noted here whether any revisions have been made in the drawing, the disposition and date of the revision, the name or initial of the person making it, an indication of approval, and the serial number on which it is to be effective. The disposition of previously made parts may also be indicated.

139

Dimensions and Limits. On the drawing of Fig. 6-35 there are spaces for the dimension of the standard stock from which the part is made. In this case the standard stock is tubing, MIL-T-6737. Limits and tolerances are shown in a supplementary block adjacent to the basic title block.

Station Numbers. A **station numbering system** can be used to help the print reader find such things as fuselage frames, wing frames, and stabilizer frames. For example, the nose of an airplane or some other point that can be easily identified is designated as the zero station, and other stations are located at measured distances in inches behind the zero station. Thus, when a print reads "Fuselage Frame-Sta. 182," that means that the frame is 182 inches (in) back of station zero. In a similar manner, the fore-and-aft center line of an airplane may be a zero station for objects on its right and left. Thus, the wing and stabilizer frames can be located as being a certain number of inches to the right or left of the airplane center line. On some drawings the firewall is a zero station, and on other drawings the leading edge of the wing can be used as a zero station for certain purposes. Always locate the zero station before looking for other stations.

Model and Next Assembly Information. When an object shown on a print is to be a part of an assembly, the next assembly number is given in the title block to indicate the drawing number of another blueprint that gives the necessary information for completing the assembly. The number of parts required, such as one for the right hand and one for the left hand, is also given. In the vicinity of these facts is found the model designation.

Material Listing. Every drawing carries information regarding the material, and the specifications are also given. For example, a drawing may have "C.M. Sh't" in the material space, that is, chromemolybdenum steel sheet. In the drawing of Fig. 6-35, the material list is shown in the upper part of the title block. The material list is sometimes referred to as the bill of materials. The specification space may show that this is SAE4130 or some other particular type of chrome-molybdenum steel sheet. In a similar manner, a drawing may show that 0.040 sheet aluminum (0.040 in thick) is required and then specify that it must be 2024-T3, which is a grade of heat-treated structural aluminum alloy.

The material specification is coded when it is for the Army, Navy, or Air Force. For example, it might read "ANQQ---," or it might have some other combination of letters and numerals and have a definite meaning to those handling military contracts.

The finish, or protective coating, may be indicated in clear language, or it may be coded, especially for a military contract.

Weight. In a space in the title block for a drawing, the weight of the object shown in the blueprint is given. This quantity may be the calculated weight, the actual weight, or both weights. This information is particularly useful for the weight and balance engineers and others who are concerned with the balance of the device involved.

Notes. Information that cannot be given completely and yet briefly in the title block is placed on the drawing in the form of **notes,** but such information does not duplicate information given elsewhere in the same drawing. Notes may be used to tell the size of a hole, the number of drill to be used in making the hole, the number of holes required, and similar information but only when it cannot be conveyed in the conventional manner without notes or when it is desirable to avoid crowding the drawing. If the notes apply to specific places on the part, they are placed on the face of the drawing. If they apply to the part in general, they go in the "General Notes" at the bottom and to the left of the title block.

Symbols and Abbreviations

Purpose. A **symbol** is a visible sign used instead of a word or words to represent ideas, operations, quantities, qualities, relations, or positions. It may be an emblem, such as a picture of a lion to represent courage or an owl to represent wisdom. Likewise the cross represents Christianity, the Star of David represents Judaism, and the crescent stands for Islam. It may be an abbreviation, a single letter, or a character. In other words, symbols constitute picture writing.

Symbols and abbreviations are used extensively in place of long explanatory notes on drawings and prints. A few of these symbols and abbreviations are common to every trade, whereas special trades have special symbols and abbreviations of their own.

Material Symbols. The ANSI has standardized certain symbols used to represent materials in section views. The military services have adopted some of these symbols and added others of their own, and these are all included in the design handbooks used by aerospace manufacturers. Figure 6-36 shows an important group of such symbols.

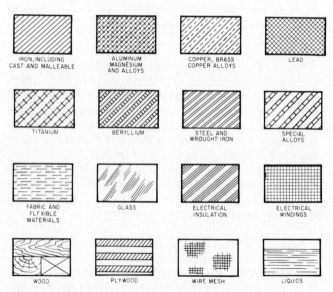

FIG. 6-36 Symbols used for materials. (MIL-STD-1A).

It must be clearly understood that, when used, these symbols are intended only for general information and not to indicate specific types of materials. For example, in the upper-right-hand corner of Fig. 6-36, there is a symbol for iron, including cast iron and malleable iron, but it does not tell the specific type of iron to be used. Such information appears elsewhere on each print.

To simplify drawing-room changes, the symbol for cast iron and malleable iron is often used by many companies to refer to all metals. The print reader must determine what metal is intended in each case.

These material symbols are not generally used on section views unless it is desired to call special attention to section parts; hence their appearance is an invitation to observe the drawing closely.

Process Code. Many manufacturers in the aerospace industry use symbols in what they call their **process code.** Examples are A, anodize; B, chromodize; C, cadmium plate; D, dichromate (Dow No. 7); H, degrease in vapor degreaser; and S, sand blast. These code letters simplify the instructions required on drawings and consume much less space than would otherwise be required.

● **REVIEW QUESTIONS**

1. Why is a copy of a drawing often called a blueprint?
2. What is the difference between a detail drawing and an assembly drawing?
3. What is the purpose of an installation drawing?
4. What is the purpose of a block diagram?
5. In what cases are shop sketches used?
6. How may a wiring diagram be used by a technician?
7. What is the function of a schematic diagram?
8. Why are logic symbols used in some drawings?
9. What is meant by a perspective drawing?
10. How does an isometric view vary from a perspective view?
11. Describe an *orthographic projection?*
12. Explain the importance of width and type of lines used for aircraft drawings?
13. Under what conditions are two-view and single-view drawings used?
14. When is a detail view used?
15. Compare location dimensions and size dimensions.
16. Where is a dimension placed when it is common to two views?
17. What unit of measurement is generally used for dimensions?
18. What is a dimension limit?
19. Explain what is meant by tolerance.
20. What information is contained in the title block of a drawing?
21. In what document are drawing standards and procedures for a particular manufacturer published?
22. Where is the drawing number placed?
23. What is the purpose of a dash number on a drawing?
24. With respect to a drawing, what is meant by scale?
25. What is the procedure for making a revision on a drawing?
26. Where is the material listing placed on a drawing?
27. Why are notes used on a drawing?
28. What is a process code?

7 WEIGHT AND BALANCE

● INTRODUCTION

Aviation has been one of the most dynamic industries since its beginning. New aircraft are continually being developed with improvements over previous models. Improvements in design have, in many cases, tended to increase the importance for the proper loading and balancing of today's airplanes. Weight-and-balance calculations are performed according to exact rules and specifications and must be prepared when aircraft are manufactured and whenever they are altered, whether the airplane is large or small. The constantly changing conditions of modern aircraft operation present more complex combinations of cargo, crew, fuel, passengers, and baggage. The necessity for obtaining the utmost efficiency from any flight has emphasized the need for a precise system of controlling the weight and balance of all airplanes.

● FUNDAMENTAL PRINCIPLES

In a previous chapter, the laws of physics were discussed. Included were discussions of specific gravity and balance, together with explanations of levers.

These principles form the basis for computing weight-and-balance data for an airplane and will be reviewed briefly.

Force of Gravity

Every body of matter in the universe attracts every other body with a certain force that is called **gravitation.** The term *gravity* is used to refer to the force that tends to draw all bodies toward the center of the earth. The weight of a body is the result of all gravitational forces acting on the body.

Center of Gravity

Every particle of an object is acted on by the force of gravity. However, in every object there is one point at which a single force, equal in magnitude to the weight of the object and directed upward, can keep the body at rest, that is, can keep it in balance and prevent it from falling. This point is known as the **center of gravity** (CG).

The CG might be defined as the point at which all the weight of a body can be considered concentrated. Thus, the CG of a perfectly round ball would be the exact center of the ball, provided that the ball were made of the same material throughout and that there were no air or gas pockets inside (see Fig. 7-1). The CG of a uniform ring would be at the center of the ring but would not be at any point in the ring itself (see Fig.

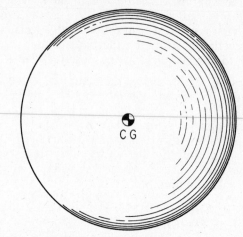

FIG. 7-1   Center of gravity of a ball.

7-2). The CG of a cube of solid material would be equidistant from the eight corners as shown in Fig. 7-3. In airplanes or helicopters, ease of control and maneuverability require that the location of the CG be within specified limits.

Location of the CG

Since the CG of a body is that point at which its weight can be considered to be concentrated, the CG of a

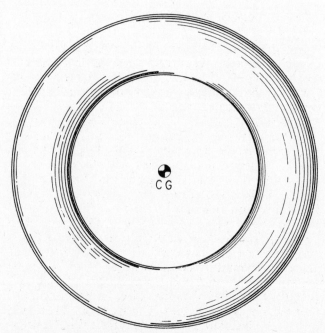

FIG. 7-2   Center of gravity of a ring.

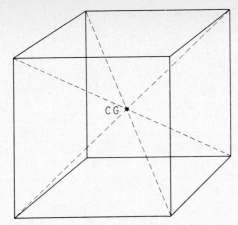

FIG. 7-3   Center of gravity of a cube.

freely suspended body will always be vertically beneath the point of support when the body is supported at a single point. To locate the CG, therefore, it is necessary only to determine the point of intersection of vertical lines drawn downward from two separate points of support employed one at a time. This is demonstrated in Fig. 7-4, which shows a flat, square sheet of material lettered *A*, *B*, *C*, and *D* at its four corners, suspended first from point *B* and then from point *C*. The lines drawn vertically downward from the point of suspension in each case intersect at the CG.

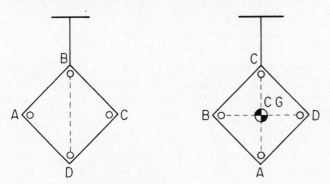

FIG. 7-4   Location of the CG.

The CG of an irregular body can be determined in the same way. If an irregular object, such as the one shown in Fig. 7-5, is suspended from a point *P* in such a manner that it can turn freely about the point of suspension, it will come to rest with its CG directly below the point of suspension *P*. If a plumb line is dropped from the same point of suspension, the CG of the object will coincide with some point along the plumb line; a line drawn along the plumb lines passes through this point. If the object is suspended from another point, which we shall call *A*, and another line is drawn in the direction indicated by the plumb line, the intersection of the two lines will be at the CG. In order to verify the results, the operation can be repeated, this time with the object suspended from another point, called *B*. No matter how many times the process is repeated, the lines should pass through the CG; hence it is evident that the CG of the object lies at the point of intersection of these lines of suspension. Therefore, any object behaves as if all its weight were concentrated at its CG.

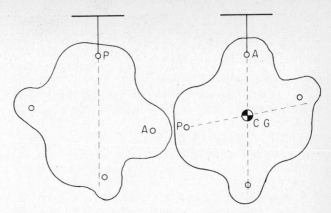

FIG. 7-5   Locating the CG in an irregular body.

Practical CG of an Airplane

If an airplane is not too heavy and can be equipped with suspension loops as in *A* and *B* of Fig. 7-6, the CG of the airplane can be determined easily by the trial method. This actually is a duplication of the process explained in the previous paragraph and illustrated in Fig. 7.5. If we suspend the airplane at point 1, draw a vertical line downward from the point of suspension, and then repeat the process with the airplane suspended from point 2, we shall find an intersection of the extended lines that will be at the CG as shown in *C* of Fig. 7-6. The foregoing is not the method normally used for locating the CG for an airplane, but it gives a demonstration of what can be

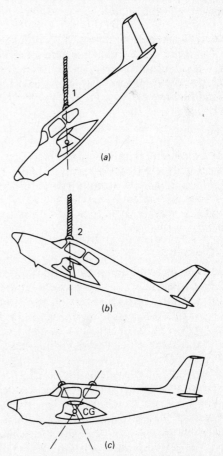

FIG. 7-6   The trial method of locating the CG of an airplane.

done and illustrates one of the basic principles of weight and balance.

The General Law of the Lever

In the chapter on the laws of physics the law of levers was explained; however, it will be repeated briefly here to show how it relates to the weight and balance of an airplane.

Wrenches, crowbars, and scissors are levers used to gain mechanical advantage, that is, to gain force at the expense of distance or to gain distance at the expense of force. A lever, in general, is essentially a rigid rod free to turn about a point called the **fulcrum.** There are three types of levers, but in the study of weight and balance we are principally interested in the type known as a **first-class lever.** This type has the fulcrum between the applied effort and the resistance as shown in Fig. 7-7.

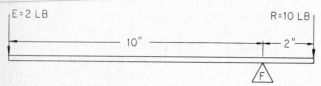

FIG. 7-7   First-class lever.

In Fig. 7-7 the fulcrum is marked F, the applied effort is E, and the resistance is R. If the resistance R equals 10 lb [4.535 kg] and it is 2 in [5.08 cm] from the fulcrum F, and if the effort E is applied 10 in [25.4 cm] from the fulcrum, it will be found that an effort of 2 lb [0.907 kg] will balance the resistance R. In other words, when a lever is balanced, the product of the effort and its lever arm (distance from the fulcrum) equals the product of the resistance and its lever arm. The product of a force and its lever arm is called the **moment** of the force.

The general law of the lever is stated as follows: **If a lever is in balance, the sum of the moments tending to turn the lever in one direction about an axis equals the sum of the moments tending to turn it in the opposite direction.** Therefore, if the lever is in balance, and if several different efforts are applied to the lever, the sum of the moments of resistance will equal the sum of the moments of effort.

Moment of a Force and Equilibrium

The tendency of a force to produce rotation around a given axis is called the **moment of the force** with respect to that axis.

The amount and direction of the moment of a force depend upon the direction of the force and its distance from the axis. The perpendicular distance from the axis to the line of the force is called the **moment arm,** and the moment is measured by the product of the force and the moment arm. Thus, a force of 10 lb acting at a distance of 2 ft [0.6096 m] from the axis exerts a turning moment of 20 ft·lb [2.765 kg·m].

In order to avoid confusion between moments tending to produce rotation in opposite directions, those tending to produce a clockwise rotation are called positive and those tending to produce counterclockwise

rotation are called negative. If the sum of the positive or clockwise moments equals the sum of the negative or counterclockwise moments, there will be no rotation. This is usually expressed in the form $\Sigma M = 0$. The symbol Σ is the Greek letter sigma, and ΣM means the sum of all the moments M, both positive and negative.

In Fig. 7-8 a moment diagram is shown with moments about the point A. M_1 acts in a counterclockwise direction, with a force of 1 lb [0.4536 kg] at a distance of 3 ft [0.9144 m]; hence the value of M_1 is -3 ft·lb [-0.4148 kg·m]. M_2 acts in a counterclockwise direction, with a force of 2 lb at a distance of 2 ft, thus producing a moment of -4 ft·lb [-0.5528 kg·m]. M_3, acting in a counterclockwise direction with a force of 1 lb at a distance of 1 ft [0.3048 m], produces a moment of -1 ft·lb [-0.1383 kg·m]. M_4 acts in a clockwise direction, with a force of 4 lb [1.814 kg] at 2 ft, which makes a moment of $+8$ ft·lb [$+1.105$ kg·m]. $-3 -4 -1 + 8 = 0$. The sum of the negative moments is equal to the positive moment; hence there is a condition of **equilibrium** and there is no rotation about point A.

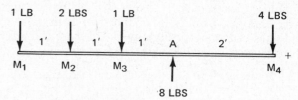

FIG. 7-8   Moment diagram.

There is a total force of 8 lb [3.629 kg] acting downward, and unless the axis is supported by an upward force of 8 lb, there will be downward movement but no rotation.

The purpose of this explanation is to show why an airplane must be designed, loaded, and operated with a constant regard for balance in order to permit it to fly safely with the minimum amount of energy required for propulsion and the minimum amount of energy expended by the pilot in the control of the airplane. When the necessary conditions of aircraft balance are not met, the pilot has difficulty in controlling the airplane even under normal flight operations. Under adverse conditions, improper balance can cause the airplane to crash.

● WEIGHT-AND-BALANCE TERMINOLOGY

Before we can proceed with explanations of the methods for computing weight-and-balance problems, we must have a good understanding of the words and terms used.

Arm (moment arm). The arm is the horizontal distance in inches from the datum to the center of gravity of the item. The algebraic sign is plus (+) if measured aft of the datum and minus (−) if measured forward of the datum (see Fig. 7-9).

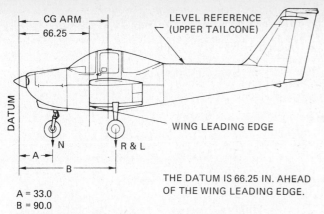

CG ARM
66.25

LEVEL REFERENCE
(UPPER TAILCONE)

DATUM

WING LEADING EDGE

N   R & L

A

B

A = 33.0
B = 90.0

THE DATUM IS 66.25 IN. AHEAD
OF THE WING LEADING EDGE.

FIG. 7-9   Leveling diagram.   *(Piper Aircraft Corporation)*

Center of gravity (CG). The CG is a point about which the nose-heavy and tail-heavy moments are exactly equal in magnitude. If the aircraft were suspended from this point it would be perfectly balanced.

Center of gravity range. The operating CG range is the distance between the forward and rearward limits within which the airplane must be operated. These limits are indicated on pertinent FAA Aircraft Type Certificate Data Sheets (see Fig. 7-10), specifications, or in aircraft weight-and-balance records, and meet the requirements of the Federal Aviation Regulations.

Datum (reference datum). The datum is an imaginary vertical plane or line from which all horizontal measurements of arm are taken (see Fig. 7-9). The datum is established by the manufacturer. Once the datum has been selected, all moment arms must be taken with reference to that point. The location of the datum may be found in the aircraft's Type Certificate Data Sheet (see Fig. 7-10).

Empty weight (EW). The empty weight of an aircraft includes the weight of the airframe, power plant, and required equipment that has a fixed location and is normally carried in the airplane. For aircraft certificated under FAR Part 23, the empty weight also includes unusable fuel and full operating fluids necessary for normal operation of aircraft systems such as oil and hydraulic fluid. For older aircraft not certificated under FAR Part 23, in place of full oil, only the undrainable oil is included in the empty weight. The current aircraft empty weight must be kept as a part of the permanent weight-and-balance records.

Empty-weight center of gravity (EWCG). The empty-weight CG is the CG of the aircraft in its empty condition and is an essential part of the weight-and-balance record that must be kept with the permanent aircraft records.

Empty-weight CG range. The EWCG range is established so that when the EWCG falls within this range, the aircraft operating CG limits will not be exceeded under standard loading conditions. The EWCG range shown for many light airplanes is listed in the Aircraft Specifications or Type Certificate Data Sheet and may eliminate further calculations by technicians making equipment changes (see Fig. 7-10).

Fleet empty weight. Fleet empty weight is an average basic empty weight which may be used for a fleet or group of aircraft of the same model and configuration. The weight of any fleet member cannot vary more than the tolerance established by the applicable governmental regulations.

Lemac. Lemac is the leading edge of the mean aerodynamic chord.

Leveling means. Leveling means are the reference points used by the aircraft technician to insure that the aircraft is level for weight-and-balance purposes (see Fig. 7-9). Leveling is usually accomplished along both the longitudinal and lateral axis. Leveling means are given in the Aircraft Specifications or the Type Certificate Data Sheet (see Fig. 7-10).

Loading envelope. Those combinations of airplane weight and center of gravity which define the limits beyond which loading is not approved.

Main-wheel center line (MWCL). A vertical line passing through the center of the axle of the main landing gear wheel.

Maximum gross weight. The maximum gross weight is the maximum authorized weight of the aircraft and its contents as listed in the Aircraft Specifications or the Type Certificate Data Sheet (Fig. 7-10).

Maximum landing weight. The maximum weight at which the aircraft may normally be landed (see Fig. 7-10).

Maximum ramp weight. Maximum weight approved for ground maneuver. (It includes weight of start, taxi, and run-up fuel.)

Maximum takeoff weight. The maximum allowable weight at the start of the takeoff run (see Fig. 7-10).

Mean aerodynamic chord (MAC). The MAC is the length of the mean chord of the wing as established through aerodynamic considerations. For weight-and-balance purposes it is used to locate the CG range of the aircraft. The location and dimension of the MAC, where used, will be found in the Aircraft Specification, the Type Certificate Data Sheet, Flight Manual, or Aircraft Weight-and-Balance Record (see Fig. 7-10).

Minimum fuel. Minimum fuel for weight-and-balance computations is no less than the quantity of fuel required for $\frac{1}{2}$ h of operation at rated maximum continuous power. It is calculated in the maximum except takeoff (METO) horsepower and is the figure used when the fuel load must be reduced to obtain the most critical loading on the CG limit being calculated. The formula usually used in calculating minimum fuel is $\frac{1}{2}$ METO hp = minimum fuel in pounds (e.g., $\frac{1}{2} \times 360$ hp = 180 lb of fuel).

Moment. The moment is the product of a weight and its arm.

Standard weights. For general weight-and-balance purposes the following weights are considered standard:

Gasoline	6 lb/gal [2.75 kg/gal]
Turbine fuel	6.7 lb/gal [3.0 kg/gal]
Lubricating oil	7.5 lb/gal [3.4 kg/gal]
Water	8.3 lb/gal [3.75 kg/gal]
Crew and passengers	170 pounds [77 kg] per person

DEPARTMENT OF TRANSPORTATION
FEDERAL AVIATION ADMINISTRATION

TYPE CERTIFICATE DATA SHEET

Engine:	Lycoming 10360 B2F with "Christen" inverted oil system, or Lycoming AEI0360-B2F fuel injected.		
Fuel:	91/96 minimum aviation grade gasoline.		
Engine Limits:	For all operations 2700 rpm (180 hp).		
Propeller:	Hoffman HO 29-180-170		
Propeller Limits:	(Utility and Acrobatic Categories) Statis rpm at maximum permissible throttle setting—2250 ± 50 Diameter 70.9 in. No cutoff permitted.		

Airspeed Limits		**Utility Category**	**Acrobatic Category**
	Never exceed	211 mph (183 kts)	211 mph (183 kts)
	Max. structural cruising	186 mps (162 kts)	186 mph (162 kts)
	Maneuvering	124 mph (108 kts)	146 mph (127 kts)
	Flaps extended	99 mph (86 kts)	99 mph (86 kts)
	See NOTE 3 for acrobatic maneuvers.		

Flight Maneuvering Load Factor (g's)	Flaps up	+4.4 −1.8	+6.0 −3.0
	Flaps down	+2.0 −1.8	+2.0 −2.0
CG Range:	Forward limit Aft limit	+10.6 in (18% MAC) +17.7 in (30% MAC)	10.6 in (18% MAC) 15.3 in (26% MAC)
Datum:	Wing leading edge at 51 in from airplane center line. (Length of wing chord at datum 59 in).		
Leveling Means:	Longitudinal: Left canopy rail Lateral: Top of bulkhead #2.		
Empty-Weight CG Range:	+8.6 in − +10.2 in		
Maximum Weight:	Takeoff Landing	1829 lb 1763 lb	1675 lb 1675 lb
No. of Seats:		2 at +22.7 in	2 at +22.7 in
Maximum Baggage:		110 lb at +55.1	None
Fuel Capacity:	Front tank (total) (usable)	19.8 (at −6.7 in) 19.0	19.8 (at −6.7 in) 19.0
	Rear tank (total) (usable)	20.8 (at +55.1 in) 20.6	0 0
	Rear tank must be empty for operations in acrobatic category. Minimum fuel quantity for acrobatics: 2.6 gal.		
Oil Capacity:	Maximum capacity: Minimum: Maximum oil quantity for acrobatics: 1.5 gal	2 gal 0.5 gal	

FIG. 7-10   Sample Type Certificate Data Sheet.

Station. A location along the airplane fuselage given in terms of distance in inches from the reference datum. The datum is, therefore, identified as station zero (see Fig. 7-11). The station and arm are usually identical. An item located at station +50 would have an arm of 50 in.

Tare. Tare is the weight of the equipment necessary for weighing the airplane (such as chocks, blocks, slings, jacks, etc.) which is included in the scale reading but is not a part of the actual weight of the airplane. Tare must be subtracted from the scale readings in order to obtain the actual weight of the airplane.

Undrainable oil. That portion of the oil in an aircraft lubricating system that will not drain out when the drain plug is removed or when the drain valve is opened with the aircraft in a level attitude.

Unusable fuel. Fuel remaining after a runout test has been completed in accordance with manufacturer's instruction. The amount and location of the unusable fuel may be found in the Type Certificate Data Sheet or the Aircraft Specifications (see Fig. 7-10).

Usable fuel. Fuel available for flight planning.

Useful load. Useful load is the weight of the pilot, copilot, passengers, baggage, and usable fuel. It is the empty weight subtracted from the maximum weight.

Weighing point. The weighing points of an airplane are those points by which the airplane is supported at the time it is weighed. Usually the main landing gear and the nose or tail wheel are the weighing points (see Fig. 7-11). Sometimes, however, an airplane may have jacking points from which the weight is taken. In any

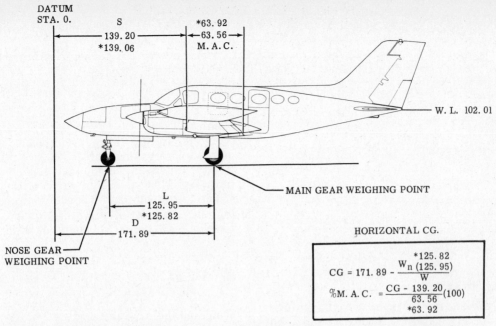

FIG. 7-11   Weighing and measuring. (Sample diagram; do not use for flight planning purposes.)   *(Cessna Aircraft Co.)*

event, it is essential to define the weighing points clearly in the weight-and-balance record.

Weight check. A weight check consists of checking the sum of the weights of all items of useful load against the allowable useful load (maximum weight less empty weight) of the aircraft.

● BALANCE FOR AIRCRAFT

Principles of Balance

The fundamental theory of weight and balance is very simple. It is merely the principle of the first-class lever and is easily demonstrated by means of an old-fashioned steelyard scale shown in Fig. 7-12. The scale shown is in a state of equilibrium when it rests on the fulcrum in a horizontal position. The weight is directly dependent on its distance from the fulcrum, and for equilibrium the weight must be distributed so that the turning effect is the same on one side of the fulcrum as it is on the other. A heavy weight near the fulcrum has the same effect as a lighter weight farther from the fulcrum.

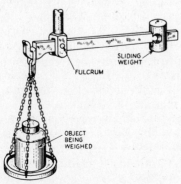

FIG. 7-12   Steelyard scale.

CG Range and Limits

The steelyard scale is in balance only when the horizontal CG is at the fulcrum. However, an aircraft can be balanced in flight anywhere within certain specified forward and aft limits if the pilot operates the trim tabs or elevators to exert an aerodynamic force sufficient to overcome any static unbalance. CG locations outside the specified limits will cause unsatisfactory or even dangerous flight characteristics.

The allowable variation within the CG range is carefully determined by the engineers who design an airplane. The CG range usually extends forward and rearward from a point about one-fourth the chord of the wing, back from the leading edge, provided that the wing has no sweepback. The exact location is always shown in the **Aircraft Specification** or the **Type Certificate Data Sheet.** Heavy loads near the wing location are balanced by much lighter loads at or near the nose or tail of the airplane. In Fig. 7-13, a load of 5 lb [2.268 kg] at *A* will be balanced by a load of 1 lb [0.4536 kg] at *B* because the moments of the two loads are equal.

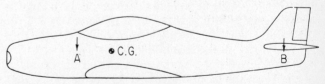

FIG. 7-13   Balancing of the load.

Since the CG limits constitute the range of movement that the aircraft CG can have without making it unstable or unsafe to fly, the CG of the loaded aircraft must be within these limits at takeoff, in the air, and on landing. In some cases, the takeoff limits and landing limits are not exactly the same, and the differences are given in the specifications for the aircraft.

Figure 7-14 shows typical limits for the CG location in an airplane. As previously stated, these limits establish the **CG range.** The CG of the airplane must fall within this range if the airplane is to fly safely; that is, it must be to the rear of the forward limit and forward of the aft limit.

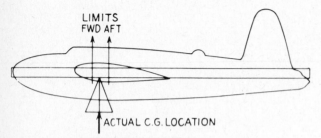

FIG. 7-14   Center-of-gravity limits.

CG and Balance in an Airplane

The CG of an airplane may be defined, for the purpose of balance computations, as an imaginary point about which the nose-heavy (−) moments and tail-heavy (+) moments are exactly equal in magnitude. Thus, the aircraft, if suspended therefrom, would have no tendency to rotate in either direction (nose up or nose down). This condition is illustrated in Fig. 7-15. As stated previously, the weight of the aircraft can be assumed to be concentrated at its CG.

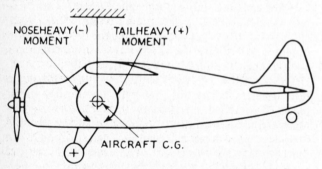

FIG. 7-15   Airplane suspended from the CG location.

The CG with the useful load installed is allowed to range fore and aft within certain limits that are determined during the flight tests for type certification. These limits are the most forward- and rearward-loaded CG positions at which the aircraft will meet the performance and flight characteristics required by the FAA. These limits may be given in percentage of the mean aerodynamic chord (MAC) or in inches forward or to the rear of the **datum line.**

The relative positions of the CG and the center of lift of the wing have critical effects on the flight characteristics of the aircraft. Consequently, relating the CG location of the **chord** of the wing is convenient from a design and operations standpoint. Normally, an aircraft will have acceptable flight characteristics if the CG is located somewhere near the 25 percent average chord point. This means the CG is located one-fourth of the total distance back from the leading edge of the average wing section (see Fig. 7-16). Such a lo-

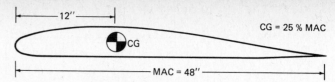

FIG. 7-16   Percent of mean aerodynamic chord.

cation will place the CG forward of the aerodynamic center for most airfoils. Technicians are seldom required to determine the MAC since this is done by the aerodynamicist in the manufacturer's engineering section. The MAC is usually given when it will be required for weight-and-balance computations; therefore the person working on the airplane is expected to have only a general understanding of its meaning.

● DETERMINATION OF EWCG LOCATION

Weighing the Aircraft

Weighing aircraft with accurately calibrated scales is the only sure method of obtaining an accurate empty weight and CG location. The use of weight-and-balance records in accounting for and correcting the aircraft weight-and-balance location is reliable over limited periods of time. Over extended intervals, however, the accumulation of dirt, miscellaneous hardware, minor repairs, and other factors will render the basic weight and CG data inaccurate. For this reason, periodic aircraft weighings are desirable; however, they are not required of aircraft operated under FAR Part 91 regulations. This is not the case for Air Taxi and Air Carrier aircraft which are required by the FARs to be periodically weighed. Aircraft may also be weighed when major modifications or repairs are made, when the pilot reports unsatisfactory flight characteristics such as nose or tail heaviness, and when recorded weight-and-balance data are suspected to be in error.

Weighing Equipment

The type of equipment which is used to weigh aircraft varies with the aircraft size. Light aircraft may be weighed on commercial-type platform scales (see Fig. 7-17).

To weigh very large aircraft, **electronic load cells** are used. These cells are strain gages whose resistance changes in accordance with the pressure applied to them. The load cell is placed between the jack and the jack point on the aircraft, with particular attention paid to locating the cell so that no side loads will be applied (see Fig. 7-18). When weight readings are taken, the entire airplane weight must be supported on the load cells.

The output of the load cells is fed to an electronic instrument that amplifies and interprets the load-cell signals to provide weight readings. The instrument is adjusted to provide a zero reading from each load cell before the aircraft is weighed. After weighing, the cells are checked again and the reading is adjusted to compensate for any change noted.

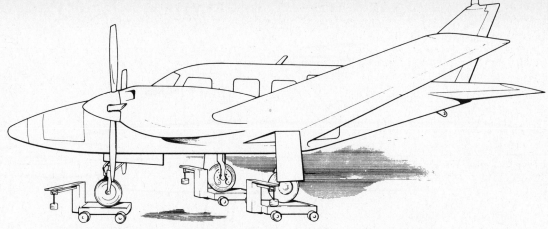

FIG. 7-17   Weighing the airplane.   *(Piper Aircraft Corporation)*

FIG. 7-18   Electronic load cells.

Electronic weighing kits are used by some airline companies for weighing large aircraft. The kit includes the load cells, electronic instruments, plumb bobs, a spirit level, a hydrometer for testing the specific gravity of the fuel, a steel tape, and a straightedge. In using the weighing kit, the technician must follow the manufacturer's instructions and also the instructions provided by the company operating the aircraft. All major airlines provide standard procedures for the handling and servicing of aircraft, and these procedures must be followed carefully to obtain accurate results and to avoid damage to the aircraft and equipment.

An airplane must be level to obtain accurate weighing information. In the case of the DC-10 airplane, an **inclinometer** consisting of a plumb bob and **grid plate** are provided in the right wheel well, and brackets for spirit levels are located in the nose-gear wheel well. In Fig. 7-19 locations of the leveling means for the DC-10 are shown.

The inclinometer indicates degrees of roll or pitch. The plumb bob is suspended by a cord and is secured in a stowage clip when not in use. During leveling operations, the plumb bob is released from the clip and is suspended by its cord over the grid plate. The level attitude of the airplane is established by the location of the plumb bob in relation to the grid-plate markings.

When a higher degree of leveling accuracy is re-

quired, spirit levels are used. The two sets of brackets provided in the nose-gear wheel well are used to support the levels in both longitudinal and lateral axes.

The DC-10 airplane can be weighed by jacking the airplane at the wing and fuselage jacking points or by jacking at the landing-gear jacking points. An electronic load cell must be placed between each jack and its associated jack point. When the airplane is jacked up from the wing and fuselage jack points, it is necessary to elevate the aircraft 26 in [66.04 cm] to give the landing gear a minimum clearance of 2 in [5.08 cm]. Jacking height may be reduced by deflating the shock struts and placing keepers on the torque arms to prevent extension of the shock struts. In jacking from the landing gear, it is necessary only to clear the wheels and tires from surface on which the airplane is standing.

Portable electronic weighing systems make it possible to find the weight and balance of large and small aircraft without jacking (see Fig. 7-20). The system consists of electronic platform scales as necessary to weigh each wheel or pair of wheels on the aircraft, signal amplifiers, a digital CG indicator, a digital gross-weight indicator, and a power panel. Each scale consists of a platform supported by strain-gage transducers usually no more than 3 in [7.62 cm] in height. Ramps are supplied with the platforms so that the aircraft can easily be towed to position on the scales. The signals from the scales provide the information that is presented on the digital CG indicator and gross-weight indicator.

Whichever type of system is selected, only weighing equipment that is maintained and calibrated to acceptable standards should be used.

Weighing Procedures

Weighing procedures may vary with the aircraft and the type of weighing equipment employed. The weighing procedures and formulas contained in the manufacturer's manuals should be followed when available. The following general instructions illustrate a common method and some of the typical precautions.

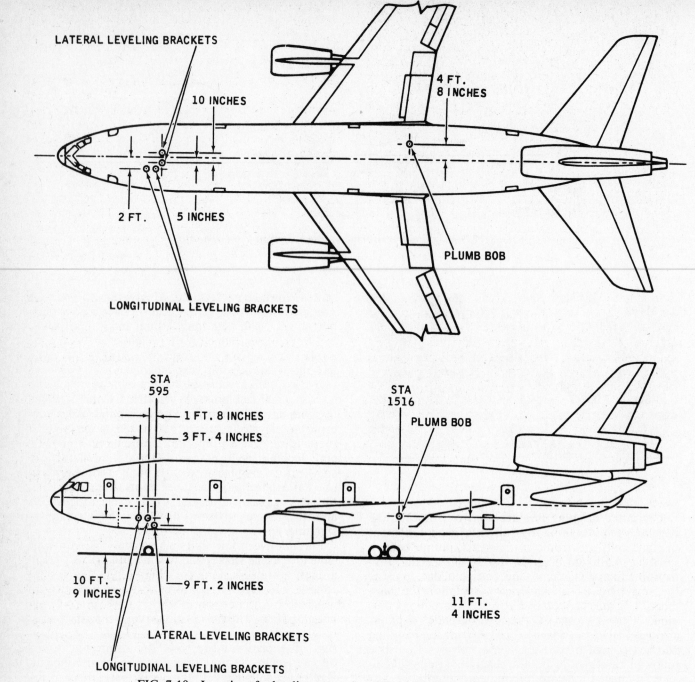

FIG. 7-19 Locations for leveling means in a DC-10 airplane. *(McDonnell Douglas Corp.)*

In order to obtain an accurate reading of the aircraft weight, the following procedures should be used:

1. The aircraft should be weighed inside a closed building to avoid errors that may be caused by wind.

2. The aircraft should be free from excessive dirt, grease, moisture, or any other extraneous material before weighing.

3. The aircraft should be weighed in the level attitude. If the main wheels are used as reaction points, the brakes should not be set because resultant side loads on the scales or weighing units may cause erroneous readings.

4. The accuracy of the scales must be established. This can be done in accordance with instructions provided by the manufacturer of the scales or by testing the scales with calibrated weights. When there is nothing on the scales, the reading should be zero.

5. All items of equipment to be installed in the aircraft and included in the certificated empty weight should be in place for weighing. Each item must be in the location that it will occupy during flight as shown in the aircraft equipment list.

6. Unless otherwise noted in the Type Certificate, the oil system and other operating fluids should be checked to see that they are full.

FIG. 7-20 Portable electronic scales. *(Evergreen Weigh, Inc.)*

7. The fuel should be drained from the aircraft unless other instructions are given. Fuel should be drained with the aircraft in the level position to make sure that the tanks are as empty as possible. The amount of fuel remaining in the aircraft tanks, lines, and engine is termed unusable fuel, and its weight is included in the empty weight of the aircraft. In special cases the aircraft may be weighed with full fuel in the tanks provided that a definite means is available for determining the exact weight of the fuel.

8. The weight of the tare should be recorded, either before or after weighing the aircraft, and the tare weight subtracted from the total weight as obtained from the scales.

9. When the aircraft is in the level position, the exact location of the weighing points must be accurately measured and recorded for use in the weight-and-balance computation.

10. The weights of the right wheel, left wheel, and the nose or tail wheel must be recorded to provide information needed for the CG determination. Several readings are taken for each reaction point and the average reading is entered on the aircraft weighing form.

11. When data for comparison is available, an attempt should be made to verify the results obtained from each weighing. Verification may be made by comparing results with a previous weighing of an aircraft of the same model.

Computing CG Location

The fundamental rule for determining the location of the CG for an airplane is as follows: **Divide the total moment of the airplane (taken from a specific reference point) by the total weight of the airplane. The result will be the distance of the CG from the reference point.**

In Fig. 7-21 a tricycle-gear airplane is weighed and it is found that the nose-wheel weight is 320 lb [145.1 kg], the right-wheel weight is 816 lb [370.1 kg], and the left-wheel weight is 810 lb [367.4 kg]. The horizontal distance between the weighing points is 75 in [190.5 cm]. These data give us all the information we need to find the location of the CG. When blocks are used on the scales, the weight of the blocks must be removed as **tare**.

To simplify the computation, we establish the center of the main wheels as the reference point from which to determine the moment. This provides a "zero" moment at the main wheels because the arm is zero. The moments and the CG location are then found as shown in Prob. 7-1.

PROBLEM 7-1

Item	Weight	× Arm =	Moment
Right wheel	816 lb	0	0
Left wheel	810 lb	0	0
Nose wheel	320 lb	−75	−24 000 in·lb
	1946 lb		−24 000 in·lb

Then $\dfrac{-24\,000}{1946}$

$= -12.33$ in $[-31.32$ cm$]$(CG dist. fwd of MWCL)

In the foregoing problem it is important to note that we are working from the **main-wheel center line (MWCL)** as the reference point. Because the nose wheel is forward of the reference point, the arm is negative and the moment of the nose wheel is negative.

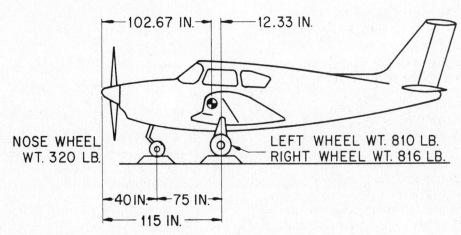

FIG. 7-21 Quantities required for determination of CG location.

The arm of the CG is negative because the CG is forward of the reference point (MWCL).

The CG location will always be the same no matter what reference point we use. To demonstrate that this is so, we shall rework the computation for the airplane in Fig. 7-21 and use the datum of the airplane as the reference point. The datum which is located at the nose of the airplane is 40 in [101.6 cm] forward of the nose-wheel center line and 115 in [292.1 cm] forward of the MWCL. (See Prob. 7-2.) We see from the computation that the CG is 102.67 in [260.78 cm] aft of the datum line (nose of airplane). This, of course, places it in the same position determined for the previous computation (102.67 + 12.33 = 115, which is the distance in inches from the datum line to the MWCL). Care must be taken to ensure that the proper sign is applied to each quantity expressed in a weight-and-balance computation.

PROBLEM 7-2

Item	Weight	×	Arm	=	Moment
Right wheel	816		+115		+93 840
Left wheel	810		+115		+93 150
Nose wheel	320		+40		+12 800
	1946				+199 790

Then $\dfrac{+199\,790}{1946} = +102.67$ in [+260.78 cm]

The weight of an airplane is always positive (+). Also, the weight of any item **installed** in the airplane is positive. The weight of any item **removed** from the airplane is negative (−). According to the standard rules of algebra, the product of two positive numbers is positive, the product of two negative numbers is positive, and the product of a positive number and a negative number is negative. This can also be stated: **The product of numbers with like signs is positive; the product of numbers with unlike signs is negative.**

When items of aircraft equipment are added or removed, four combinations are possible.

1. When items are added forward of the datum line, the signs are (+) weight × (−) arm = (−) moment.

2. When items are added to the rear of the datum line, the signs are (+) weight × (+) arm = (+) moment.

3. When items are removed forward of the datum line, the signs are (−) weight × (−) arm = (+) moment.

4. When items are removed to the rear of the datum line, the signs are (−) weight × (+) arm = (−) moment.

A simple diagram will aid in determining the effect of changes in aircraft equipment. In Fig. 7-22 a straight line represents the airplane. The nose of the airplane is shown to the left, this being the conventional method for representing aircraft in weight-and-balance diagrams. Using the CG location as a reference, we note that any item installed forward of the CG will produce a negative moment and will cause the CG to move forward. Items added to the rear of the CG produce a positive moment and move the CG rearward. Items removed in either case will have an effect opposite to that of items installed.

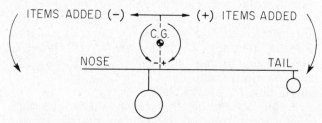

FIG. 7-22   Effects of weight changes in an airplane.

Observe that the curved arrows shown around the CG location indicate the effects of positive and negative moments. Positive moments are clockwise and cause a tail-heavy force, while negative moments are counterclockwise and cause a nose-heavy force.

Computing EWCG for a Conventional Airplane

Figure 7-23 shows a conventional airplane in position for weighing. We assume that we are weighing the airplane with 2 gal [7.57 L] of oil still in the engine and the oil arm is given as −20 in the specifications. As

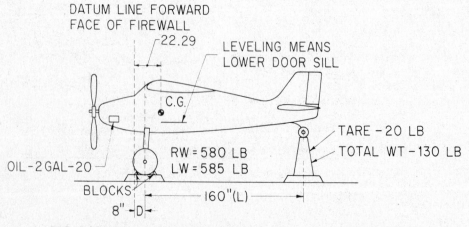

FIG. 7-23   Empty-weight CG computation for a conventional airplane.

Item	Weight	Tare	Net weight	×	Arm	=	Moment
Right wheel	580	0	580		0		0
Left wheel	585	0	585		0		0
Tail wheel	130	20	110		+160		+17 600
			1275				+17 600

shown in the illustration, the weights obtained from the scales are as follows: right wheel, 580 lb [263.1 kg]; left wheel, 585 lb [265.35 kg]; tail wheel, 130 lb [59 kg] (including tare). The computation for the EWCG can be arranged as in Prob. 7-3.

As explained previously, the CG is equal to the total moment divided by the total weight. Therefore,

$$CG = \frac{17\ 600}{1275}$$
$$= +13.8 \text{ in } [35 \text{ cm}] \quad (\text{CG aft of MWCL})$$

The result obtained in the computation is the CG location to the rear of the reference point, which is the center line of the main landing gear. To obtain the CG from the datum line, we must add the distance from the MWCL to the distance of the datum line from the MWCL. Then 13.8 + 8 = 21.8 in [55.4 cm], which is the distance of the CG from the datum line. If this is an older aircraft not certificated under FAR Part 23, the empty weight may not include the oil, in which case we have not yet completed the EWCG computation, since the CG we have obtained includes the weight of the oil that was in the aircraft at the time it was weighed. We must therefore remove the oil by computation to obtain the EWCG (see Prob. 7-4).

PROBLEM 7-4

Item	Weight	×	Arm	=	Moment
Airplane	1275		21.8		+27 795
Oil (removed)	−15		−20		+300
	1260				+28 095

$$EWCG = \frac{+28\ 095}{1260}$$
$$= +22.29 \text{ in } [56.61 \text{ cm}](\text{EWCG location})$$

Whenever possible it is desirable to use the manufacturer's weight-and-balance formulas and diagrams as shown in Fig. 7-24. If these are not available, a standard formula may be used for the EWCG computation. This formula is illustrated and explained in Fig. 7-25, taken from FAA Advisory Circular 43.13-1A. In the first diagram the datum is at the nose of the airplane, and since the airplane is of the tricycle-gear type, the CG must be forward of the MWCL. The part of the formula FL/W gives the distance of the CG forward of the MWCL. This distance must then be subtracted from the distance D to find the distance of the CG from the datum.

In the second diagram the airplane is of conven-

tional tail-wheel type, and so the CG must be to the rear of the MWCL. With the datum at the nose of the airplane, it is necessary to *add* the datum line distance D to the RL/W distance to find the EWCG from the datum line.

In the third diagram, the CG and the MWCL are both forward of the datum line; hence both distances are negative. For this reason the CG distance from the MWCL and the datum distance from the MWCL are added together, and the total is given a negative sign.

The fourth diagram shows a condition where the CG is positive from the MWCL but negative from the datum line. The datum to the MWCL is a negative distance, and the CG from the MWCL is a positive distance. Therefore, the EWCG from the datum line will be the difference between the two distances and, in this case, will carry a negative sign.

Referring back to Fig. 7-23 and disregarding the oil computation, we may solve the problem with the following formula:

$$CG = D + \frac{RL}{W} = 8 + \frac{110 \times 160}{1275}$$
$$= 8 + 13.80 = +21.80 \text{ in } [+55.37 \text{ cm}]$$

It will be noted that this answer is the same as the original computation, this is because we did not concern ourselves with the moment of the oil.

For any computation, it is always a good practice to draw a diagram of the airplane (nose to the left) with the weighing points and the datum, and from these it is easy to determine what formula should be used.

Weight-and-Balance Report

After the weight-and-balance calculations are complete, it is important that they be properly recorded and placed in the aircraft weight-and-balance records (a sample form is shown in Fig. 7-26). When a new weight-and-balance report is prepared for an aircraft, the previous report should be marked **superseded** and the date of the new document referenced. This would preclude the necessity to search for the current report.

● AIRCRAFT MODIFICATIONS

During the lifetime of many aircraft, it is often found desirable to change the type of equipment that is installed. The owner of an airplane may wish to install radio navigation equipment, an autopilot, an auxiliary fuel tank, or various other items to make the airplane more serviceable. In every case of such a change, it is

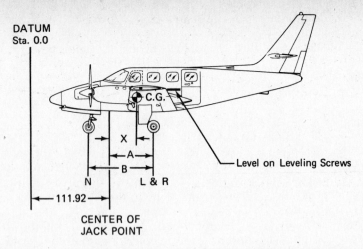

DATUM
Sta. 0.0

Level on Leveling Screws

X
A
B

N   L & R

111.92

CENTER OF
JACK POINT

Scale Position	Scale Reading	Tare	Symbol	Net Weight
Left Wheel			L	
Right Wheel			R	
Nose Wheel			N	
Sum of Net Weights (As Weighed)			W	

$$X = (A) - \frac{(N) \times (B)}{W} \; ; \; X = (\quad) - \frac{(\quad) \times (\quad)}{(\quad)} = (\quad) \text{ IN.}$$

$$\text{C.G. ARM} = 111.92 + X = \qquad \text{IN.}$$

Item	Weight (Lbs.) X	C.G. Arm (In.) =	Moment/1000 (Lbs.-In.)
Airplane Weight (From Item 5, Page 6-6)			
Add: Unusable Fuel (2 Gal at 6 Lbs/Gal)	12	162.1	1.9
Equipment Changes			
Airplane Basic Empty Weight			

FIG. 7-24   Sample airplane weighing. (Sample diagram; do not use for flight planning purposes.) *(Cessna Aircraft Co.)*

necessary to figure the effect on weight and balance. If the change in equipment should move the CG outside the limits, flight in the airplane would not be safe or legal.

Adding Equipment

Let us assume than an owner who has an airplane with an empty weight of 1220 lb [553.4 kg] and an EWCG at +25 wishes to install some radio equipment weighing 15 lb [6.8 kg]. In addition to the radio equipment, a larger generator must be installed in order to provide the additional power required to operate the radio.

The first consideration, of course, is to determine where the items of equipment are to be installed and then determine the arm of each item of equipment. This arm must be measured from the airplane datum line to the CG of the equipment to be installed. It

must be pointed out that if the CG of the item of equipment is not given in the accompanying instructions, the CG must be determined by the person making the installation. This is easily done by balancing the items of equipment at a single point in the position it will assume in the airplane. The balance point should then be marked or recorded for use in the computation.

For the purposes of the problem under consideration, we shall assume that the radio is installed at +65 and the new generator is installed at −21. These points are shown in Fig. 7-27. In order to install a new generator, the old generator must be removed. The weight and arm of the old generator are given in the Aircraft Specification as 11 lb [5.0 kg] (−21.5). The new generator weighs 14 lb [6.35 kg], and the arm is found to be −21. We now have sufficient information to make the computation. We arrange the work as

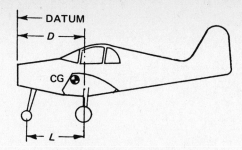

DATUM

NOSE-WHEEL-TYPE AIRCRAFT

DATUM LOCATED FORWARD OF THE
MAIN WHEELS

$$CG = D - \frac{F \times L}{W}$$

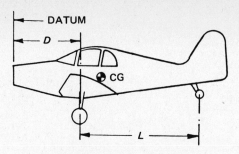

TAIL-WHEEL-TYPE AIRCRAFT

DATUM LOCATED FORWARD OF THE
MAIN WHEELS

$$CG = D + \frac{R \times L}{W}$$

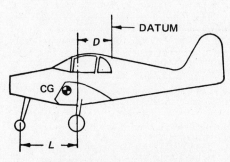

NOSE-WHEEL-TYPE AIRCRAFT

DATUM LOCATED AFT OF THE MAIN
WHEELS

$$CG = -D + \frac{F \times L}{W}$$

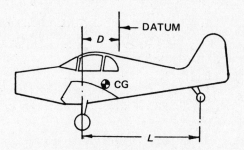

TAIL-WHEEL-TYPE AIRCRAFT

DATUM LOCATED AFT OF THE MAIN
WHEELS

$$CG = -D + \frac{R \times L}{W}$$

CG = distance from datum to center of gravity of aircraft
W = weight of aircraft at time of weighing
D = horizontal distance measured from datum to main wheel weighing point
L = horizontal distance measured from main wheel weighing point to nose or tail weighing point
F = weight at nose weighing point
R = weight at tail weighing point

FIG. 7-25   Different arrangements of the formula for EWCG.

shown in Prob. 7-5. From the computation we find that the EWCG of the airplane has moved rearward 0.38 in [0.96 cm] as a result of the new installation.

In some aircraft specifications for light airplanes, an EWCG range is given, and if the EWCG of the air-plane falls within this range, the loaded CG will also be within limits. In many cases, however, there is a wide variation of loaded conditions and an EWCG range becomes useless and will not be found in the specifications.

PROBLEM 7-5

Item	Weight	×	Arm	=	Moment
Airplane (empty)	1220		+25		+30 500
Radio	15		+65		+975
Generator (removed)	−11		−21.5		+236.5
Generator (installed)	+14		−21		−294
New empty weight	1238				+31 417.5

$$\frac{31\ 417.5}{1238} = 25.38 \text{ in [the CG (empty)]}$$

WEIGHT AND BALANCE REPORT

MAKE_____ MODEL_____ S/N_____ N_____

DISTANCE BETWEEN MAIN WHEELS AND <u>TAIL/NOSE</u> WHEEL IS _____ INCHES.

DATUM IS _____ INCHES <u>FORWARD/AFT</u> OF MAIN WHEEL CENTERLINE.

1. AIRCRAFT AS WEIGHED

POSITION	SCALE READING	TARE	NET WEIGHT
LEFT WHEEL			
RIGHT WHEEL			
TAIL OR NOSE WHEEL			

2. ITEMS ON BOARD THE AIRPLANE WHEN IT WAS WEIGHED THAT ARE NOT INCLUDED IN THE EMPTY WEIGHT

ITEM	WEIGHT	ARM	MOMENT
FUEL- GAL			
OIL- QT			
TOTAL OF ADDITIONAL ITEMS			

3. ITEMS NOT ON BOARD THE AIRPLANE WHEN IT WAS WEIGHED THAT ARE TO BE INCLUDED IN THE EMPTY WEIGHT

	WEIGHT	ARM	MOMENT
FUEL- GAL			
OIL- QT			
TOTAL OF ADDITIONAL ITEMS			

4. CENTER OF GRAVITY AS CALCULATED

	WEIGHT	ARM	MOMENT
MAIN WHEELS			
TAIL OR NOSE WHEEL			
TOTAL ITEM #2 (REMOVABLE ITEMS)			
TOTAL ITEM #3 (ADDITIONAL ITEMS)			
EMPTY WEIGHT TOTAL			

EMPTY WEIGHT CG _____   SIGNATURE _____

MAXIMUM GROSS WEIGHT _____   CERT. NO. _____

EMPTY WEIGHT _____   DATE _____

USEFUL LOAD _____

FIG. 7-26 Sample weight and balance report.

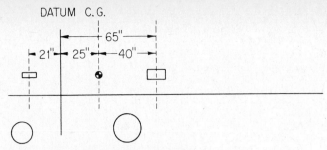

FIG. 7-27 Installation of equipment.

Removing Equipment

In removing equipment from an airplane it is just as necessary to make weight-and-balance computations as when installing equipment. Let us assume that the owner of an airplane wishes to remove flares because they are no longer necessary or required for the type of operation for which the airplane is being used. The airplane weighs 1300 lb [589.7 kg] empty (with the flares), and the flares weigh 18 lb [8.16 kg] (see Fig. 7-28). The CG of the airplane as equipped is +78, and the arm of the flares is +145 (see Prob. 7-6).

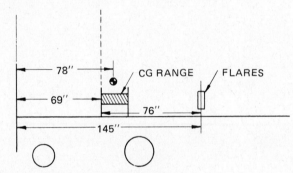

FIG. 7-28 Removal of equipment.

PROBLEM 7-6

Item	Weight	×	Arm	=	Moment
Airplane (empty)	1300		+78		+101 400
Flares (removed)	−18		+145		−2 610
New empty weight	+1282				+ 98 790

$$\frac{98\ 790}{1282} = +77.06 \text{ in (new EWCG)}$$

We observe from the foregoing computation that the removal of the flares caused the EWCG to move forward almost 1 in [2.5 cm]. It should be remembered that any removal of weight aft of the CG will cause the CG to move forward.

It is often necessary to compensate for changes in equipment by adding or removing ballast, changing the baggage-weight allowance, or some other means. If the airplane in the foregoing problem was critical for forward loading before the flares were removed, it would be necessary to correct for the change in CG caused by the removal of the flares.

It is essential that whenever equipment is added or removed from the aircraft, an entry is made in the airplane's equipment list and permanent weight-and-balance records. Many manufacturers provide a form such as the one shown in Fig. 7-29 that provides for a record of the equipment added or removed, as well as providing for a running total of the weight and balance.

EWCG Range

Some small aircraft are designed so that it is not possible to load them in a condition which will place the CG outside the fore or aft limits if standard load schedules are observed. These aircraft have the seats, fuel, and baggage accommodations located very near the CG limits. They also have **empty weight CG ranges** listed in their Type Certificates. Loads can be added to or removed from any location within the CG range with complete freedom from concern about CG movement. Such action cannot cause the CG to move beyond the CG limits of these aircraft, but maximum weight limits can still be exceeded.

Most aircraft, however, can be loaded in a manner which will place the CG beyond limits, in which case the manufacturer will be unable to establish a EWCG range. In these instances the EWCG range on the Type Certificate will be listed as "none."

● LOADING THE AIRPLANE

Improper loading reduces the efficiency of an airplane from the viewpoint of ceiling, maneuverability, rate of climb, and speed. This is the least of the harm that it can cause. The greatest danger is that improper loading may cause the destruction of life and property, even before the flight is well started, because of the stresses imposed upon the aircraft structure or because of altered flying characteristics of the airplane.

Effects of Improper Loading

Overloading. Excessive weight reduces the flying ability of an airplane in almost every respect. The most important performance deficiencies of an **overweight airplane** are:

1. Longer takeoff run
2. Lower angle and rate of climb
3. Lower ceiling
4. Reduced factors of safety for the airplane structure during rough air or takeoff from poor fields
5. Change in the flight characteristics of the airplane
6. Reduced maneuverability
7. Increase in stalling speed
8. Increased fuel consumption and loss of range

Effects of Adverse Balance. Adverse and abnormal balance conditions affect the flying ability of an airplane with respect to the same flight characteristics as those mentioned for an excess weight condition. In addition, there are two essential airplane attributes which may be seriously reduced by improper balance; these are stability and control.

WEIGHT AND BALANCE RECORD

(Continuous History of Changes in Structure or Equipment Affecting Weight and Balance)

SERIAL NUMBER			REGISTRATION NUMBER					PAGE NUMBER	

DATE	ITEM		DESCRIPTION OF ARTICLE OR MODIFICATION	WEIGHT CHANGE				RUNNING BASIC EMPTY WEIGHT	
	In	Out		ADDED (+) or REMOVED (−)					
				Wt. (lb)	Arm (In)	Moment /100		Wt. (lb)	Moment /100
			AS DELIVERED						

FIG. 7-29 Sample weight-and-balance record.

Adverse Forward Loading. When too much weight is toward the forward part of the airplane, the center of gravity (CG) is shifted forward and any one of the following conditions may exist or they may occur in combinations at the same time:

1. Increased fuel consumption and power settings
2. Decreased stability
3. Development of dangerous spin characteristics
4. Increased oscillation tendency
5. Increased tendency to dive, especially with power off
6. Increased difficulty in raising the nose of the airplane when landing
7. Increased stresses on the nose wheel

Adverse Rearward Loading. When too much weight is toward the tail of the airplane, any one of the following conditions may exist or they may occur in combination:

1. Increased danger of stall
2. Dangerous spin characteristics
3. Poor stability
4. Decreased flying speed
5. Poor landing characteristics

● EXTREME WEIGHT AND BALANCE CONDITIONS

It has been explained previously that every aircraft has an approved CG range within which the CG must lie if the aircraft is to be operated safely. In order to determine whether the loaded CG falls within the approved limits, it is necessary that two computations be made, one for **most forward loading** and one for **most rearward loading.**

Adverse-loading checks are a deliberate attempt to load an aircraft in a manner that will create the most critical balance condition while still remaining within the maximum gross weight of the aircraft.

It should be noted that when the EWCG falls within the EWCG range, it is unnecessary to perform a forward or rearward weight-and-balance check. In other words, it is impossible to load the aircraft to exceed the CG limits, provided standard loading and seating arrangements are used.

Extreme Forward Adverse-Loading Condition

To determine the conditions for most forward loading, we must examine the Aircraft Specification or Type Certificate Data Sheet to determine fuel capacity and arm, oil capacity and arm, passengers and arm, and cargo (baggage) and arm. From these we must determine which items will tend to move the CG forward and include maximum quantities of these items in our computation. Since some items may have to be included that tend to move the CG rearward, we must use a minimum of such items.

In examining a Type Certificate Data Sheet for a certain airplane, we find that the following specifications are given for weight and balance:

CG range	(+85.1) to (+95.9) at 1710 lb
	(+87.0) to (+95.9) at 1900 lb
	(+91.5) to (+95.9) at 2200 lb
Datum	78.4 in forward of wing leading edge
Leveling	Two screws left side fuselage before window
Max. weight	2200 lb
No. of seats	4 (2 at +85.5, 2 at +118)
Max. cargo	100 lb (+142.8)
Fuel capacity	50 gal (+94)
Oil capacity	2 gal (+31.7)

For the CG range we can prepare a chart such as that shown in Fig. 7-30. It will be noted that the values given on the chart agree with the specifications listed. Assuming that the empty weight of the airplane is 1075 lb [487.62 kg] and the EWCG is +84 [213.36 cm], we can load the airplane to determine whether the forward CG is within limits with maximum forward loading. Remember that we wish to load the airplane in such a manner that it will result in the most forward CG possible.

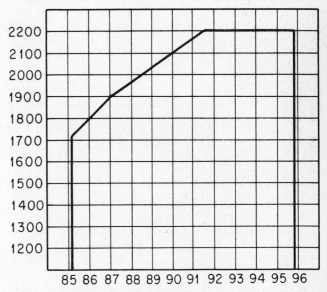

FIG. 7-30   Chart for CG limits.

The airplane must have a pilot, and so we shall load 170 lb [711 kg] at +85.5 [217.17 cm]. Since the fuel is at +94 [238.76 cm], which is substantially to the rear of the forward limit, we shall load only minimum fuel. Since the engine of this particular airplane requires 25 gal/h [94.63 L/h] at METO (maximum except takeoff) power, it is necessary that 12.5 gal [47.31 L] or 75 lb [34.02 kg] of fuel be included in the computation. This is loaded at +94 [238.76 cm]. The cargo compartment is at +142.8 [362.71 cm], and any load at this point will move the CG to the rear, hence we shall load no cargo. Full oil is required for the engine, so we must load 15 lb [6.8 kg] (7.5 lb/gal, or 3.4 kg/L) of oil at +

PROBLEM 7-7

Item	Weight	×	Arm	=	Moment
Airplane (empty)	1075		+84		+90 300
Pilot	170		+85.5		14 535
Fuel	75		+94		7 050
Oil	15		+31.7		475.5
	1335				112 360.5

Then $\dfrac{112\ 360.5}{1335}$ = 84.16 in [213.76 cm](most forward CG)

31.7 [80.51 cm]. The loading computation will then appear as in Prob. 7-7.

Checking the result of the foregoing computation against the CG limits for the airplane, we find that the CG is 0.93 in [2.36 cm] forward of the forward CG limit. To correct this condition, the airplane could carry **ballast** or a certain amount of baggage in the cargo compartment, and a warning placard should be placed on the instrument panel.

In making a check of forward CG limit, it must be remembered that we employ maximum weights for all items forward of the forward CG limit and minimum weights for all items to the rear of the forward CG limit. In the foregoing problem all items except the oil are located to the rear of the forward limit; hence minimums were used. Information required for a forward CG check is as follows:

1. Weight, arm, and moment of the empty aircraft
2. Maximum weights, arms, and moments of all items of useful load located ahead of the forward CG limit
3. Minimum weights, arms, and moments of all items of useful load located to the rear of the forward CG limit.

Extreme Rearward Adverse-Loading Condition

To check an airplane for rearward CG limit, we must use a maximum of all weights to the rear of the CG limit and a minimum of all weights forward of the rearward CG limit. Using the specifications for the same airplane as under consideration in the previous problem, we find that the cargo compartment and the two rear seats are the only items with locations to the rear of the rearward CG limit. The computation is then arranged as in Prob. 7-8.

PROBLEM 7-8

Item	Weight	×	Arm	=	Moment
Airplane (empty)	1075		+84		+90 300
Pilot	170		+85.5		+14 535
Passengers (2)	340		+118		+40 120
Fuel	75		+94		+7 050
Oil	15		+31.7		+ 475.5
Baggage	100		+142.8		+14 280
	1775				+166 760.5

Then $\dfrac{166\ 760.5}{1775}$ = 93.9 in [238.51 cm](rearward CG)

It may be noted that we could have made the CG move slightly more toward the rear by including maximum fuel, since the arm of the fuel is +94. This location is still forward of the rear limit, however, and so it could not have moved the CG beyond its rearward limit.

Information required for a rearward adverse-loading CG check is as follows:

1. Weight, arm, and moment of the empty aircraft
2. Maximum weights, arms, and moments of all items of useful load located to the rear of the rearward CG limit
3. Minimum weights, arms, and moments of all items of useful load located forward of the rearward CG limit.

● LOADING CONDITIONS

Sample loading conditions are computed as an indication of the permissible distribution of fuel, passengers, and baggage which may be carried in the aircraft at any one time without exceeding either the maximum weight or the CG range. These sample computations should be included in the aircraft's weight-and-balance records or may be posted in the form of a placard. A typical placard may be similar to the one shown in Fig. 7-31.

LOADING SCHEDULE		
FUEL	PASSENGERS	BAGGAGE
FULL	2 REAR	100 LB
40 GAL	1 FRONT AND 2 REAR	NONE
FULL	1 FRONT AND 1 REAR	FULL
INCLUDES PILOT AND FULL OIL		

FIG. 7-31 Loading schedule placard.

Correcting the CG Location

In the computation for the forward CG limit of the airplane in the previous problem, it was found that the CG was 0.93 in [2.36 cm] forward of the forward CG limit. It is therefore necessary that we find a means for correcting this condition. To start this computation, we use the forward CG limit—+85.1 in [216.15 cm]—as the reference point.

We can correct the CG by adding **fixed ballast** at a point in the rear of the airplane near the tail, or we can install fixed ballast in the cargo compartment. We can also placard the cargo compartment to the effect that a minumum amount of baggage must be carried under certain conditions. For the purpose of this computation, we shall assume that we install fixed ballast in the cargo compartment.

The cargo compartment has an arm of +142.8. Since we are now using the forward CG limit as the reference, we must subtract +85.1 from +142.8 to find the arm of the cargo compartment for this computation. In Fig. 7-32 we can see that the arm of the cargo compartment from the new reference point is +57.7 in [146.56 cm].

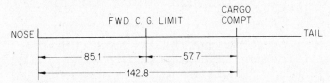

FIG. 7-32 Correction computation for forward CG limit.

The first step necessary in the computation for the correction of CG location is to determine what moment is necessary to provide the required correction. Since the CG is 0.93 in forward of the forward limit, the moment necessary for correction is 0.93 × 1335 (the weight of the forward-loaded airplane). The product is 1241.6 in·lb [14.31 kg·m], the required moment.

To determine the weight necessary to provide a moment of 1241.6 in·lb with an arm of +57.7 (the arm of the cargo compartment from the forward CG limit), we must divide 1241.6 by +57.7. The result of this division is 21.5 lb [9.75 kg], which is the weight required in the cargo compartment to correct the CG location. We can verify this by working the original computation with 22 lb [9.98 kg] installed in the cargo compartment (see Prob. 7-9).

PROBLEM 7-9

Item	Weight	×	Arm	=	Moment
Aircraft (as loaded)	1335		+84.17		+112 360.5
Ballast	22		+142.8		+3 141.6
	1357				+115 502.1

Then $\dfrac{115\ 502.1}{1357}$ = 85.1 in (forward CG limit)

If the ballast in the foregoing problem had been installed near the tail of the airplane, the weight required would have been less. The requirement was to produce a certain moment (1241.6 in·lb), and this could have been done by any combination of weight and arm that would produce this moment. Note that the moment was computed from the forward CG limit (+85.1). However, after the weight was added to the airplane, we used the original moment of the airplane and cargo compartment.

● SIMPLIFIED LOADING METHODS

CG Envelope and Loading Chart

Because of the many possible loading combinations, especially in airplanes where more than two passengers can be carried, methods have been developed whereby the pilot can quickly determine whether the

airplane is loaded within limits without going through a long process of computation. One of these methods involves the use of the **CG envelope** and the **loading chart.**

The CG envelope chart is a graph with airplane weight plotted against index units. The index unit is the moment of the airplane divided by 1000. The envelope is an area on the graph establishing the combinations of weight and index units where the CG of the airplane will be within limits. A typical CG envelope is shown in Fig. 7-33. In this chart it can be seen that there is a satisfactory range of moments for each weight of the airplane. For example, if the airplane is loaded to weight 1650 lb [748.4 kg], the moment can be from 60 000 to about 76 500 in·lb [691 to about 881 kg·m] or the index units can be from 60 to 76.5. The maximum loaded weight of the airplane is 2200 lb [997.3 kg], and the maximum index number is about 102. If we divide 2200 into 102 × 1000, we obtain the rearward CG limit. The limit is established by the line AB in Fig. 7-33.

In order to make the CG envelope simple to use, the **loading graph** is provided. This graph, illustrated in Fig. 7-34, provides for the loading of passengers, fuel, and baggage. Since passengers are loaded at two separate arms, there are two reference lines for passengers. The chart shown is designed for loading the Cessna 170 airplane. The graph plots load weight against index units. If we wish to determine the moment of any loaded item, we merely follow the weight line to the right until we intersect the position line for the item and then drop straight down to the base line and read the index unit. The index unit multiplied by 1000 is the moment.

To use the loading graph we proceed as in Prob. 7-10. We then apply the weight and index number to the CG envelope of Fig. 7-33 and find that the point is within the envelope. If we wish to operate the airplane with only a pilot, one passenger, and no baggage, the result will be as shown in Prob. 7-11. When these figures are applied to the CG envelope, we find that the CG is still within limits even though it has moved forward.

When computing the CG or loading for a particular airplane, the technician should consult the approved manual prepared for the airplane by the manufacturer. The basic CG location may be given with this manual, and the method for computing the weight and balance is explained. The charts and graphs used for the airplane CG are also included in the manual.

● CALCULATING WEIGHT AND BALANCE FOR LARGE AIRCRAFT

During the flight of a large passenger airplane, the consumption of fuel and the movement of passengers and crew members cause changes in CG location. These changes are compensated for by loading the airplane properly so the CG will not move beyond forward or rearward limits regardless of fuel quantity or movement of passengers and crew in normal situations. The CG location is calculated before the airplane takes off,

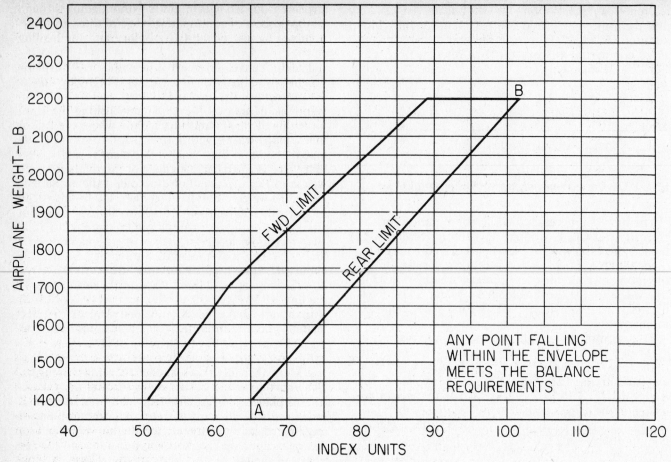

FIG. 7-33　Center-of-gravity envelope.

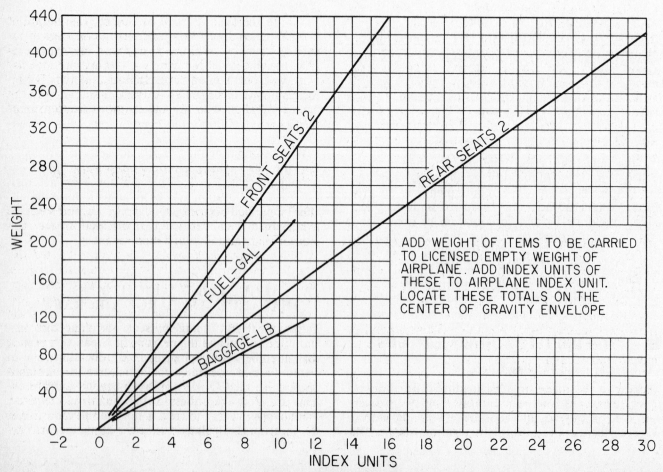

FIG. 7-34　Loading graph.

PROBLEM 7-10

Item	Weight, lb	Index units
Airplane empty weight	1210	+47.1
Oil	15	−0.3
Pilot and passenger	340	+12.2
Passengers (2)	340	+23.8
Fuel (maximum), 37 gal	222	+10.7
Baggage	70	+6.7
	2197	+100.2

PROBLEM 7-11

Item	Weight, lb	Index units
Airplane empty weight	1210	+47.1
Oil	15	−0.3
Pilot and passenger	340	+12.2
Fuel	222	+10.7
	1787	+69.7

and it is adjusted if necessary to assure that changes during flight will not cause the CG to move out of limits. Baggage, cargo, and fuel loading can be used to adjust CG location.

Computer-Calculated Weight and Balance

Different methods may be utilized in providing for proper loading of the aircraft. The system currently in use by most airlines is a computer-programmed weight-and-balance system. This computerized system has the advantage of providing improved load planning, the elimination of mathematical errors, and the ability to make adjustments for last minute changes in the number of passengers and amount of cargo loaded. A computer program system has the ability to provide the pilot with a variety of information on the load manifest. This information includes such items as the aircraft's loaded CG, ramp and take-off weights, and the payload and its distribution; it also calculates the proper trim setting for takeoff. As with many computer applications, provisions are made for a manual backup system.

For large airliners, rather complex loading charts are prepared. In use, however, these charts greatly simplify the loading process. The charts give the moment arms of the various compartments and fuel tanks and provide an easy method for determining the index of the load in any particular area. The indexes are combined by one of two or three methods, and the CG location is found on the chart. In some cases a special slide rule ("slip stick") is designed to add or subtract moments as the airplane is loaded, thus providing a quick method for computation. This special slide rule is called a **load adjuster** and operates on the principle of index units.

The weight and balance of a large airliner may be determined by means of a balance computer. This is a circular chart that is used with an overlay called the balance planning sheet. The computer and the balance planning sheet are mounted on a center peg so the overlay may be rotated over the computer. Each has a vertical index line, and these are superimposed at the start of the computation. The computer is shown in Fig. 7-35.

The upper portion of the computer has scales to the right and left of the index line representing locations and weights forward and rearward of the 21 percent MAC line of the aircraft. As each section of the aircraft is loaded with passengers and cargo, the overlay is rotated and marked. After each section is loaded, the load mark is rotated back to the vertical index line on the computer. When all passengers, cargo and fuel have been loaded, the index line on the overlay will show whether the loading is within limits. Takeoff weight and CG location in percent of MAC will be indicated.

Automatic Weight-and-Balance Systems

One of the most interesting and useful developments in the weight-and-balance field is the automatic system, or **on-board aircraft-weighing system (OBAWS)**, designed for such large aircraft as the Boeing 747. This system includes a transducer (strain-gage unit) installed inside each axle for each wheel of both the main landing-gear wheels and the nose-gear wheels, as shown in Fig. 7-36. Each transducer generates a signal that can be converted to a weight indication, because the transducer senses the shear stress on each axle. The signals from the transducers are sent to the **computer,** which integrates the weight information from all the axles and sends resulting information to the **indicator-control** unit and the **attitude sensor.** A block diagram of the weight-and-balance system is shown in Fig. 7-37.

The indicator provides a reading of gross weight of the aircraft and the CG location as a percentage of the MAC. The flight engineer is therefore always able to determine whether the weight of the aircraft and the location of the CG are within specified limits. The attitude sensor determines whether the aircraft is in the correct attitude (level) for an accurate measurement of CG location.

● WEIGHT AND BALANCE FOR A HELICOPTER

The weight-and-balance principles and procedures which have been discussed in connection with airplanes apply generally to helicopters.

Most helicopters have a much more restricted CG range than do airplanes. In some cases this range is less than 3 in [7.62 cm]. The exact location and length of the CG range is specified for each helicopter and usually extends a short distance fore and aft of the main rotor mast or the centroid of a dual-rotor system. Ideally, the helicopter should have such perfect balance that the fuselage remains horizontal while in a hover and the only cyclic adjustment required should be that made necessary by the wind. The fuselage acts as a pendulum suspended from the rotor. Any change in the CG changes the angle at which it hangs from this point of support. Many recently de-

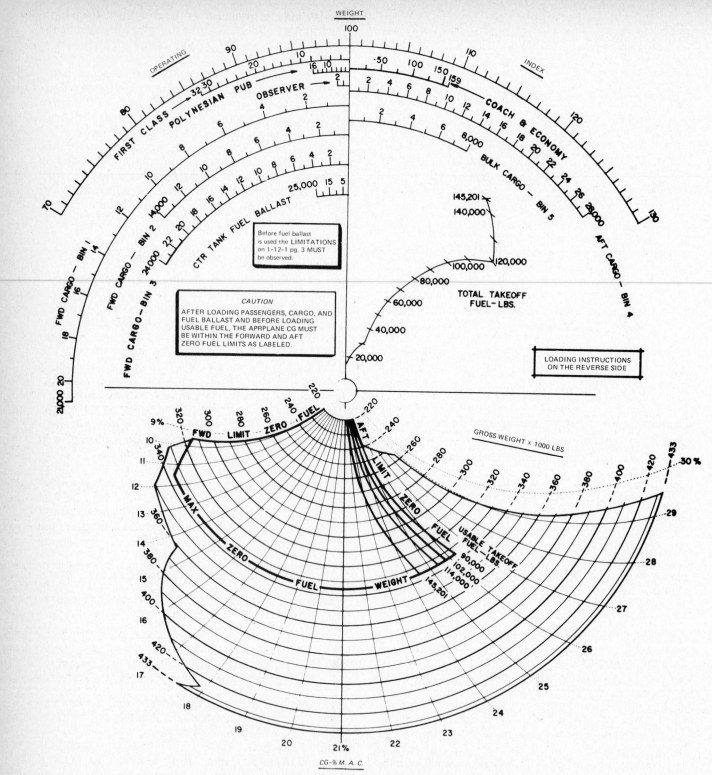

FIG. 7-35 A balance computer. *(Continental Air Lines)*

signed helicopters have loading compartments and fuel tanks located at or near the balance point.

The information in this section is typical of instructions for leveling, weighing, and computing the CG location for a helicopter. This information applies to a Bell Model 206L *Long Ranger* helicopter.

Leveling

A **level plate** is located on the cabin floor approximately 4.0 in [10.16 cm] forward of the aft seat and left of the helicopter center line. This is shown in Fig. 7-38. A slotted level plate is located directly above

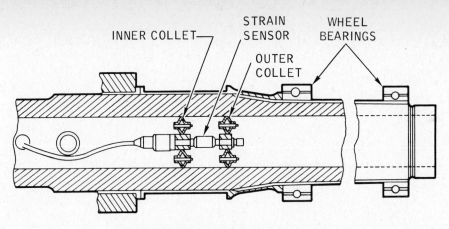

INNER COLLET
STRAIN SENSOR
OUTER COLLET
WHEEL BEARINGS

TRANSDUCER INSTALLED IN AXLE

FIG. 7-36 Transducer installed in the axle of a Boeing 747 airplane.

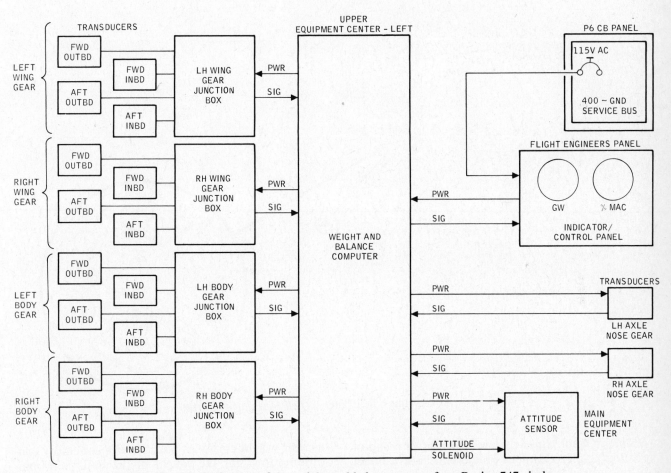

FIG. 7-37 Block diagram of the weight-and-balance system for a Boeing 747 airplane.

the level plate. The leveling procedure is then as follows:

1. Hang a plumb bob from the small hole in the slotted level plate and suspend it in such a manner that the plumb bob is just above the level plate on the cabin floor.

2. Position the helicopter on a level surface in an enclosed hangar.

3. Position three jacks under the helicopter at the jack and tie-down fittings that are permanently installed. Two forward jack fittings are located at station 55.16, and the aft fitting is located at station 204.92.

4. Adjust the aft jack at the aft jack fitting until the

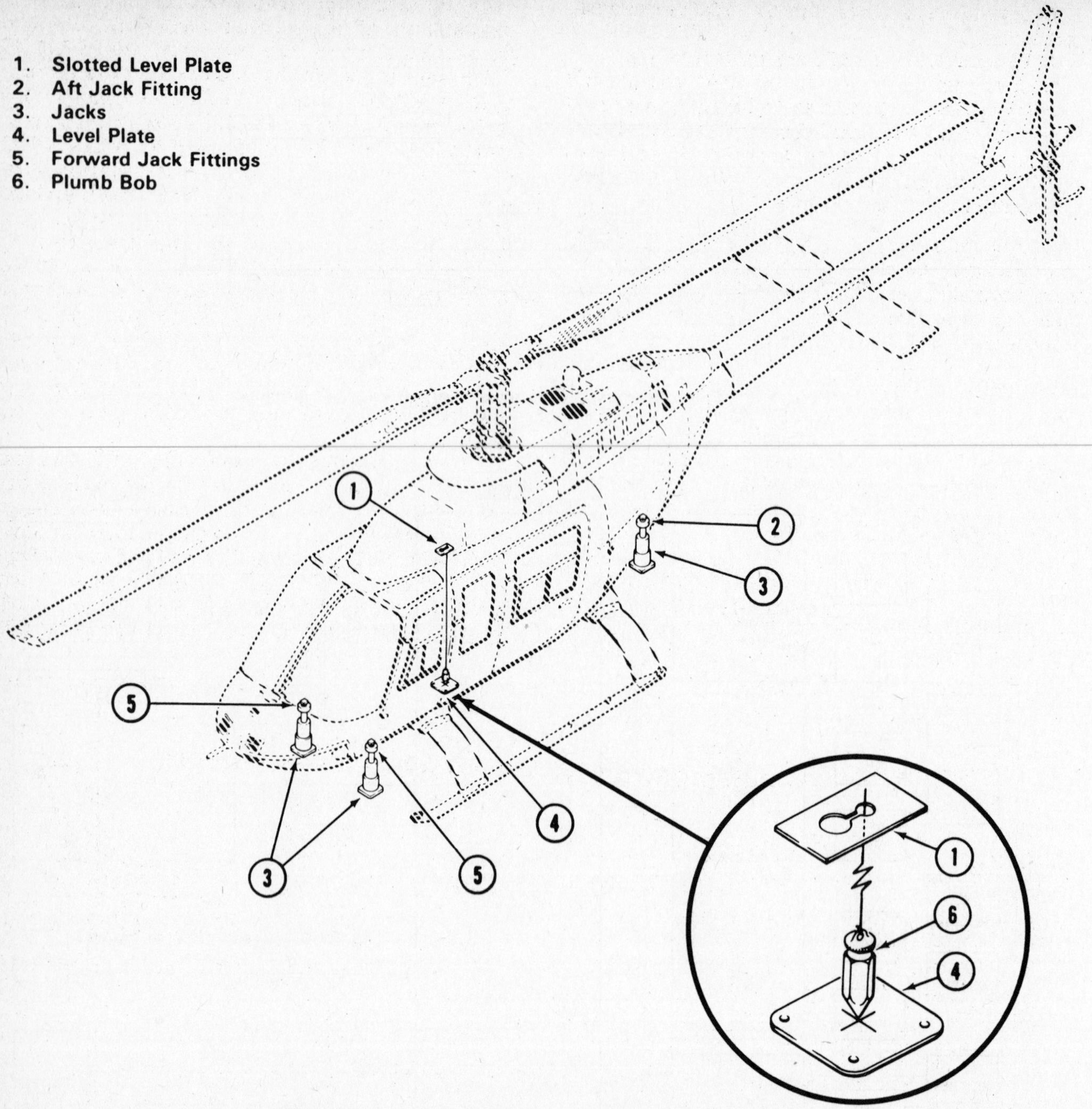

1. Slotted Level Plate
2. Aft Jack Fitting
3. Jacks
4. Level Plate
5. Forward Jack Fittings
6. Plumb Bob

FIG. 7-38   Leveling a helicopter.   *(Bell Helicopter Textron)*

helicopter is approximately level. The forward end of the landing-gear skid tubes should still be in contact with the ground.

5. Adjust all three jacks evenly until the helicopter is level, as indicated when the point of the plumb bob is directly over the intersection of the cross lines on the level plate. This is shown in Fig. 7-38.

Weighing

A helicopter may be weighed with platform scales or by means of load cells mounted on jacks. The instructions given here are for weighing with scales.

The helicopter should be weighed in a configuration as near weight-empty as possible. Weight-empty condition allows for the weight of the basic helicopter together with seats, ballast, special equipment, transmission oil, hydraulic fluid, unusable fuel, and undrainable oil. The baggage compartment should be empty. Weighing is accomplished as follows:

1. Position the scales in an approximately level area and check them for proper adjustment to the zero position. The weighing should be done in an enclosed area to avoid the adverse effects of wind, such as flapping rotors and body sway.

2. Position a scale and jack under each jack pad and raise the helicopter clear of the floor.

3. Level the helicopter with the jacks as explained previously.

4. Balance each scale and record its reading.

5. Lower the helicopter to the floor surface, and weigh the jacks, blocks, and any other equipment used between the scales and the helicopter. Deduct this tare weight from the scale readings to obtain net scale readings. The total of the net scale readings is the **as-weighed** weight of the helicopter.

A typical example of net weights is 513 lb [232.7 kg] for the forward left scale, 522 lb [236.8 kg] for the forward right scale, and 1063 lb [482.2 kg] for the aft scale. The as-weighed weight is then the sum of the net scale weights, or 2098 lb [951.6 kg].

Determining CG Location

The CG location for a helicopter is determined in the same manner as for other aircraft. In the case of the Bell Model 206L, the datum line is at the 0.0 fuselage station, which is just forward of the nose of the helicopter, as shown in Fig. 7-39. The CG location aft of the datum line is found as follows:

$$\text{CG location} = \frac{\substack{\text{moment of forward weights} \\ + \text{ moment of rear weights}}}{\text{total net weight}}$$

The location of the forward weighing point is 55.16 in [140.1 cm] aft of the datum line at FS 55.16, and the location of the rear weighing point is at FS 204.92.

The sum of the weights indicated by the forward scales is 1035 lb [469.9 kg], and the moment is 1035 × 55.16 = 57 090.6 in·lb [657.84 kg·m]. The moment of the aft weight is 1063 × 204.92 = 217 829.95 in·lb [2509.9 kg·m]. The total moment is then 274 920.55 in·lb [3167.8 kg·m]. When this is divided by the total net weight of the helicopter, the CG location is found to be 131.04 in aft of the datum line.

If a helicopter when weighed does not include all the equipment required for the weight-empty condition, these items must be added. The weights must be added to the as-weighed weight, and the moments must be computed and added to the original computed moment. The result is a total weight known as the "derived weight" and a slightly different CG location.

If the finally computed empty-weight CG location does not fall within the limitations set forth in the empty-weight CG location chart, ballast plates are installed either forward or rearward in specified locations. Ballast is never added in both forward and rearward locations. The forward ballast location in the Bell Model 206L helicopter is at +13 (FS 13.0), and the rearward ballast location is at +377.18 (FS

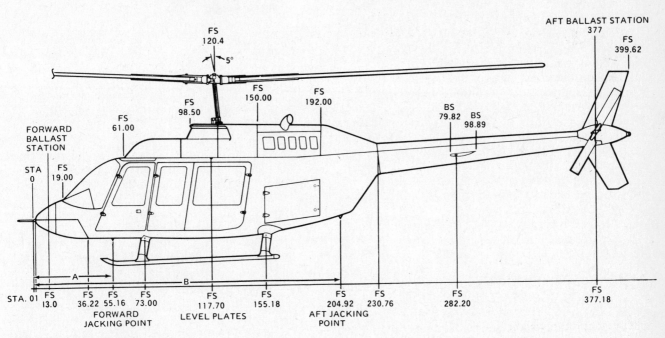

BS = BOOM STATION
CG = CENTER OF GRAVITY
STA = STATION — FUSELAGE
A = FORWARD ARM 55.16 INCHES
B = AFT ARM 204.92 INCHES
C = FORWARD SCALE READING (NET)
D = AFT SCALE READING (NET)
AC = FORWARD MOMENT
BD = AFT MOMENT
C+D = TOTAL WEIGHT

$$\frac{AC + BD}{C + D} = \frac{\text{CG FROM}}{\text{STA 0}}$$

FIG. 7-39   Datum line and weighing points for a helicopter. *(Bell Helicopter Textron)*

377.18) (see Fig. 7-39). The ballast requirement is computed in the way that was described earlier in this chapter.

● REVIEW QUESTIONS

1. Define *center of gravity.*
2. What is a *first-class lever?*
3. What conditions exist when a lever is in balance?
4. Define *arm and moment.*
5. What is the purpose of the *datum line?*
6. Define *empty weight* as used for weight-and-balance computations.
7. Define *maximum weight* for an aircraft.
8. How is the *useful load* of an aircraft determined?
9. What is the *CG range?*
10. If the wings of an airplane have no sweepback, at approximately what point should the CG be with respect to the chord of the wing?
11. Under what conditions must the MAC of an airplane be taken into consideration in establishing the correct CG location?
12. Give a brief description of an electronic weighing kit.
13. What precaution must be taken in placing the electronic load cells in position for weighing an aircraft?
14. Describe the procedure for weighing an airplane.
15. How is the *leveling means* for an aircraft determined?
16. What is *tare?*
17. What is meant by *weighing point?*
18. Explain the importance of the *EWCG range.*
19. When is it necessary to compute weight-and-balance extreme condition problems?
20. How is *minimum fuel* determined for weighing purposes?
21. List some of the dangers that may exist as a result of overloading an aircraft.
22. What undesirable conditions may exist when an airplane is loaded with too much weight forward or to the rear?
23. Why is the proper loading of a helicopter particularly critical?
24. In preparing to determine the CG location for a certain light airplane, the following conditions were found:

Right-wheel weight	875 lb
Left-wheel weight	880 lb
Tail-wheel weight	120 lb
Distance, MWCL to TWCL	200 in
Distance, datum line forward of MWCL	30 in

Find the CG location relative to the MWCL. Find the CG location relative to the datum line.

25. Determine the CG location relative to the datum line for a tricycle-gear airplane when the conditions are as follows:

Right-wheel weight	1200 lb
Left-wheel weight	1190 lb
Nose-wheel weight	400 lb
Distance of MWCL to nose-wheel CL	85 in
Distance of datum line forward of MWCL	40 in

26. Compute the most forward loading conditions for an airplane with the following specifications:

Empty weight	1200 lb
EWCG	+11.0 in
CG limits	+9.0 to +17.0
Seats	2 at +16 and 2 at +50
Maximum cargo(baggage)	120 lb at +75
Fuel	40 gal at +18
Oil	2 gal at −30
METO hp	160

27. Compute the most rearward loading conditions for the airplane in the previous problem.
28. What correction can be made when an airplane CG is outside the approved limit?
29. The EWCG of an airplane is +25.00, and the empty weight is 1100 lb. When a 15-lb generator is installed at −21 and a 12-lb battery is installed at −40, what is the new CG location?

8 AIRCRAFT MATERIALS

● INTRODUCTION

Many materials have been used in aircraft. Early aircraft were lightweight assemblies of wood and fabric kept aloft with engines that produced marginal power. Steel tubing and wood structural elements with a covering of cotton or linen fabric were used for the first practical aircraft. The development of aluminum alloy structures, beginning in the late 1930s, resulted in all-metal designs still in use today. By the early 1950s aircraft development was focused on power plants. Future design aircraft were limited by power considerations rather than structural problems.

As more powerful engines were developed, the use of existing materials was pushed to the limits. Development of supersonic aircraft resulted in the need for structural materials that would provide the needed strength at high temperatures, yet be light enough to get the aircraft off the ground. The use of titanium, corrosion-resistant steels, and metal honeycomb developed as part of this need.

Once again structural materials are the center of attention from the research and development standpoint. Emphasis is being placed upon lighter weight with the development of new metal alloys. Synthetic fibers and resins are being combined to produce composites with very favorable strength-to-weight ratios.

The remaining years of the twentieth century will see a major change in the materials used for aircraft structures. A wide variety of material types and designs will be used. Technicians will no longer be able to perform their jobs by memorizing a few aluminum alloy designations or the head marking on rivets. However, aircraft built with wood, steel tubing, fabric, and conventional aluminum alloy structures will still be flying and requiring maintenance. The aircraft maintenance technician, unlike those in some other industries, cannot ignore the old and move on to the new. The future aviation technician must have a basic knowledge of the properties of a variety of materials.

● AIRCRAFT MATERIALS

Early aircraft made extensive use of wood. Wood was used for both structural elements and as a cover or skin. Wood offered a material that was low cost, lightweight, and easily worked. Used within design limitations, wood has a high strength. The structure and surface of the Hughes HK-1 Hercules (*Flying Boat*) was built entirely of laminated wood. A large number of all-wood training aircraft were built during World War II. The need to continually protect the wood against the elements to prevent decay was a drawback to its continued use for aircraft. The demise of wood as a structural material was also affected by the development of metals for aircraft use.

Fabric materials are used for covering aircraft structures made from wood or metal. The structural strength and airfoil shapes are formed by the structural elements. The fabric forms a continuous cover over these parts. The fabric must be treated with a resinous material called aircraft dope. The dope stiffens the fabric and helps to protect it from the elements. Early fabrics were organic cottons and linens. The strength of cotton will deteriorate with age. The need for careful inspection and periodic replacement of the fabric cover was a drawback to its use. Synthetic fabrics with a longer service life are replacing cotton as an aircraft covering.

Metal structures became popular with the development of aluminum alloys usable for aircraft. Their light weight and formability made aluminum products a natural replacement of wood as a structural material. The weight of iron-based metals has limited their use to areas where high strength is required such as engine mounts or the steel-tube structure covered with lightweight fabric. Most of the steel, other than engine parts, that the technician will encounter will be in the form of steel tubing. Titanium, relatively new to aircraft use, has weight, strength, and temperature characteristics favorable for high-performance aircraft. Metals exhibit a number of properties, such as formability, that enhance their use in aircraft design. Other properties, like susceptibility to corrosion, limit their use. All metals are subject to, and must be protected from, corrosion in varying degrees. Magnesium is a strong lightweight metal, but it is very susceptible to corrosion. Its need for added protection from, and inspection for, corrosion has affected the use of magnesium as an aircraft material. The future use of metals for aircraft structures is heavily dependent upon the development of synthetic and composite materials.

Plastics are a group of synthetic, or man-made, materials. Plastics are made by taking apart the basic elements of a material. The basic elements are then recombined in a manner which produces a new material with its own properties. Plastics are widely used in aircraft. Familiar products would include the transparent window materials, fiberglass (polyester resin) wing tips, ABS type fairings, and Dacron fabric. Not so apparent are the synthetic materials used for the resins in paints, Teflon hoses, the dope used on fabric, various adhesives and sealants, and the pulleys used for control cables.

A **composite material** is one which is made of a combination of two or more materials or of a material in two different forms. A major advantage of a composite material is its light weight. A second advantage is the ease with which complex parts can be formed. The use of composite materials is not new. Laminated composite materials have been used for a number of years. For example, control cable pulleys were made from a laminated material of phenol-formaldehyde resin reinforced with linen cloth. In recent years the use of laminated and sandwich composites has shown a possibility of revolutionizing the material used for aircraft.

● PROPERTIES OF MATERIALS

Materials exhibit a number of characteristics, or **properties.** The degree to which various properties exist in a material will determine its suitability for a specific use. The properties of materials can be categorized as mechanical, physical, or chemical.

Mechanical Properties

Plasticity is the ability of a material to be deformed without rupture or failure. Plasticity can take two different forms, ductility and malleability. The ductility of a material refers to its plasticity under a tension or pulling load. **Ductility** allows materials like aluminum and copper to be drawn into very small wires. **Malleability** is the plasticity exhibited by a material under a pounding or compression load. The rolling of metal into a thin sheet is made possible by its malleability. A material can have high ductility or high malleability or both. **Brittleness** is the opposite of plasticity. A brittle material is one that cannot be visibly deformed and will shatter or break.

Elasticity describes the ability of a material to deform under load and return to its original shape when the load is removed. The **elastic limit** is the maximum amount of deformation that can occur and the material still return to its original shape. When the elastic limit is exceeded, permanent, or "plastic," deformation will occur in the material. The elastic limit is also called the **proportional limit.** It has been proven that the amount of deformation of a material is proportional to the stress that caused it as long as the elastic limit is not exceeded. The proportional principle is referred to as Hooke's law.

Hardness is not a fundamental property but is related to the elastic and plastic properties of a material. An operational definition of hardness is the resistance to penetration. In some materials, like steel, the hardness is directly related to the tensile strength. A hardness test can be used to estimate the strength of such materials. There is no direct and accurate method to measure the **toughness** of a material. Toughness is a desirable characteristic of a material to resist tearing or breaking when it is bent or stretched. Closely related to plasticity, toughness also involves the magnitude of the force causing the deformation.

One of the most important characteristics of a material is **strength.** A simple definition of strength is the ability of the materials to resist deformation. The type of load on a material will affect the strength it exhibits.

Stress. Stress may be defined as an internal force that tends to resist the deformation of a material resulting from an external load. The different types of stress are called **tension, compression, bending, torsion,** and **shear.** The five types are illustrated in Fig. 8-1. Tension stress is the result of a load which tends to pull apart or stretch the material. Compression stress is present when the load presses together or tends to crush an object.

Shear stress is a result of two layers of a material being pulled apart. Shear stress can develop when two pieces of material are bolted or riveted together. If a force is applied such that the two plates tend to slide over one another, shear stress develops in the bolt. If

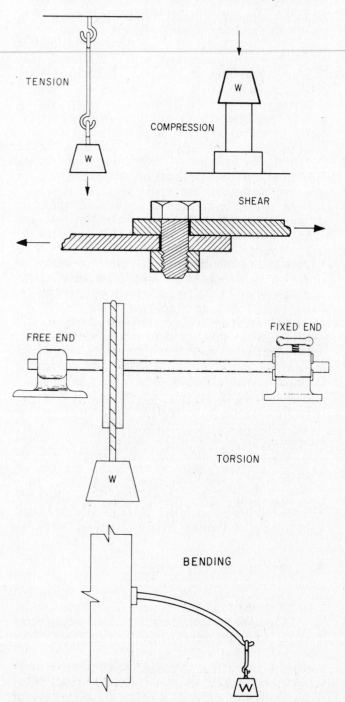

FIG. 8-1 **The five kinds of stress.**

the stress becomes greater than the shear strength of the bolt, it will be cut as with a pair of shears.

Bending produces three types of stress in a member. The material on the outside of the bend will tend to stretch and have a tension stress. Material on the inside of the bend will be compressed. Tension stress and compression stress, acting opposite to each other, will create a shear stress where they meet.

Torsion stress is the result of a twisting force. In the illustration a shaft is clamped solidly on one end and a pulley is mounted on the other end. A cable around the pulley is attached to a weight so that the weight tends to turn the pulley. This action applies a twisting or torsion force to the shaft. Stress caused by a torsion force is, like bending, a combination of tension, compression, and shear stress.

The amount of stress developed is a function of the force or load and the area upon which the force acts. Stress is expressed in units of pounds per square inch (psi). The amount of stress can be determined by the formula:

$$f = \frac{P}{A}$$

where f = stress, psi
P = force, lb
A = area, in^2

Tension stress tends to pull a member apart. In a member under tension the force acts upon a plane that is at right angles to the line of the force. A member under compression will also have the affected plane, or area, at right angles to the line of force. Stress components caused by a load acting normal, or perpendicular, to the plane of the structural member are known as **normal stress.** Tension and compression are both examples of a normal stress.

Shear stress tends to cause the material to divide into layers. In a shear stress the load is applied parallel to the surface, or plane, affected.

Strain. When a stress is applied to a piece of material, there is always some deformation of the material, even though it may be very small. This deformation is called **strain.** If the applied stress does not exceed the elastic limit of the material, it will return to its original shape as soon as the stress is removed. A tension stress will result in the strain being elongated, or in stretching of the material. Strain is measured in units of inches per inch (in/in) by the formula:

$$e = \Delta\frac{L}{L}$$

where e = strain, in/in
ΔL = change in length, in
L = the original length

If a piece of material measured 6.000 in at rest and 6.006 in under load, the strain would be 0.001 in/in. Strain is also expressed in terms of percentage of elongation.

Stress and strain can be measured by various types of testing equipment. Tensile testing is done as shown in Fig. 8-2. Material of a known cross-sectional area is gripped in the machine and a known force is applied, tending to pull it apart. The strain is measured by marking off a fixed distance before the metal is put under load. The distance is then measured under load. The amount of elongation can be measured and the strain calculated with the above formula. The load could be increased until failure of the metal occurred. This would provide a measure of strength under tension load, or tensile strength, for the material.

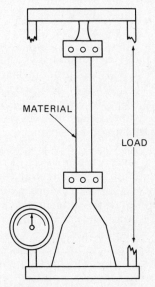

FIG. 8-2   Tensile testing.

A material could be tested under compression with similar equipment. Material under compression load will usually fail by **buckling** or bending (see Fig. 8-3) before it reaches its compressive strength. The point at which a member will buckle under compression will depend on the length of the member, its cross-sectional area, its cross-sectional shape, etc. As a result, compression test data is not widely used as a measure of the strength of a material. Most materials will ex-

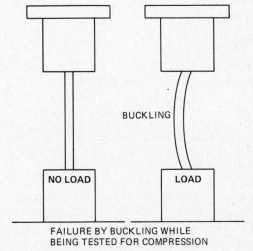

FAILURE BY BUCKLING WHILE BEING TESTED FOR COMPRESSION

FIG. 8-3   Compression testing.

hibit their greatest strength when loaded in tension. The amount of shear stress a material can take will always be less than the tensile stress.

Stress-Strain Diagram. Plotting the results of a tensile test produces a graph similar to the one in Fig. 8-4. The vertical axis of the graph shows the amount of stress (psi) and the horizontal axis is calibrated in units of strain. The amount of strain is plotted against the amount of stress causing it. The resulting curve, called a stress-strain diagram, contains considerable useful information. The diagram begins as a straight line indicating that the stress and strain are proportional. As stress increases, the elastic limit will be reached. This point, also known as the proportional limit, will be indicated by the end of the straight line portion of the diagram (*A*). Exceeding the elastic limit causes permanent deformation to take place. The point at which permanent deformation begins is known as the **yield point**, or **yield stress** (*B*). For some materials, when the yield stress is reached, the strain will continue to increase with little if any increase in stress. This is illustrated by the line in Fig. 8-4 that has a hook in it at the yield point. Increasing stress beyond the yield point causes a rapid rate of strain or deformation to occur until the material ruptures or fails (*X*). The point just prior to failure is known as the ultimate stress point or the **ultimate tensile strength** (*C*). The portion of the diagram from the origin (*O*) to the elastic limit (*A*) is known as the **elastic, or proportional, range** of the material. From the yield point (*B*) to the ultimate tensile strength (*C*) the material is said to be in the **plastic range**.

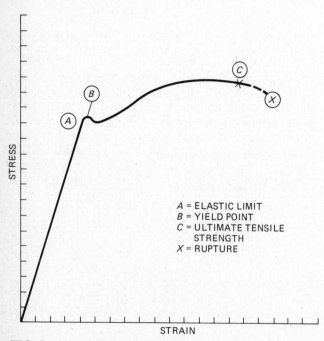

A = ELASTIC LIMIT
B = YIELD POINT
C = ULTIMATE TENSILE
    STRENGTH
X = RUPTURE

FIG. 8-4   A stress-strain diagram.

Some more ductile materials will have a curve similar to that shown in Fig. 8-5. The material does not exhibit an abrupt increase in strain at the yield point. The stress-strain diagram does not have a definite end

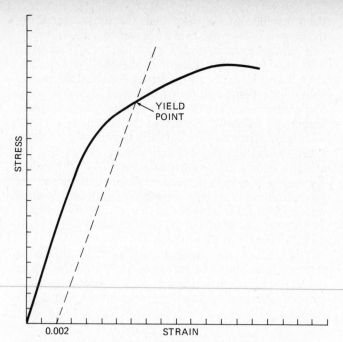

FIG. 8-5   A stress-strain diagram for a ductile material.

to the straight line portion. For a material of this type the yield point has been arbitrarily determined to be the point where a permanent strain of 0.002 occurs.

The angle formed by the straight line portion of the diagram and the horizontal axis indicates the **modulus of elasticity.** The modulus of elasticity is the value obtained by dividing the stress by the strain. Values for the modulus of elasticity are quite large. For example, the modulus for aluminum is 10×10^6, and for steel it is 30×10^6.

Stress-strain diagrams can be used to provide information about a specific material or to compare several materials. They can also be used to graphically demonstrate the mechanical properties covered in this section. Diagrams for four materials with widely varying properties are shown in the graph of Fig. 8-6.

The plasticity of a material is graphically represented by the portion of the diagram to the right of the yield point. Material *A* would have no plasticity. The abrupt failure of the material with little strain would classify it as brittle. Materials *B* and *C* have virtually the same elastic range and limits. However material *C* fails shortly after exceeding the yield point and has little plasticity. Material *D* has a lower elastic limit than *B* but has a similar value of plasticity.

Material *B* would be described as **stiffer** than material *D* because it has a higher modulus of elasticity. Material *B* also can withstand a higher stress before exceeding its elastic limit. Material *D* can have a higher strain, or more stretch, before exceeding the elastic limit and would generally be described as more elastic than *B*.

Toughness is sometimes defined as the total area under the curve of a stress-strain diagram. Using this definition, material *B* would have more toughness than material *D*. Even though they have similar plasticity, material *B* takes a higher stress. Because of the

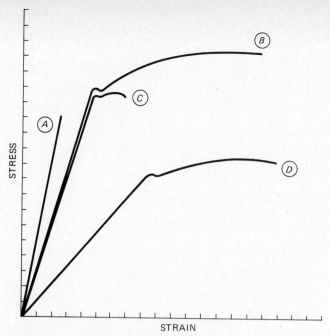

FIG. 8-6 Comparison of four materials.

low plasticity of material C, it could be considered as having less toughness than material D.

The stress-strain diagrams shown have been typical of those used for structural materials. Figure 8-7 shows a stress-strain diagram for a highly elastic material like rubber.

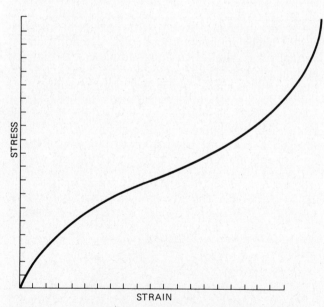

FIG. 8-7 A stress-strain diagram for an elastic material.

The strength of a material is usually reported in terms of the tensile yield stress or ultimate tensile stress. In Fig. 8-6 material B has the highest yield strength and could be described as the strongest material. However, many properties need to be considered before any material is stated as being the "strongest," "toughest," etc.

Physical Properties

Physical properties of materials that are of interest to the aircraft technician would include density, conductivity, and thermal expansion.

The **density** of a material is it's weight per unit volume, such as pounds per cubic inch. The density combined with the strength characteristics of a material produces what is known as the **strength/weight ratio** of a material. An application of the strength/weight ratio is shown in the following example.

Example. Given two materials, A with a tensile yield strength of 120 000 psi and a density of 0.28 lb/in^3 and B with a tensile yield strength of 60 000 psi and a density of 0.10 lb/in^3, provide a rod 8 in long of each material capable of supporting a load of 10 000 lb.

Solution. In the prior section we used the formula $f = P/A$ to find stress. In this example stress and load are known. To find the area, the formula will be expressed as $A = P/f$. Carrying out the calculations for the two materials finds that the cross-sectional area required for each material to be

$$A = 0.083 \text{ in}^2 \qquad B = 0.167 \text{ in}^2$$

Multiplying the area times the length (8 in) times the density of each material will give the weight of the required rod:

$$A = 0.186 \text{ lb} \qquad B = 0.136 \text{ lb}$$

Even though rod B will be larger in cross section, it will still weigh less than rod A because it has a better strength/weight ratio. Dividing the tensile strength by the density for each material provides the following values:

$$A = 4.3 \times 10^6 \qquad B = 6.0 \times 10^6$$

As illustrated in this example, material B has a better strength-to-weight ratio.

Specific gravity is used to compare weights of materials. The specific gravity of a substance is its weight divided by the weight of an equal volume of water. Since equal volumes are involved, specific gravity also is a ratio of densities.

Conductivity of a material may refer to either electricity or heat. **Thermal conductivity** is the property of a material to conduct heat. Thermal conductivity is desirable in a material used to conduct away excess heat. Insulating materials have a low thermal conductivity. **Electrical conductivity** is a measure of the material's ability to have electron flow.

Thermal expansion refers to the dimensional change that occurs as materials become hotter. Thermal expansion is of interest to those working with aircraft because of the temperature extremes, seasonal, performance, and altitude-related, in which an aircraft operates.

Chemical Properties

The chemical properties of a material refer to its atomic structure and basic elements. While they may not appear to be of interest or significance for the aircraft technician, the behavior of metals is greatly affected by chemical properties. The chemical properties of a metal determine its susceptibility to corrosion. The formability and hardness of a metal are related to chemical properties as is how it will respond to thermal treatment. Chemical properties of metals will be covered in a later section of this chapter.

● AIRCRAFT WOOD

Three forms of wood are commonly used in aircraft: solid wood, plywood, and laminated wood. To be used in an aircraft the wood must be of **aircraft quality.**

Solid Wood

The standard species of solid wood for aircraft use is Sitka spruce. Other species of wood that may be substituted for spruce include Douglas fir, noble fir, western hemlock, northern white pine, white cedar, and yellow poplar. The substitution of these materials for spruce should only be done under the guidelines established by the FAA in Advisory Circular AC43.13-1a.

A tree grows by developing new fibers around its circumference each year. During the spring the growth is more rapid and is distinguished by fibers of a larger size and thinner walls. The wood formed during this period is called **spring wood** and is lighter in color than the smaller thicker-walled fibers formed during summer growth. Spring wood is also weaker than summer wood. The dark colored layers of summer growth form what are called **annual rings.** The **grain pattern** of a wood refers to the directional orientation of the fibers. Since annual rings are layers of fibers, they also show the grain pattern. The grain pattern of annual rings are important criteria for evaluation of wood as aircraft quality.

Specification AN-W-2 for Sitka spruce specifies that the slope of the grain shall not be steeper than 1 in 15 and that the wood must be sawn vertical grain and shall have no fewer than six annular rings per inch. Measuring the slope of grain is shown in Fig. 8-8. Sawn vertical grain refers to the log being sawn in such a manner that the annual rings form an angle of 45 to 90° with the face of the board. This is accomplished by a process known as **radial sawing** as compared to plain sawing. Requiring no less than six annual rings ensures that the wood does not have an excess of the weaker spring wood.

The strength of wood will vary directly with the density. Aircraft spruce must have a specific gravity of at least 0.36. A defect called "compression wood" has the appearance of an excessive growth of summer wood. Wood with this defect would have a high specific gravity and should be rejected.

Wood must be kiln dried to be aircraft quality. As the moisture of a wood increases, the strength decreases. The use of kiln drying ensures that the wood has been brought to an acceptable moisture content. Wood fibers continually exchange moisture with the surrounding air depending upon the humidity. As moisture is absorbed or released, the fibers will expand and contract. Provisions must be made for the dimensional change which occurs. The dimensional change

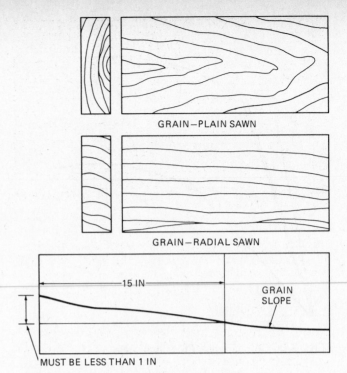

FIG. 8-8   Radial and plain sawn wood.

of a board will be the greatest across the fibers and parallel to the growth rings, somewhat less across the fibers and perpendicular to the growth rings, and negligible in a longitudinal direction. In essence, the greatest dimensional change (due to moisture) will occur to the thickness of a radial-sawn board.

Certain defects in wood will affect its use for aircraft. **Checks** are longitudinal cracks extending across annual rings. **Shakes** are longitudinal cracks between annual rings. Checks and shakes are formed during tree growth as shown in Fig. 8-9. **Splits** are longitudinal cracks caused by an artificially induced stress. Wood containing checks, shakes, or splits is not aircraft quality. A **spike knot,** shown in Fig. 8-10, runs through a beam perpendicular to the annual rings. Spike knots are not allowed in aircraft-quality wood.

Certain other defects may exist if they are within

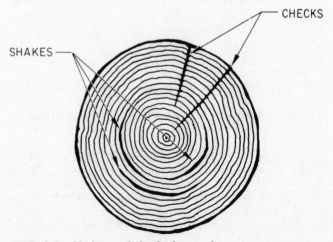

FIG. 8-9   Shakes and checks in wood.

FIG. 8-10 A spike knot.

limitations as listed in AC43.13-1a. These include wavy grain, hard knots, pin knot clusters, pitch pockets, and mineral streaks. Any form or evidence of decay is cause for the wood to be rejected.

Laminated Wood

Laminated wood is made up of a number of layers glued together. Unlike plywood, laminated wood has the grain running the same direction in all layers. The wood may have been laminated in order to form it to a shape, to use materials that have had defects removed, or to get a specific size. The FAA allows solid spruce spars to be replaced with laminated spars or vice versa provided the material is of the same quality.

Plywood

Aircraft grade plywood is made of imported African mahogany or American birch veneers laminated to cores of poplar or basswood with waterproof glue. Plywood made to specification MIL-P-6070 calls for shear testing of the plywood after immersion in boiling water for 3 h. Conventional plywood has the grain of alternate layers running at 90°. Aircraft plywood is also made in a style with grain at 45° angles in alternate plies. Aircraft plywood is used for skin or reinforcing plates.

● AIRCRAFT FABRICS

The fabric discussed in this section is used for covering aircraft wings, control surfaces, and fuselages. Fabric is a material of various types of threads woven at right angles to each other. The **warp** threads are those running parallel to the length of the cloth. Threads running across the warp are called the **fill**. The fabric as it comes from the mill has each edge bound by the fill so it will not unravel. This is called the **selvage edge** and can be used to identify which are warp threads. The manufacturer will normally print identification information on the fabric at the leading edge.

Specifications for fabric will normally reference the **thread count,** or the number of threads per inch. The thread count will be specified for both warp and fill, as they may be different. The fabric may have a weight specified which will usually be stated in ounces per square yard.

Aircraft covering materials were originally made

from cotton. Three specifications exist for the manufacture of cotton fabric. **Grade A cloth** (AMS3806) has a minimum of 80 and a maximum of 84 threads warp and fill. It is required on fabric-covered airplanes with wing loading in excess of 9 lb/ft^2 or a never-exceed speed in excess of 160 mph. Grade A cloth must have a tensile strength of 80 lb/in, warp and fill, when new. It is allowed to deteriorate to 56 lb/in (70 percent) in use before being replaced. Fabric is tested by putting a 1-in-wide strip in a tensile tester and pulling it to failure. Note that the figures are given as units of pounds per inches of width and not pounds per square inch.

Aircraft other than those requiring grade A fabric may use AMS3804 **airplane cloth,** sometimes called intermediate grade. Fabric made to this specification has a minimum of 80 threads and a maximum of 94 threads warp and fill. New strength for intermediate fabric is 65 lb/in. Gliders with a wing loading of less than 8 lb/ft^2 and a never-exceed speed of 135 mph or less can use AMS3802 cotton cloth. **Glider cloth** has a maximum of 110 threads per inch warp and fill with a new strength of 50 lb/in. **Aircraft linen** is manufactured to British specifications and can be substituted for grade A cloth.

The installation of a fabric cover requires several other types of materials. A waxed cord normally called rib-lacing cord is used to lace the fabric to the ribs and other parts of the structure. **Reinforcing tape** is a heavy cotton tape used over ribs between the fabric and the rib-lacing cord. Its purpose is to keep the cord from cutting through the fabric. **Surface tape,** also called **finishing tape,** is usually made in various widths from the same material as that used for the cover. The purpose of surface tape is to cover seams and rib-lacing cords.

The cover is not complete until the required finishes have been applied. The finishes for fabric covers include nitrate dope, cellulose nitrate dope, and cellulose acetate butyrate dope. Both clear and pigmented dope are used. The cover must be installed and finished in accordance with set standards. AC43.13-1A provides more details on fabric covers.

Synthetic materials have been developed as replacement for cotton fabrics. The most common synthetic is a **Dacron cloth** sold under several trade names. Dacron cloth has the advantage of being longer lasting than cotton. It can be heat shrunk to size and bonded seams can be used, making the application of the cover easier than cotton. Fiberglass has also been used as a cotton replacement. Virtually all fabric-covered aircraft were originally built with cotton fabric. Replacement of the cover with another material is a major alteration. The manufacturer of a synthetic material may obtain a supplemental type certificate (STC) to allow the installation of their product. If a synthetic material is being used, the instructions supplied with the STC must be strictly followed.

● METALS

General Properties of Metals

Crystalline Structure. Chemical elements may be roughly categorized into three groups; metals, nonmetals, and inert gases. Metals, in a solid state, are characterized by the following properties:

1. Crystalline structure
2. High thermal and electrical conductivity
3. Ability to be deformed plastically
4. High reflectivity

All elements are made up from various combinations of atoms. The atoms are bonded together by various forces. Temperature affects the energy levels within the bonds so that at cool temperatures the atoms are closely packed and form a solid material. As the temperature rises, the energy level between the atoms increases and the solid becomes a liquid and eventually a gas. The arrangement of the atoms affects the properties of a metal. During the process of solidification the atoms will arrange themselves in an orderly manner called a **space lattice.** The smallest unit having the same arrangement as the crystal is called the unit cell. The unit cell adds atoms in an orderly fashion and grows into a large crystal. The number of atoms in a small crystal may number in the billions. The number of atoms in the unit cell are few depending on the crystalline system. Seven systems of atom arrangement are known to exist but the important metals form either a cubic or hexagonal system. The cubic shape can be either **face-centered-cubic** (F.C.C.) or **body-centered-cubic** (B.C.C.). Figure 8-11 shows the atom arrangement of the three common systems and typical metals for that structure. Some metals will crystallize in one form and upon cooling change to another form. This type of change is called an allotropic change and is important to the processing of some metals. Metals with the face-centered lattice lend themselves to a ductile, plastic, workable state. Metals with the hexagonal lattice exhibit a general lack of plasticity and rapidly lose what they do have upon shaping such as cold forming. Metals with the body-centered lattice have properties between these two groups.

Figure 8-12a and b shows the formation of a metal crystal. From random unit cells (a) crystal growth takes place in three dimensions and continues until it is stopped by surrounding crystals (b). Crystals in a newly solidified metal will have random shapes and sizes as shown in Fig. 8-12c. The crystals found in commercial metals are commonly called grains. While

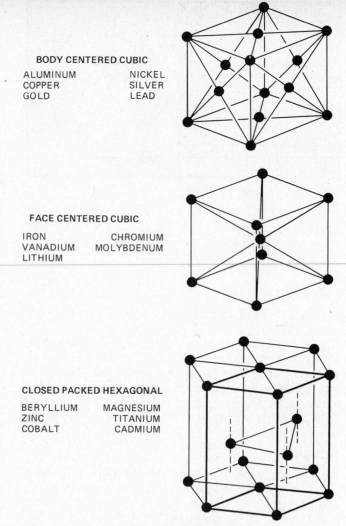

BODY CENTERED CUBIC

ALUMINUM	NICKEL
COPPER	SILVER
GOLD	LEAD

FACE CENTERED CUBIC

IRON	CHROMIUM
VANADIUM	MOLYBDENUM
LITHIUM	

CLOSED PACKED HEXAGONAL

BERYLLIUM	MAGNESIUM
ZINC	TITANIUM
COBALT	CADMIUM

FIG. 8-11   Common unit cell type for metals.

each grain has a random external shape, it is composed of layers of atoms arranged as previously discussed.

Within a crystal certain planes exist which are called **slip planes.** When an external force is applied to the crystal, the atoms along the slip planes move in relation to one another. If the force is less than the elastic limit of the material, the atoms will return to their original position after the force is removed. If the force

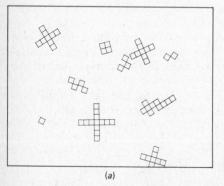
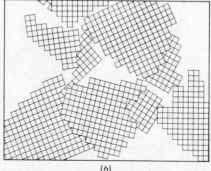
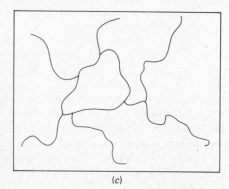

(a) (b) (c)

FIG. 8-12   Crystal growth as metal solidifies.

exceeds the elastic limit, plastic deformation will occur. The shape of the crystal will be permanently altered. As plastic deformation takes place, the slip planes are said to be used up. If the force continues, other slip planes will be brought into action or the metal will rupture. Figure 8-13 illustrates the grain changes taking place as a metal is deformed by being rolled to a thinner size.

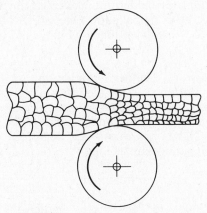

FIG. 8-13   Grain size and shape being altered by rolling.

Control of **grain size** is important to commercial metals. A fine-grain metal is usually tougher and stronger than one with a coarse grain. A coarse-grain metal will normally be easier to form to shape than one with a fine grain. The size of grain will obviously depend upon the use of the material. The grain sizes of metal are affected by the rate of cooling of molten metal. Slow cooling results in the formation of a small number of large grains. Fast cooling causes the formation of a large number of small grains.

Cold Working and Annealing. As a metal is formed, or cold worked, slip planes are used up in the crystals. As the slip planes are used up, the material becomes harder and stronger. This is known as **work hardening, strain hardening,** or **cold working** of the metal. If greater strength is desired, the effect of work hardening is good. If more forming is to be done on the part, work hardening is not good. The effects of work hardening can be removed by a process known as **recrystallization,** or **annealing.** Annealing is performed by heating the material followed by slow cooling. The recrystallization process takes place with the metal in a solid state. As the metal cools, new crystals are formed which have a new set of slip planes. The metal is once again easily deformed or shaped.

All metals are subject to cold working and annealing in varying degrees.

Alloys. An alloy is a mixture of two or more metals. Alloys consist of a **base metal** with small percentages of other materials. Alloys are used to enhance the properties of the base metal such as corrosion resistance, tensile strength, or workability. The elements of an alloy and base metal combine as solutions, compounds, or mixtures. Alloying elements often allow the base metal to be hardened or made stronger by heat treatment.

Corrosion. Corrosion is a problem for all metals. Corrosion is the decomposition of metallic elements into compounds such as oxides, sulfates, hydroxides, and chlorides. It is brought about by direct chemical action and electrolytic action. Chemical corrosion is brought about by an acid, salt, or alkali in the presence of moisture. Water provides the vehicle through which the chemical action takes place. Electrolytic action takes place when metals that have a different level of chemical activity are touching or in close proximity in the presence of moisture. The two dissimilar metals form the poles of a galvanic cell, an electric current flows, and the active metal is decomposed. The susceptibility of a metal to corrosion depends upon its position in the electromotive series. Corrosion protection and prevention is a major task in aircraft maintenance. Processes used for corrosion control will be covered in Chap. 9.

Aluminum

Aluminum is the principal structural metal for aircraft. A unique combination of properties makes aluminum a very versatile engineering and construction material. Light weight is perhaps aluminum's best known characteristic. With a specific gravity of 2.7, the metal weighs only about 0.1 lb/in^3 [2.8 g/cm^3] as compared with 0.28 lb/in^3 [7.8 g/cm^3] for iron and 0.32 lb/in^3 [8.86 g/cm^3] for copper. Commercially pure aluminum has a tensile strength of about 13 000 psi [89.6 MPa]. Its usefulness as a structural metal in this form is somewhat limited although its strength can be approximately doubled by cold working. Much larger increases in strength can be obtained by alloying it with other metals. Aluminum alloys having tensile strengths approaching 100 000 psi [689.6 MPa] are available. Aluminum and its alloys lose their strength at elevated temperatures, although some will retain good strength at temperatures as high as 400°F [204°C]. At subzero temperatures their strength increases without loss of ductility. Aluminum in general is considered as having good corrosion resistance. The ease and versatility of fabrication of aluminum surpasses that of virtually any other material.

Aluminum products are made in two forms, **cast** and **wrought.** Cast aluminum is formed into a particular shape by pouring into a mold of the required shape. Wrought aluminum alloy is made by mechanically working the metal into the form desired by rolling, drawing, and extruding it.

Wrought Aluminum Code. Wrought aluminum and aluminum alloys are designated by a four-digit system, with the first digit of the number indicating the principal alloying element:

1000 series. When the first digit is 1, the material is 99 percent, or higher, pure aluminum.

2000 series. Copper is the principal alloying element in this group. The addition of copper allows aluminum to be heat-treated to high strengths, but it also reduces its corrosion resistance.

3000 series. Manganese is the principal element in

this series. It is added to increase hardness and strength. Alloys of this series are non-heat-treatable.

4000 series. The major alloying element of this group is silicon, which can be added in sufficient quantities to cause substantial lowering of the melting point without producing brittleness. The principal use of 4000 series alloys is for welding wire.

5000 series. Magnesium is one of the most effective and widely used alloys for aluminum. When it is used as the major alloying element, the result is a moderate- to high-strength non-heat-treatable alloy. Alloys in this series have good welding characteristics and high corrosion resistance.

6000 series. Alloys in this group contain silicon and magnesium in approximate proportions to form magnesium silicide, thus making them heat-treatable. Though not as strong as the 2000 and 7000 alloys, the magnesium-silicon (or magnesium silicide) alloys possess good formability and corrosion resistance with medium strength.

7000 series. Zinc is the major alloying element of this group and, when coupled with a smaller percentage of magnesium, results in a heat-treatable alloy of very high strength. Usually other elements such as copper and chromium are added in small quantities.

The second digit of the code number indicates any modifications to the original alloy. In the 2xxx to the 8xxx numbers, the last two digits have little significance other than to identify the alloys and the sequence of development. In the 1000 series the last two digits indicate the amount of pure aluminum above 99 percent in hundredths of 1 percent. Aluminum identified with the number 1240 would be 99.40 percent pure aluminum. In the 1xxx series a modification is for the purpose of controlling one or more of the impurities. Examples of the alloy code are shown in Fig. 8-14.

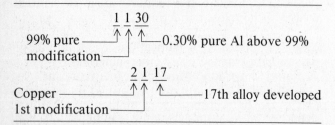

FIG. 8-14 Aluminum alloy identification.

Cast Aluminum Code. Cast aluminum also uses a four-digit identification system with the first digit indicating the alloy group. The numbers used are not the same as for wrought alloys. Cast alloy groups are:

1—aluminum, 99 plus percent
2—copper
3—silicon, with copper and/or magnesium
4—silicon
5—magnesium
6—not used
7—zinc

8—tin
9—other elements

The second two digits identify the aluminum alloy or indicate the aluminum purity. The last digit is separated from the other three by a decimal point and indicates the product form, i.e., castings or ingot. A modification of the original alloy is indicated by a serial letter before the numerical designation. Alloy A514.0 indicates an aluminum alloy casting with magnesium as the principal alloy. One modification to the original alloy has been made as indicated by the letter A.

Hardness and Temper Designation. An important factor for aluminum alloy identification is the **temper,** or **hardness,** value. A letter and number combination is placed after the alloy code to indicate the processes that have taken place and the degree of hardness. Basic designations are as follows:

F—as fabricated (no treatment)
O—annealed
H—cold worked or strain hardened (wrought products)
W—unstable condition (temporary condition while the material ages after solution heat treatment)
T—solution heat treated

The H designations are used only for the non-heat-treatable alloys. These are generally the alloys in the 1000, 3000, and 5000 series. These materials can only be hardened through the effects of cold working. The T designations are used after alloys that are capable of being hardened by thermal treatment. Heat-treatable alloys contain elements such as copper, magnesium, silicon, and zinc. Alloys from the 2000, 6000, and 7000 series use T designations. The W designation would apply only to those alloys capable of heat treatment. The annealed designation, O, and the as-fabricated designation, F, could apply to any of the alloys.

The H tempers are further subdivided to indicate the specific combination of basic operations. For example:

H1—strain hardened only
H2—strain hardened and partially annealed
H3—strain hardened and stabilized

A number following the H1, H2, or H3 indicates the degree of strain hardening of the alloy. The number 8 indicates the maximum degree of strain hardening has occurred. The number 2 indicates one-quarter hard, 4 indicates one-half hard, and 6 indicates three-quarters hard.

The T designation is followed by a number that indicates specific sequences of basic treatments. Frequently used numbers are:

T3—solution heat treated, cold worked
T4—solution heat treated
T6—solution heat treated and artificially aged

Cold working, or **strain hardening,** is any process applied at room temperature that stretches, compresses, bends, draws, or otherwise changes the shape of the metal to any appreciable degree.

Solution heat treating is a thermal process for hardening aluminum. It involves heating the material to a specified temperature causing the chemical structure of the material to be changed. The material is then quickly cooled or quenched. Upon quenching, the material is soft and unstable. Further chemical changes within the metal take place at room temperature until a stable condition is reached. This is known as **aging.** Acceleration of the aging process by additional thermal treatment is known as **artificial aging.** The material will not have developed full strength until it has stabilized through aging. Heat treatment of aluminum is covered in more detail in Chap. 9.

Corrosion. Aluminum is considered to be highly corrosion resistant under the majority of service conditions. When aluminum surfaces are exposed to the atmosphere, a thin invisible oxide skin forms immediately which protects the metal from further oxidation. Unless this coating is destroyed, the material remains fully protected against corrosion. Aluminum is highly resistant to weathering and is corrosion resistant to many acids. Alkalis are among the few substances that attack the oxide skin and are therefore corrosive to aluminum.

High-strength alloys containing copper are less resistant to corrosion than the other alloys. Alloys of this type often have a thin layer of pure aluminum rolled on each side. The pure aluminum acts as a barrier between the environment and the less resistant alloy. Aluminum of this type is known as **clad aluminum.** (Alclad is a trade name used by ALCOA for this product.)

While highly corrosion resistant by itself, aluminum is very susceptible to galvanic corrosion resulting from contact with other materials. Among the structural metals, aluminum is second only to magnesium on the electromotive series. Galvanic corrosion with any structural metal, other than magnesium, would result in aluminum being the material corroded or decomposed.

Workability. Aluminum can be cast by any method known. It can be rolled to any desired thickness including foil thinner than paper. It can be stamped, drawn, spun, or roll formed. The metal can be hammered or forged and there is almost no limit to the different shapes into which it may be extruded. Aluminum can be turned, milled, bored, or machined in other manners at the maximum speed at which the majority of the machines are capable. Almost any method of joining is applicable to aluminum—riveting, welding, brazing, soldering, or adhesive bonding.

Alloys Used for Aircraft. The majority of the aluminum products that the technician encounters will consist of the following alloys.

Non-Heat-Treatable Alloys

1100. Pure aluminum which is soft, ductile, and low strength. Its use on aircraft will be limited to nonstructural application. It is also used for making soft rivets.

3003. This alloy is similar to 1100 but has about 20 percent greater strength. It is used on aircraft for fluid lines and has a limited application for fairings or cowlings.

5052. The highest strength of the non-heat-treatable alloys, 5052 has high corrosion resistance and high fatigue strength. Having excellent workability, it is widely used for aircraft cowlings, fairing, and other nonstructural parts requiring forming. Fluid lines are another application on aircraft.

5056. This alloy is used to make rivets for riveting magnesium sheet.

Heat-Treatable Alloys

2017. Used more frequently on older aircraft, 2017 alloy will today be seen, if at all, only in aluminum rivets.

2117. A "modification" of alloy 2017, this alloy is used exclusively for the manufacture of aluminum rivets. The **2117** rivet is not as strong as a 2024 rivet, but it can be driven with no special treatment.

2024. 2024 is probably the "standard" structural metal as well as the most used metal for aircraft. It will be found in virtually every form available including sheet stock, extrusions, bar stock, standard hardware, and tubing. Heat treatable to high strengths, its ability to be cold worked is good to excellent for everything except rivets. Like most heat-treatable alloys, 2024 is not recommended for welding. To improve corrosion resistance, 2024 sheet stock is available as a clad material. 2024 ages rapidly after heat-treatment and most forms will be in the T3 condition. Alloy 2024 is highly susceptible to **intergranular corrosion** if heat treatment is done improperly.

6061. An easily worked metal, 6061 has a strength only two-thirds that of 2024. A good general-purpose material, it can be welded, offers high corrosion resistance, and can be worked by almost any means. To develop full strength after heat treatment it is normally artificially aged and will have a T6 designation.

7075. This is one of the highest strength aluminum alloys available and also is one of the more difficult aluminum alloys to work. Parts should be formed in the annealed state or at an elevated temperature. Arc and gas welding is not recommended. 7075 is available in a clad form to improve corrosion resistance. This alloy is normally artificially aged and carries the T6 designation.

Future Alloys. New materials are not limited to synthetic composites. An aluminum lithium alloy has been developed and is currently being tested. Aluminum lithium alloy is about 10 percent lighter than conventional aluminum and about 10 percent stiffer. It is estimated that new designs that take advantage of the alloy's greater stiffness could reduce aircraft weight by 15 percent. A number of problems in handling and fabrication are yet to be solved. Widespread use of the alloy by 1995 is predicted by those involved with the development.

Research has also resulted in the development of a high-temperature aluminum alloy. The still experimental alloy would more than double the present temperature limits for aluminum. Such an alloy would offer a lightweight alternative to many applications now requiring titanium.

Iron and Steel

Ferrous metals are those whose principal content is iron (Latin, *ferrum*), such as cast iron, steels, and similar products. Because of the vast number of different steels and steel alloys, we shall not attempt to describe more than a few of the more commonly used types.

A large percentage of the steels used for general aircraft work are of the wrought type and are designated as shown in Table 8-1. In addition to the standard group of wrought-carbon and alloy steels, a substantial number of heat- and corrosion-resistant steels are used in aircraft and missiles. The principal designations for these steels are given in Table 8-2.

In Table 8-1, the first digit of each number indicates the general classification of the steel, that is, carbon, nickel, etc. The number 1 indicates a carbon steel. The

TABLE 8-1 SAE Identification for Wrought Steels

Carbon steels	
10xx	Nonsulfurized carbon steel (plain carbon)
11xx	Resulfurized carbon steel (free machining)
12xx	Resulfurized and rephosphorized carbon steel
Alloy steels	
13xx	Manganese 1.75% (1.60–1.90%)
23xx	Nickel 3.50%
25xx	Nickel 5.00%
31xx	Nickel-chromium (Ni 1.25%, Cr 0.65%)
32xx	Nickel-chromium (Ni 1.75%, Cr 1.00%)
33xx	Nickel-chromium (Ni 3.50%, Cr 1.50%)
40xx	Molybdenum 0.25%
41xx	Chromium-molybdenum (Cr 0.50 or 0.95%, Mo 0.12 or 0.20%)
43xx	Nickel-chromium-molybdenum (Ni 1.80%, Cr 0.50 or 0.80%, Mo 0.25%)
46xx	Nickel-molybdenum (Ni 1.75%, Mo 0.25%)
47xx	Nickel-chromium-molybdenum (Ni 1.05%, Cr 0.45%, Mo 0.20%)
48xx	Nickel-molybdenum (Ni 3.50%, Mo 0.25%)
50xx	Chromium 0.28 or 0.40%
51xx	Chromium 0.80, 0.90, 0.95, 1.00, or 1.05%
5xxxx	Chromium 0.50, 1.00, or 1.45%, Carbon 1.00%
61xx	Chromium-vanadium (Cr 0.80 or 0.95%, V 0.10 or 0.15%)
86xx	Nickel-chromium-molybdenum (Ni 0.55 or 0.05 or 0.65%, Mo 0.20%)
87xx	Nickel-chromium-molybdenum (Ni 0.55%, Cr 0.50%, Mo 0.25%)
92xx	Manganese-silicon (Mn 0.85%, Si 2.00%)
93xx	Nickel-chromium-molybdenum (Ni 3.25%, Cr 1.20%, Mo 0.12%)
98xx	Nickel-chromium-molybdenum (Ni 1.00%, Cr 0.80%, Mo 0.25%)

TABLE 8-2 AISI Identification for Heat- and Corrosion-Resistant Steels

2xx	Chromium-nickel-manganese (nonhardenable, austenitic, nonmagnetic)
3xx	Chromium-nickel (nonhardenable, austenitic, nonmagnetic)
4xx	Chromium (hardenable, martensitic, magnetic)
4xx	Chromium (hardenable, ferritic, magnetic)
5xx	Chromium (low chromium, heat-resisting)

second digit of the number indicates the approximate percentage of the principal alloying element; for example, a 2330 steel contains more than 3 percent nickel. The last two digits of the number indicate the approximate amount of carbon in hundredths of 1 percent.

One of the most important considerations for ordinary carbon steel is the quantity of carbon it contains. A low-carbon steel contains 0.10 to 0.15 percent carbon. Medium-carbon steels contain 0.20 to 0.30 percent carbon. The higher the carbon content of steel, the greater its hardness and also its brittleness. High-carbon steels are used for cutting tools, springs, etc. For general purposes, low- or medium-carbon steels are best because they are more easily worked, they are tougher, and they have a much greater impact resistance. Such steels were used for many years in the manufacture of aircraft structures and fittings.

Probably the most commonly used steel for general aircraft structural purposes today is **SAE 4130 chromium-molybdenum (chrome-moly) steel.** When properly heat-treated, it is approximately 4 times as strong as 1025 mild-carbon steel. Depending on heat treatment, the tensile strength of 4130 steel will range from 90 000 psi [620.6 MPa] to more than 180 000 psi [1241.4 MPa]. SAE 4130 chrome-molybdenum steel is easily worked, readily weldable by any method, hardenable, heat-treatable, easily machined, and well adapted to high-temperature conditions of service. The technician should remember the name and number of this steel because it is commonly used for the tubular steel structures of aircraft.

Nickel steels, SAE 23xx and 25xx, contain from 3.5 to 5 percent nickel and a small percentage of carbon. The nickel increases the strength, hardness, and elasticity of the steel without appreciably affecting the ductility. Nickel steel is used for making various aircraft hardware including nuts, bolts, clevis pins, and screws.

Nickel-chromium and chromium-vanadium steels are used where still greater strength, hardness, and toughness are required. Such steels are often found in highly stressed machine parts such as gears, shafts, springs, and bearings.

It is not essential that technicians know the type of steel used in a particular factory-made part. However, if they need to fabricate a steel part which is no longer available, they must make sure that it is made from a material as good or better than the original and that it is properly stress-relieved and heat-treated if such treatment is necessary to provide the correct degree of hardness and strength. The nature and temper of the original part can often be determined by means of a Rockwell or Brinnell hardness tester.

Corrosion-Resistant (Stainless) Steels

Since the 1940s the term **stainless steel,** also designated corrosion-resistant steel (CRES), has become a household word because of its many applications in consumer items as well as in aircraft and missiles. The development of stainless steel has made possible many of the outstanding advances in aircraft, gas-turbine engines, and rockets. The most important characteristics of stainless steels are corrosion resistance, strength, toughness, and resistance to high temperatures. Stainless steels can be divided into three general groups based on their structures: **austenitic, ferritic,** and **martensitic.**

The austenitic steels are chromium-nickel (Cr-Ni) and chromium-nickel-manganese alloys. They can be hardened only by cold working, and heat treatment serves only to anneal them. They are nonmagnetic in the annealed condition, although some may be slightly magnetic after cold working.

Austenitic steels are formed by heating the steel mixture above the critical range and holding to form a structure called **austenite.** A controlled period of partial cooling is allowed followed by a rapid quench just above the critical range.

Ferritic steels contain no carbon; hence they do not respond to heat treatment. They contain a substantial amount of chromium and may have a small amount of aluminum. They are always magnetic.

Martensitic steels are straight chromium alloys that harden intensely if they are allowed to cool rapidly from high temperatures. They differ from the two preceding groups because they can be hardened by heat treatment.

The most widely used stainless steels for general use are those in the 300 series, called 18-8 because they contain approximately 18 percent chromium and 8 percent nickel. Typical of these types are 301, 302, 321, and 347.

Although stainless steels have many advantages, there are certain disadvantages that must be faced by the fabricator and designer: (1) Stainless steels are more difficult to cut and form than many materials. (2) Stainless steels have a much greater expansion coefficient than other steels, and they conduct heat at a lower rate; this makes welding more difficult. (3) Many of the stainless steels lose their corrosion resistance under high temperatures.

In the use of corrosion-resistant steels for aircraft, the technician must assure that the proper type is selected for the part of the aircraft involved. In most cases a damaged part can be replaced by a factory-made part identified by the part number; however, there are situations where it is preferable to repair a part by patching or welding. In these cases, the correct type of corrosion-resistant steel (CRES) must be chosen.

In welding CRES, inert-gas arc welding is preferred, because this process causes less deformation due to heat expansion of the metal and it prevents oxidation. The expansion of stainless steel due to temperature increases may be more than twice that of ordinary carbon steels.

Because of its toughness, stainless steel is more difficult to cut, form, shear, machine, or drill than is ordinary steel. For this reason the technician who is to work with this material successfully must be experienced in the necessary processes or must be directed by an experienced technician.

Magnesium

Magnesium alloys are used frequently in aircraft structures in cast, forged, and sheet form. The greatest advantage of magnesium is that it is one of the lightest metals for its strength. The disadvantages in the use of magnesium are that it is more subject to corrosion than many metals, it is not easily worked at room temperatures, and if it becomes ignited, it is extremely difficult to extinguish.

When magnesium is used in an airplane structure, it can often be recognized by the fact that it has a yellowish surface due to the chromate treatment used to prevent corrosion and furnish a suitable paint base. When technicians encounter magnesium in an aircraft, they must know that it cannot be cut easily but is likely to tear, it cannot be bent or otherwise worked under normal temperatures, it is subject to corrosion and therefore should be treated with the proper coating, and it presents a certain degree of fire hazard.

When standard parts are made of magnesium, this fact will usually be stated in the manufacturer's overhaul and service manuals. Also in the manuals will be directions for proper treatment of such parts.

Because of magnesium's tendency to corrode easily, it is incumbent upon the technician to make sure that the correct hardware items, such as the correct rivets, bolts and screws, are used with any magnesium parts; for example, rivets used with magnesium should be made of 5056-H aluminum alloy. Any metal part used with magnesium should be of a compatible metal or there should be no metal-to-metal contact.

Titanium

The use of **titanium** as a structural material has become prevalent only during the past three decades. Before then, the methods for refining and working titanium had not been developed to an extent that would make the use of the metal economically feasible.

Titanium was discovered in 1790 at Cornwall, England, by William Gregor, a priest who was also an amateur mineralogist. Gregor isolated the oxide of titanium from black magnetic beach sand. A few years later, Gregor's findings were confirmed by Martin Klaproth, a German chemist, who gave the name "Titan metal" to the new element. Later the name was expanded to titanium.

The first isolation of pure metallic titanium in sufficient quantity for practical study was accomplished by Dr. M. D. A. Hunter in 1906. Dr. Hunter was searching for a suitable material for electric-lamp filaments, and because of the reported high melting point of titanium, he believed it would be an ideal metal for the purpose.

The Kroll process, which has been widely used for

extracting titanium metal, was developed by Wilhelm Kroll, a Luxembourg scientist, in 1932. This process was improved and employed by the United States Bureau of Mines, which began in 1946 to produce titanium sponge in 100-lb batches. Since that time, continued improvement has taken place; titanium is produced today in relatively large quantities in rod, bar, sheet, and other forms for use in the manufacture of a wide variety of metal products.

Titanium and its alloys are used widely in the aerospace industry because of its high strength, light weight, temperature resistance, and corrosion resistance. The weight of titanium is approximately 56 percent of the weight of steel, but its strength is equal to that of steel.

The strength of titanium is maintained to temperatures of more than 800°F [427°C]; hence it is useful in the cooler sections of gas-turbine engines, for cowling and baffling around engines, and for the skin parts of aircraft which may be subjected to elevated temperatures that would be damaging to aluminum alloys. Supersonic transport airplanes utilize titanium extensively for the skin because of the atmospheric heating which occurs at high supersonic speeds. Titanium is also used for the manufacture of supersonic military aircraft.

Titanium may be worked by many of the methods employed for steel and stainless (corrosion-resistant) steel. It can be sheared, drawn, pressed, machined, routed, sawed, and nibbled. The operator handling titanium must be familiar with its peculiarities and special characteristics in order to obtain good results. The cutting dies and shear blades used in cutting titanium must be of good-quality steel and must be kept very sharp.

When titanium is exposed to high temperatures, 1000°F [538°C] and above, it must be protected from the atmosphere, because at these temperatures it combines rapidly with oxygen. The usual method of protection is to heat the metal in an atmosphere of argon or helium gas. One of the most satisfactory methods for welding titanium is, therefore, inert-gas welding.

One of the most outstanding properties of titanium is its resistance to corrosive substances, including some of the most troublesome industrial chemicals. It is uniquely resistant to inorganic chloride solutions, chlorinated organic compounds, chlorine solutions, and moist chlorine gas. It also has excellent resistance to oxidizing acids such as nitric or chromic acids. Strong reducing acids, however, will attack titanium. The resistance of titanium to corrosion by natural environmental substances is unequaled by other structural metals. It is completely inert when exposed to stagnant water, urban atmosphere, marine atmosphere, saltwater spray, and seawater.

Titanium has excellent properties in the pure form and also with the addition of various alloying elements. The pure form may have small amounts of carbon and nitrogen with maximums of 0.10 and 0.05 percent, respectively. These maximums are also a requirement for alloyed types.

The addition of 8 percent manganese to the Republic Steel RS-110A titanium alloy brings about an increase of tensile strength. The pure material may have

a tensile strength of 50 000 to 90 000 psi [344.75 to 620.55 megapascals (MPa)], and the addition of manganese brings this up to as high as 139 000 psi [958.4 MPa]. Aluminum in amounts of 3 to 7 percent is commonly used as an alloying element. Other alloying elements include molybdenum, tin, iron, chromium, and vanadium. Alloying and heat treating have made it possible to develop titanium products with more than 180 000 psi [124a MPa] tensile strength.

Two of the titanium alloys are Ti-6A1-4V and Ti-8A1-1Mo-1V. The latter of these, also called 8-1-1, is employed to a large extent on supersonic aircraft. This alloy was chosen because of its high creep resistance at high temperatures and its stiffness. Because of these qualities, it is more difficult to work than many of the other titanium alloys; however, its strength qualities outweigh the disadvantages of its workability. Extensive research has been carried out to discover the best methods for cutting, forming, and drilling the material, and it is now reasonably economical for the manufacturing of aircraft parts.

Titanium has a very low coefficient of thermal expansion; much lower than that of other structural metals such as Monel metal or stainless steel. The thermal conductivity is approximately the same as that of stainless steel. The low thermal-expansion coefficient simplifies the design of complex structures made with titanium because it is unnecessary to make such large allowances for expansion as those required for the metals with high expansion coefficients. Table 8-3 gives some characteristics of titanium alloys.

Titanium is used extensively in both military and commercial aircraft and in missiles because of its high strength/weight ratio, its freedom from stress corrosion and cracking, its ability to withstand high operating stress, and its high-temperature resistance. It is used for major aircraft structures, engines, and numerous small parts and components.

An example of the use of titanium in spacecraft is a space capsule manufactured by the McDonnell Aircraft Company, a division of McDonnell Douglas Corporation. Selected for the internal skin of the capsule was commercially pure titanium with usable strengths up to 900°F [482°C]. Hat-section stringers are fabricated from titanium alloy Ti-5A1-2.5Sn, a medium-strength alloy providing excellent weldability, resistance to oxidation up to 1200°F [649°C], and a fatigue endurance limit of approximately 60 percent of its ultimate strength of 125 000 psi [861.8 MPa].

In current spacecraft and jet aircraft, including supersonic types, titanium is extensively used wherever light weight, high strength, corrosion resistance, and heat resistance are important factors in the performance of the structure.

Illustrating the adaptability of titanium to precision operating equipment is the use of Ti-6A1-4V alloy for turbine wheels in a U.S. Navy missile. This application calls for a high-strength, heat-resistant material with low inertia forces because the wheel must attain a speed of 60 000 rpm [6281.4 rad/s] in a fraction of a second.

The working of titanium can be accomplished in much the same manner as that employed for sheet or stainless steel. It can be sheared, sawed, stretched,

TABLE 8-3 Titanium Alloy Properties

Specification No.		Producers				Composition, % max, (Bal. Ti)	Forms*	1000 psi [6.89 MPa]		% E. in 2 in [5.08 cm]
AMS	Military	Crucible	M-STC	Republic	TMCA			Y.S.	T.S.	
4900A		A55	MST-55	RS-55	T00025	Unalloyed	S,B,E,W	55	65	18
4901B	(MIL-T-7993-CL1)	A70	MST-70	RS-70	T00035	Unalloyed	S,B,E,W	70	80	15
4902		A40	MST-40	RS-40	T00020	Unalloyed	S,B,E,W,T	40	50	20
4908A	(MIL-T-9046-CL1)	C110M	MST-8Mn	RS-110A		8Mn	S	110	120	10
4911	(MIL-T-009046-CL2)	C120AV	MST-6A1-4V	RS-120A	T34615	6A1-4V	S,B,E,W	120	130	10
4921A	(MIL-5-9047-CL1)	A70	MST-70	RS-70	T00035	Unalloyed	B,F	70	80	15
4923	(MIL-T-9047-CL4)				T96035	2Fe-2Cr-2Mo	S,B,E,W,F	120	130	15
4925A	(MIL-T-9047-CL6)	C130AM	MST-4A1-4Mn	RS-130		4A1-4Mn	B,W,F	130	140	10
4926		A110AT	MST-5A1-2.5Sn	RS-110C	T00820	5A1-2.5Sn	S,B,W	110	115	10
4927	(MIL-T-9047-CL3)		MST-3A1-5Cr			3A1-5Cr	B,F	135	145	10
4928	(MIL-T-9047-CL5)	C120AV	MST-6A1-4V	RS-120A	T34620	6A1-4V	B,F	120	130	10
4929					T94520	5A1-1.5Fe-1.5Cr-1.20Mo	B,F	135	145	10
4941		A40	MST-40	RS-40	T00020	Unalloyed	T (welded)	40	50	20
4951		A40		RS-110C	T00020	Unalloyed	W		50	
4953		A110AT		RS-140	T00820	5A1-2.5Sn	W		115	
......						5A1-2.75Cr-1.25Fe	B	140	150	10
......		C105A		RS-110		3Mn-1.5Al	S,B,W	100	110	12
......		C115A		RS-110B		3.25Mn-2.25Al	S,B,W	110	120	10
......			MST-2.5Al-16V			2.5Al-16V	S,B,W	55	90	12
......					Ti-4Al-3Mo-1V	4Al-3Mo-1V	S	90	125	16
......					Ti-6.5Al-3Mo-1V	6.5Al-3Mo-1V	B	150	155	17
......		C130AMO				6.5Al-3.75Mo	B	152	162	16
......			MST-821			8Al-2Cb-1Ta	S,B	120	127	
......					Ti-8Al-1Mo-1V	8Al-1Mo-1V	B	132	137	18
......		C120VCA			Ti-13V-11Cr-3Al	13V-11Cr-3Al	S,B,W	120	125	10
......			MST-881			8Al-8Zr-1(CB + Ta)	S,B,F	125	135	16
......						7Al-4Mo	B,F	130	140	10

B: bar and billet   E: extrusions   F: forgings   *S: rolled flat products—sheet, strip, plate   T: tube   W: wire punched, and formed into a variety of shapes. It is not what would be called "easy" to work, but with care and proper tools and techniques it can be handled satisfactorily.

Two precautions must be observed while working with titanium. Both of these are necessary because of the strong affinity which titanium has for oxygen and other elements at high temperatures. At about 1950°F [1065°C], titanium will ignite in the presence of oxygen and burn with an incandescent flame. Its affinity for nitrogen is even more pronounced, because it will ignite at about 1500°F [815°C] with nitrogen.

When titanium is being cut or ground in any appreciable quantity, it is necessary to have fire-extinguishing equipment immediately available. The hot sparks from a grinding wheel can ignite an accumulation of titanium dust and chips to produce an extremely hot fire. It is recommended that liquid coolants of the proper type be used during grinding to avoid the possibility of such a fire.

As explained previously, if titanium is heated to temperatures above 1000°F [538°C] the metal should be protected by a surrounding atmosphere of inert gas. Otherwise, it will combine with the oxygen and nitrogen in the air at rates depending upon the temperature.

Copper and Its Alloys

Copper is one of the comparatively plentiful metals and has been used by human beings for thousands of years. It is easily identified by its reddish color and by the green and blue colors of its oxides and salts. It is very ductile in the annealed state but hardens with cold working. A primary use for comparatively pure copper is as an electrical conductor. Because of its conductivity, copper is used extensively for electrical wire. Before the discovery and development of aluminum as a useful metal, copper was used for tubing and in many other applications where aluminum is used today.

The principal alloys of copper are bronze, brass, and beryllium copper. Bronze is primarily a blend of copper and tin, the tin content being from 10 to 25 percent. Brass is an alloy of copper with 30 to 45 percent zinc plus small amounts of other metals. Beryllium copper is approximately 97 percent copper, 2 percent beryllium, and 1 percent other metals.

Bronze and brass are used extensively for bushings, bearings, valve seats, fuel metering valves, and numerous other applications.

Beryllium copper is heat-treatable and can be brought up to a tensile strength of 200 000 psi [1379 MPa]. Beryllium copper is used for precision bearings, bushings, spring washers, diaphragms, ball cages, and other applications where its qualities of wear resistance, toughness, strength, and elasticity are desirable.

Copper is alloyed with aluminum, manganese, silicon, iron, nickel, and other metals to make a variety of "bronzes." These are not true bronzes in the original sense of the word because they do not contain tin. Among these bronzes are aluminum bronze, silicon bronze, and manganese bronze. These alloys are available in sheet, bar, rod, plate, and other standard shapes.

The alloys of copper are developed to increase strength, corrosion resistance, and other qualities not possessed by pure copper. Designers select the alloy best suited for the purposes required, taking into consideration hardness, strength, wear resistance, and corrosion resistance. A technician finding that a part must be made should make sure to use the correct alloy as specified by the manufacturer.

Monel Metal

The value of Monel metal lies principally in its strength and corrosion resistance. It is a nickel alloy of approximately two-thirds nickel and one-third copper. Small amounts of other metals such as iron and manganese may also be included.

Monel metal is nonmagnetic in all conditions. It is easily worked in a manner similar to steel and has comparable strength. K-Monel includes a small amount of aluminum and is heat treatable to develop maximum strength. It is particularly useful in manufacturing durable parts that are or may be subjected to corrosive conditions.

High-Temperature Alloys

Because of the need for metals that can withstand the extremely high temperatures found in gas-turbine engines, afterburners, thrust reversers, etc., and because of the high temperatures generated by air friction at high supersonic speeds, it has become necessary to develop metal alloys which retain their strength even though hot. The products of high-temperature metal research have led to the development of **superalloys,** which utilize a wide variety of metal elements in combinations to produce the desired results.

High-temperature superalloys are those containing high percentages of nickel, cobalt, chromium, molybdenum, titanium, and other alloying elements that make them particularly resistant to heat and corrosion and retain high tensile strength at elevated temperatures such as 1000 to 2200°F [538 to 1205°C]. The qualities given consideration in the selection of high-temperature alloys for use in the hot parts of gas-turbine engines include thermal stability, tensile strength at elevated temperatures, low-cycle fatigue strength, stress rupture properties, hot-corrosion resistance, and oxidation resistance.

Hundreds of high-temperature alloys have been developed and tested; however, relatively few of these are in extensive use. Some of these are described briefly as follows:

Hastelloy alloy X. This alloy consists of approximately 50 percent nickel (Ni), 22 percent chromium (Cr), 9 percent molybdenum (Mo), and 0.07 percent carbon (C), with small amounts of other elements. It is used extensively for combustion liners and transition ducts in turbine engines. This alloy is manufactured by the Stellite Division of the Cabot Corporation.

Hastelloy alloy S. This is another Stellite Division alloy used in gas-turbine engines for hot sections where low thermal expansion is a requirement. This alloy has a high nickel content with 15.5 percent Cr and 14.5 percent Mo, together with small amounts of other alloying elements. The desirable properties of this alloy are low thermal expansion, good high-temperature strength, excellent surface stability, excellent thermal stability, and excellent fabricability.

Haynes alloy No. 25 (also known as Haynes stellite No. 25). This alloy has had extensive use in the gas-turbine industry for combustion liners, nozzle vanes, shrouds, spray bars, and flameholders. It is a cobalt-base alloy that is particularly well adapted to temperatures in the range of 1800 to 2100°F [982 to 1150°C]. It contains approximately 50 percent or more of cobalt (Co), which makes it quite expensive.

Haynes alloy No. 188. This is one of the newer cobalt-base alloys manufactured by the Stellite Division of the Cabot Corporation. It contains about 40 percent Co, 22 percent Cr, 14 percent Ni, 14 percent tungsten (W), plus small amounts of other elements. It is considered by some to be the best combustor alloy available. The high-temperature strength, surface stability, and fabricability of this alloy are all excellent. Because of its high cobalt and tungsten content, it is somewhat expensive. Typical components of gas-turbine engines produced from alloy No. 188 are combustion liners, transition ducts, afterburner liners, turbine nozzle baffles, turbine midframe liners, and turbine air seal rings. It can operate satisfactorily in temperatures up to 1800°F [982°C].

Haynes alloy No. 625. This is a nickel-base alloy containing approximately two-thirds Ni with substantial quantities of Cr, Mo, and columbium (Cb). It is used for the hot parts of gas-turbine engines and in the cast condition is used for turbine blades and vanes.

Inconcel alloy No. 617. This is a nickel-base alloy developed by Huntington Alloys, Inc., a subsidiary of International Nickel Co., Inc. (INCO), for use in extremely high-temperature situations. It is more than one-half Ni and also has substantial quantities of Cr, Co, and Mo. It is used for combustion liners, transition ducts, and other hot sections of turbine engines.

Haynes alloy No. 263. This alloy was developed by Rolls-Royce, Ltd. and was licensed to the Stellite Division of the Cabot Corporation for manufacture. This is a nickel-base alloy with more than 50 percent Ni plus Cr, Co, and Mo, with smaller amounts of titanium (Ti) and aluminum (Al). It is used primarily for the hot parts of Rolls-Royce engines.

The alloys listed in the foregoing paragraphs are used primarily for hot parts made of sheet alloy. In addition to these are alloys developed especially for casting to manufacture turbine blades and nozzle vanes. In earlier engines the nickel alloys with substantial quantities of cobalt were used for turbine blades and nozzles vanes. Two of the alloys were known as Waspalloy and René 41, developed by Pratt & Whitney Aircraft and General Electric, respectively. Currently, cobalt-base alloys such as MarM-509, AMF-5382, and X-40 are used for turbine blades and nozzle vanes. Nickel-base alloys for the same purpose are René 80, IN-713, IN-100, Udimet-41, Udimet-700, PWA-655, PWA-1455, and PWA-1422.

The foregoing alloys are only a few of the special alloys designed for use under high-temperature conditions in aircraft, jet engines, and missiles. However, their descriptions give an indication of the extensive developments that have been made by manufacturers to meet the requirements of modern technology. Figure 8-15 shows the ultimate tensile strength of some high-temperature alloys.

In the graph of Fig. 8-15, the stress values in the SI metric system are given in megapascals. One megapascal (1 000 000 pascals) is equal to approximately 145 psi. One ksi is equal to 1000 psi. See the Appendix for conversion factors.

● PLASTICS

The word *plastic* is derived from the Greek word *plastikos,* meaning "to form." In a broad sense, any material that can be formed into various shapes can be called a plastic; however, common usage of the term *plastic* limits it to synthetic materials that have been developed for industrial use and consumer products. Most of the materials called plastics are often termed "synthetic resins." Natural resins are usually produced from the sap or pitch of plants or from certain insects. Formerly, natural resins were used in the manufacture of varnishes and lacquers; however, synthetic resins are now usually used.

Classified according to how they react to heat, plastics are called **thermosetting resins** or **thermoplastic resins.** Thermosetting resins harden or set when heat of the correct value is applied. This type of plastic cannot be softened and reshaped after having been solidified. Thermoplastic resins can be softened by heat and reshaped or reformed many times without changing composition, provided that the heat applied is held within proper limits.

Thermosetting Resins

Common thermosetting resins include *phenolics, epoxies, polyurethanes, polyesters,* and *silicones.*

Phenolic resin is based on phenol and formaldehyde. These resins are resistant to heat, moisture, chemicals, and oils and are excellent insulators. They are therefore used extensively for various parts and insulators in electrical devices.

Epoxy resins find many uses in aircraft. They have excellent heat resistance, insulation qualities, dimensional stability, chemical resistance, and moisture resistance. When applied in a liquid state, they have outstanding adhesive qualities. Epoxy resins are used for potting, encapsulating, casting, reinforced laminates, adhesives, and protective coatings. Epoxies are usually supplied in two parts. The resin component is in a syrupy liquid state, and the curing agent may be a liquid or a powder. The curing agent is mixed with the resin just before the material is used. The time allowable between mixing and solidification is stated on the container. It is incumbent on the technician using the product to avoid mixing a larger quantity than can be used within the specified time.

In some modern aircraft, epoxy resin is used as an adhesive agent for metal bonding. The faying surfaces (surfaces to be joined) of metal sheet are coated with the resin or the resin is applied in tape form. The seams are then cured under heat and pressure. The resulting bond has excellent strength and durability.

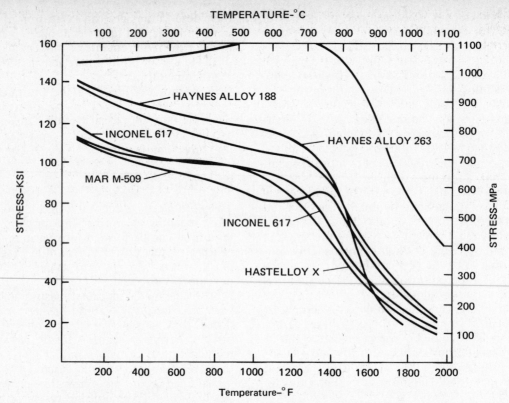

TEMPERATURE-°C

FIG. 8-15   Comparison of tensile strengths of high-temperature superalloy.

Another common use for epoxy resins is in coatings for aircraft. Epoxy primers provide an effective base for other finishes, such as polyurethane.

Polyurethane plastics may be used for rigid or flexible structures. They are commonly used as foams that, when solidified, make very light, heat-resistant, and thermal-insulating materials. Polyurethane can also be used as the internal structure for small flight control surfaces.

Polyurethane enamels make superior finish coatings for aircraft. When properly applied, the finish has a high gloss and there is no need for sanding, rubbing, polishing, or waxing. These finishes are weather-resistant and retain good appearance and quality for several years.

Polyester resins are commonly used for glass fiber laminates. In aircraft these are found as fairings, tail cones, antenna housings, radomes, cowling, wheel pants, and similar items.

When it is desired to make a particular glass fiber part, the first step is to make a wooden or plaster form or die in the shape of the item desired. This form is coated with a type of varnish that will prevent the plastic material from sticking.

The resin is usually a clear, syruplike liquid that will become hard after the addition of a suitable catalyst (curing agent). The catalyst can be obtained from the same source as the resin. The type and amount of catalyst will usually determine how soon the resin will "set up." In any case, the manufacturer's directions must be followed closely to obtain the desired result.

When the form has been made and the proper finish applied, the actual construction of the part may begin. The first step is to coat the form or die with a thick layer of the resin to which the proper type and amount of catalyst have been added. Strips of glass cloth are then carefully applied to the surface, completely covering the form over the area delineating the part to be made. After the layer of glass cloth is firmly in place, another coating of the resin is applied by brushing or spraying. A second layer of glass cloth is then applied and pressed firmly into place. This is followed by a coating of the resin. The number of layers of glass cloth and resin is determined by the thickness required.

If it is desired to have a perfectly smooth finish inside and outside the part, it is necessary that male and female dies be provided. The glass cloth and resin are laid up on one die, and then the other die is mated to it and the two sections are firmly pressed together. They are kept in this position until the resin has set.

After removal of the part from the form or die, it is usually necessary to trim the part and smooth the rough spots. The trimming can be done with several types of cutting tools, such as shears, a saw, or a knife. Smoothing can be accomplished with sandpaper, a sanding machine, or other means.

Silicone resins can be used for reinforced laminates with glass fiber, fibrous graphite, and other materials. The silicones are superior in heat resistance and for this reason are used in products exposed to high temperatures. Among high-temperature silicone materials are oils, greases, rubbers, and reinforced sheet.

Thermoplastic Resins

Among the thermoplastic resins that may be encountered by the aviation technician are **cellulose acetate,**

polyethylene, polypropylene, vinyls, polymethyl methacrylate (acrylic resin), polytetrafluoroethylene (Teflon), and **nylon.**

Cellulose acetate is used for transparent film and sheet. In aircraft, cellulose acetate is often used for windows and windshields.

Polyethylene resin is made in low-density and high-density qualities. Low-density polyethylene is made in thin, flexible sheet or film and is used for plastic bags, protective sheeting, and electrical insulation.

High-density polyethylene is used for containers such as fuel tanks, large drums, and bottles.

Vinyls are manufactured in a variety of types and have a wide range of applications. Their use in aircraft includes seat coverings, electrical insulation, moldings, and tubing. They are flexible and resistant to most chemicals and moisture.

Acrylic resin is a water-clear plastic that has a light transmission of 92 percent. This property, together with its weather and moisture resistance, makes it an excellent product for aircraft windows and windshields.

Polytetrafluoroethylene (Teflon) is encountered in nonlubricated bearings, tubing, electrical devices, and other applications. It is extremely tough and almost frictionless. It has good resistance to such high temperatures as 500°F [260°C] and remains flexible at low temperatures.

Windshields and Windows

The most suitable clear plastic material for windows and windshields is acrylic (polymethyl methacrylate) sheet. One of the best known brand names for this material is Plexiglas, manufactured by the Rohm & Haas Co. This material is manufactured in almost any size and thickness desired and can be used for many purposes other than windows and windshields.

Great care should be exercised in the handling and storing of acrylic sheet. The surface can be easily scratched if it is allowed to rub against any kind of rough surface. In handling the material, it is best that soft gloves be worn and that the sheets be stored on edge in a specially prepared rack. The new sheet is often protected with a paper mask, which is held to the surface by an adhesive. This masking paper can be easily peeled off when the material is installed. Acrylic sheet can be stored horizontally with the sheets stacked together, provided that the supporting surface is perfectly flat and smooth and that no particles of metal, wood chips, sand, or other foreign material are on the surface. If the sheet is masked, there will not be as much danger of damage as otherwise. When the material is stacked horizontally, large sheets should be placed at the bottom of the stack and then smaller sheets can be stacked in order of their size. This is to assure that all sheets have full support and will not sag and become deformed.

In the replacement of windows and windshields in aircraft, it is essential to make sure that the type of material to be installed is of the same quality as that being replaced. There are many types of transparent plastics, and their properties vary greatly, particularly with respect to expansion characteristics, brittleness under low temperatures, strength, etc. It may be noted here that acrylic plastics are stronger and more durable than the cellulose acetate types. Furthermore, the acrylics have a lower expansion coefficient.

When an acrylic plastic panel is installed, it should never be forced into place to make it fit. If the fit is poor, the panel should be trimmed or a new one should be obtained. When an acrylic plastic sheet is clamped or bolted in place, care must be taken that the material is not placed under excessive stress. If it is held in place by a nut and bolt or machine screw, the nut must not be turned up tight. The correct method is to tighten the nut to a firm fit and then back it off one turn. The purpose of this procedure is to allow for expansion and contraction of the material and to avoid crushing the point of attachment. In many cases where bolts are used to hold plastic sheet in place, the bolts are used with spacers or stops, which prevent overtightening. In the replacement of panels, the spacers, washers, and other parts should be installed as in the original configuration. The edges of plastic sheet should be mounted between rubber, cork, or protective material to reduce the effects of vibration and distribute compressive stresses on the material.

Acrylic plastics expand and contract about 3 times as much as the metal channels in which they are installed. It is therefore necessary that adequate provision be made to allow for this expansion and contraction. Clearances of $\frac{1}{8}$ in [0.317 cm] minimum should be allowed around the edges of small panels, and larger clearances around the edges of large panels. Where holes are drilled in the plastic material for bolts or screws, the holes should be oversize by $\frac{1}{8}$-in diameter and centered so that there will be no binding or cracking at the edges of the holes. Slotted holes are also recommended.

Panels of plastic must be mounted in the channels to a sufficient depth to prevent the panels from coming out when they shrink as the result of cold temperatures. Cellulose acetate panels are mounted in the same manner as acrylics; however, allowance must be made for greater expansion and contraction.

Plastic panels may be repaired when cracked or damaged, but when the damage is extensive, it is best to replace the panel with a new one. Small cracks can be stopped with $\frac{1}{8}$-in holes drilled at the end of the crack. Cracks and holes may be patched in accordance with directions given by the manufacturer of the plastic or according to directions provided by the FAA in Advisory Circular 43.13-A.

Clear plastic panels should be cleaned by washing with soap and water. Solvents and chemical cleaners should not be used, because there is danger that the cleaner will attack or soften the surface of the plastic. If, after dirt and grease are removed, no great amount of scratching is visible, the plastic should be finished with a good grade of commercial wax. The wax should be applied in a thin, even coat and brought to a high polish by rubbing lightly with a soft cloth. If the surface of a plastic panel has small scratches, it can be polished with a fine grade of polish and a soft cloth or buffing wheel. Care must be taken that the surface is not heated appreciably, because overheating will cause the material to soften and damage may result.

● COMPOSITE MATERIALS

Structural materials known as composites are made of many different materials and in a variety of forms. The use of plastic resins, as described earlier in this chapter, has made it possible to develop nonmetallic aircraft structures that are often superior to metals in strength/weight ratio, corrosion resistance, ease of fabrication, and cost. Aircraft have been developed that utilize composite materials for all their structural parts. The use of composite materials for parts such as wing tips, cowlings, fairings, flaps, spoilers, and ailerons is a common practice.

For purposes of this text composite materials will be divided into three categories: **solid laminates, honeycomb sandwich,** and **solid-core sandwich.** All three of these types of materials have a common function, to produce a lightweight material with high tensile strength. The material must also have adequate stiffness for compression and bending loads without requiring additional reinforcement.

Solid Laminates

Solid laminates are made by laying up resin-saturated cloth fabric or other reinforcing material over or within a mold to produce a desired shape. The laminate will be made up of a number of individual layers, called plies, of the reinforcing material. For full strength to be developed, the reinforcing fibers must be completely encapsulated by the resin. It is essential that each individual layer is in contact with the layers next to it with no void or air pockets. For individual parts, or those with complex shapes, the layups are usually done manually.

On laminated items such as tubing piping, pressure vessels, and cylindrical containers, fabrication may be done by machine using continuous bundles of fibers called rovings or tows. The roving is drawn through a reservoir of activated resin and is then machine-wound on a form. The windings will be directionally oriented so that successive layers will have the rovings running at different angles. This is done to provide bidirectional strength.

Much of the stiffness or compressive strength in a part made from a solid laminate is derived from the shape of the part and is not necessarily an inherent property of the materials used. An advantage of solid laminates is that a complex shape can be made in one piece.

Solid laminates are used in several forms for sandwich composites.

Honeycomb Sandwich Materials

A section of honeycomb sandwich material is shown in Fig. 8-16. The core is made up of corrugated material assembled in a way that resembles the honeybee's honeycomb. The face or skin is bonded to the honeycomb core. The facings provide high tensile strength. Bonding the two facings and core into one unit produces a lightweight material with the desired stiffness.

Honeycomb core can be made of solid laminates,

FIG. 8-16   Honeycomb core material.

paper, or metal. Honeycomb core made of stainless steel and titanium is used for high-temperature applications. The honeycomb core is specified by the type of material, the thickness of the material, and the size of the hexagon-shaped cell. Honeycomb material is manufactured in blocks, allowing the core material to be cut to the desired thickness or shape.

The facings can be metal or solid laminate materials. The combination of materials chosen for a honeycomb sandwich will depend upon the conditions in which the component will be used. Aircraft parts made from laminated core material are often made by shaping the core and bonding on the skins.

Solid-Core Sandwich Composites

A solid-core sandwich material is similar to honeycomb core materials in design. Instead of the open cell honeycomb material, the core is a low-density solid material such as balsa wood or styrofoam. Bonding the skins to the core provides the necessary stiffness to complement the tensile strength of the facing material. Numerous combinations of material can be used for solid-core composites. Because of the ease of use in forming shapes, the use of solid laminates is very popular.

Figure 8-17 illustrates the construction of a wing made entirely of solid laminates and solid-core sandwich construction. The wing is made up of molded upper and lower halves. Each half is made up of a sandwich of styrofoam with solid laminate facings. The use of laminated materials allows each half to be laid up in its mold. An I-beam-shaped main spar and a rectangular rear spar with connecting ribs are all made of sandwich construction with a foam core. When all the parts are laminated in place, the wing in effect becomes one piece with no rivets or other fasteners that may come loose. The wing in the drawing is made with fiberglass, but any composite fabric would be suitable.

Figure 8-18 shows the cross section of a wing built of similar materials but built in what is known as a "moldless" method of construction. In this case the whole wing is a solid core. Styrofoam, or a similar ma-

⬛⬛⬛⬛ = FOAM CORE SANDWICH STRUCTURE
(MIN. 3 PLIES FIBERGLASS EACH SIDE)

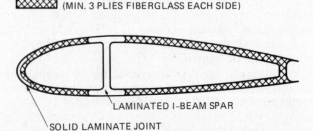

LAMINATED I-BEAM SPAR

SOLID LAMINATE JOINT

FIG. 8-17   A wing design with sandwich construction.

terial, is first cut to the desired airfoil shape. All lines, wires, cables, etc., that will pass through the wing are either put in place in the styrofoam or a passage is made for them. The entire core is then covered with the necessary number of layers of resin-impregnated material and finished. Moldless construction is widely used by builders of experimental home-built aircraft.

SOLID FOAM CORE COVERED
WITH FIBERGLASS LAMINATE

FIG. 8-18   A wing design for moldless construction.

Laminate Materials

In-depth coverage of the various materials being used and developed for laminated composites far exceeds the scope of this text. The following is an overview of the common materials in use today for laminated structure.

Two types of material are involved in advanced composites, the fibers or reinforcing material and the resin. The fibers are usually woven into a cloth. Like the aircraft fabric covered earlier in the chapter, composite fabrics are specified by thread counts (warp and fill), tensile strength per inch of width, and weight in ounces per square yard.

The **weave patterns,** or methods in which the warp and fill threads interlace, are also a part of the specification. A number of weaves are available, including the three shown in Fig. 8-19. The **plain weave** consists of a warp thread woven over and under alternate fill threads. The plain weave is characterized by fabric stability with minimum pliability except at low thread counts. The **crowfoot weave** has a warp thread going

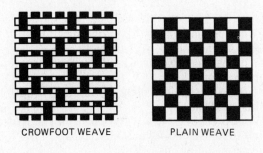

CROWFOOT WEAVE   PLAIN WEAVE

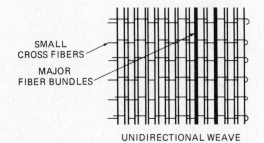

SMALL
CROSS FIBERS

MAJOR
FIBER BUNDLES

UNIDIRECTIONAL WEAVE

FIG. 8-19   Bidirectional plain and crowfoot weave and unidirectional weave.

over three and under one fill thread. The crowfoot is characterized as being more pliable than a plain weave and being better able to conform to complex shapes or compound curves. Both the crowfoot and plain weave are **bidirectional** meaning that they have virtually equal strength whether across the warp or the fill threads. A **unidirectional** fabric has most of the fibers running parallel to the warp and will have its strength in that direction. Unidirectional fabric is used in areas where the primary loads are only in one direction such as in the cap strip of a spar.

Fibers used for composite materials are made of two types of fiberglass (E glass and S glass), Kevlar, graphite (carbon), and ceramics; they are defined below. Table 8-4 compares these materials in terms of cost, weight, stiffness, heat resistance, toughness, and impact resistance. As with most materials, the one selected for a task will often be a compromise.

TABLE 8-4   Comparison of Composite Fabrics (Best = 1, Worst = 5)

Material	E glass	S glass	Kevlar	Graphite	Ceramic
Cost	1	2	3	4	5
Weight (density)	4	3	1	2	5
Stiffness	5	3	2	1	4
Heat	3	2	4	5	1
Toughness	3	2	1	5	4
Impact resistance	3	2	1	5	4

E glass. Plain glass fibers. Most inexpensive and one of heavier materials.

S glass. The chemical formulation is different than that for E glass. S glass is 30 percent stronger and 15 percent stiffer and will retain its properties up to 1500°F.

Kevlar. Kevlar is the registered trademark of Dupont for an aramid fiber. Kevlar ranks high is stiffness, toughness, and impact resistance. It is also lightweight. Kevlar can be identified by its yellow color. An unusual benefit is that cowlings made of Kevlar transmit less noise and vibration than one made of glass or graphite. The use of polyester resin with it is not recommended because of poor bonding. Vinyl-ester resin has been found to be the system most compatible. Kevlar is a difficult material to cut.

Graphite. High strength, high stiffness, and low density are strengths of graphite fabrics.

Ceramic. The main advantage of ceramic cloth is its ability to withstand temperatures of almost 3000°F. Otherwise it is heavy, very expensive, and comparable with S glass in strength.

The fabrics listed above are made in many different weaves, weights, and strengths. Table 8-5 shows values for an example of each type of fabric.

The resin systems used with the above fabrics will probably be based on an epoxy, a polyester, or a vinyl-ester resin. A number of different systems are manufactured. The manufacturer's directions should be carefully followed.

TABLE 8-5 Composite Fabrics

Material	Weight, oz/yd^2	Thickness, in	Thread count, W × F	Weave	Tensile, warp	Strength, fill
E glass	3.70	0.0055	24 × 22	Plain	160	135
S glass	3.70	0.0050	24 × 22	Plain	205	175
Kevlar	5.00	0.0100	17 × 17	Crow	630	650
Graphite	5.70	0.0070	12.5 × 12.5	Plain	1704	1704
Ceramic	7.50	0.0090	48 × 47	Crow	>200	>200

● REVIEW QUESTIONS

1. Discuss the five types of *stress*. Give two common examples of each type.
2. Why are materials normally not rated in terms of compressive strength?
3. What is the difference between *elasticity* and *plasticity?*
4. How can you estimate the *modulus of elasticity* from looking at a stress-strain diagram?
5. Name some common structural materials used for aircraft.
6. What characteristics must materials for modern, high-performance aircraft possess?
7. What types of woods are satisfactory for aircraft structures?
8. What defects in aircraft woods make them unsuitable for use?
9. Describe some commonly used aircraft fabrics.
10. Give the *AMS numbers* that describe fabrics approved for aircraft use.
11. How can a technician determine whether a synthetic fabric is approved for use on aircraft?
12. Why is *aluminum alloy* a good structural material for aircraft?
13. What is the principal alloying element for *2024 aluminum alloy?*
14. Describe the *aluminum alloy 7075-T6*.
15. How are steels designated according to SAE numbers?
16. What is the SAE number for *chrome-molybdenum steel?*
17. Name some parts for aircraft that are usually made from nickel steel.
18. What alloying elements are used in *18-8 stainless steel?*
19. What AISI numbers are used to designate 18-8 stainless steel?
20. What difficulties are encountered in the use of stainless (corrosion-resistant) steels?
21. What type of welding process is preferred for *CRES?*
22. What are the advantages of magnesium for aircraft parts?
23. What are the disadvantages of magnesium?
24. What are the principal advantages of titanium as a structural material for aircraft?
25. Why is titanium used extensively for the skin of supersonic aircraft?
26. Discuss the methods for working titanium.
27. What precautions must be taken when titanium is heated to a temperature above 1000°F [538°C] in welding the materials?
28. What are the principal uses of copper in aircraft?
29. What are some alloys of copper?
30. For what purpose is *beryllium copper* used?
31. Describe *Monel metal*.
32. Discuss *high-temperature superalloys* and name the basic metals of which they are composed.
33. What are the principal characteristics desired in *high-temperature alloys?*
34. Discuss *Hastelloy alloy X,* its composition, and its use.
35. Describe *Haynes alloy No. 188*.
36. List some of the high-temperature alloys used for turbine blades and nozzle vanes.
37. What is the difference between thermosetting and thermoplastic plastics?
38. Name a few common thermosetting resins.
39. Discuss uses of *epoxy resins* in aircraft.
40. List some of the uses of *polyurethane resins*.
41. What are the advantages of *polyurethane enamels?*
42. For what purpose are polyester resins commonly used?
43. Describe a process by which a part may be made from fiberglass-reinforced polyester resin.
44. Discuss the uses of *polyethylene*.
45. Why is *Teflon* a useful resin for aircraft applications?
46. What materials are used for aircraft windshields and windows?
47. What precautions must be taken in the handling and installation of windshield plastic?
48. What installations practices are designed to deal with expansion characteristics of acrylic plastics?
49. Describe *honeycomb sandwich material*.
50. What materials may be used in a honeycomb sandwich?
51. What parts of aircraft may be made of honeycomb sandwich?
52. How are solid laminates constructed?

9 FABRICATION TECHNIQUES AND PROCESSES

PROCESSES APPLIED TO AIRCRAFT MATERIALS

This section describes various processes utilized in the production and preparation of metals and alloys used in the manufacture of aircraft and explains processes employed by technicians in the maintenance of aircraft. Detailed processes of fabrication are not covered in this section but are presented in the chapters dealing with the specific area of repair involved.

Basic processes applied to aircraft metals include heat treating, case hardening, hardness testing, nondestructive testing, and treating to prevent corrosion.

MILL PRODUCTS

Metal materials are supplied in a number of shapes and sizes. A few of interest to the technician are described below:

Bar. A solid product that is long in relation to its cross section, which is square or rectangular (excluding plate or flattened wire) with sharp or rounded corners or edges, or is a regular hexagon or octagon and in which at least one perpendicular distance between parallel faces is $\frac{3}{8}$ in [9.5 mm] or greater.

Foil. A rolled product rectangular in cross section of thickness less than 0.006 in [0.15 mm].

Forging. A metal part worked to a predetermined shape by one or more processes such as hammering, upsetting, pressing, rolling, etc.

Pipe. Tube in standardized combinations of outside diameter and wall thickness, commonly designated by nominal pipe sizes and ANSI schedule numbers.

Plate. A rolled product rectangular in cross section and form with thickness of 0.250 in [6.35 mm] or more and either sheared or sawed edges.

Rod. A solid product that is long in relation to its cross section, which is $\frac{3}{8}$ in [9.55 mm] or greater in diameter.

Shape. A wrought product that is long in relation to its cross-sectional dimensions and has a cross section other than that of sheet, plate, rod, bar, tube, or wire.

Sheet. A rolled product rectangular in cross section and form with a thickness of 0.006 through 0.249 in [0.15 through 6.32 mm] with sheared, slit, or sawed edges.

Tube. A hollow wrought product that is long in relation to its cross section, which is round; is a regular hexagon, a regular octagon, elliptical, square, or rectangular with sharp or rounded corners; and has uniform wall thickness except as affected by corner radii.

Wire. A solid wrought product that is long in relation to its cross section, which is square or rectangular with sharp or rounded corners or edges, or is round; is a regular hexagon or a regular octagon; and has diameter or greatest perpendicular distance between parallel faces (except for flattened wire) that is less than $\frac{3}{8}$ in [9.5 mm].

Aluminum alloy sheet as it comes from the manufacturer is usually marked with letters and numbers in rows about 6 in [15.24 cm] apart. These identification symbols may include a federal specification number, the alloy number with temper designation, and the thickness of the material in thousandths of an inch. Standard methods for marking aluminum sheet are shown in Fig. 9-1. Note that the rows of letters and figures are parallel to the grain of the metal. Methods for marking foil sheet and other shapes are shown in Fig. 9-2. Items like rivets are too small for conventional markings and are identified by symbols and numbers in the metal.

HEAT TREATING

Heat treating is any method employed for the controlled heating and cooling of metals in order to develop the desired hardness or softness, ductility, tensile strength, and grain structure. Annealing, normalizing, tempering, and hardening are all heat-treating processes. Quenching (rapid cooling) in oil, water, brine, or air is a part of the various heat-treating processes.

It must be noted that not all metals and alloys can be heat-treated. Ferrous metals such as iron and steel can usually be heat-treated; however, many of the corrosion-resistant (stainless) steels cannot be heat-treated. Some of the alloys of aluminum can be heat-treated, but others must be hardened by cold working. The high-temperature superalloys can be heat-treated in varying degrees, depending upon their composition. The temperatures involved vary considerably from alloy to alloy, and it is essential that the exact temperatures and times specified by the manufacturer be employed to attain the qualities desired.

Events in Heat Treating

The general cycle of events in heat treating includes the following processes:

191

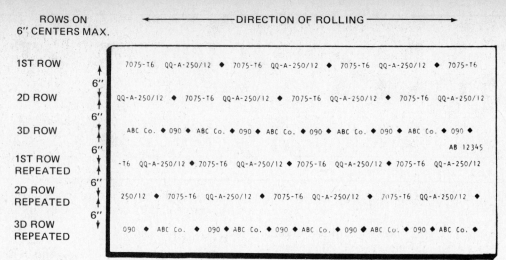

FLAT SHEET AND PLATE – 0.012 THROUGH 0.375 IN THICKNESS
6 THROUGH 60 IN WIDTH 36 THROUGH
200 IN LENGTH

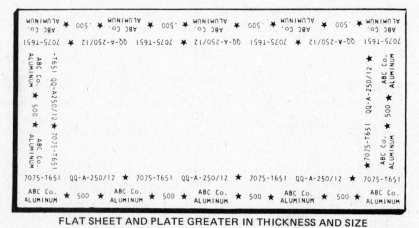

FLAT SHEET AND PLATE GREATER IN THICKNESS AND SIZE
THAN SHOWN ABOVE.

FIG. 9-1   Identification markings for aluminum alloy sheet.   *(Aluminum Association, Inc.)*

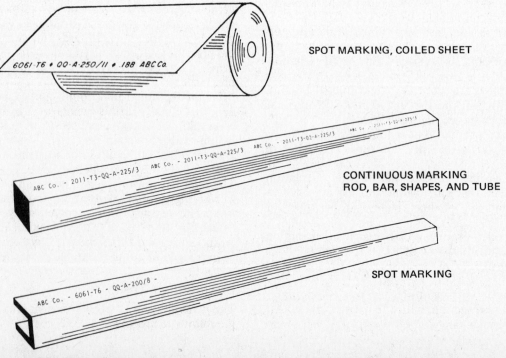

SPOT MARKING, COILED SHEET

CONTINUOUS MARKING
ROD, BAR, SHAPES, AND TUBE

SPOT MARKING

FIG. 9-2   Markings for coiled and other shapes.   *(Aluminum Association, Inc.)*

1. **Heating** a metal to a temperature within or above its critical temperature under conditions that are carefully controlled. The **critical temperature** is the temperature level where a phase change takes place in the crystalline structure of the metal.

2. **Soaking,** or **holding,** is the process of keeping a metal at an elevated temperature for a definite period of time so that it can become thoroughly saturated with heat and the necessary changes in grain structure can take place.

3. **Cooling** (quenching) is returning the metal to room temperature by various methods including air cooling or immersion in a liquid bath.

Heat-Treating Aluminum Alloys

Basic Purposes. The basic purposes of heat-treating aluminum alloys are to increase their strength, improve their corrosion resistance, and to improve their workability. The latter effect is accomplished by annealing.

Steps in Heat-Treating Aluminum Alloys. As mentioned previously, there are two principal steps in a heat-treating process: (1) heating the material to a required temperature for a specified time and (2) cooling the metal in a prescribed manner. An additional conditioning process takes place automatically for certain alloys. This is **age hardening,** which occurs over a period of hours or days after the material is quenched.

Methods for Heating Metal. There are two principal methods for heating the metal during a heat-treating process: (1) a furnace, illustrated in Fig. 9-3 and (2) a molten salt bath. The liquid salt bath has the advantage of rapid heating and uniformity of temperature, but the hot-air furnace is more flexible in operation and is not as hazardous as the salt bath. The salt bath usually consists of molten sodium nitrate, potassium nitrate, or a combination of the two. The use of the salt-bath method requires additional washing of the parts after quenching.

Since close control of the temperature is necessary during heat treatment, an automatic control should be used with a recording device that produces a permanent record of the time and temperature relations. The temperature of the furnace is detected by means of a pyrometer in connection with a recording device. The pyrometer is usually of the thermocouple type.

The furnace must be arranged so that the parts can be immediately transferred to the quench. This is im-

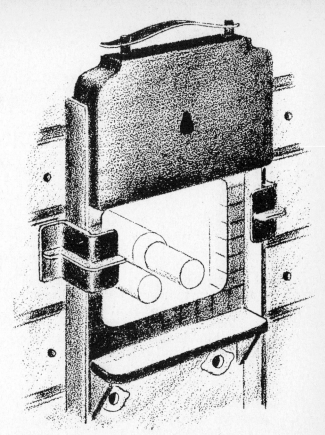

FIG. 9-3   A heat-treating furnace.

portant, because the parts must remain above the required temperature until they are quenched. If the metal cools below the required temperature before quenching, its corrosion resistance will be substantially reduced.

Conditions for Heat Treatment of Aluminum Alloys. Table 9-1 gives the temperatures required for **solution-heat-treating** common aluminum alloys together with quenching and aging data.

The time required for soaking a particular type of alloy depends upon the alloy and the thickness of the parts. Those parts clad with pure aluminum should not be soaked longer than necessary because diffusion of the alloying elements into the pure-aluminum coating takes place and can lead to a reduction in corrosion resistance. Table 9-2 gives suggested times for soaking of heat-treatable aluminum alloys. Solution

TABLE 9-1   Aluminum-Alloy Heat-Treating Data

Alloy	Temperature		Quench	Aging temperture		Time of aging
	°F	°C		°F	°C	
2014T	930–950	499–510	Hot water	335–345	168–174	10 h
2017T	930–950	499–510	Cold water	Room	Room	4 days
2117T	930–950	499–510	Cold water	Room	Room	4 days
2024T	910–930	488–499	Cold water	Room	Room	4 days
5053T	960–980	515–527	Water	312–325	155–163	18 h
6061T	960–980	515–527	Water	315–325	157–163	18 h
7075T	860–930	460–499	Cold water	345–355	174–179	6–10 h

TABLE 9-2 Aluminum-Alloy Soaking Times

Alloy	Soaking time, min			
	Less than 0.032 in [0.081 cm] thick	0.032–0.125 in [0.317 cm] thick	0.125–0.250 in [0.635 cm] thick	More than 0.250 in [0.635 cm] thick
2014T	20	20	30	60
2017T	20	30	30	60
2117T	20	20	30	60
2024T	30	30	40	60
2024T (clad)	20	30	40	60
5053T	20	30	40	60
6061T	20	30	40	60
7075T	25	30	40	60
7075T (clad)	20	30	40	60

heat treating involves heating the metal to a level that allows the alloying elements to go into solid solution with the base metal and soaking at the required temperature until the solid solution is complete.

It will be noted in the table that the thinner sheets of clad material have slightly reduced soaking periods. This is to reduce the diffusion of alloying elements into the pure-aluminum coating.

Precipitation Heat Treatment. The process of heating and quenching metal for heat treating is called **solution heat treating.** This simply means that alloying elements are in **solid solution** with the base metal. Solid solution means that the alloying elements are evenly dispersed throughout the material.

The process of **artificially aging** an alloy after heat treating to increase its strength is called **precipitation heat treatment.** When the temperature of the alloy is raised to an intermediate level as given in Table 9-1, certain alloying elements tend to precipitate out of the solid solution and form particles throughout the material. These microscopic particles give additional strength to the material. The time involved in the precipitation heat treatment is given in the table. Table 9-3 lists the common temper designations for aluminum-alloy materials.

Intergranular Corrosion. One of the effects of improper heat treating is **intergranular corrosion.** This is a condition wherein corrosion takes place between the grains of the metal. If an alloy is not quenched quickly enough during the heat-treating process, undesired precipitation of alloying elements takes place and this causes formation of tiny electrolytic cells within the metal. Electrolytic action then takes place and brings about destruction of the aluminum.

Intergranular corrosion is observed as blisterlike flaws or irregularities on the surface of the metal. Penetration of the raised spots will reveal a white powder and metal flakes (exfoliation) where the alloy has decomposed.

Annealing. The **annealing process** for softening aluminum and aluminum alloys is a form of heat treatment requiring the heating and cooling of the material in accordance with specific instructions. The temperature range for a full anneal is 750 to 800°F [399 to 427°C]. The full-annealing process is sometimes called **recrystallization.** After annealing, the metal is in the "0" temper condition, which is dead soft. The soaking time and cooling rates for annealing depend upon the type of alloy. In general, the metal should be soaked at the annealing temperature for at least 1 hour (h), but it may require more than this for some materials. The cooling rate should be not more than 50°F [28°C] per hour until the material has reached a temperature of less than 500°F [260°C]. For non-heat-treatable alloys or pure aluminum, the cooling rate is not important.

Stress Relief. When aluminum or aluminum alloy is formed into various shapes from the flat-sheet condition, the formed product will usually contain areas where internal stresses exist. These stresses may cause distortion, cracking, or corrosion during the life of the part. It is therefore necessary that these stressed areas be eliminated, and this is one function of heat treatment. The stresses in aluminum and its alloys can be reduced by solution heat treatment or by partial annealing at temperatures between 700 and 800°F [371

TABLE 9-3 Aluminum-Alloy Temperature Designations

T1	Cooled from an elevated-temperature shaping process and naturally aged to a substantially stable condition.
T2	Cooled from an elevated-temperature shaping process, cold-worked, and naturally aged to a substantially stable condition.
T3	Solution heat-treated, cold-worked, and naturally aged to a substantially stable condition.
T4	Solution heat-treated and naturally aged to a substantially stable condition.
T5	Cooled from an elevated-temperature shaping process and then artificially aged.
T6	Solution heat-treated and then artificially aged.
T7	Solution heat-treated and stabilized.
T8	Solution heat-treated, cold-worked, and then artificially aged.
T9	Solution heat-treated, artificially aged, and then cold-worked.
T10	Cooled from an elevated-temperature shaping process, cold-worked, and then artificially aged.

and 427°C]. If the finished part is to attain full strength, a full heat treatment should be applied.

Heat Treatment of Ferrous Metals

Ferrous metals are those that contain large percentages of iron. The element that makes it possible to harden and toughen iron is carbon absorbed into the iron during the smelting process. Depending upon the process, the iron becomes low-carbon steel, medium-carbon steel, or high-carbon steel. Percentages of carbon in steel range from about 0.10 percent to more than 0.80 percent.

Heat-treating processes are used to rearrange the atoms of the carbon and iron to alter the strength, toughness, and hardness of the steel. Pure iron cannot be heat-treated because there is no element to change the structure of the metal. The presence of carbon in steel makes it possible to form several different combinations of iron and carbon that affect the characteristics of the steel.

When carbon steel is heated to a temperature of 1341°F [727°C], a solid solution of iron and carbon is formed that consists of 0.76 percent carbon. This is called **austenite.** If the steel contains less than 0.76 percent carbon, the result will be a mixture of **ferrite** and austenite at temperatures above 727°C. Ferrite is iron in which a maximum of 0.025 percent carbon is dissolved.

The solid solution of iron and carbon is a situation in which the carbon is actually dissolved in the iron just as sugar can be dissolved in water. When austenite is cooled below 727°C, a precipitate of iron carbide called **cementite** is formed at the boundaries of the crystals. As cooling continues at a slow rate, the austenite continues to decompose into alternate platelets of ferrite and cementite in a form called **pearlite** because of its appearance. The amount of pearlite formed and the nature of its structure depend upon the rate of cooling and the carbon content of the steel.

Very rapid cooling of austenite precludes the formation of pearlite, and the austenite changes immediately to a structure called **martensite.** In this form, the steel attains a maximum of hardness, particularly when the carbon content is high. Various conditions of hardness, toughness, ductility, and brittleness of steel depend upon the mixture and arrangement of ferrite, cementite, pearlite, and martensite in the steel as brought about by heat treating.

Alloy steels can develop greater strength and toughness than carbon steels because of the effects of the alloying elements such as chromium, molybdenum, tungsten, nickel, and vanadium. Heat treatment of the alloy steels must be tailored to the particular type of steel being treated in accordance with the elements it contains.

Hardening is done by heating the metal slightly above its critical temperature and then rapidly cooling it by quenching in oil, water, or brine. This produces a fine-grain structure, great hardness, maximum tensile strength, and minimum ductility. Material in this condition is usually too brittle for most uses, but this treatment is the first step in the production of high-strength steel. This method has been used for many years by blacksmiths for hardening certain tools. The tool is heated to a cherry red and then plunged into cold water. This produces a maximum of hardness. If more toughness is desired, the heated tool is plunged into the water for a moment and then removed. The color of the filed edge changes from gray to straw to blue. If the tool is placed back in the water when the filed surface turns to the straw color, the edge will be reasonably hard but will not be brittle.

The hardening of steel in industry is a carefully controlled process. The rate of heating, the time of soaking, and the rate of cooling are all regulated to achieve the desired results. The more rapidly a steel is cooled from the critical temperature, the harder the steel will be. It is usually necessary to temper the steel to achieve the necessary toughness and to reduce brittleness. Table 9-4 gives information on temperatures for heat treating and tempering of steel.

Tempering (drawing) is a process generally applied to steel to relieve the strains induced during the hardening process. It consists of heating the hardened steel to a temperature *below* the critical range, holding this temperature for a sufficient period, then cooling in water, oil, air, or brine. The degree of strength, hardness, and ductility obtained depends directly upon the temperatures to which the steel is raised. When high temperatures are reached in tempering, the ductility is improved at the expense of hardness, tensile strength, and yield strength.

Annealing is a form of heat treatment that consists of heating and cooling operations for the purpose of removing gases and stresses; of inducing softness; of altering ductility, toughness, electrical resistance, or magnetic properties; or of refining the grain structure. Annealing is done by gradually heating the material to a point above the critical temperature, soaking (holding) it at this temperature for a definite length of time, then cooling it slowly according to the method prescribed for the specific material being annealed. Annealing differs from other forms of heat treating in the slow cooling of the metal.

Normalizing is a process of heating iron-base metals above their critical temperature to obtain better solubility of the carbon in the iron, followed by cooling in still air. The normalizing process reduces stresses in the metal that have been caused by forging, machining, cold working, or other process. It improves the grain structure, toughness, ductility of the metal. Welded steel parts should always be normalized to eliminate stresses caused by uneven heating and to improve the grain structure of the weld. When weld metal solidifies, the result is a cast grain structure that does not match the wrought grain structure of the base metal.

Work hardening is simply any mechanical process that sets up a condition of hardness. It consists of repeatedly applying a mechanical force, such as rolling, hammering, bending, and twisting. This sets up stresses that resist outside forces.

Case-Hardening Treatments

Case-hardening treatments are given to iron-base alloys to produce a hard, wear-resisting surface and, at

TABLE 9-4 Temperatures for Heat Treatment and Tempering of Steel

Steel No.	Temperatures, °F			Quenching medium[n]	Tempering (drawing) temperatures for tensile strength (psi), °F				
	Normalizing air cool	Annealing	Hardening		100 000	125 000	150 000	180 000	200 000
1020	1650–1750	1600–1700	1575–1675	Water	—	—	—	—	—
1022 (x1020)	1650–1750	1600–1700	1575–1675	Water	—	—	—	—	—
1025	1600–1700	1575–1650	1575–1675	Water	a	—	—	—	—
1035	1575–1650	1575–1625	1525–1600	Water	875	—	—	—	—
1045	1550–1600	1550–1600	1475–1550	Oil or water	1150	—	—	n	—
1095	1475–1550	1450–1500	1425–1500	Oil	b	—	1100	850	750
2330	1475–1525	1425–1475	1450–1500	Oil or water	1100	950	800	—	—
3135	1600–1650	1500–1550	1475–1525	Oil	1250	1050	900	750	650
3140	1600–1650	1500–1550	1475–1525	Oil	1325	1075	925	775	700
4037	1600	1525–1575	1525–1575	Oil or water	1225	1100	975	—	—
4130 (x4130)	1600–1700	1525–1575	1575–1625	Oil[c]	d	1050	900	700	575
4140	1600–1650	1525–1575	1525–1575	Oil	1350	1100	1025	825	675
4150	1550–1600	1475–1525	1500–1550	Oil	—	1275	1175	1050	950
4340 (x4340)	1550–1625	1525–1575	1475–1550	Oil	—	1200	1050	950	850
4640	1675–1700	1525–1575	1500–1550	Oil	—	1200	1050	750	625
6135	1600–1700	1550–1600	1575–1625	Oil	1300	1075	950	800	750
6150	1600–1650	1525–1575	1550–1625	Oil	d, e	1200	1000	900	800
6195	1600–1650	1525–1575	1500–1550	Oil	f	—	—	—	—
NE8620	—	—	1525–1575	Oil	—	1000	—	—	—
NE8630	1650	1525–1575	1525–1575	Oil	—	1125	975	775	675
NE8735	1650	1525–1575	1525–1575	Oil	—	1175	1025	875	775
NE8740	1625	1500–1550	1500–1550	Oil	—	1200	1075	925	850
30905	—	g, h	i	—	—	—	—	—	—
51210	1525–1575	1525–1575	1775–1825 [j]	Oil	1200	1100	k	750	—
51335	—	1525–1575	1775–1850	Oil	—	—	—	—	—
52100	1625–1700	1400–1450	1525–1550	Oil	f	—	—	—	—
Corrosion resisting (16–2)	—	—	—	—	m	—	—	—	—
Silicon chromium (for springs)	—	—	1700–1725	Oil					

[a] Draw at 1150°F for tensile strength of 70 000 psi.
[b] For spring temper draw at 800 to 900°F Rockwell hardness C-40-45.
[c] Bars or forgings may be quenched in water from 1500–1600°F.
[d] Air-cooling from the normalizing temperature will produce a tensile strength of approximately 90 000 psi.
[e] For spring temper draw at 850 to 950°F Rockwell hardness C-40-45.
[f] Draw at 350 to 450°F to remove quenching strains. Rockwell hardness C-60-65.
[g] Anneal at 1600 to 1700°F to remove residual stresses due to welding or cold work. May be applied only to steel containing titanium or columbium.
[h] Anneal at 1900 to 2100°F to produce maximum softness and corrosion resistance. Cool in air or quench in water.
[i] Harden by cold work only.
[j] Lower side of range for sheet 0.06 in and under. Middle of range for sheet and wire 0.125 in. Upper side of range for forgings.
[k] Not recommended for intermediate tensile strengths because of low impact.
[l] AN-QQ-S-770.—It is recommended that, prior to tempering, corrosion-resisting (16 Cr-2 Ni) steel be quenched in oil from a temperature of 1875 to 1900°F, after a soaking period of ½ h at this temperature. To obtain a tensile strength at 115 000 psi, the tempering temperature should be approximately 525°F. A holding time at these temperatures of about 2 h is recommended. Tempering temperatures between 700 and 1100°F will not be approved.
[m] Draw at approximately 800°F and cool in air for Rockwell hardness of C-50.
[n] Water used for quenching shall not exceed 65°F. Oil used for quenching shall be within the temperature range of 80–150°F.

the same time, to leave the core of the metal tough and resilient. Three common methods are **carburizing, nitriding,** and **cyaniding.**

Carburizing consists of holding the metal at an elevated temperature while in contact with a solid, liquid, or gaseous material that is rich in carbon. Time must be allowed for the surface metal to absorb enough carbon to become high-carbon steel.

Nitriding is accomplished by holding special alloy steels containing small amounts of chromium, molybdenum, and aluminum at temperatures below the critical point in anhydrous ammonia. Nitrogen from the ammonia is absorbed into the surface of the steel as iron nitride and produces a greater hardness than carburizing, but the hardened area does not reach as great a depth as it does in carburizing.

Cyaniding is a fast method of producing surface hardness on an iron-base alloy of low carbon content. The steel may be immersed in a molten bath of cyanide salt, or powdered cyanide may be applied to the surface of the heated steel. During this process, the temperature of the steel must range from 1300 to

1600°F [538 to 871°C], the exact temperature depending upon the type of steel, the depth of the case hardening desired, the type of cyanide compound used, and the time that the steel is exposed to the cyanide. In using sodium cyanide or potassium cyanide, great care must be taken to avoid getting any of the cyanide into the mouth, eyes, or any other part of the body. These materials are deadly poisons.

Heat-Treating Techniques

As mentioned previously, heat treating is accomplished by heating metal to a certain prescribed temperature and then cooling it according to given directions depending upon the type of metal and the desired result. The heating is done in a furnace which may be gas-heated, oil-heated, or electrically heated. The metal is placed in the oven in suitable racks or containers, the heat of the oven is regulated to the required temperature, and the time which the metal is exposed to the heat is carefully controlled. After the heat has been applied, it is necessary to cool the metal according to the specified process. The cooling process must start the moment that the metal is removed from the heat. If the metal is to be quenched in a liquid, the methods shown in Figs. 9-4 and 9-5 may be used. Figure 9-4 shows a part being immersed with the part perpendicular to the bath. In Fig. 9-5 the part is immersed at an angle to the bath. The method of quenching, in any case, should follow the specified process.

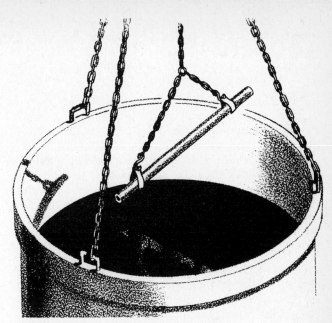
FIG. 9-5   Quenching metal at an angle.

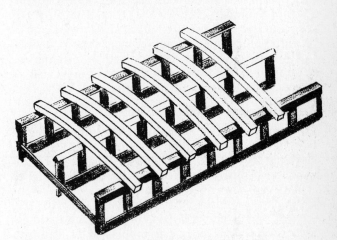
FIG. 9-6   Support for small parts in a furnace.

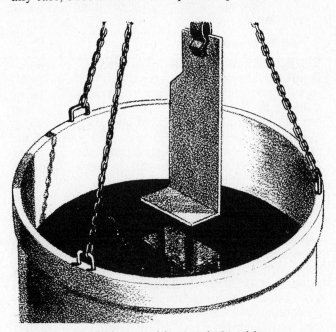
FIG. 9-4   Quenching metal in a vertical position.

When metal parts to be heat-treated are placed in a furnace, it is necessary that they be properly supported. Because the metal may be required to soak for as long as 1 h, there is a tendency for the material to sag. This causes warping and distortion of the parts. Figure 9-6 shows a method for supporting small parts.

The size and shape of parts to be heat-treated will often determine the methods of handling and supporting. Large sheet-metal parts require more careful support than small parts which have a thicker cross section. In any event, the operator must use care in the arrangement of parts in the furnace and the supporting structures used to hold the parts.

Heat Treatment of Titanium

Requirements for Heat Treatment. Heat treatment for titanium is required for some alloys but not for others. If a part is to be made, the technician must make certain to use the specified material and then to determine whether the material should be heat-treated. Some of the alloys that can be strengthened by heat treating are RS-120A, RS-135, RS-140, Ti-6A1-V, and Ti-8A1-1Mo-1V.

Temperatures for Heat-Treating Titanium. The temperatures employed for the heat treatment of titanium range from 1450 to 1850°F [788 to 1010°C]. At these temperatures the atmosphere in the furnace must be inert, that is, argon gas should displace all air in the furnace to protect the material from combining with oxygen and nitrogen. The soaking time is specified according to the material. For 8-1-1, the soaking time is 8 h at 1450°F [788°C]. The material is then air-cooled. For RS-120A, 135, and 140, a water quench is used and this is followed by artificial aging at 1000 to 1100°F [538 to 593°C].

Annealing of Titanium. As for other alloys, the primary reason for annealing titanium alloys is to make the metal more workable. This increases the ductility and machinability and improves the dimensional and structural stability at high temperatures. Annealing of titanium alloys must be accomplished according to the specifications for the particular alloy being treated. A typical annealing procedure is to heat the metal to 1350°F [732°C] and soak it at this temperature for 1 h. The metal is then removed from the furnace and allowed to air cool.

Stress Relieving. Titanium alloys are stress-relieved to remove stress concentrations developed during cold working. Actually, stress relieving could be considered a partial annealing. The alloy is heated to a lower temperature and soaked for a shorter period of time before it is air-cooled. The stress-relieving temperature is under 1000°F [538°C], and the alloy is soaked at this temperature for about 30 min.

In stress relieving and annealing, a scale may form on the metal. This can be removed by pickling the metal in a bath consisting of 15 percent nitric acid, 2 percent hydrofluoric acid, and water.

Heat Treatment of Stainless Steels

The chrome-nickel stainless steels, commonly referred to as 18-8 steels, cannot be hardened by heat treating. These are types 302 through 347. They can be annealed at temperatures of 1850 to 2050°F [1010 to 1121°C] and can be work-hardened.

The chromium stainless steels, types 410, 416, 420, and 431, can be heat-treated to increase their hardness and strength. Hardening is accomplished by heating to a range of 1750 to 1900°F [954 to 1038°C] and quenching in oil.

Heat Treatment of Magnesium Alloys

Magnesium alloys may be heat-treated in much the same manner as that employed for aluminum alloys; however, the heating times, soaking times, and cooling rates will vary in accordance with the type of alloy. The majority of magnesium parts are castings, and these are readily heat-treatable.

Magnesium alloy castings are solution-heat-treated to improve such characteristics as tensile strength, ductility, and shock resistance. The temperatures to which magnesium alloys are heated for heat treatment are a little less than those used for aluminum alloys, being in a range of 730 to 780°F [388 to 416°C]. The actual temperature for heat treatment depends upon the particular alloy involved. Specification MIL-H-6857 gives temperatures for heat treating of various magnesium alloys.

Precipitation heat treatment (artificial aging) is applied to some magnesium alloys after solution heat treatment. This treatment improves hardness and yield strength and also increases corrosion resistance.

Temperatures employed for precipitation heat treating are in the range of 325 to 500°F [163 to 260°C].

Heating of magnesium alloys must be undertaken with careful control to avoid overheating. At high temperatures, magnesium oxidizes rapidly and will ignite if the temperature is too high. Magnesium fires are very difficult to extinguish, particularly where large quantities of chips and shavings may have accumulated. To avoid oxidation and the danger of fire at high temperatures, magnesium can be heated in an atmosphere of inert gas.

● HARDNESS INSPECTION

Importance of Hardness Testing

All materials required for the various structural parts of an airplane must be examined to determine their hardness as an indication of strength as specified either by the appropriate drawing or by material specifications. This is accomplished by means of various types of instruments, all of which enable the operator of the instrument to determine the tensile strength of the material.

Hardness testing is usually the function of the process laboratory in a manufacturing plant; however, there are times when a technician may be required to determine the hardness of a material to find out whether it is satisfactory for a particular repair. Hardness testing is also a method for the identification of metals and alloys.

Brinell Hardness Test

The **Brinell hardness tester** shown in Fig. 9-7 is an instrument commonly used on aluminum-alloy castings, forgings, and billets before machining. The check is made by forcing a steel ball of a known diameter into the material by a specified and known pressure, as suggested by Fig. 9-8. Three distinct pressures are provided for ferrous, nonferrous, and soft materials. The reading is taken by measuring the width of the impression made by the steel ball in the material, by means of a microscope. The reading thus obtained is compared with an established comparison chart which gives the tensile strength for each reading.

Before testing any material with the Brinell instrument, the operator must make sure that the surface of the material is clean, free from scale, flat, and fairly smooth, in order to obtain an accurate reading. The work must be adequately supported to avoid twist or movement when applying the test load. A decarbonized surface might cause a low reading even though the hardness is correct. If the reading is much lower than expected, the surface should be filed or ground to remove decarbonization, and another test made. The instrument is never used where excessive vibration is present or on thin material where the indentation will show through.

The Rockwell Hardness Test

The **Rockwell hardness tester** shown in Fig. 9-9 is another instrument used in examining the hardness of metals. Its operation is similar to that of the Brinell

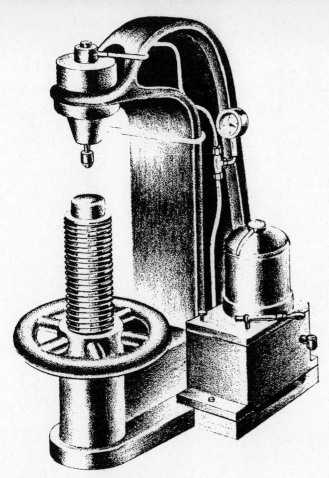

FIG. 9-7   Brinell hardness tester.

FIG. 9-8   Forcing the steel ball into the surface of the metal.

instrument because the test is made by the depth of penetration of a diamond point, or of a ball of fixed size, under predetermined load. The reading is made directly from a calibrated dial showing a given number for depth of penetration. This number is then checked against the comparison chart to find the tensile strength. Table 9-5 gives readings of Brinell, Rockwell, and Vickers instruments for a wide range of hardness values.

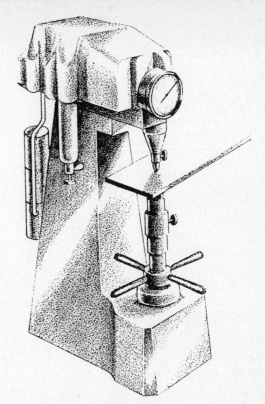

FIG. 9-9   Rockwell hardness tester.

There are three ranges of material covered by the diamond point, or "brale," as it is called on the Rockwell instrument, each using a specified pressure. Other ranges of softer material are checked by means of a given-size steel ball and predetermined pressure. Each range, for both ball and brale, is covered by an alphabetically designated scale shown on the calibrated head.

When testing with the Rockwell instrument, the operator must be certain that the surface of the material is clean and smooth. The material should be held square, as shown in Fig. 9-10, and in all cases where the material is large or heavy it should be adequately supported to avoid any movement of the material while the reading is taken.

Figure 9-11 shows what will happen if too small a piece is tested with the Rockwell instrument. If the metal can flow away from the ball or brale as shown, a false reading will be obtained. False readings will also result from taking two readings too close together. This is illustrated in Fig. 9-12.

Other precautions must also be observed. Operators should not strike the penetrator when removing or replacing the anvil. They should not raise the jackscrew with the anvil in place and force it against the penetrator. They should not overrun the set point in applying the minor load, and they should not back off to the set point if it has been overrun and then attempt a reading. The jackscrew, anvil seat, and penetrator seat should be clean to avoid a "cushioning" effect.

Vickers Hardness Tester

The Vickers tester operates on a principle similar to that of the Brinell and Rockwell machines; however,

TABLE 9-5 Rockwell, Brinell, and Vickers Hardness Values

Rockwell hardness number				Vickers hardness number, diamond-pyramid penetrator	Brinell hardness number, 3000-kg load, 10-mm standard ball	Approx. tensile strength, 1000 psi
A scale 60-kg load, brale penetrator	B scale 100-kg load, $\frac{1}{16}$-in [0.16 cm] diam. ball	C scale 150-kg load, brale penetrator	D scale 100-kg load, brale penetrator			
78.5		55	66.9	595		287
78.0		54	66.1	577		278
77.4		53	65.4	560		269
76.8		52	64.6	544	500	262
76.3		51	63.8	528	487	
75.9		50	63.1	513	475	245
75.2		49	62.1	498	464	239
74.7		48	61.4	484	451	232
74.1		47	60.8	471	442	225
73.6		46	60.0	458	432	219
73.1		45	59.2	446	421	212
72.5		44	58.5	434	409	206
72.0		43	57.7	423	400	201
71.5		42	56.9	412	390	196
70.9		41	56.2	402	381	191
70.4		40	55.4	392	371	186
69.9		39	54.6	382	362	181
69.4		38	53.8	372	353	176
68.9		37	53.1	363	344	172
68.4	(109.0)*	36	52.3	354	336	168
67.9	(108.5)	35	51.5	345	327	163
67.4	(108.0)	34	50.8	336	319	159
66.8	(107.5)	33	50.0	327	311	154
66.3	(107.0)	32	49.2	318	301	150
65.8	(106.0)	31	48.4	310	294	146
65.3	(105.5)	30	47.7	302	286	142
64.7	(104.5)	29	47.0	294	279	138
64.3	(104.0)	28	46.1	286	271	134
63.8	(103.0)	27	45.2	279	264	131
63.3	(102.5)	26	44.6	272	258	127
62.8	(101.5)	25	43.8	266	253	124
62.4	(101.0)	24	43.1	260	247	121
62.0	100.0	23	42.1	254	243	118
61.5	99.0	22	41.6	248	237	115
61.0	98.5	21	40.9	243	231	113

*Values in parentheses are beyond normal range and are given for information only.

the penetrator is a diamond pyramid. The Vickers tester is particularly useful for testing the hardness of very hard steels. The scale of hardness numbers is shown in Table 9-5.

The Scleroscope Hardness Test

The **Scleroscope hardness tester,** shown in Fig. 9-13, is another instrument used for testing the hardness of metals. This instrument tests not by indentation but by rebound. It has a diamond-pointed ball that drops through a glass tube onto the material being tested. This tube is mounted with a graduated scale, and the rebound of the ball, or point, is caught by the eye at its peak, and the corresponding graduation is then read. The instrument is also made with a direct-reading head which records the amount of rebound. A soft

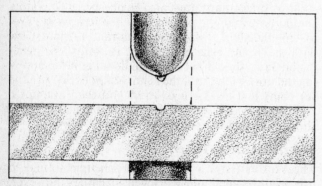

FIG. 9-10   Position of material being tested for hardness.

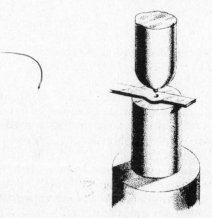

FIG. 9-11   Result of testing too small a part.

FIG. 9-12 Incorrect reading from tests too close together.

FIG. 9-13 The Scleroscope.

twice a day when working continuously with the same pressures. Continuous work with the Rockwell tester requires a calibration check at least twice in any 8-h shift. The Scleroscope, on production testing, should have a calibration of at least once each hour. An erratic reading on any machine always calls for a calibration check. Any change of materials, beyond the scale range, requires a change in the machine setup, such as penetrators and loads. Such changes are made by authorized personnel who are directly responsible for the care and upkeep of the instruments.

The Shore Durometer for Plastics and Rubber

The **Shore durometer,** shown in Fig. 9-14, is used to determine the hardness of treated or molded rubber, plastics, and allied materials. Readings are taken directly from a reading head, actuated by the upward movement of a calibrated pin that is forced against the material to be tested. The pin is spring-mounted under a predetermined pressure; hence its travel is governed by the degree of hardness of the material being tested. Until this pressure is reached by the resistance of the material, the pin will penetrate the material up to the point where the spring pressure is overcome.

FIG. 9-14 The Shore durometer.

ball is provided for the testing of nonferrous metals, and readings are taken from the same scale.

When using the Scleroscope, the surface of the metal must be clean and smooth, the glass must be perpendicular, and the material square with the glass. The hammer must not be dropped twice in the same place because this may chip the diamond point and would give a false reading.

Calibration of Hardness Testers

The accuracy of the readings on the hardness testers already described should be checked frequently, the frequency of checking depending upon the extent of their use. Checking for accuracy, or calibration, is done by means of test blocks supplied by the manufacturers of the instruments. These blocks have been tested to positive readings against master blocks. The Brinell instrument should have a calibration check

The Webster Hand-Type Hardness Tester

The **Webster hand-type hardness tester** is shown in Fig. 9-15. This is a simple pliers-type unit with an anvil on one jaw and a series of inclined indenters of increasing crest area on the other jaw. This instrument is used for testing aluminum and its alloys. When the indenters are forced into the metal by the action of the pliers, the number of indentations appearing on the surface is an indication of the hardness. An improved type of Webster instrument has a dial indicator that is read directly during the plier action, thus eliminating the necessity of looking at the part to count the indentations. Care must be taken in applying the indenter jaw to be sure it is at right angles to the surface being tested, because any inclination or rotation will give inaccurate readings.

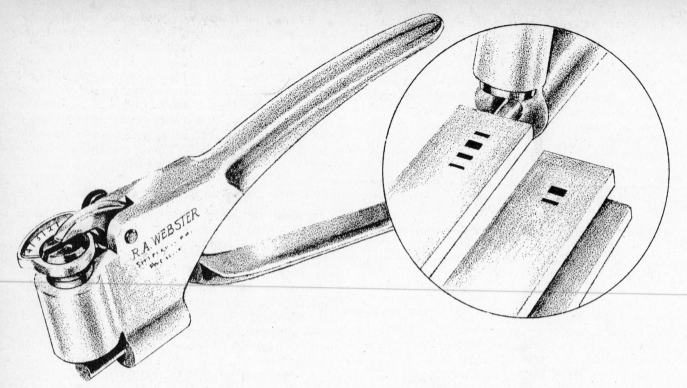

FIG. 9-15   Webster hand-type hardness tester.

The Barcol Hardness-Testing Machine

Like the Webster tester, the Barcol machine is used for testing aluminum and its alloys. This machine is shown in Fig. 9-16. It has a small spring-loaded needle point extending from the face of a small housing. The dial registers the amount the spring is compressed in forcing the needle into the material until the housing face contacts the material. As more pressure is exerted in forcing the needle into harder metal, the dial registers the increased pressure exerted. The instrument may be used as a production test for heat treating, after obtaining direct readings with the Rockwell instrument.

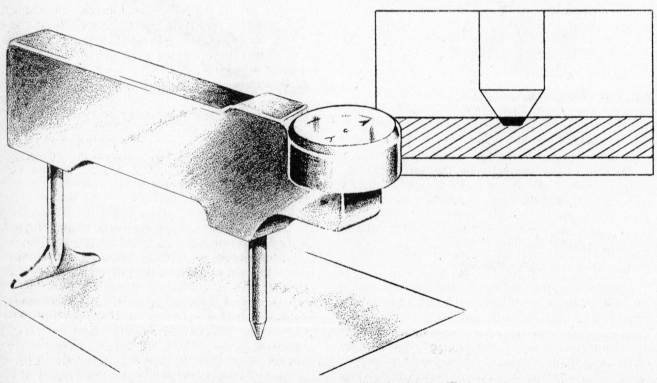

FIG. 9-16   Barcol hardness-testing machine.

The Barcol tester may be adjusted for any homogeneous material to give an identical "one-point" reading equivalent to any Rockwell scale reading made on the same material. On clad aluminum and other non-homogeneous materials where the coating thickness varies with materials, the Barcol tester must be calibrated against the Rockwell tester for each gage tested. The penetrating needle on the Barcol tester is very hard and brittle. It must be applied at right angles to the surface being tested, with no sliding force, or it will break. Broken, flattened, or bent needles must be replaced at once. Replacement is made only by the authorized personnel, because it requires disassembly and resetting of the spring. Resetting consists of preloading the spring so that the instrument will be centered to give direct readings over the entire aluminum-alloy range. In either the Webster or the Barcol machine, the indicator must be periodically examined to detect whether or not soft material is adhering to the jaws.

Figure 9-17 shows the proper method for testing hollow tubing. Testing without an internal support, as shown in Fig. 9-18, will result in a false reading, because the tubing will tend to flatten and permit increased movement of the testing needle.

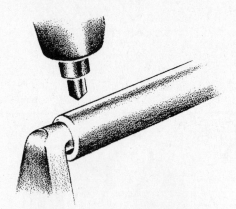

FIG. 9-17   Testing hollow tubing for hardness.

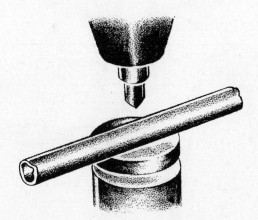

FIG. 9-18   Improper method for testing hollow tubing.

When a curved rod is tested for hardness, the rod must be placed as shown in Fig. 9-19. If the rod is placed in the convex position as in Fig. 9-20, a certain amount of give will take place with the result that a false reading will be obtained.

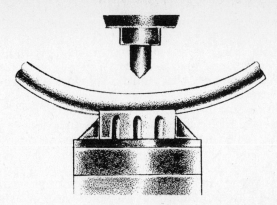

FIG. 9-19   Correct method for testing a curved rod.

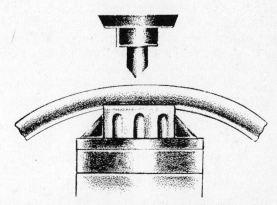

FIG. 9-20   Improper method for testing a curved rod.

● NONDESTRUCTIVE TESTING

Nondestructive testing is employed to detect flaws in aircraft parts without damaging or destroying the parts. Because aircraft parts are manufactured from a variety of metals, alloys, and nonmetallic materials, several different types of inspection and inspection processes are employed, and the method chosen for a particular part depends upon the shape of the part and the material from which it is made.

Technicians employing nondestructive testing methods must be well trained in the methods and procedures in order to obtain satisfactory results and to avoid damaging the parts being tested. Methods and procedures to be used are often specified in maintenance and overhaul manuals. Manufacturers of testing equipment provide information and instructions for the use of their equipment.

Magnetic Inspection

The purpose of magnetic inspection is to render visible to the naked eye defects in magnetic metals, such as cracks, inclusions, and miscellaneous faults. This type of inspection can be used only on ferrous materials which are magnetic.

A thorough inspection is required on machined parts and welded structures made from magnetic materials, except on those subjected to very low stresses or those excepted by the specifications in effect at the time the job is done. Parts that require such inspections are so noted on the drawing or specification. The percentage of bolts, nuts, etc., requiring magnetic inspection is determined by the specifications for the job.

To prepare for a magnetic inspection, the parts must be thoroughly cleaned and freed from dirt, grease, and foreign matter that might produce erroneous indications.

In performing a magnetic inspection, the part to be inspected is placed between the faces of the testing machine and clamped tight by means of air pressure. Figure 9-21 shows a part in place for testing. The switch for current is pressed, thus sending a strong current through the part. This current produces magnetic lines of force around the part as shown in Fig. 9-22. These lines of force flow within and around the part in a uniform manner if there is no defect. If the part is cracked or otherwise damaged, the lines of force will be broken. This is indicated by the drawing of Fig. 9-23. Where the crack or other discontinuity exists, the lines of force will leave the part and form a local field near the surface. When iron oxide powder, either dry or in a liquid, is applied to the part, the magnetic field around the cracked area will attract the magnetic particles and reveal the location of the crack.

To detect cracks that are approximately perpendicular to the longitudinal axis of a part, it is best to mag-

netize the part by means of a coil. The field of a coil is shown in Fig. 9-24. When a part is placed in the coil as shown in Fig. 9-25, the magnetic lines of force will pass through the part from end to end as shown in Fig. 9-26. A crack will distort the lines of force, and this in turn will cause magnetic particles of iron oxide powder to collect in the cracked area. This, of course, makes the position of the crack visible.

Magnaflux. The name **Magnaflux** is a copyrighted word used to denote the magnetic process and equipment developed by the Magnaflux process.

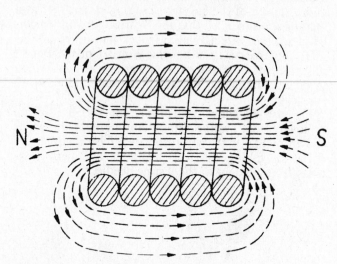
FIG. 9-24   Magnetic field of a coil.

FIG. 9-21   A part in position for magnetizing and inspection.

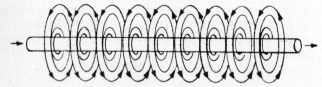
FIG. 9-22   Magnetic field around a conductor.

FIG. 9-25   Parts being magnetized in a coil.

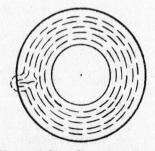
FIG. 9-23   How a defect affects the magnetic field.

FIG. 9-26   Magnetic field passing through a rod.

Many shops whose operations require the magnetic inspection of parts will send the parts to certified or approved Magnaflux stations to have the inspections performed. After the parts are inspected and found satisfactory, they are indelibly stamped or marked to show that the inspection has been performed. An inspection certificate is usually issued to show that the parts have been subjected to approved Magnaflux inspection. In the case of aircraft or aircraft engine parts, the Magnaflux certificate should be retained with the overhaul records.

The technician using magnetic inspection equipment must take particular care to prevent the formation of electric arcs during the magnetizing process. High currents are applied to the parts, and if the contact area is not adequate, arcs can form and burn pits in the parts being tested. These pits can cause the parts to be unserviceable. Woven wire pads are often used at the contact points to increase the contact area.

Magnaglo. Magnaglo is a process developed by the Magnaflux Corporation in which the parts are magnetized as for magnetic testing but the magnetic powder used to detect the flaw is fluorescent and will glow under ultraviolet (black) light. The process is the same as for Magnaflux, but the part is examined under ultraviolet light to detect the presence of flaws.

Fluorescent-Penetrant Inspection

Fluorescent-penetrant (Zyglo) inspection is used for both magnetic and nonmagnetic materials. The part to be inspected is placed in a penetrant solution after having been thoroughly cleaned. After soaking in the solution for a specified period of time, the part is washed off with the designated solvent and dried thoroughly. It is then placed in the developing powder in a manner that assures that all parts of the item being tested are covered with powder. The part is then dusted off and placed under an ultraviolet light in a darkened cabinet. If there are cracks or other discontinuities in the surface, a bright line will show up. This line is the result of the penetrant being drawn from the crack by the developer and creating a fluorescent indication at the crack.

The appearance of a fluorescent line or spot does not always indicate a flaw. If a fluorescent indication appears, the technician should investigate further to determine whether the indication was caused by a flaw or by a rough spot on the surface of the material.

Dye-Penetrant Inspection

The **dye-penetrant process** involves the use of a highly penetrating dye which seeps into cracks or other defects. A white developer is applied, and the dye coming out of the defect through the film of developer reveals the defect as a bright red spot or line.

Before applying the dye, the part must be thoroughly cleaned to remove oil, rust, scale, and other extraneous material. The part to be inspected must be dry and at a temperature of over 70°F [21°C] when the dye is applied. This is done by dipping the part in the dye, brushing, or spraying. The dye is allowed to "soak" on the part for 2 min or more to be sure it has penetrated all possible cracks or fissures. It is then washed off with a special cleaner or with warm water, after which the part is dried. The developer is an extremely fine white powder suspended in a solvent, and is sprayed or brushed on the part. The part can also be dipped in the developer liquid. The developer is allowed to dry, and the dye penetrant will then reveal defects as red spots or lines. After inspection, the developer is washed off with solvent.

In the use of any particular dye-penetrant process, the technician should follow carefully the instructions provided by the manufacturer. These instructions are usually on the container for the penetrant or the developer.

Ultrasonic Inspection

Ultrasonic inspection utilizes high-frequency sound waves to reveal flaws in metal parts. The element transmitting the waves is placed on the part, and a reflected wave is received and registered on an oscilloscope. If there is a flaw in the part, the reflected wave will show a "blip" on the oscilloscope trace. The position of the blip indicates the depth of the flaw.

Ultrasonic inspection can be accomplished satisfactorily by a well-trained and experienced technician. The technician must be familiar with the responses of the equipment and be able to interpret all of the indications observed accurately.

Eddy-Current Inspection

Eddy-current inspection is used to "see" inside metal parts in a manner similar to ultrasonic inspection except that electromagnetic waves rather than sound waves are used to penetrate the material. The transducer in eddy-current equipment emits high-frequency electromagnetic waves that penetrate the metal being tested. If there are no flaws in the material, the eddy currents generated in the material will be uniform in their flow and produce a consistent output signal at the dial indicator. If the eddy currents are blocked by a flaw inside the material, the output will be changed and this will be indicated on the output meter.

As with ultrasonic equipment, the operator must be an experienced technician in order to interpret the output signals of the instrument accurately. Eddy-current inspection can be used for both magnetic and nonmagnetic metals.

X-Ray Inspection

X-ray inspection, or **radiography,** is often used for the inspection of metal parts. A very powerful x-ray machine is used to produce the rays necessary to penetrate metal. The rays pass through the metal and impinge upon a photographic plate. Flaws in the metal will be revealed as shadows in the picture of the part.

X-ray inspection is performed by persons certificated for this type of work. There is danger from the high voltages required and from inadvertent exposure to the powerful radiation. Furthermore, the technician must know what type of photographic film to use and what voltages to apply, depending upon the type of material being tested and the thickness of the material.

Radiography inspection is also accomplished by means of a small cobalt "bomb" that is placed inside parts where it would be impossible to use an x-ray machine. The cobalt is radioactive and continuously emits gamma radiation. This is the same as x-rays and penetrates the material being tested. The photographic film is placed on the side of the material opposite the cobalt and reveals the nature of the interior of the material by the shadows on the developed film.

The use of radioactive cobalt for nondestructive testing is subject to numerous regulations to assure that no one is exposed to the radioactive material. Special permits must be obtained from government agencies before this type of inspection can be used.

● CORROSION CONTROL

Essentially, corrosion is the decomposition of metallic elements into compounds such as oxides, sulfates, hydroxides, and chlorides. It is brought about by direct chemical action and by electrolytic action. Chemical corrosion is brought about by an acid, salt, or alkali in the presence of moisture. Water provides the vehicle through which the chemical action takes place. Electrolytic corrosion takes place when metals that have a different level of chemical activity are touching or are in close proximity in the presence of moisture. The two dissimilar metals form the poles of a galvanic cell, an electric current flows, and the more active metal is decomposed. This is a proces similar to electroplating.

Types of Corrosion

The types of corrosion most commonly encountered in aircraft are surface corrosion, dissimilar-metals (electrolytic) corrosion, fretting corrosion, stress corrosion, and intergranular corrosion. Corrosion is noted as discoloration on a metal surface, rust on steel parts, greenish deposits on brass and copper, white or gray powder or deposits on aluminum and magnesium, or other colors, depending upon the chemical combination involved.

Surface corrosion of aluminum occurs most often on bare aluminum alloy that is not painted or otherwise treated to prevent corrosion. Clad aluminum is much more corrosion-resistant than bare aluminum; however, clad aluminum will also suffer corrosion if allowed to become dirty, especially in a humid climate or exposed to sea air. Surface corrosion on aluminum generally appears as white blotches if the corrosion is general on the surface and as small, dark gray lumps if the corrosion has penetrated below the clad surface and is attacking the interior of the metal.

Surface corrosion on metal that has been painted is evidenced by peeling or blistering of the paint. In the case of steel, the reddish color of rust will often show up on the outer surface of the paint.

Dissimilar-metals corrosion is caused when metals with different chemical activity are in contact in the presence of moisture. As explained previously, this situation causes galvanic action that decomposes the metal most easily oxidized.

Metals have been classified in four groups in accordance with their tendency to oxidize. The most active metals are in group I, and the least active are in group IV. The groups are as follows:

Group I: Magnesium and its alloys
Group II: All aluminum alloys, cadmium, and zinc
   Subgroup A: Aluminum alloys 1100, 3003, 5052, 6061, 220, 355, 356, all clad alloys
   Subgroup B: Aluminum alloys 2014, 2017, 2024, 7075, 7079, 7178, 195
Group III: Iron, lead, tin, and their alloys except stainless steels
Group IV: Stainless steels, titanium, chromium, nickel, and copper, and their alloys, and graphite

The aluminum alloys in subgroups A and B should be considered as dissimilar metals with respect to corrosion prevention. This is particularly true when a large area of an alloy in subgroup B is in contact with a small area of an alloy in subgroup A. Severe corrosion of the alloy from subgroup A may be expected.

In the design of aircraft metal structures, every effort is made to prevent contact between metals of different groups. This is the reason that aircraft bolts are usually cadmium-plated. It will be noted that cadmium is in group II along with aluminum alloys. Copper electric terminals are cadmium-plated for the same reason.

Intergranular corrosion is caused by improper heat treatment. If an alloy is not quenched rapidly enough following precipitation heat treatment, alloying elements will precipitate along grain boundaries and form minute galvanic cells. The aluminum alloy 2024 is particularly vulnerable to intergranular corrosion because of its copper content. The zinc in the 7000 series alloys leads to intergranular corrosion unless the heat treatment is accurately controlled. Intergranular corrosion also occurs in some corrosion-resistant steels.

Intergranular corrosion shows up as raised surface flaws unless it is deep inside a casting or forging. In this case, ultrasonic or eddy-current inspection techniques produce good results. If apparent blisters are noted under the surface of bare or painted aluminum alloy, intergranular corrosion may be present. Scraping or piercing the raised area will reveal the nature of the damage. Intergranular corrosion will show up under the surface of the metal, and the outer surface may not show any corrosion. Under the surface will be white powder mixed with aluminum particles. When the aluminum breaks up or delaminates, the condition is called **exfoliation.**

Stress corrosion results when a metal part is overstressed over a long period of time under corrosive conditions. Parts that are susceptible to stress corrosion are overtightened nuts in plumbing fittings, parts joined by taper pins that are overtorqued, fittings with pressed-in bearings, and any other assembly where the metal is stressed almost to the yield point. Stress corrosion is not easy to detect until cracks begin to appear.

One type of stress corrosion is known as corrosion fatigue. This occurs where cyclic stresses are applied to a part or assembly. These stresses not only affect the metal but also produce pores and minute cracks in the surface coating, thus allowing moisture to penetrate.

Fretting corrosion occurs when there is slight movement between close-fitting metal parts. The movement prevents the formation of oxides that inhibit corrosion and produces fine particles of metal and oxide that tend to absorb and retain moisture. This further aggravates the corrosion. The particles also act as an abrasive that keeps the bare metal exposed to the corrosive conditions. Fretting corrosion is sometimes called "false brinelling" because of the appearance of parts that are affected.

Repair of Corroded Parts

The repair of parts that are affected by corrosion but are not rendered unserviceable is simply a matter of removing the existing corrosion products and applying a coating or finish that will prevent further corrosion. Care must be taken to assure that the parts are not damaged by the removal process and that the correct type of refinish is applied.

Parts made of carbon or alloy steel and not highly stressed can be cleaned with buffers, wire brushes, sandblasting, steel wool, or abrasive papers. If corrosion remains in pits and crevices, it may be necessary to use chemical inhibitors to prevent corrosion. Unpainted steel parts are often coated with rust-inhibiting oil or greases. Steel parts that are highly stressed must not have appreciable material removed during cleaning and must not have pits or appreciable surface damage. Parts that are satisfactory after corrosion removal can be primed and painted in accordance with the original finish.

When removing corrosion from aluminum parts, the use of steel or wire brushes, steel wool, emery cloth, or other harsh abrasives should be avoided. Steel brushes and steel wool are likely to leave metal particles embedded in the surface, and these will lead to electrolytic corrosion. Mild abrasives or polish meeting Specification MIL-P-6888 can be used on clad alumimun alloy but must not be used for cleaning anodized alloys. The anodized film is so thin that it can be removed by polishing. When it is necessary to remove severe corrosion that cannot be removed by polishing, a number of chemical cleaners are available. One such process utilizes a 10 percent solution of chromic acid to which a small amount of sulfuric acid has been added. This solution can be brushed on with a stiff bristle brush and allowed to remain for more than 5 min. If the existing corrosion is removed, the solution can be rinsed off with water. The cleaned surface should be repainted within a few hours after cleaning.

If an anodized film has been partially removed from an aluminum-alloy surface, a protective coating can be partially restored by treating with a chromic acid solution. After treatment, the part should be primed and painted as soon as possible.

Structural aluminum-alloy parts that have suffered severe intergranular corrosion must usually be replaced because of the loss of strength in the parts. Sometimes a small amount of intergranular corrosion can be removed from the outer surface of a part, and

if there is no evidence of additional penetration, the part can be treated chemically and then refinished.

Since magnesium is the most chemically active of the metals employed in aircraft, it is also the most critical with respect to corrosion. Magnesium parts are chemically treated to produce a film coating that prevents corrosion; however, this film can be scratched or worn off with the result that corrosion starts immediately if any moisture is present. When the protective coating on magnesium is damaged, it should immediately be repaired by chemical treating with a 10 percent chromic acid solution to which has been added a small amount of sulfuric acid. This acid can be obtained in the form of lead-acid battery electrolyte. The recommended amount for the chromic acid solution is 20 drops of the electrolyte to 1 gal of the solution. The solution should be brushed on the cleaned magnesium part and allowed to remain for 10 to 20 min.

● FINISH AND SURFACE-ROUGHNESS SYMBOLS

The surface of a metal part is "finished" by performing a machining, coating, or hand-finishing operation on that surface. Scraping, file-fitting, reaming, lathe turning, shaping, and grinding are some finishing operations.

On many existing blueprints the symbol for a finished surface is a letter V with its point touching the surface to be finished, drawn with an angle of 60° between the sides of the V. Numbers may be placed within the angle formed by the sides of the V to represent the type of finish to be applied to that particular surface. When a part is to be finished on all surfaces, the abbreviation F.A.O. is sometimes used to represent "finish all over."

Some factories use symbols on shop orders, but not necessarily on drawings, according to what they call a "finish code." Examples taken at random are as follows: 1, zinc chromate primer; 2, darkened primer, 3, aluminized primer, and 9, dope. However, it is a more common practice to write a finish specification to show how each material and portion of a particular assembly finished.

Many manufacturers in the aerospace industry have adopted the root-mean-square (rms) microinch system of surface-roughness designation. This system has been standardized by the National Aerospace Standards Committee in Specification NAS30 and is also set forth in MIL-STD-10A. All new drawings of machined castings, machined forgings, and other machined parts will use this method of specifying surface finishes. **Surface roughness** is a term used to designate recurrent or random irregularities that may be considered as being superimposed upon a plane or wavy surface. On "smooth-machined" surfaces, these irregularities generally have a maximum crest-to-crest distance of not greater than 0.010 in and a height that may vary from 0.000 001 to 0.000 05 in. "Waviness" should not be confused with roughness, as the crest distances are much greater, generally running from 0.04 to 1.00 in [0.10 to 2.54 cm], and the height as much as several thousandths of an inch.

The need for a simple control of the surface quality of a machined part by means of production drawings has long been apparent. Dimension tolerances as well as process notes such as "rough machine," "smooth machine finish," "grind," and "polish" limit the surface characteristics in a general way but are not sufficiently specific to describe the desired result. Certain machining processes, such as grinding, could produce several degrees of smoothness; hence the decision as to which degree was intended generally was made in the past by the shop or the vendor.

By means of the rms system of surface-roughness designation, it becomes possible for the engineering department of any company to specify precisely the degree of finish required and for the shop to produce the specified finish without resorting to judgment. The rms average is a unit of measurement of surface roughness and is expressed in microinches. The microinch is one-millionth (0.000 001) part of the U.S. Customary System linear inch. The rms average is chiefly affected by the highest and lowest deviations from a mean surface and is a mathematical indication of average surface roughness.

The National Aerospace Standards Committee has selected a series of preferred roughness numbers that cover the range of aircraft requirements. These numbers are 1, 2, 5, 10, 20, 40, 100, 250, and 500. All these numbers, with the exception of number 1, are used by manufacturers. They indicate maximum allowable or acceptable roughness of the surface on which they are specified in rms microinches.

Engineers have a fairly clear conception of finish characteristics when expressed in terms of the machining process used to produce a surface finish. It is therefore necessary for the designer to designate this surface condition as an rms finish number. Figure 9-27 is a table of roughness numbers for surface-finish designation.

In order to allow the machine shop latitude in developing the specified finish, drawings specify the maximum roughness allowable. Any smoother finish will be acceptable as long as economy is not sacrificed. The shop may develop the specific finish by whatever method is the most practical. Surface defects, flaws, irregularities, and waves that generally occur in only a few places will be handled by the inspection department.

The shop and engineering departments must use the standard reference samples for comparison with machined parts. By visual inspection and by feel, they can check the completed surfaces. In borderline cases the surface can be checked by instruments called the **profilometer** and the **Bush analyzer.** These are tracer-point instruments that measure the surface roughness when drawn along the surface to be measured.

The rms numbers on drawings always are called out by the use of the standard symbol illustrated in Fig. 9-28. The roughness number must always be on the left side of the long leg close to the horizontal bar as indicated by *xx* in the illustration. The roughness of natural surfaces is not specified unless such surfaces are critical because of functional or manufacturing requirements.

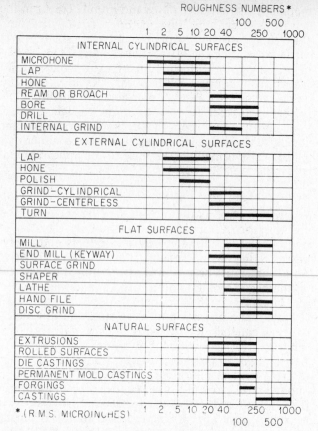

FIG. 9-27   Table of roughness numbers.

The finish of such items as drilled holes, reamed holes, and spot faces is not usually specified if the maximum roughness to be produced will be acceptable. The roughness of fillets and chamfers conforms to the rougher of the two connected or adjacent surfaces, unless otherwise indicated.

Unless otherwise specified, a symbol used on a

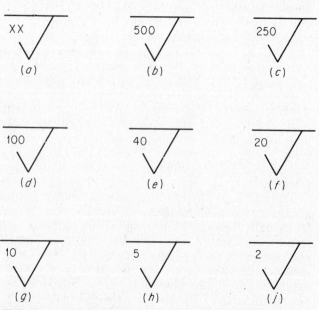

FIG. 9-28   Root-mean-square (rms) finish numbers and symbols.

plated or coated surface always signifies that a control applies to the parent-metal surface before plating or coating.

Lay may be defined, for the purpose of this discussion, as the direction of tool marks or the grain of the surface roughness. Waviness and tool lay designations also covered in the National Aerospace Standards Committee Specification NAS30 are not yet adopted by all manufacturers.

The symbol in Fig. 9-28b indicates rms 500. This is a very rough, low-grade machine surface resulting from heavy cuts and coarse feeds in milling, turning, shaping, and boring, as well as from rough filing and rough disk grinding. This is also the natural finish of some forgings and sand castings. It is not used for aluminum alloys or other soft metals, for surfaces in tension, or where notch sensitivity is a factor. It may be used on secondary items but is generally not called out as a finish for aircraft or missile parts.

Figure 9-28c is the symbol for rms 250. This is a medium machine finish and is fairly inexpensive to produce. Figure 9-28d is the symbol for rms 100. This finish is generally known as smooth machine finish and is the product of high-grade machine work in which relatively high speed and fine feeds are used in taking light cuts with well-sharpened cutters.

Figure 9-28e is the symbol for rms 40. This is a fine machine finish produced by a carbide or diamond bore, a medium surface or cylindrical grind, a rough emery buff, ream, burnish, and similar operations.

Root-mean-square 20, a very fine finish, is indicated by the symbol of Fig. 9-28f. This finish is produced by fine cylindrical or surface grind, very smooth ream, smooth emery buff, and coarse to medium lap or hone. The extremely smooth finishes are indicated by Fig. 9-28g, h, and j. These finishes are produced by honing, lapping, microhone, polishing or buffing.

Root-mean-square 10 and the finer finishes may have either a dull or bright appearance, depending upon the method used to produce them. The surface appearance must not be considered in judging quality, but the degree of smoothness must be determined by "feel" or roughness-measuring instruments.

● METAL SURFACE TREATMENTS

A number of surface treatments for metals have been developed to reduce or eliminate corrosion. Treatments for aluminum alloys include Anodizing, Alodizing, Iridite, and treatments with chromic acid and sodium dichromate. All these processes provide a protective film, and parts so treated must be handled carefully to avoid damaging the film. The aviation maintenance technician is not usually required to apply the more complicated treatments because they require equipment not generally available in the field. Usually such treatments are applied by manufacturers and large overhaul bases.

Magnesium is usually treated with a chrome-pickle or sodium-dichromate process. These processes produce an excellent protective film that can be repaired by brushing chromic acid solution on damaged areas.

Steel parts are protected by cadmium plating, chrome plating or by phosphate processes such as Parkerizing, Bonderizing, Parco Lubrizing, or Granodizing. These processes effectively protect the surface from oxidation, and the parts may be painted over after preparation as specified by the manufacturer of the product applied.

Cadmium Plating

Cadmium plating is a nonporous, electrolytically deposited layer of cadmium that offers high corrosion resistance for steel. Plating is per specification MIL-P-416A. Three types of cadmium plating are considered in this specification:

Type I: Pure silver-colored cadmium plate, without supplementary treatment. This type of cadmium coating was used on all steel aircraft hardware in the past.

Type II: This consists of type I plating followed by a chromate treatment. Type II plating is a light to dark gold color. It has improved corrosion resistance. Procurement specifications for aircraft now specify type II plating.

Type III: This is type I coating followed by a phosphate treatment. It is used mainly as a paint base.

Anodizing

This finish, applied to aluminum by an acid plating process, hardens the surface, reduces porosity, increases abrasion resistance, and has high dielectric strength. Anodized aluminum can be dyed almost any color.

Specification MIL-A-8625B covers three types of anodizing:

Type I: Chromic anodize coating will vary from a light to a dark gray color depending on the alloy. Coating is given a chromate treatment to seal the surface.

Type II: Sulfuric anodize coating is the best coating for dying. Nondyed coating will have dull yellow-green (gold) appearance when sealed with a chromate treatment.

Type III: Hard anodize coating can be used as an electrical insulation coating or as an abrasion-resisting coating on devices such as hydraulic cylinders and actuating cams.

● REVIEW QUESTIONS

1. What is meant by *heat treating*?
2. Describe a typical heat-treating process.
3. How are temper designations indicated in an alloy number?
4. What is meant by *solution heat treatments* of aluminum alloys?
5. What is *artificial aging*?
6. How is aluminum alloy marked for identification?
7. How is maximum hardness attained in carbon steel?
8. What is meant by *tempering* of steel?
9. Describe *annealing* of steel.

10. What is the purpose of *normalizing?*
11. Describe how case hardening of steel is accomplished.
12. What equipment and materials are required for heat-treating steel?
13. By what methods is aluminum alloy heated during the process of heat treating?
14. How is the temperature of a heat-treating furnace accurately controlled?
15. Describe *solution heat treating* of aluminum alloys.
16. Explain the purpose of *precipitation* heat treatment.
17. How is annealing of aluminum alloy accomplished and in what condition does it leave the alloy?
18. What is the purpose of *stress relief?*
19. Compare the heat treatment of magnesium with that of aluminum alloys.
20. What precautions must be observed in heating magnesium?
21. What temperature ranges are utilized in the heat treatment of titanium?
22. What types of stainless (corrosion-resistant) steels can be heat-treated?
23. Explain the principles of hardness testing with typical testers.
24. What does a hardness index reveal about a metal?
25. Compare Rockwell, Brinell, and Vickers hardness testing.
26. What problem may be experienced if a hardness test is made on a very small part?
27. What must be done when testing a section of tubing for hardness?
28. How are hardness testers calibrated?
29. What is meant by *nondestructive testing?*
30. Briefly explain *magnetic inspection.*
31. If a part is suspected of having cracks approximately perpendicular to the longitudinal axis, what type of magnetization should be employed?
32. What damage may result from magnetic inspection?
33. Explain *fluorescent-penetrant inspection.*
34. How is dye-penetrant inspection accomplished?
35. What is the principle of *ultrasonic inspection?*
36. For what type of defect is *eddy-current inspection* particularly useful?
37. What precautions must be taken when using x-ray and other radiographic-inspection techniques?
38. What are causes of *corrosion?*
39. Explain *electrolytic corrosion.*
40. Name and describe five types of corrosion.
41. How can dissimilar-metal corrosion be prevented?
42. What is the cause of *intergranular corrosion?*
43. What should be done to repair corrosion where no appreciable damage has occurred?
44. How should corrosion be removed from aluminum alloys?
45. Why is steel wool not approved for use in cleaning aluminum alloys?
46. Describe a chemical process that is suitable for treating the surface of aluminum alloys.
47. What precaution must be taken when cleaning aluminum alloys that have been anodized?
48. What is the likely result of intergranular corrosion?
49. What processes are used to prevent corrosion on magnesium parts?
50. Name some treatments that are used to provide a protective film on the surface of aluminum alloys.
51. How are steel parts protected from corrosion?

10 AIRCRAFT HARDWARE

● INTRODUCTION

In the design, production, and maintenance of aircraft and space vehicles or any other device for which specific quality and performance requirements are established, it is necessary that standards be selected and specifications be prepared to assure that the aircraft or other device will meet its requirements. Generally speaking, standards and specifications establish quality, size, shape, performance, strength, finish, materials used, and numerous other conditions for the manufacture and design of aircraft and their components. Because of the almost infinite number of sizes, shapes, materials, etc., involved in mechanical devices, a wide variety of standards and specifications have been developed covering hardware, metals, plastics, coatings, nonmetallics, and manufactured components.

In order to provide measures of uniformity, the military services, technical societies, manufacturers, and other agencies have attempted to establish uniform standards that would be universally acceptable for particular materials, products, dimensions, etc.

This chapter is concerned with the use of standards for parts known as hardware. Standard aircraft hardware includes such items as fastener assemblies, control cable fittings, fluid line fittings, and electrical wiring components. Fluid line fittings are covered in Chap. 12 of this text and wiring components are covered in a separate text of this series.

A listing of all the aircraft hardware for fasteners, cables, and miscellaneous applications would require a publication much larger than this text. The hardware covered in this chapter has been chosen as being representative of the types of hardware that the technician will encounter and that will also be encountered on a large number of aircraft. Much of the information in this chapter relates to standards and designation codes. Learning this information will enable the technician to use technical information available for hardware not covered in this text.

The emphasis in this chapter will be on identification and function of the various hardware items. It is impossible to provide the student with specific instructions relating to the installation of all types of fasteners and other items of hardware that a technician may encounter. In various sections of the texts of this series, specific information is given relating to the items of hardware involved in particular operations. For example the installation of rivets is discussed with structural repair.

It must be stressed that all items must be installed or applied in accordance with the manufacturer's instructions. Numerous items of hardware apply only to certain makes and models of aircraft. These should be handled as specified by the manufacturer of the product.

● STANDARDS

A **standard** is variously defined as (1) something established for use as a rule or basis of comparison in measuring or judging capacity, quantity, content, extent, value, quality, etc., (2) a level or grade of excellence, and (3) any measure of extent, quality, or value established by law or by general usage or consent.

In the normal performance of their duties, technicians will encounter an extensive array of standards establishing the characteristics of the materials and components that they will use from day to day in repair and maintenance work. Among these standards are **AN** (Air Force and Navy), **AND** (Air Force–Navy Aeronautical Design), **MS** (Military Standard), **NAS** (National Aerospace Standard), **NAF** (Naval Aircraft Factory), and **AS** (Aeronautical Standard).

Hydraulic fittings manufactured before World War II were manufactured to AC (Air Corps) standards. Some of these fittings may still be found on older aircraft. Such fittings are now covered by AN or MS numbers.

The most widely used standards for aircraft hardware are the AN and MS standards. These standards have been established by the Air Force and the Navy Bureau of Aeronautics. Items manufactured according to these are not limited to use by the Air Force and Navy but are found in all classifications of aircraft including those certificated by the FAA. Examples of AN parts are shown in Fig. 10-1, and some MS parts are shown in Fig. 10-2.

In recent years, all items approved for military applications have been manufactured according to MS drawings and standards. Many former AN parts are now produced as MS parts; for example, the universal head rivet listed under AN standards is now produced under MS standards.

Hardware items that have not yet been approved for military aviation but have been proven satisfactory by the industry are usually given NAS (National Aerospace Standard) numbers. After such items have been approved by military agencies, the MS numbers are assigned.

In addition to standard parts, manufacturers often design their own hardware items and these are given manufacturers' parts numbers. Many of the parts are

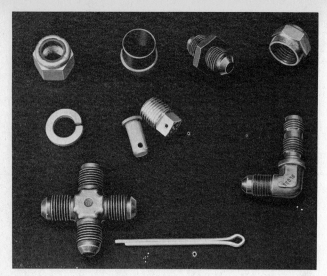

FIG. 10-1   Some AN standard parts.

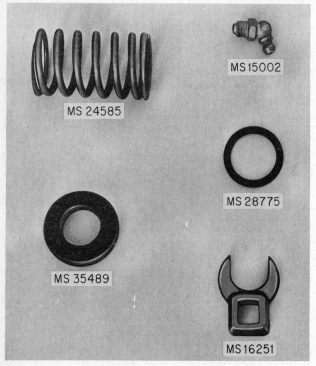

FIG. 10-2   Typical MS standard parts.

identical or almost identical to standard parts. The technician must be careful to assure that only approved parts are installed, regardless of whether they are standard parts or parts designated by manufacturers' parts numbers. In many cases, parts will be standard but will have manufacturers' parts numbers assigned.

The NAF standards are those that were developed and approved for use by the Naval Aircraft Factory. Items or parts manufactured under these standards are almost all covered by AN or MS standards at the present time. If an NAF number is called out for a particular part in an older airplane, it is likely that a comparable part can be found under an AN or MS stan-

dard; hence there should be no difficulty in obtaining a replacement.

AND standards are generally of more interest to the engineer than the technician because they are concerned principally with design. It is well, however, that the technician know the meaning and purpose of AND standards.

AS (Aeronautical Standards) have been established by the Society of Automotive Engineers (SAE). They include design standards, parts standards, and specifications that have not already been assigned AMS (Aeronautical Materials Specifications) numbers. Materials manufactured under AMS specifications are often required for use on civil aircraft by the FAA; for example, grade A fabric used for covering aircraft is designated AMS3806.

Table 10-1 lists a few of the AN, MS, and NAS standard numbers and the items to which they apply. This listing is only a sample, and it should be understood that there are hundreds of items covered by these standards. The agency for which the aircraft maintenance technician works should have the complete listing among its technical publications.

TABLE 10-1   Aircraft Hardware Standards

Sample AN, MS, and NAS numbers	
AN214	Pulley, control, plain bearing
AN253	Hinge
AN501	Screw, machine, fillister head
AN175	Bolt, close tolerance
AN960	Washer
AN815	Union
MS20365	Nut
MS20667	Fork end, cable
MS20995	Safety wire
MS20470	Rivet
MS24693	Screw, flat head
NAS1134	Bolt, pan head, close tolerance
NAS697	Nut, plate
NAS501	Bolt, hex head
NAS2007	Lock bolt, pan head
NAS1291	Nut, lightweight

Industry Standards

In addition to the standards previously described, industry organizations have also developed standards and specifications that are not necessarily covered by other standards. The American Society for Testing and Materials (ASTM) is one of the most active organizations in the establishment of material standards.

The American National Standards Institute (ANSI) is a federation of other organizations. Its function is to serve as a clearinghouse for standards. When a standard is established by the ANSI, the item standardized is generally accepted by all groups concerned with the particular item.

One of the leading organizations concerned with standards for iron and steel products is the American Iron and Steel Institute (AISI). Standards established for iron and steel products are therefore designated by AISI numbers. Many iron and steel products are also covered by standards issued from the SAE, ANSI, and ASTM. Special types of products are also standardized by the Alloy Casting Institute (ACI), the Investment Casting Institute (ICI), the Gray Iron Founders' Society (GIFS), the Steel Founders' Society of America (SFSA), and the Metal Powder Association.

The SAE, ANSI, and ASTM are concerned with almost all types of materials used in the aircraft field. Other organizations concern themselves with limited fields such as iron and steel, nonferrous metals, plastics and rubber, nonmetallics, and finishes and coatings.

● SPECIFICATIONS

A **specification** may be defined as a particular and detailed account or description of a thing, specifically, a statement of particulars describing the dimensions, details, or peculiarities of any work about to be undertaken, as in architecture, building, engineering, or manufacturing. It may also be stated that a specification sets forth the standards of quality and performance that a particular aircraft or aircraft component must meet to be acceptable for the purpose intended. Products of many types are manufactured according to specifications and then tested to assure that the specifications are met.

Military Specifications

Of greatest concern to persons working in the aircraft field are the military specifications for materials and products. Such specifications are designated MIL followed by a letter and a number. For example, the specification for standard aircraft cable, 7 by 19, carbon steel, is MIL-C-5424. MIL is the abbreviation for military, C is the first letter of the first word in the title of the specification (cable), and 5424 is the basic serial number. The number could be followed by a letter, which would indicate a revision in the basic specification.

A few of the products produced to a MIL specification that the technician may encounter are listed in Table 10-2.

TABLE 10-2 Military Specifications

Sample MIL-specifications	
MIL-R-5521	Ring, hydraulic fitting
MIL-P-5315A	Packing, O-ring
MIL-P-7034	Pulley
MIL-R-5674	Rivet
MIL-B-6946	Bronze bar stock
MIL-S-5000	Steel, annealed

FAA Specifications

The FAA requires all aircraft, aircraft engines, and propellers to be certificated. As part of the certification process, specifications covering all parts and components of the aircraft will be prepared. These specifications may include reference to parts built under AN, MS, or NAS standards.

Replacement or modification parts for a certificated aircraft must come from one of several sources:

1. Those produced under the aircraft type or product certificate which would primarily involve those made by the manufacturer.
2. Parts produced under a Technical Standard Order (TSO) of the FAA.
3. Parts produced by the holder of FAA Parts Manufacturer Approval (PMA). PMA parts are usually components or parts other than hardware.
4. Standard parts (such as bolts and nuts) conforming to established industry or United States Specifications, i.e., AN, MS, NAS.

FAA certification procedures for products and parts are given in part 21 of the Federal Aviation Regulations.

● THREADED FASTENERS

Fasteners are hardware items used to fasten two or more parts together. **Threaded fasteners** allow the parts to be taken apart and reassembled as necessary. In most cases threaded fasteners are reusable. Fasteners are essential to the structural integrity of the aircraft. It is essential that the aircraft technician be able to determine that the proper fastener is, or has been, used.

Threaded fasteners, for the purposes of this chapter, consist of bolts, nuts, and associated items, such as washers or locking devices.

Designation Codes

All standard hardware will have a designation code that identifies the basic item, dimensions, material, and special features. This code will usually relate to an AN, MS, or NAS standard. In some cases the manufacturer may have its own code that supplements the standard. While some codes look quite complex, most consist of a meaningful assortment of letters and numbers. The technician must have some knowledge of the basic part (bolt, screw, rivet) to develop an understanding of the code. The code structure for a few common items should be committed to memory. Because of the number of items of standard hardware, it will be necessary in many cases for the technician to make reference to technical literature or distributor's catalogs.

Machine Screw Thread

The winding groove around a bolt or screw or in the hole of a nut forms what is called a **screw thread.** Screw threads can take a variety of shapes such as on a wood screw, sheet metal screw, or a bolt. The thread on bolts or screws designed for nuts is referred to as **machine screw thread.**

A number of machine screw thread designs are in use. Aircraft hardware uses a design based upon a 60° thread. In Fig. 10-3 it can be seen that the angles cut into the bolt or nut form a series of equilateral trian-

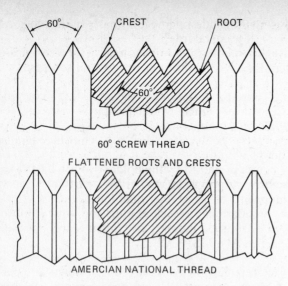

60° SCREW THREAD

FLATTENED ROOTS AND CRESTS

AMERCIAN NATIONAL THREAD

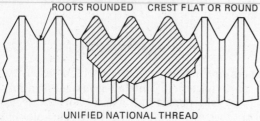

UNIFIED NATIONAL THREAD

FIG. 10-3 Thread form.

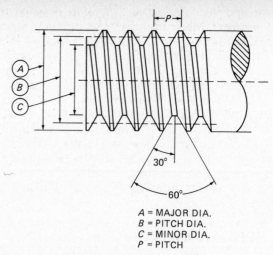

A = MAJOR DIA.
B = PITCH DIA.
C = MINOR DIA.
P = PITCH

FIG. 10-4 Thread dimensions.

gles. Sharp points at both the crest and the root of the thread created several problems. These were overcome by a slight flattening of the crest and filling of the root. This modification reduces the thread depth by about 25 percent. This thread form was standardized around the time of World War I as the **American (national) Form.** In 1948 the United States, Canada, and Great Britain agreed to adopt the **unified national form thread** to provide for interchangeability of threaded parts in these countries. This form is essentially the same as the national form except that the roots must be rounded and the crests may be flat or rounded. Fasteners using national and unified national thread forms are interchangeable except for some of the close tolerance fits. The European metric thread also uses a similar 60° design. The difference in the systems of measurement prevents interchangeability of metric and unified national threads.

Figure 10-4 shows some of the dimensions that are applied to screw threads. Threads are classified as internal and external depending upon application. The **major diameter** of an external thread is the diameter measured across the thread crest. The **minor diameter** is the diameter at the root of the thread. For a given size, the major and minor diameter will be the same for both the internal and external threads. It should be noted that in relation to internal threads the major diameter applies to the root and the minor diameter to the crest. Pitch diameter is a standard value for a given size and is approximately half way between the major and minor diameter.

The **lead** of a screw refers to the distance that the screw will advance into another threaded object with one revolution. This is the same as the **pitch,** or distance, between crests for most bolts. Most threads are described in terms of the number of **threads per inch** (tpi) or the number of crests in a length of 1 inch (in). It should be apparent that tpi and pitch are related in that the reciprocal of tpi (1/tpi) equals pitch.

Unified national threads are made in two series, unified national fine (UNF) and unified national coarse (UNC). The basic difference for these two series is in the number of threads per inch for a given diameter. For example a $\frac{1}{4}$-in-diameter bolt that is UNC will have 20 tpi while a $\frac{1}{4}$-in UNF will have 28 tpi. The advantages and disadvantages of coarse or fine threads are beyond the scope of this chapter.

Threaded fasteners smaller than $\frac{1}{4}$-in diameter are dimensioned by screw sizes. Fasteners larger than $\frac{1}{4}$ in are dimensioned in fractions of an inch. The $\frac{3}{16}$-in bolt is an exception to this practice. **Machine screw sizes** are shown in Table 10-3 and range from 0, the smallest, to 12, the largest. It should be noted that the dimension of a No. 10 machine screw is 0.190 in or very close to $\frac{3}{16}$ in (0.1875). The screw threads for a $\frac{3}{16}$-in-diameter bolt will use the threads specified for a No. 10 machine screw. Screw thread sizes may be better visualized by remembering that a No. 10 is approximately $\frac{3}{16}$-in diameter and that a No. 5 is $\frac{1}{8}$-in diameter. A further association that can be made is that the small screws used for switch and plug plates in house wiring are No. 6. The technician will encounter extensive use of Nos. 8 and 10 screws in aircraft maintenance.

In Table 10-3, the number of threads per inch for both UNC and UNF is shown for the various sizes as well as the decimal equivalent of each size. Technicans will find that their work is often enhanced by a knowledge of the threads per inch for the various sizes.

External threads are cut with a **die.** Some dies are adjustable so that the depth of thread can have slight variation. To cut an internal thread with a **tap** requires that a hole be drilled for the tap to go into. The size of the drill used is called the **tap drill.** The tap drill must be as large as the minor diameter. Tap drill sizes are given in Table 10-3.

TABLE 10-3 Thread Chart

National coarse-thread series, medium fit, class 3 (NC)					National fine-thread series, medium fit, class 3 (NF)				
Size and threads	Dia. of body for thread	Body drill	Tap Drill Pref'd dia. of hole	Tap Drill Nearest stand'd drill size	Size and threads	Dia. of body for thread	Body drill	Tap Drill Pref'd dia. of hole	Tap Drill Nearest stand'd drill size
					0–80	0.0600	52	0.0472	$\frac{3}{64}$
1–64	0.0730	47	0.0575	No. 53	1–72	0.0730	47	0.0591	No. 53
2–56	0.0860	42	0.0682	No. 51	2–64	0.0860	42	0.0700	No. 50
						0.0990			
3–48	0.0990	37	0.0780	$\frac{5}{64}$	3–56		37	0.0810	No. 46
4–40	0.1120	31	0.0866	No. 44	4–48	0.1120	31	0.0911	No. 42
5–40	0.1250	29	0.0995	No. 39	5–44	0.1250	25	0.1024	No. 38
6–32	0.1380	27	0.1063	No. 36	6–40	0.1380	27	0.1130	No. 33
8–32	0.1640	18	0.1324	No. 29	8–36	0.1640	18	0.1360	No. 29
10–24	0.1900	10	0.1472	No. 26	10–32	0.1900	10	0.1590	No. 21
12–24	0.2160	2	0.1732	No. 17	12–28	0.2160	2	0.1800	No. 15
$\frac{1}{4}$–20	0.2500	$\frac{1}{4}$	0.1990	No. 8	$\frac{1}{4}$–28	0.2500	F	0.2130	No. 3
$\frac{5}{16}$–18	0.3125	$\frac{5}{16}$	0.2559	No. F	$\frac{5}{16}$–24	0.3125	$\frac{5}{16}$	0.2703	I
$\frac{3}{8}$–16	0.3750	$\frac{3}{8}$	0.3110	$\frac{5}{16}$ in	$\frac{3}{8}$–24	0.3750	$\frac{3}{8}$	0.3320	Q
$\frac{7}{16}$–14	0.4375	$\frac{7}{16}$	0.3642	U	$\frac{7}{16}$–20	0.4375	$\frac{7}{16}$	0.3860	W
$\frac{1}{2}$–13	0.5000	$\frac{1}{2}$	0.4219	$\frac{27}{64}$ in	$\frac{1}{2}$–20	0.5000	$\frac{1}{2}$	0.4490	$\frac{7}{16}$ in
$\frac{9}{16}$–12	0.5625	$\frac{9}{16}$	0.4776	$\frac{31}{64}$ in	$\frac{9}{16}$–18	0.5625	$\frac{9}{16}$	0.5060	$\frac{1}{2}$ in
$\frac{5}{8}$–11	0.6250	$\frac{5}{8}$	0.5315	$\frac{17}{32}$ in	$\frac{5}{8}$–18	0.6250	$\frac{5}{8}$	0.5680	$\frac{9}{16}$ in
$\frac{3}{4}$–10	0.7500	$\frac{3}{4}$	0.6480	$\frac{41}{64}$ in	$\frac{3}{4}$–16	0.7500	$\frac{3}{4}$	0.6688	$\frac{11}{16}$ in
$\frac{7}{8}$–9	0.8750	$\frac{7}{8}$	0.7307	$\frac{49}{64}$ in	$\frac{7}{8}$–14	0.8750	$\frac{7}{8}$	0.7822	$\frac{51}{64}$ in
1–8	1.0000	1.0	0.8376	$\frac{7}{8}$ in	1–14	1.0000	1.0	0.9072	$\frac{49}{64}$ in

The fit between internal and external threads has been standardized into five classes ranging from No. 1, loose, to No. 5, tight. Aircraft bolts utilize a class 3 thread, while aircraft screws may have either a 2 or a 3 thread.

Figure 10-5 shows a code used to specify screw threads. The letters designate it as unified national coarse or unified national fine. This is followed by the diameter expressed as a screw size or fraction of an inch. Following a dash number are the number of threads per inch. A number from 1 to 5 following the tpi will give the standards of thread fit. A letter A following the number indicates an external thread and a B indicates an internal thread.

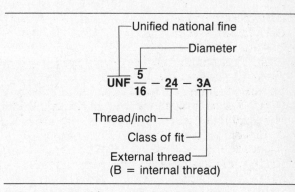

FIG. 10-5   Machine-screw thread designation.

Bolts

A **bolt** is designed to hold parts together. It may be loaded in shear or tension or both. Bolts are designed to be used with a nut and have a portion of the shank that is not threaded, which is called the **grip.** Machine screws and cap screws have the entire shank threaded.

Dimensions for a bolt are shown in Fig. 10-6. The diameter is the diameter of the shank, and the length is the distance from the bottom of the head to the end of the bolt. Bolt sizes are expressed in terms of diameter and length. The grip length should be the same as the thickness of the material being held together. The grip length can be determined from a reference chart for AN bolts.

Bolt heads are made in a variety of shapes with the hexagon (hex)-shaped head being the most common.

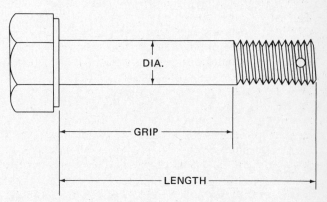

FIG. 10-6   Bolt dimensions.

General-Purpose Bolts. An all-purpose structural bolt used for both tension and shear loading is made under AN standards 3 through 20. The bolt diameter is specified by the AN number in sixteenths of an inch. For example:

AN3 $\frac{3}{16}$ in diameter
AN11 $\frac{11}{16}$ in diameter

These bolts have hex heads with drilled shanks, are made from alloy steel, and have UNF (fine) threads.

The length is given with a dash number. Bolts increase in lengths by eighths of an inch. The dash number will be one or two digits. If there is one digit, it will give the length in eighths. A two-digit dash number will always be 1 in or longer. The first number gives inches of length and the second gives the eighths. For example:

AN3-7 $\frac{7}{8}$ in long
AN3-15 $1\frac{5}{8}$ in long

The lengths stated will be nominal lengths; the actual length will be from $\frac{1}{32}$ to $\frac{3}{32}$ in longer than shown.

The bolts in this series were designed to be used with a castellated nut and cotter pin. Thus the standard bolt has a shank drilled for a cotter pin. If fiber-type self-locking nuts are used, a plain or undrilled shank should be used. To specify the bolt with a plain shank, the letter A is placed after the dash number. For example:

AN3-7   Drilled shank
AN3-7A   Plain shank

In some installations it is necessary to secure the bolt with a safety wire through the head. A drilled head is indicated with the letter H placed before the dash number. For example:

AN3-7   Undrilled head
AN3H-7   Drilled head

The standard bolt is made of alloy steel (2330). AN all-purpose bolts are also available in aluminum alloy and corrosion-resistant steel. To specify aluminum alloy the letter DD is inserted in front of the dash number. Corrosion-resistant steel is specified with the letter C. For example:

AN3-7   Alloy steel
AN3DD-7   Aluminum alloy
AN3C-7   Corrosion-resistant steel

The bolt material can be identified by head markings as shown in Fig. 10-7. Alloy steel bolts will have a cross, aluminum-alloy bolts will have two raised dashes, and the corrosion-resistant steel bolts will have one raised dash. Alloy steel bolts smaller than $\frac{3}{16}$-in diameter and aluminum-alloy bolts smaller than $\frac{1}{4}$-in diameter are not to be used in aircraft primary structure. Aluminum-alloy bolts should not be used where they will be repeatedly removed for purposes of maintenance and inspection. Alloy steel bolts should be cadmium plated and aluminum-alloy bolts should be anodized.

Close-Tolerance Bolts. AN4 bolts have a specified diameter of 0.249 (+0.000/−0.003) in. This tolerance is adequate for all applications except where the joint is subject to frequent load reversals or severe vibration. A close-tolerance bolt is made for this type of application under AN standards 173 through 186. AN174 bolts have a specified diameter of 0.2492 (+0.0000/−0.0005) in. The close-tolerance bolt has a triangle marking (see Fig. 10-7) on the head in addition to the standard material markings. The designation code and options are the same for the close-tolerance bolt as for the general-purpose bolt, with the exception of the AN number itself. The AN173 bolt is $\frac{3}{16}$-in diameter and the diameter increases by $\frac{1}{16}$ in for each AN number increase. An AN181 would be $\frac{11}{16}$-in diameter. For example:

AN5-7A   General-purpose bolt, $\frac{5}{16}$-in diameter, $\frac{7}{8}$ in long, plain shank.
AN175-7A   Close-tolerance bolt, $\frac{5}{16}$-in diameter, $\frac{7}{8}$ in long, plain shank.

FAA Advisory Circular 43.13-1A allows a general-purpose bolt to be used in place of a close-tolerance bolt if the bolt has a light-drive fit in the hole. One definition of a light-drive fit is a maximum clearance between the bolt and the hole of 0.0015 in. A second definition is a fit that requires the bolt to be pushed into the hole. A tight-drive fit, by comparison, would require a sharp blow from a 12- to 14-oz hammer.

Drilled-Head Engine Bolts. Drilled-head bolts have a deeper, or thicker, head than standard AN bolts. The extra material is required because there are three holes drilled through the head for safety wire (see Fig. 10-8). Drilled-head bolts were designed to be installed into threads tapped in the engine. The bolt can only be secured by wires through the head. Drilled-head bolts are identical to AN bolts 3 through 20 in terms of shear and tensile strength. A major difference is that the drilled-head bolts are made in both fine and coarse

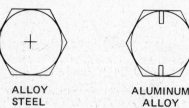

ALLOY
STEEL

ALUMINUM
ALLOY

CORROSION-
RESISTANT
STEEL

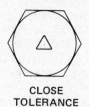

CLOSE
TOLERANCE

FIG. 10-7   Code marks for aircraft bolts.

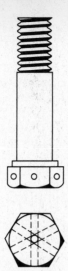

FIG. 10-8 A drilled head bolt.

threads. Drilled-head bolts were originally produced under AN73 through AN81 with a code similar to that for the bolts described above. The coarse-thread bolt was identified by a letter A in front of the dash number. The shank is not drilled for a cotter pin. For example:

AN74A-7   Drilled-head bolt, $\frac{1}{4}$-in diameter, $\frac{7}{8}$ in long, UNC thread

AN74-7   Drilled-head bolt, $\frac{1}{4}$-in diameter, $\frac{7}{8}$ in long, UNF thread

The AN standards for AN73 through 81 have been superseded by MS standards. Drilled-head bolts are produced under MS20073 for fine threads and MS20074 for coarse threads. Two dash numbers are used for diameter and length.

Clevis Bolts. Clevis bolts (AN21 through 36) are designed to be used only for shear load applications. The slotted, domed head results in this bolt often being mistaken for a machine screw. Closer examination (see Fig. 10-9) reveals that, unlike a machine screw, it has a long plain shank with a short threaded portion. Since it is only loaded in shear, a thin shear nut is used. This design provides more clearance for moving parts at each end than does a regular bolt. The clevis bolt is alloy steel with a cross mark on the head. The threaded shank is drilled for a cotter pin unless

FIG. 10-9 A clevis bolt.

the letter A appears after the dash number in the designation code. Designation of diameter follows the method used for other AN bolts. AN23 clevis bolts are $\frac{3}{16}$ in in diameter. Unlike standard bolts, clevis bolts are made in diameters smaller than $\frac{3}{16}$ in. An AN21 has the diameter of a No. 6 machine screw. A significant difference is that the clevis bolt's length increases in increments of $\frac{1}{16}$ in. The dash number of the code states the nominal length in sixteenths of an inch. For example, AN26-14A is a clevis bolt, $\frac{3}{8}$-in diameter, $\frac{14}{16}$ or $\frac{7}{8}$ in long, with a plain shank.

MS and NAS Bolts. A large number of standard bolts are produced under AN and MS numbers. These include high-strength bolts, high-temperature bolts, and internal wrenching bolts as shown in Fig. 10-10. Being standard items, these bolts will have a code for identification. The particulars of these codes may be found in various technical publications relating to MS and NAS standards, as well as in hardware distributors' catalogs.

FIG. 10-10 AN internal wrenching NAS bolt.

Nuts

Nuts are used on bolts to hold them in place and to provide the necessary clamping force to make a strong joint. The strength of a joint depends upon a bolt and nut being tightened to a specified torque. To ensure that the torque is maintained, various means are used to "lock" the bolt and nut together. Nuts are locked or safetied to the bolt by several means. One method involves mechanically locking the two together with safety wire or a cotter pin. A second method is to provide adequate friction between the threads of the nut and the bolt to, in effect, lock them together. The amount of friction can be increased in several ways. One method is the use of a spring-type lock washer. A second method is the use of a second nut on the bolt called a check nut. A third method is to use a nut with special design features that make them self-locking.

The designation code for nuts includes information similar to that for the bolt: the basic part, the size, the material, and the thread size.

The nut designed for use with AN bolts is the **AN310 castellated nut** shown in Fig. 10-11. The

FIG. 10-11 Common AN standard nuts.

AN310 nut is shaped for a cotter pin to be used to lock the nut and bolt together. The standard material for the AN310 nut is alloy steel; aluminum and corrosion-resistant steel are also available. The size of the AN310 nut is specified with a dash number that is the same as the number of the bolt it fits. The type of material is designated with the letter D before the dash number for aluminum, the letter C for corrosion-resistant steel, and no letter for plain steel. For example, AN310-5 is a steel castellated nut for a $\frac{5}{16}$-in-diameter bolt. All AN310 nuts have UNF class 3 fit threads.

Similar to the AN310 nut is the **AN320 shear nut.** The AN320 shear nut is designed for use with a cotter pin but is only about one-half to two-thirds as thick as the AN310. It is designed for use with the clevis bolt. The code for the AN320 is the same as for the AN310. The shear nut is made in dash numbers from 1 to 20. (-1 for No. 6 and -2 for No. 8 machine screw size). AN320 nuts are only made with fine threads.

The **AN315 plain nut** is shown in Fig. 10-11. In this case the name is very descriptive as the AN315 has no special features except that it is made in both left-hand and right-hand threads. The code for size and material is the same as for the AN310. Right- or left-hand thread is designated with an L or R following the dash number. For example:

AN315-6R   Plain nut with right-hand threads for a $\frac{3}{8}$-in bolt.
AN315C6L   Plain nut with left-hand threads for a $\frac{3}{8}$-in bolt from stainless steel.

Standard aircraft hardware uses right-hand threads. Plain nuts are used on control rods and wires which have one end threaded right-hand and the other end threaded left-hand.

The plain nut can be locked with a lock washer or a check nut. The **AN316 check nut** is a thinner version of the plain nut. To lock a plain nut, the check nut is tightened against it. This loads the plain nut in such a way that the bolt and nut threads are pushed tightly together. The designation code for the check nut is the same as for the plain nut, including identification of right-hand and left-hand threads.

Self-Locking Nuts. Self-locking nuts are made with a nonmetallic insert or with the top two or three threads distorted. As the bolt screws into the nut and encounters the insert, or distorted threads, a downward force is placed on the nut. The force removes all the axial play between the threads of the nut and bolt and creates adequate friction to prevent the nut from vibrating loose.

Self-locking nuts are made in both coarse and fine threads. The type of thread is designated in the code by listing the size and number of threads per inch as the dash number. A $\frac{1}{4}$-in bolt (-4) and a No. 4 machine screw will have dash numbers starting with 4. The technician must know the number of threads per inch to designate the correct size (see Table 10-3). For example:

-440   No. 4 machine screw (NC)
-448   No. 4 machine screw (NF)

-420 $\frac{1}{4}$-in bolt (NC)
-428 $\frac{1}{4}$-in bolt (NF)

Earlier in this chapter it was stated that a $\frac{3}{16}$-in-diameter bolt uses a No. 10 machine screw thread. The self-locking nut used with an AN3 bolt has a dash number of 1032.

One of the first self-locking nuts was the **AN365** which uses a nonmetallic insert to provide the locking force. Original inserts were of an elastic, fibrous material giving the nut the name of **fiber locknut,** or **elastic stop-nut.** The insert is currently made from nylon. The bolt does not cut threads in the insert but forces its way into the elastic material. As long as the insert material retains its elasticity, the nut may be reused. The **AN364 nut** is a shear nut version of the AN365. The AN364 is used with clevis pins or in other applications where the bolt is loaded primarily in shear. The nonmetallic insert will soften and melt under high temperatures. Both the AN364 and the AN365 are limited to operating temperatures below 250°F. Materials for locknuts include steel, aluminum, and brass. The letters D and B before the dash number designate aluminum and brass as the material. For example:

AN365-1032   Steel locknut for AN3 bolt
AN365B1032   Brass locknut for AN3 bolt
AN365D1032   Aluminum locknut for AN3 bolt

Standards for AN364 and AN365 have been superseded by MS20364 and MS20365. The material and size designations of the code remain the same.

The need for a nut to operate in higher temperatures resulted in the **AN363 metallic locknut.** The locking action in these nuts is caused by slightly squeezing the top threads of the nut. As the bolt is screwed into these threads, a downward force is placed upon the nut, pushing it against the bolt threads and holding it tight. The AN363 nut is usable to 550°F [288°C]. The AN363C, made of corrosion-resistant steel, can be used up to 800°F [427°C]. The standards for AN363 have been superseded by MS20363. The size designation is the same as that used for AN365 nuts.

Two types of lightweight, self-locking nuts usable to 450°F [232°C] have been developed under NAS1291 (MS21042) and NAS679A (MS21040). The NAS1291 is an all-metal hexagon design. An identifying feature is that the wrench size is much smaller than that used on conventional nuts. For example, an AN310-4 will use a $\frac{7}{16}$-in wrench, an NAS1291 for a $\frac{1}{4}$-in bolt will use a $\frac{5}{16}$-in wrench. The NAS679A is of conventional size but has been stamped to shape. Both of these nuts have the threads slightly squeezed to provide the self-locking force.

Plate Nuts. Nuts which are made to be riveted in place in the aircraft are called **plate nuts.** The purpose of such nuts is to allow bolts and screws to be inserted without having to hold the nut. Self-locking plate nuts are made under a number of standards and in a variety of shapes and sizes. Pictured in Fig. 10-12 is an AN366 two-lug plate nut with a nonmetallic insert. Also shown is an NAS680A lightweight, all-metal, 450°F [232°C] plate nut.

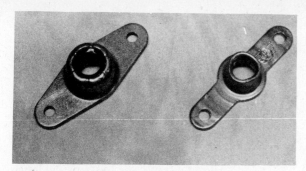

FIG. 10-12   Self-locking plate, or anchor, nuts.

Washers

Washers serve up to three functions when used with a bolt and nut. One function is to protect the material being fastened from being marred or crushed. A second function is to take up the excess grip length on the bolt and allow the nut to tighten before reaching the ends of the threads. The grip length of the bolt should be equal to, but never less than, the thickness of the metal. Since bolts vary in length by eighths of an inch, the grip length will usually be slightly longer than the thickness of the material. This extra length is compensated for by adding washers. A third function of a washer is to provide a locking force between the nut and the bolt.

The **AN960 flat washer** is a general-purpose washer for use under the heads of bolts or nuts. The AN960 is available in cadmium-plated steel, corrosion-resistant steel, or aluminum. The washer is made in two thicknesses, regular and light, to provide more variation in positioning the nut on the threads. The thin washer is one-half the thickness of the regular washer. The AN960 washer is sized by the screw or bolt that it fits. The dash number for use with a screw is the same as the screw size. For an AN bolt the dash number of the bolt followed by 16 (e.g., 416, 516) is used. The exception is the AN3 bolt which uses a -10 washer. Material designation is similar to that for nuts. A washer from the light, or thin, series is designated by placing the letter L after the dash number. For example:

AN960-4   Steel flat washer for a No. 4 screw
AN960-416   Steel flat washer for an AN4 bolt
AN960D416L   Aluminum flat washer for an AN4 bolt one-half the regular thickness

Figure 10-13 shows the dimensions of an AN960 washer.

The **AN970 large-area flat washer** was designed to be used with AN bolts in wood structures. Figure 10-14 shows an AN970 washer for an AN4 bolt compared to an AN960 washer for an AN8 bolt. The washer spreads the clamping force over a large area and keeps the relatively soft wood fibers from being crushed. Although originally designed for wood, the AN970 will work equally well in any installation of a similar nature. Since the AN970 is only made in sizes for AN bolts, the dash number of the size code will be

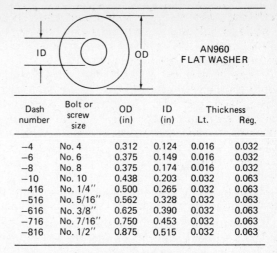

Dash number	Bolt or screw size	OD (in)	ID (in)	Thickness Lt.	Thickness Reg.
−4	No. 4	0.312	0.124	0.016	0.032
−6	No. 6	0.375	0.149	0.016	0.032
−8	No. 8	0.375	0.174	0.016	0.032
−10	No. 10	0.438	0.203	0.032	0.063
−416	No. 1/4″	0.500	0.265	0.032	0.063
−516	No. 5/16″	0.562	0.328	0.032	0.063
−616	No. 3/8″	0.625	0.390	0.032	0.063
−716	No. 7/16″	0.750	0.453	0.032	0.063
−816	No. 1/2″	0.875	0.515	0.032	0.063

FIG. 10-13   Dimensions of AN960 washers.

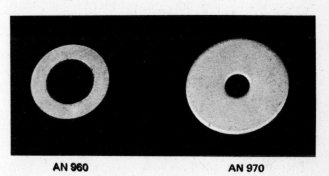

FIG. 10-14   A general-purpose and large-area washer.

the same as that for the bolt. For example, AN970-5 is a steel large-area washer for an AN5 bolt.

Lock washers are made in two designs. The **AN935 split-ring lock washer** is made of a twisted piece of steel. As the nut turns on the thread, the steel is flattened. The spring action of the steel provides a friction force between the threads that keeps the nut tight. The size code under AN935 standards uses the same dash number scheme as for AN960 washers. The AN935 standard has been superseded by an MS35338 standard. The size in the MS35338 code is given by a sequential number. For example:

AN935-416   Split-ring lock washer for an AN4 bolt
MS35338-40   Split-ring lock washer for a No. 4 machine screw
MS35338-44   Split-ring lock washer for an AN4 bolt

A second design of a lock washer is the **AN936 shakeproof lock washer.** The AN936 is a relatively thin washer with a number of twisted teeth that are flattened as the nut is tightened. As shown in Fig. 10-15, there are two design variations. The type A design has internal teeth and the type B design has external teeth. The locking forces generated by this washer are less than those generated by an AN935. The size code for AN936 follows that of the AN935 with the letter A or B being inserted before the dash number to identify the teeth locations. AN936 has

FIG. 10-15   Shakeproof washers.

been superseded by MS35333 for the type A, internal-teeth, washer and MS35335 for the type B, external teeth. As with the split-ring washer, the size is designated by a sequential number under the MS standard. For example:

AN936-A416   Shakeproof washer for AN4 bolt internal teeth
AN936-B416   Shakeproof washer for an AN4 bolt external teeth
MS35333-40   Shakeproof washer for an AN4 bolt internal teeth
MS35335-33   Shakeproof washer for an AN4 bolt external teeth.

Cotter pins

Cotter pins are used to lock castellated nuts onto drilled bolts or to secure plain-shank pins in a hole. Figure 10-16 illustrates the proper use of cotter pins.

FIG. 10-16   Applications of cotter pins.

AN380 is the standard for cadmium-plated steel cotter pins, with AN381 being the standard for those made from corrosion-resistant steel. The size of cotter pins under the two AN numbers is designated with two dash numbers. The diameter of both types is expressed in terms of thirty-seconds of an inch. The second dash number for the AN380 gives the length in eighths of an inch. The length for AN381 cotter pins is given in sixteenths of an inch. Both AN standards for cotter pins have been superseded by a combined standard, MS24665. Diameter, length, and type of material are represented by a single series of sequential numbers under MS24665 specifications. For example:

AN380-2-4   Steel cotter pin $\frac{1}{16}$ by $\frac{1}{2}$ in
AN381-2-8   Corrosion-resistant steel cotter pin $\frac{1}{16}$ by $\frac{1}{2}$ in
MS24665-132 Steel cotter pin $\frac{1}{16}$ by $\frac{1}{2}$ in
MS24665-151 Corrosion-resistant steel cotter pin $\frac{1}{16}$ by $\frac{1}{2}$ in

The size of cotter pin to use with AN3, 4, and 5 bolts is the AN380-2-2 ($\frac{1}{16}$ by $\frac{1}{2}$ in). AN380-3-3 cotter pins ($\frac{3}{32}$ by $\frac{3}{4}$ in) are used for AN6, 7, and 8 bolts.

Safety Wire

Safety wire is used in some cases to lock castellated nuts to drilled bolts or to secure a bolt with a drilled head. Stainless-steel safety wire is made in accordance with standard MS20995. The size of the safety wire is specified by a dash number representing the diameter of the wire in one-thousandths of an inch. Safety wire is available in diameters of 0.021, 0.025, 0.035, 0.041, and 0.051 in. The designation code contains the letter C to indicate that the material is corrosion-resistant steel. For example, MS20995-C41 is 0.041 stainless-steel safety wire. Safety wire is available for purchase in 5-lb spools or 1-lb dispensing packages.

Aircraft Screws

Aircraft use a large number of machine and self-tapping screws. Screws are used to fasten inspection panels, cowlings, fairings, and similar components which do not require high-strength fasteners. Screws are designed to be installed into a threaded object. On the aircraft they are often used with plate nuts or regular AN nuts.

A screw normally has the shank threaded all the way to the head as shown in Fig. 10-17. The machine screws specified for aircraft use normally have coarse threads with a class 2 fit. An exception is made for the No. 10 screw which has a fine thread compatible with nuts used with AN3 bolts. In essence, the common screws used for aircraft are the 6-32, 8-32, and 10-32. AN standards exist for Nos. 6 and 8 fine-thread screws and some may be encountered by the technician.

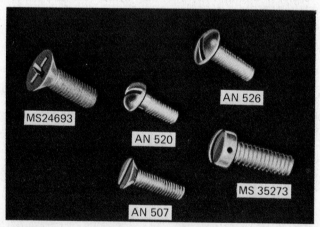

FIG. 10-17   Typical machine screws.

The head is designed to accept the installation tool. The slotted head for a plain screwdriver is the original standard design. A head made for use with a Phillips screwdriver is called a recessed head and is available for all AN screws. Because the Phillips design provides a better grip, the recessed head has, in effect, become the standard head used for aircraft.

Machine screws for aircraft are made in a number

of shapes and are used to identify the type of screw. Figure 10-17 shows five head styles. They are:

MS24693   Flat head (recessed) 100°
AN507   Flat head (slotted) 100°
AN520   Round head (slotted)
AN526   Truss head (slotted)
MS35273   Fillister head (slotted)

Other commonly used machine screw heads, shown in Fig. 10-18, are the AN 525 washer head and the MS35206 pan head.

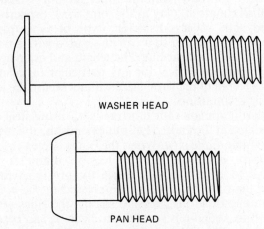

FIG. 10-18   Washer head and pan head screw.

The **AN507 flat head** (superseded by MS24693) requires that the material in which it is placed have a 100° recessed or countersunk area. Some flat head screws are designed for an 82° recess and should not be used in place of the AN507. The flat head screw is used where flush surfaces are desired.

The **AN526 truss head** is widely used as a protruding head machine screw. Its large head area provides a good clamping force on the sheet-metal parts with which it is commonly used. The AN526 and 507 are the two most common screws encountered in general aviation aircraft.

The **AN520 round head screw** has been replaced in most cases by the truss and the pan head screws.

The **pan head** (MS35206, coarse thread; MS35207, fine thread) screw is used for protruding head applications where more strength is required than is provided by the truss head.

The **fillister head screw** has a deep head with a small diameter when compared to other screws. This screw is widely used for components such as carburetors and magnetos. The head design allows a large clamping force to be applied. The head is usually drilled for a safety wire. The fillister screw was originally made to one of four standards. The AN500 screw has a coarse thread, class 2 fit and is made of plain carbon steel. The AN501 is identical to the AN500 except that it has fine threads. The AN502 is made of alloy steel with a class 3 fit, fine thread. The AN503 is identical to the AN502 except that it has a coarse thread. 502 and 503 screw heads are marked with an x to indicate their composition is alloy steel. The AN500 and 501

standards have been replaced with several different MS standards.

With the exception of the structural screws and some fillister head screws, most screws are made from plain carbon steel. Specifications also provide for screws made from aluminum alloy, brass, and corrosion-resistant steel.

Structural Screws. Certain machine screws are made from alloy steel and have a portion of the shank that has no threads to provide a grip length. These screws are known as structural screws and can be used in a manner similar to a bolt. Structural screws include the AN509 (100° flat head), AN525 (washer head), and the MS27039 (pan head). The AN525 is called a washer head because the head shape appears to have a washer on it. The AN509 standard has been superseded by MS24694. Structural screws have a grip length which must be considered in choosing screw length.

Screw Designations. AN screws use two dash numbers to indicate screw size. The first number is the screw diameter (6, 8, 10) and the second gives the length in sixteenths of an inch. The letter R before the second dash number designates a recessed (Phillips) head. The letter B, C, or D before the first dash number indicates material to be brass, corrosion-resistant steel, or aluminum.

Some AN standards have been superseded by MS standards. The MS standards use sequential numbers to designate size that are similar to those used for cotter pins. Because of the simplicity of the AN code, it is still in widespread use. For example:

AN507-8-R6   Flat head No. 8 $\times$ $\frac{3}{8}$ in screw with Phillips head
MS24693-S48   Flat head No. 8 $\times$ $\frac{3}{8}$ in screw with Phillips head
MS24693-C48   Flat head No. 8 $\times$ $\frac{3}{8}$ in with Phillips head made of corrosion-resistant steel
AN526-10-8   Truss head No. 10 $\times$ $\frac{1}{2}$ in screw with slotted head
AN526C8-R6   Truss head No. 8 $\times$ $\frac{3}{8}$ in screw with slotted head made of corrosion-resistant steel

Self-Tapping Screws. Self-tapping screws are also called sheet-metal screws and PK screws. At one time a commonly used self-tapping screw was the Parker-Kalon screw, thus the use of the term "PK." The term "sheet-metal screw" is derived from the fact that the self-tapping screw is designed to fasten sheet-metal parts together. The large coarse threads pull their way into sheet-metal parts that have not been tapped. AN standards exist for self-tapping screws but those in common use are made to industry standards. Self-tapping screws are only used in nonstructural applications.

Self-tapping screws are made in type A with sharp points (see Fig. 10-19) and type B with blunt points. The sharp-point style can start threading into smaller hole. Head styles for self-tapping screws include flat

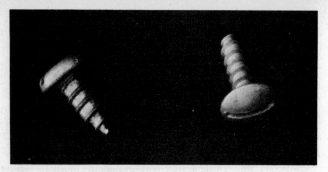

FIG. 10-19 Self-tapping screws.

head (82 and 100°), truss head, and pan head similar to the shapes found in machine screws. The heads can be slotted or have a recessed design for a Phillips screwdriver. The screws are made of steel with cadmium or nickel plating or of stainless steel. The diameters are specified in terms of machine-screw sizes. Lengths are available from $\frac{1}{4}$ in up, depending upon the diameter. To properly specify a self-tapping screw you should designate head style, screwdriver style, material, type of point, diameter, and length.

● NONTHREADED FASTENERS

With the exception of the pins, the fastener systems covered in this section are not designed to be taken apart. To disassemble parts held together with these systems requires the destruction of the fastener.

Pins

Metal pins of various shapes are used in certain locations in aircraft where their characteristics make their use beneficial. A flathead pin, taper pin, and roll pin are shown in Fig. 10-20.

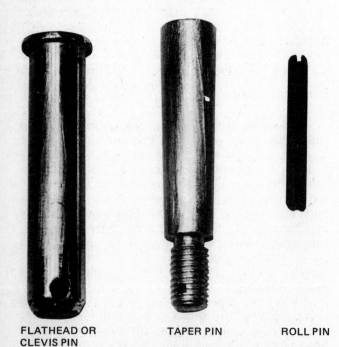

FLATHEAD OR
CLEVIS PIN   TAPER PIN   ROLL PIN

FIG. 10-20 Flat head, taper, and roll pins.

A flat head pin, also called a clevis pin, for use in aircraft is covered by MS20392. Pins of this type are used to join rod ends to bellcranks, secondary-control-cable terminals to control arms or levers, and in other, similar situations where the control is not in continuous operation. This type of pin is installed with the head up to reduce the possibility of its dropping out if the safety wire comes loose through wear. The pin is safetied with a cotter pin or safety wire.

A taper pin is designed to carry shear loads in a situation where a rod and tube are telescoped or where two tubular members of different diameters are telescoped to form a rigid joint. Since the pin is tapered, it will eliminate all play in the joint when properly installed. The most satisfactory type of taper pin is threaded on one end so it can be secured with a taper-pin washer and shear nut. The length and diameter of a taper pin are critical for any particular installation because a pin of the wrong size will not secure the joint in a rigid condition.

The roll pin is a split tube made of spring steel and chamfered at the ends. The split extends the full length of the pin on one side. When the pin is driven into an undersize hole, the pin reduces in diameter just enough to enter the hole. Since the pin is normally larger than the hole, when compressed in the hole it will maintain strong pressure against the sides of the hole, thus keeping it securely in place. It can best be removed with a pin punch.

Rivets. Rivets are metal pin-type fasteners designed primarily for shear-type loads. Thousands of rivets are used in aircraft. The majority of the rivets for aircraft are aluminum-alloy material. Aluminum rivets made to AN standards had the following head shapes (see Fig. 10-21):

AN426 Flush head, 100° countersink
AN430 Round head
AN442 Flat head
AN455 Brazier head
AN456 Modified brazier
AN470 Universal head

The AN standards have been superseded by two MS standards. The MS20470 universal head rivet can replace any other protruding head rivet. The MS20426 flush head rivet supersedes the AN426.

Aircraft rivets are made from five aluminum alloys. The material from which a rivet has been made can be determined by the head marking on the rivet. A one- or two-letter code identifies the alloy used in the rivet designation code. The alloys used, their head marking, and letter code are:

Alloy 1100	A	Plain head
Alloy 2017	D	Raised dot
Alloy 2117	AD	Dimple
Alloy 2024	DD	Raised double dash
Alloy 5056	B	Raised cross

Alloy 1100 type A rivets are made from pure aluminum. The rivets are soft and low strength. They are only used for nonstructural purposes.

MATERIAL	HEAD MARKING		AN MATERIAL CODE
1100	PLAIN	◯	A
2117T	RECESSED DOT	◉	AD
2017T	RAISED DOT	◉	D
2024T	RAISED DOUBLE DASH	⊖	DD
5056T	RAISED CROSS	⊕	B

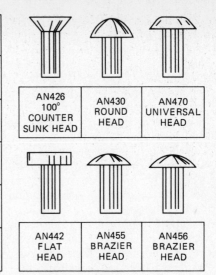

AN426 100° COUNTER SUNK HEAD | AN430 ROUND HEAD | AN470 UNIVERSAL HEAD

AN442 FLAT HEAD | AN455 BRAZIER HEAD | AN456 BRAZIER HEAD

FIG. 10-21   Rivet heads and markings.

Alloy 2024 type DD rivets are the strongest of the aluminum rivets. The material in the 2024 rivets age-hardens very rapidly after heat treatment and quenching. In a very short time the material becomes too hard to "drive" without internal stresses causing the rivet to crack. These rivets must be heat-treated and quenched immediately prior to driving. An alternate method is to keep the rivets refrigerated after quenching until they are to be driven. The cold temperature slows the aging process. For this reason 2024 rivets are called **icebox rivets.**

Alloy 2017 type D rivets have approximately 85 percent of the strength of 2024 rivets. Although not as much of a problem as 2024 rivets, the type D rivets also age-harden rapidly. Larger sizes of these rivets must be kept refrigerated or heat-treated immediately prior to use.

Alloy 2117 type AD rivets are made from a modification of the 2017 alloy that allows the rivets to be driven "off the shelf" at any time. The type AD rivet has 77 percent of the strength of 2024 rivets. The 2117 rivet, identified by a dimpled head, is the "standard" rivet for aluminum aircraft structure.

Alloy 5056 type B rivets are for riveting magnesium sheet. Magnesium is too hard to be used for rivets. The 2117 rivet used for aluminum contains copper as a major alloy. Since copper presents a potential for corrosion in the proximity of magnesium, 2117 rivets should not be used for magnesium. Aluminum alloy 5056 has magnesium as a major alloying element and provides adequate strength for rivets in magnesium structures.

Some other materials that aircraft rivets may be made of and the letter that would identify the material are:

Copper   C
Stainless steel F
Monel   M

Rivets are sized by the diameter of their shanks and their length. Rivets are made in diameters that increase in $\frac{1}{32}$-in increments. Common sizes of rivets used for aircraft are $\frac{3}{32}$- to $\frac{3}{16}$-in shank diameter. A protruding-head rivet length is the distance from the bottom of the head to the end of the shank. The length of a flush-head rivet is the overall length of the rivet. The designation code for rivets uses two dash numbers for the size. The first dash number gives the diameter of the rivet in thirty-seconds of an inch. The length of the rivet is stated in sixteenths of an inch by the second dash number.

The complete designation code for rivets includes the basic AN number, which gives the head shape. This is followed by the letter code indicating the material. Finally the size is given by diameter and length. For example, MS20470AD-3-7 is a universal head rivet of 2117 aluminum alloy, $\frac{3}{32}$ in in diameter, and $\frac{7}{16}$ in long. Most rivets are anodized or given a zinc-chromate treatment by the manufacturer. The color of the rivet provides no information in terms of material.

Detailed instructions for rivet installation are given in the associated text *Aircraft Maintenance and Repair.* Information is also available in FAA Advisory Circular 43.13-1A.

Special Fasteners

In addition to the standard nuts, bolts, screws, and rivets, there are many special fasteners that have been developed to join parts or structures where the more common fasteners are not usable or do not provide adequate strength. Special fasteners are used in installations where only one side of the material is accessible. Fasteners used for this purpose are called blind fasteners. Other special fasteners are used because of the need for high strength. Still others are used because they provide high strength with less weight than conventional hardware.

Many of the special fasteners use manufacturer's part numbers even though they are made to MS or NAS standards. Before installing a special fastener, the manufacturer's technical data should be checked to make sure the fastener is approved. When making the installation, the manufacturer's instructions should be closely followed.

Blind Rivets. Blind rivets, or fasteners, are designed to be installed where access to both sides of a sheet assembly or structure is not possible or practical. The blind rivet usually consists of a tubular sleeve in which a stem having an enlarged end is installed. The heads of such rivets are made in standard configurations such as brazier, universal, and flush. The rivet and stem are inserted in a correctly sized hole, and the stem is drawn into the sleeve by means of a special tool. The bulb or other enlargement on the end of the stem expands the end of the rivet and locks it into the hole. A typical blind rivet is the Cherrylock shown in Fig. 10-22. This rivet includes a locking collar that locks the stem plug in the rivet.

In addition to the Cherrylock rivet shown in Fig. 10-22, the Townsend Division of Textron, Inc., has developed more advanced and effective blind rivets. These are the bulbed Cherrylock rivet and the Cherrymax rivet. The installation of a bulbed Cherrylock rivet is shown in Fig. 10-23. This rivet conforms to NAS1738 and NAS1739.

The Cherrymax rivet is shown in Fig. 10-24. This rivet provides a high clamp up and a high shear strength plus a simplified installation system. Further description of this rivet is provided in the associated text *Aircraft Maintenance and Repair.*

A **rivnut,** shown in Fig. 10-25, is a hollow blind rivet that also serves as a nut. It is manufactured by the B. F. Goodrich Company and was originally designed for the attachment of deicer boots to aircraft wings and other parts of the aircraft subject to icing. When the rivnut is installed, a threaded spindle inside the rivnut is turned by means of the installing tool. As the spindle is turned, it pulls the shank toward the head and forms a collar on the side of the sheet opposite the head. This holds the rivnut securely in the hole and also holds the metal sheets together when installed through two or more sheets. After installation, the spindle is unscrewed and removed so that the rivnut can be used as a nut into which attaching screws can be placed.

Blind Bolts. A blind bolt, like a blind rivet, is one that can be completely installed from only one side of a structure or assembly. The blind bolt is used in place of a blind rivet where it is necessary to provide high shear strength. The bolt is usually made of alloy steel, titanium, or other high-strength material. A typical blind lock bolt is shown in Fig. 10-26, consisting of a nut, sleeve, and screw. Selecting the bolt of the correct length and diameter for the particular installation is very important, and the specification sheet should be followed closely.

Installation consists of inserting the sleeve in the hole, ready for pull-up by the mandrel of the installation gun. The mandrel pulls the nut toward the sleeve and causes the nut to form a collar against the workpiece. After setting the nut, the mandrel is withdrawn

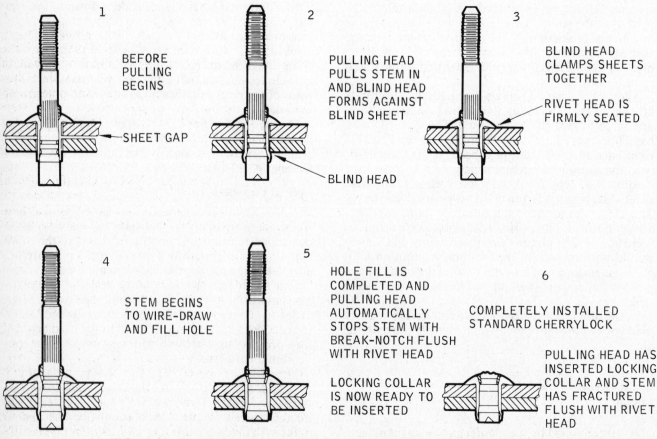

FIG. 10-22   Installation of a cherrylock rivet.   *(Townsend Division of Textron, Inc.)*

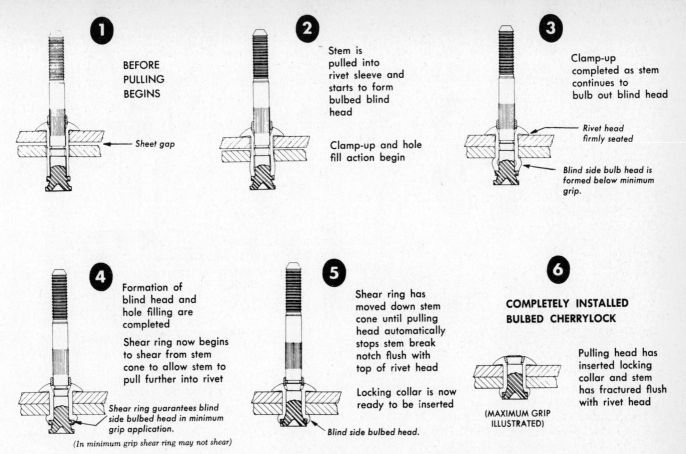

1 BEFORE PULLING BEGINS

Sheet gap

2 Stem is pulled into rivet sleeve and starts to form bulbed blind head

Clamp-up and hole fill action begin

3 Clamp-up completed as stem continues to bulb out blind head

Rivet head firmly seated

Blind side bulb head is formed below minimum grip.

4 Formation of blind head and hole filling are completed

Shear ring now begins to shear from stem cone to allow stem to pull further into rivet

Shear ring guarantees blind side bulbed head in minimum grip application.

(In minimum grip shear ring may not shear)

5 Shear ring has moved down stem cone until pulling head automatically stops stem break notch flush with top of rivet head

Locking collar is now ready to be inserted

Blind side bulbed head.

6 COMPLETELY INSTALLED BULBED CHERRYLOCK

Pulling head has inserted locking collar and stem has fractured flush with rivet head

(MAXIMUM GRIP ILLUSTRATED)

FIG. 10-23   Installation of a bulbed cherrylock rivet.   *(Townsend Division of Textron, Inc.)*

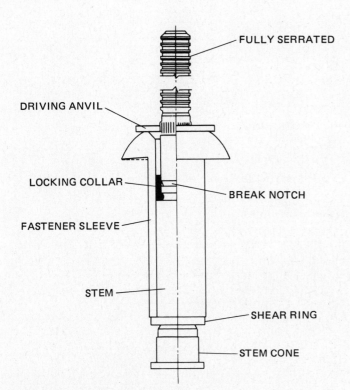

FULLY SERRATED

DRIVING ANVIL

LOCKING COLLAR

BREAK NOTCH

FASTENER SLEEVE

STEM

SHEAR RING

STEM CONE

FIG. 10-24   Drawing of a cherrymax rivet.   *(Townsend Division of Textron, Inc.)*

from the sleeve. The core bolt is then inserted in the sleeve and torqued to the level set forth in installation instructions. A blind bolt similar to the one described here is called the **Jo-Bolt.** There are many sizes, types, and designs of blind bolts, and the technician must make certain to use the size, type, and material that has been approved for the repair or replacement that is being made.

Hi-Shear Fasteners. Hi-Shear rivets, stump bolts, and **lock bolts** are designed for quick, permanent installations where it is desired to reduce weight and installation time. Such rivets or bolts can be used only where they can provide adequate strength and where their use has been approved by the appropriate authority.

The Hi-Shear rivet is made of steel and employs a swaged aluminum collar to hold it in place. The collar is driven on to the end of the rivet by means of a special tool in a conventional pneumatic rivet gun. The installation of a Hi-Shear rivet is shown in Fig. 10-27.

The Hi-Lok bolt is illustrated in Fig. 10-28. The installation of the Hi-Lok fastener is completed on one side of the assembly after the bolt has been inserted through the hole from the other side. The hexagonal wrench tip of the installing tool is inserted into a recess in the bolt, which holds the pin (bolt) while the tool turns the collar (nut). As the collar is tightened to

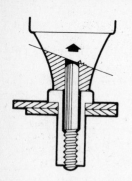 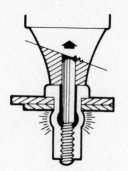 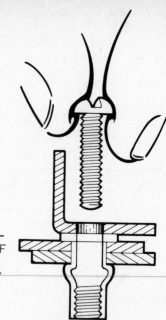

RIVNUT ON HEADER TOOL
MANDREL, INSERTED IN
DRILLED HOLE READY
FOR INSTALLATION.
ARROW INDICATES
DIRECTION OF MANDREL
MOVEMENT AS TOOL IS
OPERATED.

MANDREL RETRACTS PULLING THREADED POR-
TION OF RIVNUT SHANK TOWARD BLIND SIDE OF
WORK, FORMING BULGE AROUND UNTHREADED
SHANK AREA. RIVNUT IS CLINCHED SECURELY
IN PLACE. UNTHREADED THE TOOL MANDREL
LEAVES INTERNAL RIVNUT THREADS INTACT,
UNHARMED.

INSTALLED RIVNUTS ALSO
SERVE AS BLIND NUT PLATES
FOR SIMPLE SCREW ATTACH-
MENTS. WHEN IMPERATIVE
THAT ATTACHED PART BE
FLUSH FIT, COUNTERSUNK
RIVNUT HEADS ARE USED
INSTEAD OF FLAT HEADS.

FIG. 10-25   Installation of a rivnut.

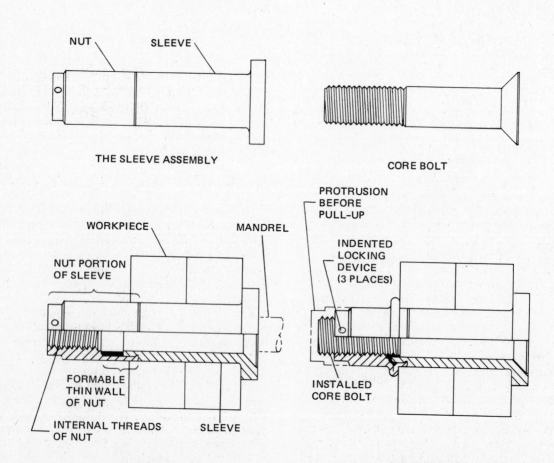

THE SLEEVE ASSEMBLY

CORE BOLT

BEFORE PULL-UP

AFTER PULL-UP
(COMPLETED INSTALLATION
WITH CORE BOLT INSTALLED)

FIG. 10-26   A blind lock bolt.

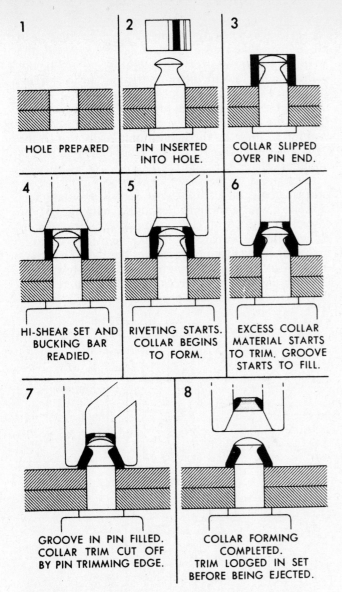

1	2	3
HOLE PREPARED	PIN INSERTED INTO HOLE.	COLLAR SLIPPED OVER PIN END.
4	5	6
HI-SHEAR SET AND BUCKING BAR READIED.	RIVETING STARTS. COLLAR BEGINS TO FORM.	EXCESS COLLAR MATERIAL STARTS TO TRIM. GROOVE STARTS TO FILL.
7	8	
GROOVE IN PIN FILLED. COLLAR TRIM CUT OFF BY PIN TRIMMING EDGE.	COLLAR FORMING COMPLETED. TRIM LODGED IN SET BEFORE BEING EJECTED.	

FIG. 10-27 Installation of a Hi-Shear rivet. *(Hi-Shear Corp.)*

the design torque level built into the collar, the hex portion of the collar is sheared off automatically by the driving tool. This leaves the installation with the correct amount of torque and preload.

A **lock bolt** is similar to the Hi-Shear rivet in function. These fasteners are manufactured in two principal types. The **pull-type** lock bolt includes a grooved pintail by which the bolt is pulled into place by means of a pneumatic installation gun. When the collar is firmly in place, the force of the gun fractures the pintail at the breakneck groove.

The **stump-type** lock bolt is installed in locations where the clearance is such that the installation gun for the pull-type bolt cannot be used. Pull-type and stump-type lock bolts are shown in Fig. 10-29.

Lock bolts are used where permanent assemblies are made, because they save weight, provide strength equivalent to standard bolts, and are easy to install. They provide excellent shear strength, and the tension lock bolts provide good tension strength. The tension

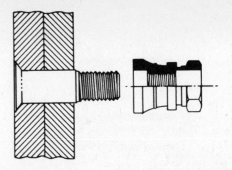

THE INSTALLATION OF THE HI-LOK FASTENER IS COMPLETED ON ONE SIDE OF THE ASSEMBLY AFTER THE PIN HAS BEEN INSERTED THROUGH THE HOLE FROM THE OTHER SIDE.

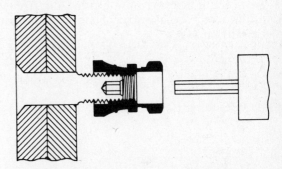

FOR NON-INTERFERENCE-FIT APPLICATIONS, THE HEX WRENCH TIP OF THE POWER DRIVER IS INSERTED INTO THE HEX RECESS OF THE PIN. THIS KEEPS THE PIN FROM ROTATING AS THE COLLAR IS DRIVEN.

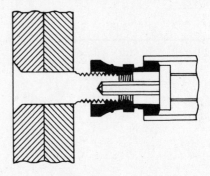

PROGRESSIVE TIGHTENING TAKES PLACE AS TORQUE IS APPLIED.

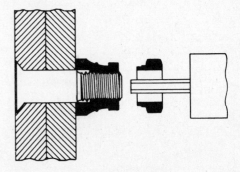

AT THE DESIGNED TORQUE LEVEL, BUILT INTO THE HI-LOK COLLAR, THE HEX PORTION OF THE COLLAR IS SHEARED OFF AUTOMATICALLY BY THE DRIVING TOOL. REMOVAL OF THE INSTALLATION TOOL FROM THE HI-LOK PIN COMPLETES THE INSTALLATION.

FIG. 10-28 Installation of a Hi-Lok fastener. *(VOI-Shan Division of VSI Corp.)*

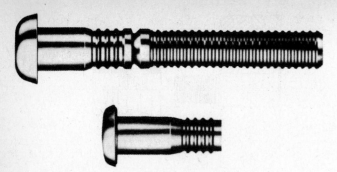

FIG. 10-29   Pull-type and stump-type lock bolts.

lock bolts have four collar-locking grooves, while the shear-type lock bolts have only two locking grooves. In selecting a lock bolt, the shear and tension loads must be known.

A pull-type lock bolt is installed as shown in Fig. 10-30. The pin is inserted from one side of the work, and the locking collar is placed over the projecting lock-bolt pintail. The gun is then applied, and the chuck jaws automatically engage the pull grooves of the projecting pintail. Depressing the gun trigger causes a pull to be exerted on the pin. The reaction of the pull is taken against the collar by the swaging anvil, which draws the work tightly together. The pin is thus pulled into the hole. As the pull on the pin increases, the anvil of the tool is drawn over the collar, swaging the collar material into the locking grooves of the pin to form a rigid, permanent lock. Continued buildup of force by the gun automatically breaks the lock-bolt pin at the breakneck groove. The pintail is automatically ejected.

Installation of the stump bolt is accomplished by means of a pneumatic rivet gun (air hammer) equipped with a swaging set and a bucking bar, as shown in Fig. 10-31. The stump bolt is inserted or driven into the prepared hole, and the locking collar is placed over the locking grooves of the bolt. The swaging set, driven by the air hammer, is forced down over the locking collar and forces the collar material into the locking grooves, thus providing a positive lock.

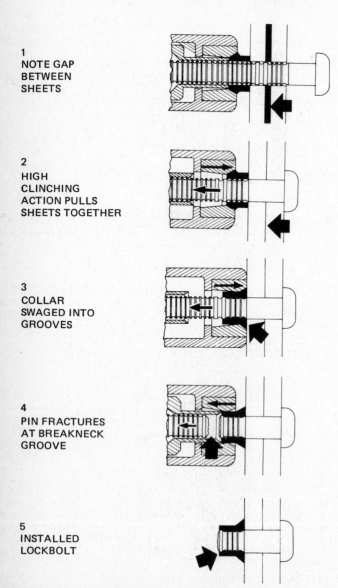

1
NOTE GAP
BETWEEN
SHEETS

2
HIGH
CLINCHING
ACTION PULLS
SHEETS TOGETHER

3
COLLAR
SWAGED INTO
GROOVES

4
PIN FRACTURES
AT BREAKNECK
GROOVE

5
INSTALLED
LOCKBOLT

FIG. 10-30   Installation of a pull-type lock bolt.

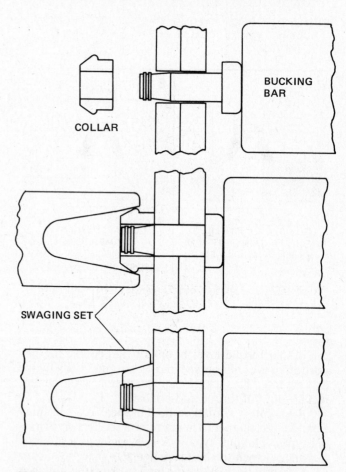

COLLAR

BUCKING
BAR

SWAGING SET

FIG. 10-31   Installation of a stump-type lock bolt.

Lock bolts of the type described here were formerly called Huck lock bolts. They are made of alloy steel, aluminum alloy, or stainless steel and carry NAS numbers from NAS1414 to NAS1562.

● PANEL AND COWLING FASTENERS

Panel and cowling fasteners that may be disengaged quickly are important in the inspection and servicing of an aircraft. For small aircraft, fasteners such as the Dzus, Camloc and Airloc are suitable; these are illustrated in Fig. 10-32. For high-performance large aircraft, a panel fastener is needed that not only will hold a panel in place but also will pull the panel in to a secure fit such that the panel becomes a part of the load-bearing structure. This type of fastener is called a **structural panel fastener** and is illustrated in Fig.

10-33. With this type of fastener, the sleeve bolt is retained in the panel by means of a retainer ring. The receptacle assembly is riveted to the structure.

● CABLE FITTINGS

Cable fittings are required to connect a cable to control arms, to other fittings, to turnbuckles, and to other

DZUS

CAMLOC

AIRLOC

FIG. 10-32   Panel and cowling fasteners: Dzus, Camloc, Airloc.

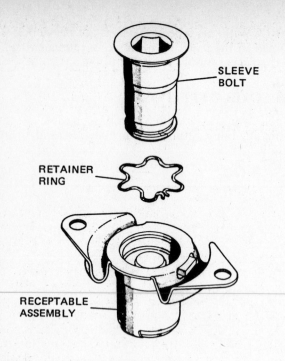

SLEEVE BOLT

RETAINER RING

RECEPTABLE ASSEMBLY

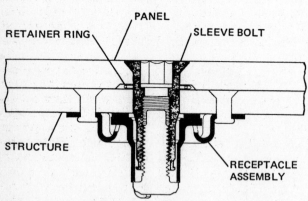

PANEL

RETAINER RING

SLEEVE BOLT

STRUCTURE

RECEPTACLE ASSEMBLY

FIG. 10-33 A structural panel fastener. *(VOI-Shan Division of VSI Corp.)*

sections of cable. When it is necessary to attach a cable to a turnbuckle or other device and swageable fittings are not available, the AN100 cable thimble is used. This thimble is illustrated in Fig. 10-34 and is attached by the swaged Nicopress sleeve.

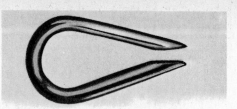

FIG. 10-34 A cable thimble.

The Nicopress sleeve and a spliced cable fitting are shown in Fig. 10-35. The Nicopress sleeve is composed of copper and is pressed (swaged) on the cable by means of a special tool. The specifications for a properly installed sleeve are established by the manufacturer.

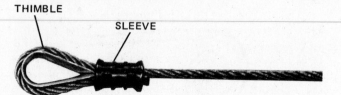

THIMBLE

SLEEVE

FIG. 10-35 A Nicropress cable splice.

When swaging equipment is available, it is desirable to use swaged fittings because these fittings develop the full cable strength when properly installed. The AN664 ball end, the AN666 stud end, the AN667 fork end, and the AN668 eye end are shown in Fig. 10-36.

To install a swaged fitting, the cable is inserted to the full depth of the barrel in the fitting and is held firmly in this position while the swaging operation is completed. The swaging can be done either with a hand machine or with a power swaging machine. The machine presses the metal of the fitting barrel into the cable to the extent that there is no visible division between the fitting and the cable when a cross-sectional cut is made through the swaged portion. After the swaging operation is completed, the swaged barrel should be checked with a go no-go gage to make sure that the proper degree of swaging has been accomplished. It is also advisable to mark the cable with adhesive tape when it is inserted into the barrel of the

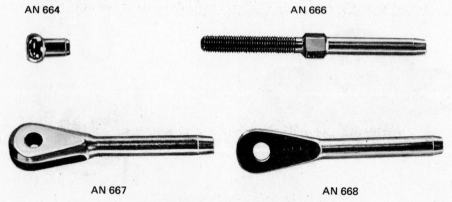

AN 664

AN 666

AN 667

AN 668

FIG. 10-36 Swaged cable fittings.

fitting to make sure that it does not slip during the swaging operation. It is good practice to paint the junction of the terminal or fitting and the cable with red paint to provide a means of detection for slipped cable at later inspections.

● TURNBUCKLES

Turnbuckles are commonly used for adjusting the tension of control cables. A standard turnbuckle consists of a **barrel** and two steel ends, one end having a right-hand thread and the other having a left-hand thread. Thus, when the barrel is rotated, the ends are moved together or away from each other. The end of the barrel having the left-hand threads is marked with a groove completely around the end of the barrel.

Typical turnbuckles are illustrated in Fig. 10-37. The AN-type turnbuckle is safetied with safety wire, and the MS-type turnbuckle is safetied with an MS21256 safety clip. The barrel (MS21251) of this turnbuckle has a groove in the threads on both ends to receive the straight end of the safety clip. The end fittings also have the threads grooved for the clip. The clip locks the threads of the barrel and fittings together to prevent any rotation.

Turnbuckles may be supplied with several different types of ends. Some of these are illustrated in Fig. 10-38. Figure 10-38*a* is a **cable eye** for use with a cable thimble; Fig. 10-38*b* is a **fork** by which the turnbuckle can be attached to a flat fitting; Fig. 10-38*c* is a **pin eye** to be inserted into a forked or double-sided fitting; and Fig. 10-38*d* is a **swage fitting** by which the cable can be attached to the turnbuckle after having been swaged into a sleeve. The barrel of the turnbuckle is made of brass and, as stated previously, is grooved around one end to indicate the left-hand thread. The hole through the center of the turnbuckle is used for the purpose of turning the barrel and for safetying.

Two principal factors must be considered in the installation and adjustment of turnbuckles: (1) when the turnbuckle is tightened, not more than three threads must show outside the barrel at each end; (2) the turnbuckle must be properly safetied.

Typical turnbuckle parts are designated by MS and AN standards as follows:

Barrel	MS21251	AN155
Fork	MS21252	AN161
Pin eye	MS21254	AN165
Cable eye	MS21255	AN170
Swaging terminal		AN669
Safety clip	MS21256	

● SAFETY BELTS

Although safety belts for civil aircraft may not be considered as aircraft hardware, they do involve hardware and must meet rigid standards. For this reason, they are included in this section.

All seats in an airplane which may be occupied during takeoff or landing must be equipped with approved safety (seat) belts. The requirements for safety belts are set forth in Technical Standard Order C22d, issued by the FAA. Models of safety belts manufactured for installation on civil aircraft since November 30, 1960, have had to meet the standards of National Aerospace Standards Specification (NAS) 802, with certain exceptions. Any safety belt meeting all the requirements of NAS802 may be approved for use.

The principal difference between the requirements of TSO C22d and NAS802 is in the strength of the safety-belt assembly. TSO C22d states that the safety-belt strength need be only 1500 lb [680 kg] for a single-person belt and 3000 lb [1360 kg] for a two-person belt. NAS802 requires 3000 and 6000 lb [2720 kg], respectively. Any safety belt meeting the requirements of TSO C22d may be used on civil aircraft.

Safety belts must be designed so as to be easily adjustable. Each belt must be at least $1\frac{15}{16}$ in [4.92 cm] wide and equipped with a quick-release mechanism designed so that it cannot be released accidentally. A safety belt may be approved for one person or for two adjacent persons, depending upon its strength. A belt for one person must be capable of withstanding a load of 1500 lb [680 kg] applied in alignment with the anchored belt. The quick-release mechanism must be capable of withstanding this load without undue distor-

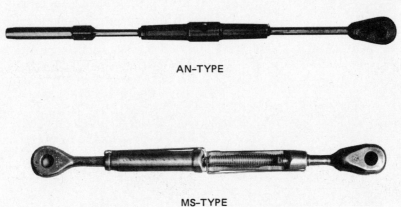

AN–TYPE

MS–TYPE

FIG. 10-37   AN and MS type turnbuckles.

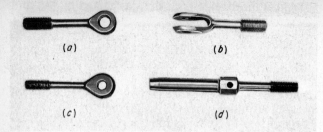

FIG. 10-38   End fitting for turnbuckles: (a) cable eye, (b) fork, (c) pin eye, (d) swage fitting.

tion and must be easily releasable under a load simulating a person hanging on the belt.

A safety belt approved for two adjacent persons must be capable of withstanding a load of 3000 lb [1360 kg] applied in alignment with the anchored belt. The quick-release mechanism must be easily releasable as described previously. After a test under extreme load, the release mechanism must be releasable with a pull of not more than 45 lb [20 kg].

The strength of a safety belt is determined by a test as specified in NAS802. The static testing of the belt and its attachments must be accomplished under conditions simulating the belt pulling against a human body. Each half of an approved safety-belt assembly must have legibly and permanently marked on or attached to it a nameplate or identification label with the following information: (1) manufacturer's name and address, (2) equipment name or type or model designation, (3) serial number and/or date of manufacture, and (4) applicable TSO or NAS number.

● KEEP CURRENT

It must be emphasized that new standards are constantly being issued and that older standards become obsolete. Nevertheless, many of the older standards are still effective, and it is up to the maintenance technician to use only approved parts and materials. It is always safe to select materials and parts that are specified in the manufacturer's overhaul or maintenance manual for a particular airplane, powerplant, or accessory.

● REVIEW QUESTIONS

1. Define a *standard*.
2. List the meanings of the following standard designations: AC, AF, AN, AS, NAS, NAF.
3. What type of hardware is covered by *AN5; AN310; AN470?*
4. What standard has supplanted the AN standard?
5. What type of part is covered by *MS20426; MS21252?*
6. Under what conditions is an NAS standard applied?
7. In case of doubt, should a technican install a standard part or a part designated by a manufacturer's part number?
8. What is the meaning of *AMS?*
9. What is the meaning and purpose of *ASTM?*
10. Who establishes standards for iron and steel?
11. What is the meaning of *SAE?*
12. Define *specification.*
13. What is an *MIL specification?*
14. What is the product covered by *MIL-C-5424?*
15. What agency establishes aircraft, engine, and propeller specifications?
16. How is the material from which a bolt is made indicated on an MS or AN standard bolt?
17. Describe an *MS internal wrenching bolt.*
18. What is the purpose of a *close-tolerance bolt?*
19. What is meant by a UNF class 3 thread?
20. How is the material of an aircraft standard nut indicated in the standard identification number?
21. How is a plain nut safetied?
22. Describe a *castle nut* and how it may be safetied.
23. Explain the principle of a *fiber locknut.*
24. What precaution with respect to temperature must be taken in the selection of fiber locknuts?
25. What is the purpose of a *shear nut?*
26. What precaution must be taken in the use of *check nuts* or *shear nuts?*
27. Why are *plate nuts* used on aircraft?
28. What is the difference between a standard machine screw and a structural screw?
29. What is the purpose of a *lock washer?*
30. Describe a *self-tapping screw* and its purpose.
31. Explain the use of *cotter pins.*
32. What does the number *MS20470-AD-3-4* indicate?
33. How are rivet heads marked to indicate material?
34. What is an *"icebox" rivet?*
35. For what purpose is a *taper pin* used?
36. Describe a *roll pin.*
37. Explain the purpose of a *turnbuckle.*
38. How may a turnbuckle be safetied?
39. How is aircraft control cable attached to turnbuckle fittings? Give three methods.
40. Describe a *Nicopress sleeve.*
41. Describe and explain the purpose of a *blind rivet.*
42. Describe a *rivnut* and explain its purpose.
43. What is the purpose of a *blind bolt?*
44. Describe a typical blind bolt.
45. Describe the installation of a *Hi-Shear rivet.*
46. What is the difference between a pull-type lock bolt and a stump-type lock bolt?
47. What are the advantages of *lock bolts?*
48. How is a *stump-type lock bolt* installed?
49. Describe three types of panel and cowling fasteners.
50. What is a structural panel fastener?
51. Give the requirements for a safety belt in aircraft.
52. To what instructions should a technician refer before installing units of aircraft hardware?

11 HAND TOOLS AND THEIR APPLICATION

● INTRODUCTION

Technicians are often judged by their knowledge of tools and by the manner in which they care for their tools. A variety of hand tools are necessary for the maintenance of aircraft. It is essential that the aviation maintenance technician be well informed and skilled in the use of these tools.

Common hand tools have very broad application and are used by people with a wide variety of training and experience. As a result hand tools are often taken for granted and misused by many. Many users are surprised to find there are many details affecting the design, construction, and use of the various tools. Tools are designed for specific purposes and for working with specific types of materials. Using a tool in a manner for which it was not designed can result in damage to the part or the tool as well as serious injury to the user. The proper and safe use of hand tools is very important to the aircraft technician.

Because of the number of tools in existence, this chapter concentrates on the frequently used hand tools that all technicians would have in their personal toolboxes. Specialized tools for use in areas such as sheet metal or dope and fabric are covered in the appropriate sections of other textbooks in this series. The emphasis of this chapter is placed upon the ability to identify the tool and understand its function.

● MEASUREMENT AND LAYOUT

One of the most important considerations in the manufacture, maintenance, and overhaul of machinery is measurement. The concept of interchangeable parts depends upon precision measurement. Repair of an aircraft structure requires measurement of damage and replacement material.

Layout refers to the process of transferring measurements from a drawing to materials from which parts or repairs will be fabricated. The tools used for measurement and layout are closely related.

Rules and Scales

The most commonly used measuring instrument is the rule. A rule is a straight-edged piece of material, either steel, wood, or plastic, and marked off in units of length. The **scale** of the rule refers to how the units of length are marked off. The markings of the scale are referred to as **graduations.** The graduations may divide the rule into fractional parts of an inch ($\frac{1}{2}$, $\frac{1}{4}$, $\frac{1}{8}$, $\frac{1}{16}$, $\frac{1}{32}$, and $\frac{1}{64}$), decimal parts of an inch (0.1 and 0.01), or metric graduations of centimeters and millimeters.

Rules exist in a wide variety of sizes and shapes. One of the most useful scales for an aviation technician is the 6-in **flexible steel rule** illustrated in Fig. 11-1. Rules of this size are available with a number of different graduations. The one shown has scales graduated in tenths and one-hundredths of an inch on one side and thirty-seconds and sixty-fourths of an inch on the other. A similar rule that has widespread use is pictured in Fig. 11-2 and is a part of a combination square set. The rule of a combination square is usually 12 inches (in) long and has four scales. The types of scales available for this rule are similar to those described for the 6-in rule.

An instrument for taking measurements up to several feet in length is the flexible steel tape shown in Fig. 11-3. The **flexible steel tape** is equipped with a hook on one end so that it will hold to a corner or ledge, thus making it possible for the rule to be used by one individual.

Examples of specialized rules are shown in Figs. 11-4 and 11-5. The triangular-shaped engineering scale has six scales graduated in tenths, twentieths, thirtieths, fortieths, fiftieths, and sixtieths of an inch. The graduations on the six scales are numbered so that such an instrument may be used to scale down a drawing to any desired proportion of an actual measurement. The **shrink rule** shown in Fig. 11-5 is used for those making patterns for use in casting. The pattern for the part must always be made a certain proportion larger than the actual part, depending upon the metal to be cast. This is because the metal will shrink a small amount as it solidifies in the cavity of the mold. For this reason a shrink rule, made to conform to the shrinkage of the metal, is used in laying out and measuring the dimensions of the pattern. Many other types of specialized scales exist.

FIG. 11-1   A 6-in flexible rule.

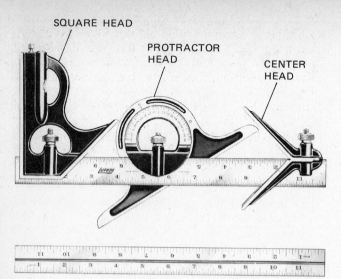

SQUARE HEAD

PROTRACTOR HEAD

CENTER HEAD

FIG. 11-2 A rule from a combination square.

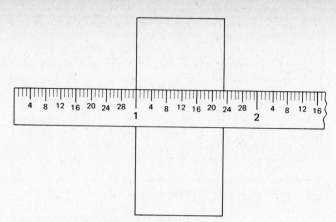

FIG. 11-6 Reading a measurement.

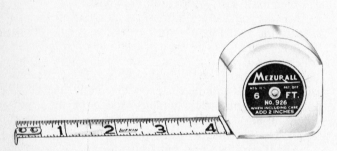

FIG. 11-3 Flexible steel tape.

FIG. 11-4 Engineering scale.

FIG. 11-5 A shrink rule.

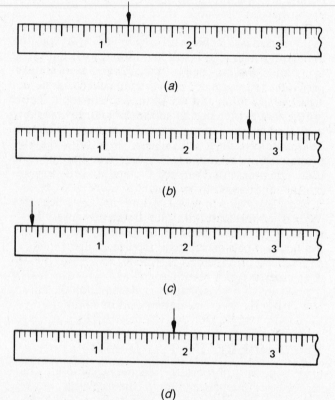

(a)

(b)

(c)

(d)

FIG. 11-7 Scale readings.

Reading a Rule

The first step in reading a rule is to know the value of the graduations on the rule. A careful study of the rule prior to taking measurements will make it possible for the technician to read the measurements quickly and accurately. Figure 11-6 shows a measurement being taken with a steel rule graduated in thirty-seconds of an inch. Examination of the drawing will show that the measurement being taken is $\frac{23}{32}$ in. Observe that the measurement is taken from the 1-in point on the scale, rather than from the end of the scale, which would be more difficult to align with the edge of the part to be measured. The method shown is more likely to be accurate than if the measurement is taken from the end of the scale.

The reading of a rule is not difficult; however the mechanic or technician must know the units into which the rule is divided and also know the subdivisions of these units. Figure 11-7 shows four different edges of rules to illustrate given measurements. The rules shown are subdivided into sixteenths of an inch. The ruler in Fig. 11-7a shows a measurement at the arrowhead of $1\frac{1}{4}$ in. This could also be read $1\frac{2}{8}$ in or $1\frac{4}{16}$ in. In b, the ruler shows a measurement of $2\frac{5}{8}$ in at the arrowhead, in c, $\frac{3}{16}$ in, and in d, $1\frac{13}{16}$ in.

A decimal rule is shown in Fig. 11-8. The user of this rule should note the fact that the divisions of the inches are in tenths rather than eighths. If a person

FIG. 11-8 A decimal rule.

were not careful, it would be easy to read a measurement as $1\frac{5}{8}$ instead of $1\frac{6}{10}$. Decimal rules are marked with divisions as small as $\frac{1}{100}$ in.

To obtain accurate measurements with a rule or scale, the readings should be taken very carefully, and the rule or scale should be kept in good condition. A rule should not be thrown into a box with other tools, because in this way it may become scratched, nicked, or otherwise damaged, and also it may become so dirty and worn that it is not possible to read the scales accurately. A steel scale should be wiped clean and dry and from time to time should be wiped with a cloth and a light oil.

Calipers

In some cases it is not possible, or is inconvenient, to take measurements with a rule, such as when measuring inside and outside diameters or the width of a slot. A caliper can be used for these measurements. A basic caliper has two parts that can be moved in relation to one another allowing different dimensions to be set. There are many types of calipers. Figure 11-9 shows two **spring calipers** commonly used for inside or outside measurement. The distance between the ends of the legs are adjusted with a screw working against a spring. The calipers shown require the use of a rule or other direct reading device to determine the units of measurement. Calipers can be used as a gage when machining a part to size by first setting it, with the aid of a rule, to the desired dimension and comparing it to the part as work progresses.

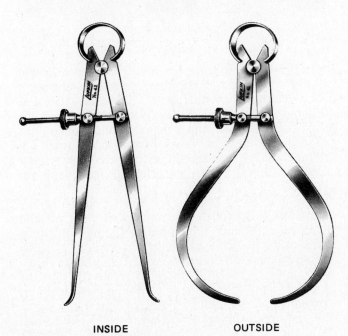

INSIDE   OUTSIDE

FIG. 11-9 Spring calipers.

Similar in appearance to the caliper is the **divider** shown in Fig. 11-10. The legs of the divider are straight and have sharp points. Dividers are designed to transfer a desired dimension from a scale to a piece of material. One setting will allow a line to be divided into a number of segments of equal length. They can be used as a compass for drawing arcs and circles for layout work. The divider can also be used in a manner similar to a caliper to measure thickness, etc. Calipers and dividers are sized by the length of the legs. Common sizes used by technicians are 3 in and 6 in.

FIG. 11-10 Dividers.

Figure 11-11 shows a **slide caliper**. This device has a scale, located on a beam, attached to one fixed jaw. A second jaw is free to slide up and down the beam. The ends of the jaws are formed so that either inside or outside measurements can be made. The sliding jaw has two reference lines marked inside and outside. The distance between the jaws is the value on the scale opposite the appropriate reference line.

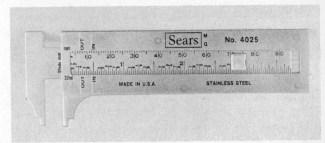

FIG. 11-11 Slide caliper.

Precision Measurement

The ability of the human eye to accurately read the graduations of a scale is limited. The need for precision measurement as small as ten-thousandths of an inch has necessitated the development of equipment capable of this function. Two ways of obtaining more precise measurements are use of the micrometer principle and the vernier scale.

Micrometers. The **micrometer** is a device to make small measurements. The term comes from micro plus meter, or one-millionth of a meter. In terms of English units the basic micrometer is designed to accurately measure to one-thousandth of an inch. A micrometer caliper is shown in Fig. 11-12. The basic micrometer unit is made up of three parts, a **hub,** or **barrel,** a **spindle,** and a **thimble.** Adding a C-shaped frame and an anvil completes the caliper. The anvil is fixed to the hub which also has the scale. The spindle serves the same function as the sliding jaw on the slide caliper. Although the basic micrometer mechanism is used on a number of different devices, the micrometer caliper is the most common. Micrometers are designed for a range of measurement of 1 in. The size of the frame determines if the micrometer will measure from 0 to 1 in, 1 to 2 in, etc. The ability of the micrometer to measure to 0.001 in is based on the fact that it has a spindle threaded with 40 threads per inch. The spindle screws into internal threads in the barrel. The thimble is attached to the spindle and moves with it. By rotating the thimble the spindle is caused to move to and from the anvil at the rate of $\frac{1}{40}$ in per turn of the thimble. Since $\frac{1}{40}$ in is equal to 0.025 in, each full turn of the thimble will move the spindle 0.025 in either in or out, depending on the direction of rotation. The barrel or sleeve of the micrometer is marked in divisions of 0.025 in each. The beveled edge of the thimble which surrounds the barrel is marked in divisions of one twenty-fifth of a revolution. Thus each division on the thimble represents a distance of 0.001 in of spindle travel.

The graduations on the barrel and on the thimble of a micrometer are shown in Fig. 11-13. The reading on

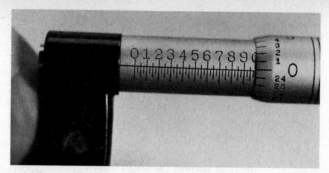

FIG. 11-13   Graduations on a micrometer.

the micrometer shown is 0.975 in. The figures on the barrel of the micrometer are spaced 0.10 in apart; hence the figure 1 represents a space of 0.10 in, the figure 2 represents 0.20 in, etc. The micrometer scale in Fig. 11-14 is set to read 0.203 in. Observe that the thimble is barely outside the graduation at the figure 2 on the barrel. Then note that the 0 mark on the thimble is three spaces below the 0 line of the barrel. If the 0 on the thimble were aligned with the 0 line on the barrel, the reading would be 0.200 in. However, since the 0 is three graduations below the 0 line, 0.003 in must be added to the reading to make a total of 0.203 in.

FIG. 11-14   Reading a micrometer scale.

In Fig. 11-15 the scales for four different micrometer readings are shown. Examine these settings and see if your readings show that *a* is 0.235, *b* is 0.036, *c* is 0.121, and *d* is 0.762. To develop proficiency in the use of a micrometer, the technician should obtain one and practice taking readings on a variety of small objects. These readings should be compared with similar readings taken by an experienced technician for accuracy.

It will be noted that the micrometer shown in Fig. 11-12 has a part called the cam lock at the frame and a ratchet at the end of the thimble. The purpose of the cam lock is to lock the spindle in place after a setting has been made so that there will be no change in the reading when the micrometer is removed from the work. The purpose of the ratchet is to provide a uniform pressure for taking readings. Figure 11-16 shows the micrometer being used to take a reading on a small part, and the pressure of the spindle is applied through the ratchet. The ratchet and spindle are turned to the left by means of the ratchet itself, and when the proper

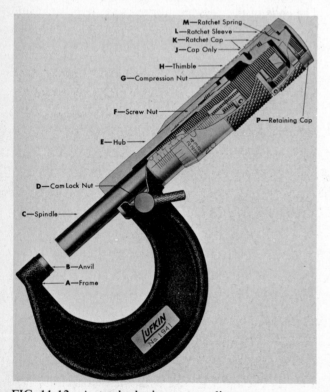

M—Ratchet Spring
L—Ratchet Sleeve
K—Ratchet Cap
J—Cap Only
H—Thimble
G—Compression Nut
F—Screw Nut
E—Hub
D—Cam Lock Nut
C—Spindle
P—Retaining Cap
B—Anvil
A—Frame

FIG. 11-12   A standard micrometer caliper.

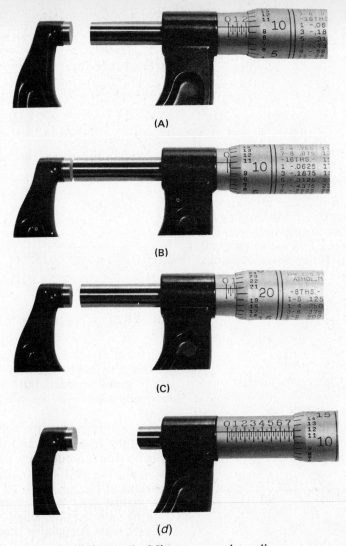

(A)

(B)

(C)

(d)

FIG. 11-15   Micrometer scale readings.

FIG. 11-16   Taking a measurement with a micrometer.

FIG. 11-17   Special micrometer for tubing wall thickness.

pressure is reached, the ratchet will slip, thus preventing any further pressure from being applied to the spindle. Thus the ratchet provides uniform pressure for each reading taken with the micrometer. Special micrometers are made to take readings which cannot be made with the standard micrometer. Figure 11-17 shows a micrometer designed to measure the distance between a concave and a convex surface,

such as the wall thickness of a tube. Figure 11-18 shows a micrometer which has a 60° angle on the spindle and anvil to make it suitable for measuring screwthread diameters.

An inside-measuring micrometer is shown in Fig. 11-19. The **inside micrometer** consists of a basic micrometer mechanism with extension rods of various lengths to provide a range of measurements. The mi-

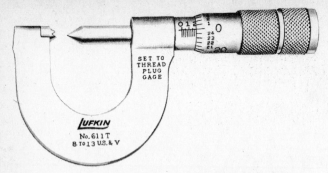

FIG. 11-18   Micrometer for measuring screw threads.

crometer unit shown measures 0 to 0.5 in. Adding the extension rods shown will allow it to measure up to 6 in. For example, the addition of a 2-in rod will provide measurements of 2.0 to 2.5 in. Inside micrometers are used to measure inside diameters and widths of slots.

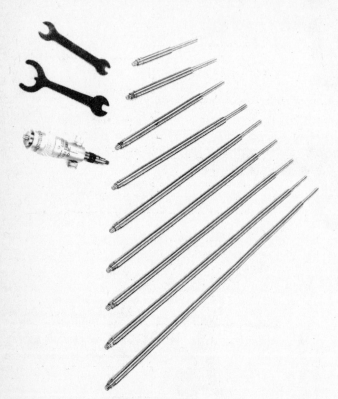

FIG. 11-19   An inside micrometer set.

A **micrometer depth gage** can be used to measure the depth of a hole or slot or the range of travel between two parts. As shown in Fig. 11-20, it has a micrometer unit mounted on a base. As the thimble is turned, the spindle will advance or retract into the base. The distance it has extended from the base can be read on the micrometer scale. Depth micrometers have interchangeable spindles of various lengths to provide for measurements greater than 1 in.

Vernier Scales. When it is necessary to make measurements smaller than 0.001 in, it will be necessary to use a micrometer with a vernier scale. A **vernier scale** is an auxiliary scale on a measurement tool that divides the smallest graduation on the main scale into even smaller increments. On the micrometer a vernier

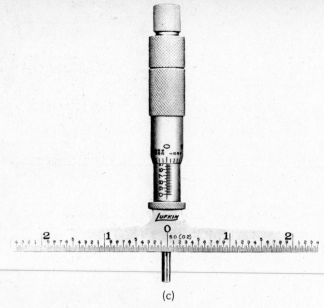

(c)

FIG. 11-20   A depth gage micrometer.

scale will divide each unit of 0.001 into 10 parts, or 0.0001. On other tools it may divide an increment of 0.025 into 25 parts (0.001). A fractional scale with $\frac{1}{16}$-in graduations may be subdivided in increments of $\frac{1}{128}$ in.

An instrument with a vernier scale will have two scales, the main scale and the vernier scale, located side by side but not connected. The vernier scale will have slightly smaller divisions than the main scale. Looking at Fig. 11-21, it can be seen that the scale of the vernier has been divided into four parts. It can also be seen that the length of four divisions on the vernier scale is equal to three divisions on the main scale. The vernier scale will always have one more division than an equal length on the main scale. The number of divisions on the vernier scale determines how much the main scale increment is subdivided. With four vernier divisions each main scale division can be divided by 4. To divide a scale increment into 25 parts would require a vernier scale having 25 divisions equal in length to 24 divisions on the main scale.

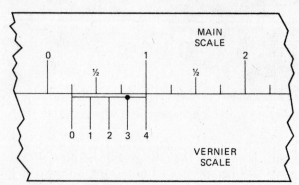

FIG. 11-21   A vernier scale.

Figure 11-22 shows a micrometer with a vernier scale. The scale is located on the barrel adjacent to the scale on the thimble. The scale on the barrel will have 10 divisions compared to 9 on the thimble. Each thim-

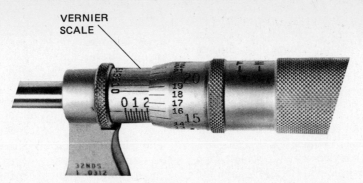

FIG. 11-22   Micrometer with a vernier scale.

ble division (0.001) can be broken down into units of 0.0001. To read the vernier micrometer you first read the barrel and thimble as with a standard micrometer. In Fig. 11-22 the reading appears to be between 0.215 and 0.216. The vernier scale will tell exactly how far it is between these two values.

Figure 11-23 illustrates an enlarged view of a vernier scale (A) and a thimble scale (B) from a micrometer. In looking at Fig. 11-23 it can be seen that only one of the marks (0 through 9) on the vernier scale will line up with a mark on the thimble. In number 1 the 0 mark of the vernier scale is the one that lines up. This would indicate that the reading in this case is in even thousandth measurements. In number 2 it is the 5 mark that lines up, indicating that the measurement is five-tenths of the way between the two one-thousandths marks. Using the reading from Fig. 11.22, we add 0.0005 to the 0.215 value and get a final reading of 0.2155.

To summarize, in reading a vernier scale look for the line on the vernier scale that lines up with a line on the main scale. The value of the vernier scale line will be the amount to add to your base value. Figure 11-24 represents a micrometer barrel scale (A), thimble scale (B), and vernier scale (C). Determine what value is shown. Your answer should be 0.1724.

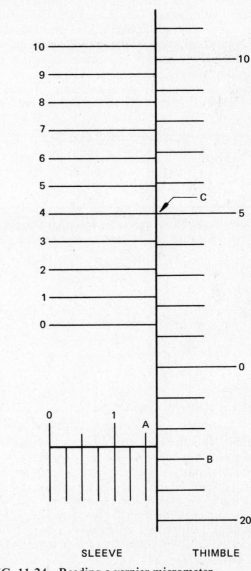

FIG. 11-24   Reading a vernier micrometer.

FIG. 11-23   Principle of the vernier scale.

Care of Micrometers. The micrometer is a delicate, precision instrument and must therefore be handled with great care. The technician should always practice using a light touch and avoiding excessive clamping

239

pressures. The micrometer should be kept in its own case or wrapped in a soft cloth and put in a special compartment. It should not be dropped or thrown into the toolbox. The technician should check the micrometer periodically for accuracy. Better quality micrometers are adjustable and have instructions for calibration.

Vernier Calipers. A slide caliper with vernier scale is commonly called a **vernier caliper.** The vernier caliper shown in Fig. 11-25 has two separate scales, each with its own vernier. One scale will normally have the capability of measuring to 0.001 in. The second scale may be decimal inches, fractions, or metric. A decimal scale area is illustrated in Fig. 11-26. Most vernier calipers will measure up to 6 in, although other sizes are available. The graduations on the main scale are similar to those on the barrel of the micrometer. The largest number represents inches. The smaller number is tenths of an inch. Each tenth is divided into four parts, or 0.025, per graduation. The vernier has 25 divisions and thus will divide each graduation on the main scale in 0.001. The vernier value is read in the same way as the micrometer. The jaws of the vernier caliper are designed so that the scale reading will be the same for both inside and outside measurements. Therefore only a single reference mark is needed to read the scale. The reference mark is the 0 mark on the vernier scale. The scale in Fig. 11-26 has a reading of:

Main scale	1.600
plus	0.050
vernier scale	0.004
	1.654

The vernier caliper is a precision tool that should be cared for in the same manner as the micrometer.

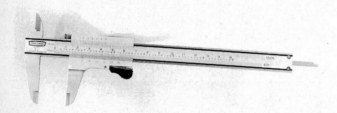

FIG. 11-25   A vernier caliper.

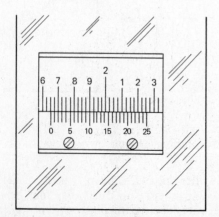

FIG. 11-26   A vernier caliper scale.

Gages

The inside micrometer and vernier caliper can make inside measurements. However, there are many occasions where they are difficult to use. Various gages have been developed to be used with micrometer calipers that, in general, provide more convenient and accurate inside measurements.

A set of telescoping gages is shown in Fig. 11-27. The telescoping gage is shaped like a T, with one fixed arm and one telescoping arm. The fixed arm is hollow and contains a coiled spring which extends into the telescoping arm. The gage has a screw stop arrangement by which the telescoping arm may be held at any desired point of extension.

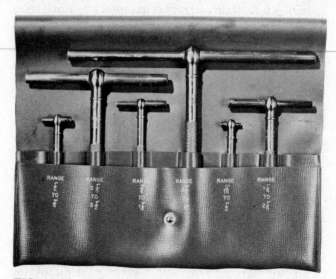

FIG. 11-27   Telescoping gages.

To use the telescoping gage, the gage with the correct size range must be chosen. For example, if it is desired to measure a hole having a diameter of 1 in, the $\frac{3}{4}$- to $1\frac{1}{4}$-in gage should be selected. The telescoping arm is depressed, and the knurled spindle is turned to the holding position. The gage is then inserted into the hole to be measured, and the telescoping arm is released. When it seems certain that the gage is in the maximum extended position which the hole will allow, the operator tightens the knurled spindle to hold the gage in a fixed position. The gage is then removed carefully and measured with a micrometer. It is best to take more than one measurement for each dimension, because there is a possibility of error if the gage is not exactly in the right position when the telescoping arm of the gage is set.

In using the telescoping gage, the gage must not be *forced* into any position and the holding mechanism must not be screwed in too tightly. As is true with any precision measuring tool, the telescoping gage must be handled with care at all times.

Small-hole gages are recommended to accurately measure holes smaller than $\frac{1}{2}$ in. A set of small-hole gages is shown in Fig. 11-28. The gage consists of a small split ball or half ball with an inner shaft upon which is mounted a cone. When the knurled knob on the end of the gage is turned, the shaft moves axially and changes the position of the cone in the split ball.

This, in turn, changes the outer dimension of the split ball. When it is desired to measure the inside of a small bearing or any other small hole requiring exact dimensions, a gage is selected which will fit easily into the hole when the gage is at its smallest dimension. The gage is inserted into the hole and expanded until it fits the hole firmly but not tightly. It is then removed, and its largest dimension is measured with a micrometer. This will provide an accurate measurement of the hole diameter.

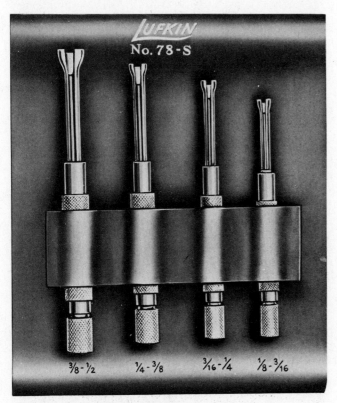

FIG. 11-28   Small-hole gages.

For certain measurements, manufacturers will sometimes specify that "go" and "no-go" gages be used. For example, if the dimension of a particular hole is specified as 0.525 ± 0.0005 in, the go gage will have a diameter of 0.5245 in and the no-go gage will have a diameter of 0.5256 in. If the go gage will fit into the hole, it is known that the hole is large enough, and if the no-go gage will not fit into the hole, the hole is not too large.

A **thickness gage** is pictured in Fig. 11-29. This gage consists of a number of metal leaves ranging from as thin as 0.001 in to as thick as 0.060 in. The thickness gage is used to determine the dimension of a gap or the clearance between two parts such as a set of breaker points.

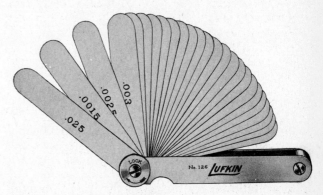

FIG. 11-29   A thickness gage.

The **radius and fillet gage** shown in Fig. 11-30 consists of a number of metal leaves which have been shaped with inside and outside radii of specific sizes. This gage can be used to check the radius on a part being fabricated or to check for stretch or other deformation in a part in use. It can also be used for laying out radii for new parts.

The **screw-pitch gage**, or **thread gage**, is shown in Fig. 11-31. It is used to measure the number of threads per inch on a threaded fastener.

Angular Measurement

The measurement of angles is involved in many aircraft maintenance operations. Many of these, such as setting propeller angles or checking control surface travel, require precision equipment beyond the scope of this chapter. General-purpose measurement of angles can be done with a protractor or combination square.

A **protractor** is an instrument for drawing or measuring angles. One type of protractor used by many

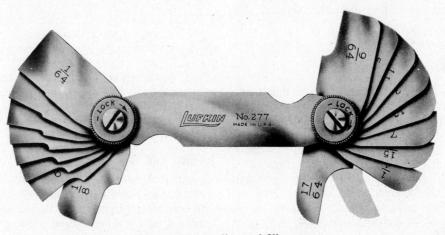

FIG. 11-30   Radius and fillet gage.

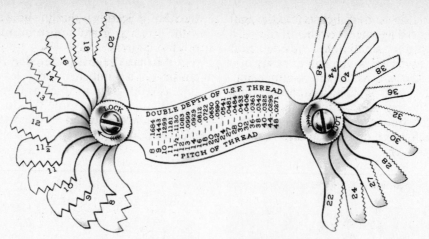

FIG. 11-31 A screw pitch gage.

technicians is shown in Fig. 11-32. The **combination square,** shown in Fig. 11-2, consists of a steel rule used with one of three heads. The **square head** is used to measure, lay out, or check 90 and 45° angles. This head can also be positioned on the scale to provide assistance in making a number of repetitive measurements or can be used as a marking gage. The **protractor head** provides for measuring or drawing angles. The **center head** can be used to locate the center of round stock.

FIG. 11-32 Protractor.

A number of other types of fixed and adjustable squares are available to the technician. Because many are of a specialized nature they are not covered in this chapter.

● WRENCHES

Wrenches are used for the installation and tightening of threaded fasteners. They are made in a variety of shapes to meet the requirements of the fastener or installation design or the physical location of the fastener assembly. The technician should always use a wrench designed for the operation to be performed. If the wrong wrench is used, the work may be damaged.

Basic aircraft hardware uses the hexagonal, or six point, shape for bolt heads and nuts. Hardware is dimensioned by the distance across flats. Wrenches are sized by the hardware they fit and are made in fractional and metric sizes. Table 11-1 lists commonly used fractional wrench sizes. The bolt and nut sizes shown are for standard AN hardware. Other types of hardware may use different sizes of wrenches. Metric-sized wrenches and sockets that are manufactured are shown in Table 11-2.

TABLE 11-1 Wrench Sizes

Wrench size	Fits
$\frac{11}{32}$	Nuts for 8-32 screws
$\frac{3}{8}$	Nuts for 10-32 screws
	AN3 bolts
$\frac{7}{16}$	AN4 bolts and all nuts
$\frac{1}{2}$	AN 5 bolts and all nuts
$\frac{9}{16}$	AN6 bolts and all nuts
$\frac{5}{8}$	AN7 bolts and all nuts
	AN8 bolts and AN363 and AN365 nuts
$\frac{1}{2}$	AN310, 315, 316, and 320 -8 nuts

*The bolt and nut sizes shown are for standard AN hardware. Other types of hardware may be different sizes.

TABLE 11-2 Metric Wrench and Socket Sizes, mm

4.0	10	17
5.0	11	18
5.5	12	19
6.0	13	20
7.0	14	21
8.0	15	22
9.0	16	23

The wrenches covered in this chapter are designed as hand tools and should be used in that manner. Specially designed wrenches for use with power or impact tools are available. Wrenches may be purchased individually or in sets.

Open-End Wrench

Open-end wrenches are commonly used for general mechanical work requiring the tightening or loosening of threaded fasteners. The **open-end wrench** is used for both square and hexagonal headed bolts and nuts. The illustration in Fig. 11-33 shows that an open-end wrench will make contact only on two faces of the bolt or nut. For this reason it is essential that the open-end wrench be precisely made to fit the head. Clearance between the wrench and the head may result in the wrench slipping and rounding off the corners of the head. An open-end wrench of good quality will have the proper fit but can become worn or deformed with use. A wrench that does not fit tightly should be rejected by the technician.

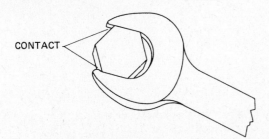

FIG. 11-33   How open-end wrenches apply force to a nut.

The jaws of an open-end wrench may be offset as shown in Fig. 11-34. A 15° offset allows the wrench to operate in a space as small as 30° when turning a hexagonal nut. Other degrees of offset are used to allow the wrench to be used in tight locations. The open-end wrench is usually double ended as shown in the figure. This allows two different wrench sizes to be on the same handle.

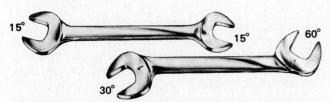

FIG. 11-34   Open-end wrenches.   *(Snap-on Tool Corp.)*

In using open-end wrenches, it is always better to pull the wrench toward you than to push it. There is less likelihood of the wrench slipping when it is pulled. Slipping of the wrench will damage the head of the bolt or nut as well as exposing the hands of the technician to injury.

Box-End Wrenches

The most effective type of wrench to use in turning a nut or the head of a bolt is the **box wrench** illustrated in Fig. 11-35. In the illustration the 12-point, or "double-hex box wrench," is shown. Box wrenches are the most effective for use in turning nuts and bolt heads because they apply pressure to six points on the nut or bolt head. Figure 11-36 shows how two different types of box wrenches apply force to the head of a bolt. It will be noted that both wrenches apply force at six points equally spaced around the head. For this reason the 12-point wrench may be used just as effectively as the 6-point wrench; however, the 12-point wrench may operate through an angle of 30°, whereas the hex

wrench requires an angle of 60° through which to operate. This makes the 12-point wrench more useful in close or restricted areas. The 12-point wrench is required by some specialized aircraft hardware using 12-point heads and nuts.

FIG. 11-35   Box wrench.   *(Proto Tools)*

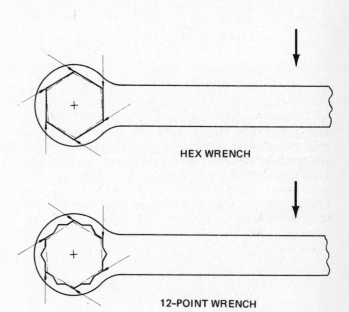

HEX WRENCH

12-POINT WRENCH

FIG. 11-36   How box wrenches apply force to a nut.

Box-end wrenches are among the most useful used by the technician. The box-end wrench should be the preferred wrench to use whenever possible. Like open-end wrenches, box-end wrenches have two different size ends. Box-end wrenches are made with a variety of angular offsets between the handle and end as shown in Fig. 11-37. A **ratcheting box-end wrench** is shown in Fig. 11-38. This wrench makes it possible to tighten a nut or bolt completely without having to remove the wrench from the nut or bolt.

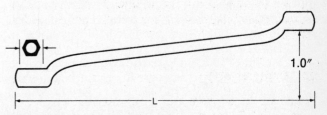

FIG. 11-37   Box wrench offset.

FIG. 11-38   A ratcheting box wrench.   *(Snap-on Tool Corp.)*

Combination Wrenches

Combination wrenches are manufactured with a box wrench on one end and an open-ended wrench on the other end. Such a wrench is illustrated in Fig. 11-39. These wrenches are quite popular with technicians because they provide the versatility necessary for many mechanical operations.

FIG. 11-39   A combination wrench.   *(Proto Tools)*

Flare-Nut Wrenches

The **flare-nut wrench** (see Fig. 11-40) looks like a box-end wrench with one side removed, forming a gap. The gap allows the wrench to be slipped over a tube and placed on a flare nut. The flare-nut wrench provides the advantages of a box-end wrench in tightening the flare nut. Flare nuts are of lower strength than standard hardware nuts and are easily damaged by tightening with an open-end wrench. Flare-nut wrenches are manufactured in both 6- and 12-point designs.

FIG. 11-40   A flare-nut wrench.

Adjustable Wrenches

An open-ended **adjustable wrench** is shown in Fig. 11-41. This wrench should be considered as an "emergency" tool and not used unless other types of wrenches are not available. Because of the open-end design, the wrench makes contact on only two sides of the head. The adjustment mechanism makes slippage of the wrench, with resulting damage to the bolt or nut, more likely. When using a wrench of this type, care should be taken to see that the jaws tightly fit the nut. Placing the thumb on the adjusting screw and applying pressure in the direction to tighten the jaws will help prevent slippage. The wrench should be turned in a direction so that the maximum stress is applied toward the inner end of the fixed jaw as shown in Fig. 11-42. Adjustable wrenches are manufactured in sizes designated by their overall length. Table 11-3 shows some sizes and the maximum opening of some common sizes of wrenches.

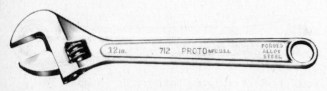

FIG. 11-41   Open-end adjustable wrench.

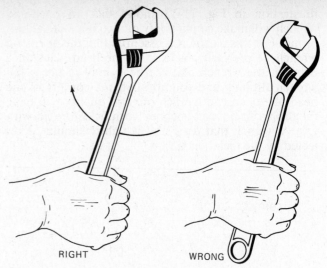

FIG. 11-42   Using an adjustable wrench.

TABLE 11-3   Adjustable Wrench Sizes

Size, in	Jaw capacity, in
4	$\frac{1}{2}$
6	$\frac{3}{4}$
8	$\frac{15}{16}$
10	$1\frac{1}{8}$
12	$1\frac{5}{16}$

Allen Wrenches

Set screws and certain aircraft bolts use an internal wrenching head. These fasteners do not have a hex head but rather a hex-shape recessed into the shank. Tightening these fasteners requires the use of a hex-shaped tool called an **Allen wrench.** Most of these wrenches are shaped as an L as shown in Fig. 11-43, although T shapes are available. Allen wrenches are made in a variety of sizes from 0.028 to $\frac{3}{4}$ in as measured across the flats.

FIG. 11-43   Allen wrenches.   *(Proto Tools)*

Socket Wrenches

One of the tools most useful to the technician is the **socket wrench.** A set of socket wrenches is shown in Fig. 1-44. Socket sets are made of a variety of types and sizes of drivers and sockets.

Socket Drivers. Socket wrenches have a square opening which accepts the driving tool. The basic size

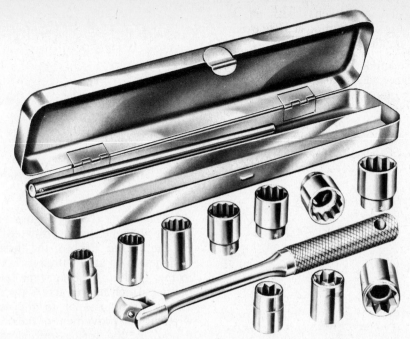

FIG. 11-44   A set of socket wrenches.

of a socket set is determined by the size of the drive. Sets are made in $\frac{1}{4}$-, $\frac{3}{8}$-, $\frac{1}{2}$-, and $\frac{3}{4}$-in drive. Adapters are made to allow different sizes of sockets and drivers to be used together. The aircraft technician will make extensive use of the $\frac{1}{4}$- and $\frac{3}{8}$-in drive sets. Metric-sized socket wrenches use standard fractional-sized drivers.

Sockets can be turned or driven by a number of tools. These include the **speed handle,** the **breaker bar,** the **sliding T handle,** and the **ratchet handle** as shown in Fig. 11-45. The speed handle is used with a socket wrench to make the rotation of the bolt or nut more rapid than with other types of handles. The breaker bar and T handle are used where it is necessary to get more torque on a bolt, such as to "break loose" a tight nut. The ratchet handle is used to rotate the socket wrench in a very restricted area where only a small amount of handle movement is available. The ratchet handle allows the socket to be left on the head of the bolt, or nut, until it is completely tightened. Not pictured is a handle that is best described as a "screwdriver-type" handle. This handle is widely used for socket wrenches in low-torque applications.

When it is necessary to reach a nut that is deeply recessed or in an extremely tight area, an extension

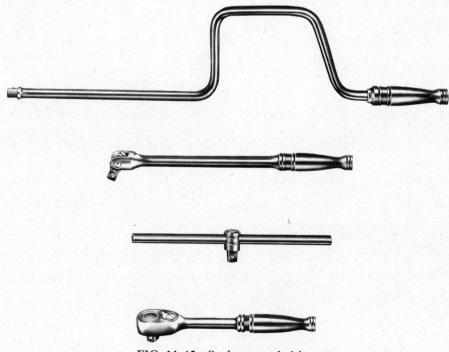

FIG. 11-45   Socket wrench drivers.

may be used. The extension attaches between the driver and the socket wrench. Common sizes of extensions are available in lengths ranging from 2 to 24 in.

Sockets. Sockets are made in a variety of shapes and sizes for use with a given bolt head or nut size. Figure 11-46 shows a **standard socket** and a **deep socket.** The standard socket and the deep socket are similar in design and use except for the amount of bolt clearance provided. Figure 11-47 shows the difference between the two types of sockets. Both types pictured are available as either a 6-point or 12-point socket.

FIG. 11-48   Crowsfoot sockets.

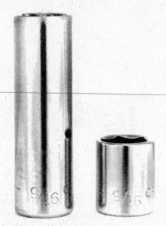

FIG. 11-46   Socket wrenches.

A variety of different tools are made to be used with socket wrench drives. Figure 11-48 shows two examples called **crowsfoot wrenches.** Crowsfoot wrenches fit on the extension of a ratchet handle and can be used on bolt heads, nuts, and fittings that would be inaccessible to normal socket wrenches.

Socket wrenches are made with Allen wrench bits for use with internal wrenching bolts. In addition virtually all types of screwdriver ends are available as bits for socket sets.

Special socket wrenches are made for use with electric or pneumatic impact drivers. These are of heavier construction than socket wrenches designed for use with hand tools.

Torque Wrenches

A **torque wrench** includes a measuring device designed to indicate the twisting force being applied to a nut or a bolt in inch-pounds (in·lb), foot-pounds (ft·lb), or newton-meters (N·m). Figure 11-49 illustrates three torque wrenches having different methods for indicating the torque. Each of the three torque wrenches is designed to be used with a socket wrench. The wrench in Fig. 11-49a is a **beam-type torque wrench.** It has a pointer which moves across the indicating scale in an amount proportional to the force exerted by the wrench. The movement of the pointer is caused by deflection of the spring-steel handle. A **"torsion-type"** torque wrench is shown in Fig. 11-49b. In this wrench the torque is indicated by a small dial gage attached to the wrench. The gage is actuated by a spring mechanism in the handle. An adjustable **toggle-type** wrench is shown in Fig. 11-49c. The operator sets the wrench at the desired torque value with a micrometer-type adjustment on the handle. When the desired torque is applied to the wrench, the wrench clicks and the handle momentarily releases. The wrench is capable of applying increased torque once this point has been reached. Technicians using this type of wrench should be sure they recognize the release action. This type of torque wrench does not have the direct reading capability of the two previously described.

Torque wrenches are available in a number of ranges such as 30 to 200 in·lb, 150 to 1000 in·lb, 5 to 75 ft·lb, and 30 to 250 ft·lb. The technician should choose a wrench that has a range of torque values compatible with the work being done. Torque wrenches are precision instruments and should be handled and stored with care. Torque wrenches must be calibrated on a periodic basis.

Torque wrenches should be used in every case where the force applied to a nut or bolt is critical. This is especially true in the assembly of components in the aircraft engine. Extensions and special attachments

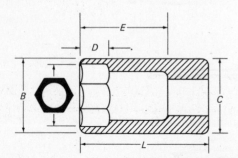

DIMENSION	STD SOCKET	DEEP SOCKET
B	23/32	23/32
C	21/32	21/32
D	5/16	13/32
E	1/2	1 1/16
L	15/16	2 1/8

FIG. 11-47   Standard and deep sockets.

are required for use of the torque wrenches in certain applications. These will change the effective arm of the wrench and a correction value must be calculated to get the correct torque setting. Methods for calculating this correction are explained in FAA Advisory Circular 43.13-1A as well as in other texts of this series.

The Pipe Wrench

The **pipe wrench** is used for assembling and disassembling pipes and fittings. A pipe wrench is illustrated in Fig. 11-50. The jaws of a pipe wrench are fitted with sharp hardened-steel teeth designed to grip pipes or rods. The wrench is designed so that when the jaws are adjusted to a pipe and the handle is pulled forward, the jaw pressure will increase, causing the teeth to indent the pipe and grip it very firmly. The pipe wrench must not be used on any pipe or rod where the finish must be preserved intact, because in every case the teeth of the wrench will mark the surface of the object being gripped. Furthermore, a pipe wrench will crush thin-walled tubing. It is unlikely that the technician will ever use a pipe wrench on an aircraft, but it is useful for other jobs around the shop. When it is necessary to turn a pipe or rod on which the surface must not be marred, a **strap wrench,** shown in Fig. 11-51, should be used.

Spanner Wrenches

Several different types of **spanner wrenches** are illustrated in Fig. 11-52. The first of these is an **adjustable-hook spanner wrench.** This wrench is designed for use with large nuts having notches cut in the periphery into which the hook may be inserted. With the adjustable hook on the wrench it is possible to accommodate several different sizes of nuts. The second wrench is called the **"pinhook" spanner wrench.** This wrench is designed for use on nuts having holes drilled in the periphery. The third wrench shown is an **adjustable pin-face spanner wrench.** This wrench is used to turn large nuts which are recessed flush with the part in which they are installed. Holes are drilled on the face of such nuts so that an adjustable pin-face spanner wrench may be used for installing or removing the nut.

● SCREWDRIVERS

The most frequently used tool for aircraft maintenance is probably the screwdriver. Hundreds of machine screws and sheet-metal screws are used in aircraft to hold cowlings, fairings, access panels, and inspection plates in place. Screwdrivers are made in a wide range of sizes and blade design.

Plain Screwdrivers

A **plain screwdriver** is shown in Fig. 11-53. When driving a particular screw, the width of the screwdriver blade should not exceed the diameter of the screw head. If the screwdriver tip is too wide, it will mar the material into which the screw is being driven. A screwdriver on which the tip has been rounded or beveled should not be used to drive a screw, because there is a

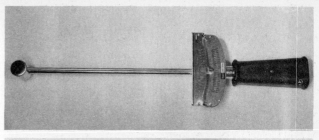

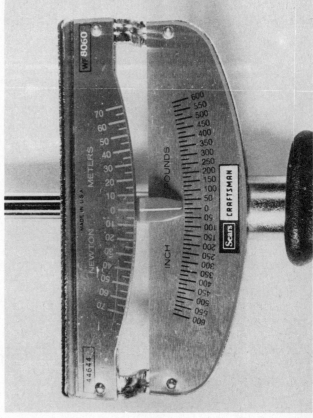

(a)

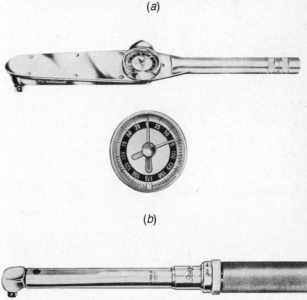

(b)

(c)

FIG. 11-49 Torque wrenches. *(Sears Roebuck & Co.; Snap-on Tools Corp.)*

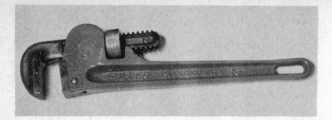

FIG. 11-50   A pipe wrench.

FIG. 11-51   A strap wrench.

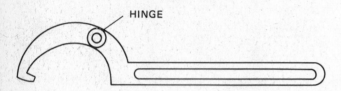

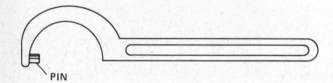

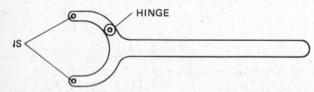

FIG. 11-52   Spanner wrenches.

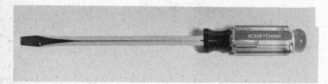

FIG. 11-53   A plain screwdriver.

danger that it may slip out of the slot in the screw head, thus damaging both the screw and the material into which the screw is driven. If the screwdriver tip is found to be worn, it should be replaced or ground to the shape shown in Fig. 11-54.

FIG. 11-54   Tip of a correctly ground screwdriver.

Plain screwdrivers are manufactured in many sizes. Size criteria includes the width and thickness of the tip and the length of the blade. The length of the blade is measured from the end of the handle to the tip. A large number of different sized screwdrivers are manufactured. The technician has excellent catalog and reference material available from the manufacturers to provide guidance in the selection of screwdrivers. The screwdriver blade may have a round or square shank. The square shank will permit a wrench to be used (see Fig. 11-55) to give more turning force to the screw. A provision for the use of a wrench on some round shanks is made with a hex shape as shown in Fig. 11-56.

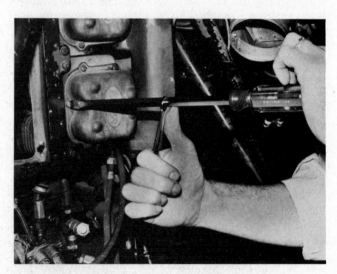

FIG. 11-55   Using a wrench with a heavy-duty screwdriver.

FIG. 11-56   Hex shape on shank for wrench.

Cross-Point Screwdrivers

Screws are manufactured with a number of designs for the driving tool in addition to the straight slot. A recessed-cross design known as a **Phillips head** provides a more positive drive than is possible with a straight slot screwdriver. The Phillips head screw has become a virtual standard for aircraft. The Phillips screwdriver tip is made with four flutes, as shown in Fig. 11-57, which fit the recess in the Phillip head screw. Phillips screwdriver tips are made in four sizes with No. 1 being the smallest and No. 4 the largest. It is important when using this screwdriver to have the size which matches the recess in the screw head. Phillips screwdrivers will wear on the ends as shown in Fig. 11-58. If this condition exists, the screwdriver should be replaced.

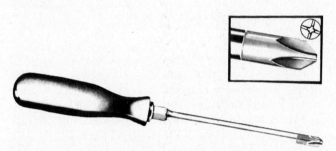

FIG. 11-57   A Phillips (cross-point) screwdriver.   *(Snap-on Tool Corp.)*

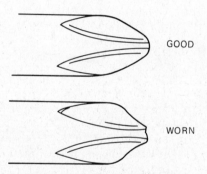

GOOD

WORN

FIG. 11-58   Phillips screwdriver with worn tip.

A second recessed-cross head design was used extensively in older aircraft. A **Reed & Prince screwdriver** is shown in Fig. 11-59. By comparing the pictures of the Phillips and Reed & Prince tips, it can be seen that the Reed & Prince has a sharper point. Each screwdriver will only work correctly with the screw head of its own design.

The majority of the screws used on aircraft use the straight slot or Phillips design. A number of other screw-head designs have been developed and may be

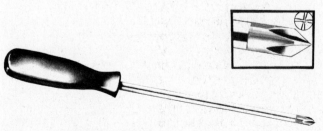

FIG. 11-59   A Reed & Prince screwdriver.   *(Snap-on Tool Corp.)*

encountered on an aircraft. An example is the **clutch head** design shown in Fig. 11-60. Some of these designs appear similar to the Phillips head. The technician must be sure that a screwdriver with the proper tip is used.

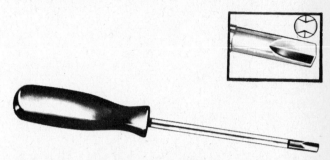

FIG. 11-60   A clutch-head screwdriver.

Special Screwdrivers

To compensate for a variety of tip shapes and sizes, screwdriver sets have been developed. The set consists of a screwdriver-handled driver with hex-shaped bits. In addition to the variety of shapes and sizes available, the replacement of worn tips is more economical with a set.

Stubby screwdrivers are designed for use where longer screwdrivers cannot be used conveniently. They have different bit designs, such as standard and Phillips as shown in Fig. 11-61. In many installations screws are placed in a position where they cannot be reached conveniently with a straight blade screwdriver. In order to get at the screw placed in such a restricted position, it is necessary to use an **offset screwdriver** such as that illustrated in Fig. 11-62.

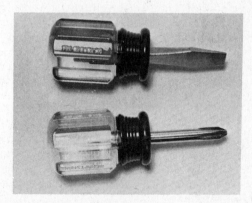

FIG. 11-61   Stubby screwdrivers.

To provide for rapid, semiautomatic rotation of a screw, the **spiral-ratchet screwdriver** (see Fig. 11-63) was developed. This type of screwdriver may be adjusted to rotate either to the right or to the left in order to install or remove screws. Pressing down on the handle of the screwdriver causes spiral grooves in the spindle to rotate the screwdriver blade. The screwdriver is also provided with a locking ring to lock the spindle so the screwdriver can be used in a conventional manner. This screwdriver is also provided with

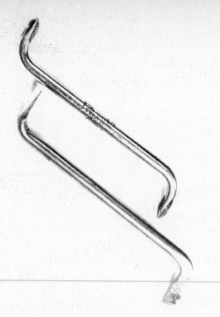

FIG. 11-62   An offset screwdriver.

FIG. 11-63   A spiral-ratchet screwdriver.   *(Snap-on Tools Corp.)*

a ratchet to allow continuous rotation without removing the blade from the screw head.

Power Screwdrivers

The use of power screwdrivers has been limited by the cost of electric and pneumatic units and the need for a power source. Lightweight battery-powered power screwdrivers have been developed that are an economical addition to the technician's toolbox. The speed that a powered unit provides will make the technician's work more time-efficient. The torque that a power screwdriver can provide should be controllable to prevent damage to screw heads from excessive force.

Use of Screwdrivers

It must be emphasized that screwdrivers are designed only for driving screws and should not be used as pry bars, as chisels, or for any other purpose which may damage the tool.

● PLIERS

The technician will need an assortment of pliers to perform aircraft maintenance. As with the other tools covered in this chapter, pliers exist in many different designs for specific functions.

A pair of **combination slip-joint pliers** is shown in Fig. 11-64. Combination pliers have jaws capable of gripping flat or round objects. Most combination pliers also have cutting capability. A slip-joint makes

it possible to enlarge the jaw width. Combination slip-joint pliers are used for general-purpose holding and gripping. Pliers are sized by their overall length. Combination pliers are made in sizes ranging from $4\frac{1}{2}$ in to almost 10 in.

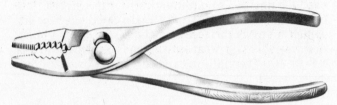

FIG. 11-64   Combination slip-joint pliers.   *(Snap-on Tools Corp.)*

Another type, **adjustable slip-joint pliers,** is shown in Fig. 11-65. These pliers are often called "water-pump pliers" because an early use of the design was to tighten the packing nut on an automotive water pump. The jaws of these pliers are designed for gripping. The shape and length of the handles are made to provide a heavy gripping force on the jaws. The slip-joint allows the jaws to be set to several widths but still stay parallel to each other.

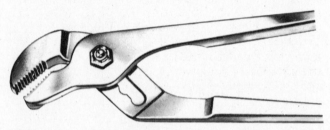

FIG. 11-65   Adjustable slip-joint pliers.   *(Snap-on Tools Corp.)*

The **Channellock pliers** shown in Fig. 11-66 are similar in appearance and function to the adjustable slip-joint set. The major difference is in the design of the adjustable joint. Curved grooves make up a series of interlocking joints. Considerable force may be exerted with these pliers, with no danger that the adjustment may slip. The adjustable slip-joint and Channel-lock pliers are made in sizes ranging from $4\frac{1}{2}$ to 20 in.

Adjustable **lever-wrench pliers** are illustrated in Fig. 11-67. **Vise-Grip,** a common name for these pliers, is

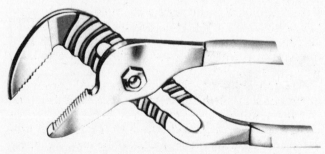

FIG. 11-66   Channellock pliers.   *(Snap-on Tool Corp.)*

a registered manufacturer's trade name. These pliers provide a powerful clamping force through the action of a cam and lever mechanism. The amount of jaw opening and clamping force is varied by turning the knurled screw on the handle. The pliers shown have the standard jaws. A variety of other jaw styles are made for gripping, cutting, and clamping. Standard lever-wrench pliers are made in sizes from 4 to 10 in. Under normal conditions, adjustable pliers should never be used as a substitute for a wrench to hold a bolt head or nut.

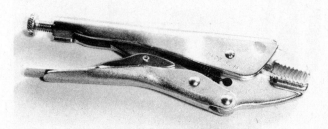

FIG. 11-67   Adjustable lever wrench pliers.

Diagonal-cutting pliers are so named because the cutting face is at an angle of approximately 15° to the plane of the handles. Diagonal-cutting pliers are designed exclusively for cutting of wire, cotter pins, nails, and other comparatively small soft-metal pins. The aircraft technician will use diagonal cutters primarily with cotter pins and safety wire. Diagonal-cutting pliers are made in a number of lengths (4 to $7\frac{1}{2}$ in) and cutting capabilities. They should not be abused by cutting hard steel or objects greater than their capacity. A pair of diagonal-cutting pliers is shown in Fig. 11-68.

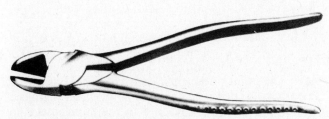

FIG. 11-68   Diagonal-cutting pliers.   *(Snap-on Tool Corp.)*

Figure 11-69 illustrates **needle-nose pliers,** which are also known as "long-nose" pliers. They are designed to reach into restricted areas. Needle-nose pliers are made in overall lengths of 4 to 7 in and jaw lengths of 1 to 3 in. The jaws of some needle-nose pliers will be bent at an angle, while others may have wire-cutting capability.

The **duckbill pliers** shown in Fig. 11-70 are similar in function to needle-nose pliers. The jaws are wide and flat resembling the bill of a duck. The design of the jaws provides a greater gripping area than that provided by the needle nose. They can be used for many applications where a heavy clamping force is not re-

FIG. 11-69   Needle-nose pliers.   *(Snap-on Tool Corp.)*

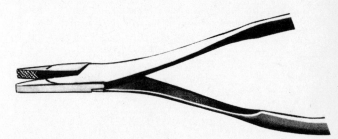

FIG. 11-70   Duckbill pliers.   *(Snap-on Tool Corp.)*

quired. The major use for the aircraft technician will be for safety wiring fasteners.

Safety wire twisters, shown in Fig. 11-71, are specially designed pliers for safety wiring aircraft fasteners. The jaws of these pliers are designed to both cut and grip the safety wire. The handles have a lock that will clamp the jaws on the wire. A helical screw arrangement will then provide twisting action to tighten the wire. The technician must understand the principles of safety wiring to properly use these pliers.

FIG. 11-71   Safety-wire twisters.

● HAMMERS

As a group of tools, hammers are a strong candidate for the title of most misused. While primarily designed as a pounding tool, hammers have a number of designs for different types of pounding as well as the material to be worked. The use of the wrong hammer can cause damage to the material and personal injury to the user.

One method of classifying hammers is by the shape of the peen end. Peen may be spelled pein or peen; however, for the purposes of this text we shall use the spelling **peen.** The peen end is the end of the hammer head opposite the face. The aircraft maintenance technician will use the **ball peen hammer** shown in Fig.

251

11-72 frequently. The ball peen hammer is also called the toolmaker's or machinist's hammer because of its all-around usefulness when working with metals. The head is made of a tough steel to permit its use in pounding against other steel parts such as punches and chisels. The combination of the flat face and rounded peen make the ball peen hammer useful for a variety of metal-forming tasks. Ball peen hammers are sized by the weight of the head. The weights range from 4 oz, for light work, to as much as 3 lb. A common size used by technicians would be 8 oz.

FIG. 11-72   A ball peen hammer.   *(Proto Tools)*

Figure 11-73 shows two other types of metal-working hammers, the **straight peen** and the **cross peen.** These hammers are used where the shape of the ball peen would not reach into a sharp corner or other restricted area.

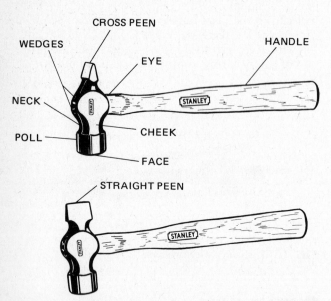

FIG. 11-73   Straight and cross peen hammer.   *(Stanley Tool Co.)*

Soft hammers are made with heads or faces of plastic, rawhide, wood, brass, copper, lead, or rubber. Soft hammers are designed to apply moderately heavy blows to various materials, including metals, without causing damage. For many jobs the soft hammer will serve as well as a steel hammer. Soft hammers are used for forming metal, separating tightly fitted parts, assembling parts where a tight fit is required, removing tight-fit bolts, and a variety of other jobs. The soft hammer may have replaceable tips as shown in Fig. 11-74 or be of construction similar to the rubber mallet in Fig. 11-75.

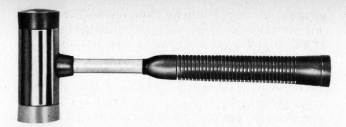

FIG. 11-74   A soft hammer with replaceable tips.

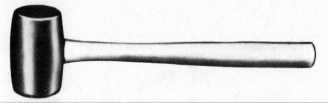

FIG. 11-75   A rubber mallet.   *(Snap-on Tools Corp.)*

● CUTTING TOOLS

Cutting is the process of a tool shearing away or removing portions of a material. The material removed can be a large piece, such as a piece of sheet metal cut into two parts. The material removed can also take the form of a number of small particles or chips made by a hacksaw. To cut a material the tool must be harder than the material it is cutting. Materials are cut both to size and shape.

Chisels

The chisel is the basic metal-cutting tool. The chisel consists of a steel tool with a hardened point. The shearing force is usually supplied by a hammer. The term "cold chisel" refers to the fact that the metal does not have to be heated to be cut. Cutting with a chisel is called chipping.

Chisels are made with four different shapes of cutting edges as shown in Fig. 11-76. The **cape chisel** is normally used for cutting grooves such as keyways or slots. The point of the chisel is narrow and has a very short cutting edge. The **diamond-point chisel** is ground to a sharp V edge. It is used for chipping metal in

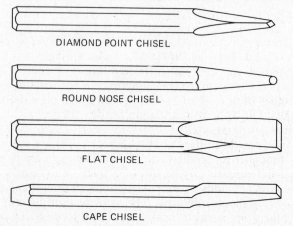

FIG. 11-76   Types of chisels.

sharp corners or for cutting V grooves. The **round-nose chisel** is used for cutting rounded or filleted corners. The **flat chisel** has a wide cutting edge. It may be used for cutting sheet metal or comparatively thin bar stock. It may also be used to "chip" a surface to a desired shape.

The chisel most commonly used by the technician is the flat chisel. The wedge-type cutting action is shown in Fig. 11-77. The cutting angle of a flat chisel should be sharpened to between 50 and 75° depending upon the type of metal. In general, as the material gets softer, the point will become sharper (closer to 50°). An angle of 60° is adequate for average work. The head of the chisel is the end opposite the point. The head is not hardened and may become "mushroomed" as shown in Fig. 11-78 during use. If mushrooming develops, the head should be ground to eliminate the possibility of injury from flying chips.

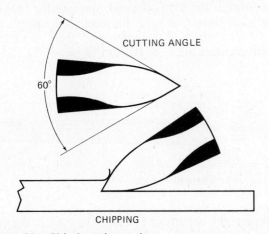

FIG. 11-77　Chisel cutting action.

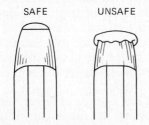

FIG. 11-78　Mushroomed chisel head.

The aviation technician's use of a chisel will be limited, with removal of rivets a possible use. Flat chisels are sized by the width of the tip. A chisel of $\frac{3}{8}$ in is adequate for most work.

Hacksaws

Handsawing of metal is done with a **hacksaw.** The hacksaw as shown in Fig. 11-79 consists of the **frame** and the **blade.** The frame is designed to hold the blade. It is adjustable for different length blades and is designed so that the blade can be installed in any of four positions. The blade can be compared to a number of small chisels, each cutting or chipping the metal. The hacksaw blade varies in length, number of teeth, and type of material.

FIG. 11-79　A hacksaw and frame.　*(Snap-on Tools Corp.)*

As with all cutting tools, the blade must be harder than the material it is cutting. Blades are made in various degrees of hardness. A harder blade will usually cut longer and straighter than a softer, more flexible blade. The harder blade will also be more brittle and break more easily if it is twisted while cutting. Some blades are made with a flexible back and hardened cutting edges.

Hacksaw blades are made with different sizes of teeth varying from 14 to 32 per inch. The number of teeth to use depends upon the thickness of the metal. The teeth must be far enough apart to provide clearance for the chips to get out. If the clearance is inadequate, the blade will be clogged with chips and stop cutting. When cutting thin material, there should be at least two teeth in contact with the work at all times to prevent the teeth being broken. The teeth of hacksaw blades have a "set" (see Fig. 11-80) that makes the cut, or kerf, slightly wider than the blade. The purpose of the set is to provide clearance for the blade as it passes through the metal. The width of the cut will decrease as the blade wears during use. A cut made with a worn blade will be narrower than that required for the teeth of a new blade to pass through. New blades should not be used to continue cuts started with a worn blade.

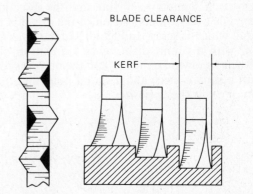

FIG. 11-80　Set of a hacksaw blade.

Hacksaw blades are commonly made in lengths of 10 and 12 in. Common sizes of teeth available are 14, 18, 24, and 32/in. A blade marked 1032 would be 10 in long and have 32 teeth per inch. A 1218 would have 18 teeth per inch and be 12 in long.

The hacksaw blade should be installed in the frame with the teeth pointing forward. Cutting action takes place on the forward movement of the saw. To prevent excessive wear, pressure on the blade should be lightened during the backstroke.

Hand Shears

Aircraft sheet metal can be cut with hand shears. The most commonly used shears for sheet-metal work are the **aviation snips** shown in Fig. 11-81. The snips shown in the photograph are a design intended for making straight-line cuts. Aviation snips are made in two other designs, **right hand** and **left hand,** with curved jaws for curved cuts. Right hand and left hand refers to the relative position of the blades. The design type can be identified by holding the snips with the tip of the blades pointing away from your body. The right-hand snips will have the lower jaw on the right side and vice versa for the left. A common practice among manufacturers is to have green handles on right-hand snips, red on left-hand snips, and yellow on straight snips.

FIG. 11-81   Aviation snips.

Making a curved cut with the snips requires that the metal on one side of the snips be distorted to provide for movement of the tool. In general the smaller the radius of the curve the more the metal will be distorted. Aviation snips are designed so that the lower jaw will push the metal on that side up and out of the way. Figure 11-82 illustrates the use of both types of snips to cut a round circle. Note that the direction of cut depends upon which portion of the metal is to be used and which is scrap. In Fig. 11-82*a* the object is to end up with a round disk and the direction of cut would be as shown. In Fig. 11-82*b* the object is to end up with a square piece and a round hole, and the direction of cut is the opposite.

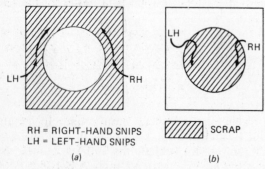

RH = RIGHT-HAND SNIPS
LH = LEFT-HAND SNIPS

▨ = SCRAP

(a)   (b)

FIG. 11-82   Directions of cut.

Aviation snips are made with serrated teeth on the jaws to provide better shearing action. If the jaws are damaged or become dull, they should be sharpened by a specialist or replaced.

Files

Files are used by technicians to cut and shape metal. Files consist of hardened metal shapes with a number of chisel-like teeth cut into them. While there are hundreds of file types, the aviation technician needs to only be familiar with a few general-purpose types of files. Files are categorized by **size, shape, type of cut,** and **coarseness of cut.** The selection of a file from these criteria will depend upon the material being filed, the nature of the cut to be made, the type of finish desired, and the amount of material to be removed in a given time.

Figure 11-83 shows the names of the parts of the file and how the length is measured. The **tang** of the file is the sharp-pointed end and is designed for a handle. The **heel** of the file is next to the tang and contains the start of the teeth. The **face** of the file is one of the wide flat sides. The side opposite the face is called the **back.** The **point** of the file is the end opposite the tang. The **length,** as shown, is from the heel to the point. Files vary in length in 2-in increments from 4 to 20 in. Common sizes are 6, 8, 10, and 12 in.

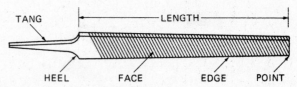

TANG   LENGTH

HEEL   FACE   EDGE   POINT

FIG. 11-83   Parts of the file.

Type and Coarseness of Cut. The type of cut refers to the shape of the teeth on the file. Figure 11-84 shows, from left to right, a **single cut,** a **double cut,** a **rasp cut,** and a **curved-tooth,** or **vixen,** cut. The single cut has a single series of teeth. The double cut has a second series of teeth cut at an angle to the first. The first set of teeth is called the overcut and is deeper than the second set, called the upcut. In general the double cut will remove metal faster but will leave a rougher finish than the single cut.

The rasp cut has a series of raised, individual teeth. The rasp makes fast but rough cuts. It is used primarily on soft materials such as aluminum and lead where fast removal of material is desired.

The curved-tooth file has relatively large spaces between the teeth. This allows the file to be used on soft materials without clogging. Curved-tooth files produce a very smooth surface on soft materials such as aluminum.

The coarseness of cut, for single- and double-cut files, is related to the spacing between the teeth. Six terms used to describe these spacings are (from most coarse to finest) **rough, coarse, bastard, second cut, smooth,** and **dead smooth.** The spacing for these cuts is not fixed but is dependent upon the length of the file. As the files get shorter, the spacing decreases and the cut becomes finer. As an example, a 14-in smooth file may be more coarse than an 8-in bastard file. The aviation technician will have the most need for bastard, second-cut, and smooth files.

FIG. 11-84   Types of file tooth cuts.

File Shape.   The **shape** of a file is its general outline and cross section. Many files are **tapered,** which means that they decrease in width and/or thickness from the heel to the point. Files that do not change in cross section are called **blunt** files. Figure 11-85 shows several shapes of files. The **flat file** has a rectangular cross section that is slightly tapered toward the point in both

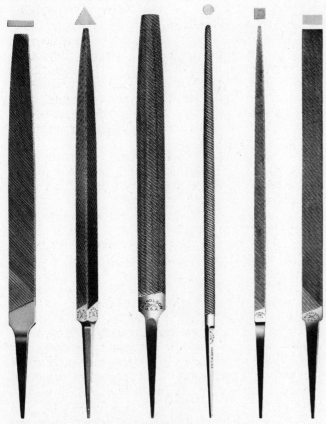

FIG. 11-85   Different types of files.

width and thickness. The flat file has double-cut teeth on both faces and the edges. A **hand file** is similar to the flat file with tapered thickness but uniform width. The hand file also has one **safe edge** which means the edge does not have teeth cut in it. The **mill file** may be tapered or blunt but has single-cut teeth and is used where smooth finishes are desired.

A **triangular file** is also called a **three-square file.** The triangular file is tapered and double cut. It is used to file corners and angles of less than 90°.

A **half-round** file is tapered with one rounded side (back) and one flat side. The flat side is always double cut. The back side is double cut except for files with a smooth cut. The half-round file is used for concave surfaces. The **round,** or **rattail, file** is tapered and can be single or double cut. The round file is used to enlarge holes and file curved surfaces.

A **square file** is square in cross section and tapered toward the point. It is principally used for filing slots and keyways. A **pillar file** is rectangular in cross section but is narrower and thicker than a hand file. It has at least one safe edge so the file may be used close to a corner without danger of filing on the side. Like the square file, the pillar file is used primarily for keyways and slots.

File Identification.   The length of a file and type of teeth are readily determined by observation. Most of the manufacturers will stamp the file shape and coarseness of cut on the heel of the file.

Use and Care of Files.   There are three basic ways in which a file can be put to work: (1) **straight filing,** (2) **drawfiling,** and (3) **lathe filing.**

Straight filing consists of pushing the file lengthwise across the metal and applying sufficient pressure to make the file cut. **Drawfiling** consists of grasping the file at each end and pushing or pulling it across the work to produce a fine finish. **Lathe filing** is the application of the file to a piece of metal turning in a lathe.

Usually, work to be filed should be held in a vise. The jaws of the vise should be covered with soft sheet metal to prevent damage to the work being filed. The vise should be at approximately elbow height for best results. If the work is of a heavy nature requiring a great amount of filing, the vise should be lower, but if it is to be fine, delicate work, it is better for the vise to be nearer eye level.

Before using a file, a handle should be installed on the tang. A typical handle is turned from wood and has a metal ferrule on the end into which the tang of the file is to be inserted. The metal ferrule prevents the handle from splitting. The handle is installed simply by inserting the tang of the file into the hole at the end of the handle and tapping the handle a few times to drive it into place.

The right and wrong ways to hold a file are shown in Fig. 11-86. In the first picture the operator is holding and moving the file in a manner which is most likely to produce a flat surface on the work. The operator in the other picture is likely to produce a rounded surface because the file will rock as it is moved across the work.

When taking a cut with the file, one should apply

FIG. 11-86   Holding a file.

It is of great importance to keep the teeth of files clean of filings and chips, which often collect as a result of use. After every few strokes the end of files should be tapped on the bench to loosen the metal particles from the teeth. The teeth of the file should be brushed frequently with a file brush or card. (See Fig. 11-87.) To remove stubborn "pinnings" from the file teeth, a sharp scorer is used. This tool is made of soft iron and is often included with the file card. **Pinnings** are the soft metal particles which clog the teeth of the file and stick in place so they are not easily removed by brushing.

FIG. 11-87   A file card.

pressure only **on the forward stroke.** On the return stroke the file should be lifted from the work unless the material being filed is soft. In this case the file may be permitted to remain in contact with the material but no pressure should be applied other than that of the file's weight.

When a light stroke is desired and the pressure demanded becomes less, the thumb and fingers of the point-holding hand may change their direction until the thumb lies at right angles with the length of the file.

In holding the file with one hand, as in filing pins, dies, and edged tools not held in a vise, the forefinger instead of the thumb is generally placed on top of the file and as nearly as possible in the direction of the file's length.

One of the quickest ways to ruin a file is to use too much or too little pressure on the forward stroke. Different materials, of course, require different touches; but, in general, just enough pressure should be applied to keep the file cutting at all times. If allowed to slide over the harder metals, the teeth of the file rapidly become dull; and if they are overloaded by too much pressure, they are likely to chip or clog.

Drawfiling is used extensively to produce a perfectly smooth, level surface. Ordinarily, a standard-mill bastard file is used for drawfiling, but where a considerable amount of stock is to be removed, a double-cut flat or hand file will work faster. The double-cut file usually leaves small ridges in the work and consequently does not produce a finished job where a smooth surface is required. In such cases the double-cut file may be used for the "roughing down," and the single-cut mill file is used for finishing.

File life is greatly shortened by improper care as well as by improper use and improper selection. Files should never be thrown into a drawer or toolbox containing other tools or objects. They should never be laid on top of each other or stacked together. Such treatment ruins the cutting edges of the teeth and causes nicks in the edges. Files should be kept separate, standing with their tangs in a row of holes or hung in a rack by their handles. They may also be placed in separate grooves made by attaching strips of wood to a base board.

Drills

The maintenance technician is often faced with the necessity of boring accurately sized holes in metal parts in order to make attachments and to join parts in an assembly. The tool usually used for boring such holes is the **spiral,** or **twist, drill.** The steel drill usually consists of a cylinder into which has been cut spiral grooves or "flutes." One end is pointed, and the other is shaped to fit a particular drilling machine such as a hand-drill motor.

The principal parts of a drill are illustrated in Fig. 11-88. The illustration shows the **shank, body, flute, land, margin,** and **lip,** or **cutting edge.**

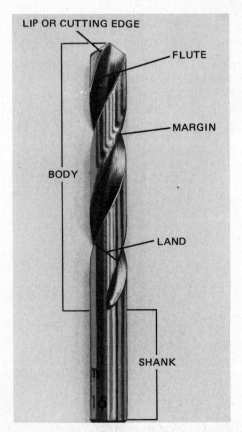

FIG. 11-88   The parts of a drill.

The **shank** of a drill is the part designed to fit into the drilling machine. It may be a plain cylinder in shape, which is the design for use in a drill chuck on a drill motor, drill press, or hand drill. The drill shank may also be tapered. The tapered drill shank is usually used in drill presses.

The **body** of a drill is the part between the point and shank. It includes the spiral flutes, the lands, and the margin. The body is slightly larger in diameter at the tip than at the shank, thus causing it to bore a hole with clearance to prevent the drill from binding.

The **point** of a drill includes the entire cone-shaped cutting end of the drill. The point includes the **cutting edges,** or **lips,** which are sharpened when the drill is ground.

The **web** is the portion of the drill at the center along the axis. It becomes thicker near the shank. The web may also be defined as the material remaining at the center of the drill after the flutes have been cut out. The web forms the dead-center tip at the point of the drill. The **dead center** is in the exact center of the tip and is on the line forming the axis of the drill.

Drill Sizes. Drills are available in three series of sizes. **Number-size drill bits** are available in sizes from No. 80 (0.0135 in) to a No. 1 (0.2280 in). **Letter-size drill bits** start with A (0.2340 in) and go to Z (0.4130 in). **Fractional sizes** start with $\frac{1}{64}$ in (0.0156) and increase in $\frac{1}{64}$-in increments to 1 in (1.0″). Table 11-4 presents all the drill sizes and their decimal conversion. A study of Table 11-4 reveals that the three series of drill bits have a total of 138 sizes between No. 80 and $\frac{1}{2}$ in. The only common sizes in the three sets occurs with the letter E and the fractional $\frac{1}{4}$-in drills. Drills are also available in sizes above 1 in. In addition to the three series of drill sizes listed, drills are available in metric sizes.

Drill-Point Angles. In order to perform correctly, the drill must be ground or sharpened to the correct angle. The angles on the drill point will vary depending upon the hardness of the material. Figure 11-89 shows correct angles for three different types of materials. The **drill-point angle** has an effect on the length of the **cutting lips.** For harder materials a shorter lip is desirable and therefore a blunter angle. Softer materials can utilize a sharper point with a longer cutting lip. The general-purpose cutting lip angle (118°) is also given as 59°, which is the angle formed by one cutting lip and the centerline of the drill bit.

The **lip clearance angle** provides clearance between the cutting lip and the rest of the bit. The cutting lips are comparable to chisels. To cut effectively, that area back of the lip must be relieved. Too little clearance results in the drill point rubbing and not penetrating. If the angle is too large, the cutting lip will be weakened. Because of the shape of the bit, lip clearance should be increased slightly as the cutting lip approaches the dead center. If the lip clearance has been correctly ground, the angle formed by the dead center and the cutting lip should be between 120 and 135°.

The angle and lip of each cutting lip must be equal. If they are not equal, the dead center will be displaced from the center line of the drill. The resulting hole will be oversized or distorted.

Drill Speeds and Feeds. A properly sharpened drill will not function properly if the correct drill speed and feed is not used. Drill speed refers to the speed at which the cutting lips pass through the metal. Cutting speed varies depending upon the material being drilled and the metal from which the drill is made. Cutting speed is stated in terms of feet per minute (fpm). The speeds of drill bits are normally measured in terms of revolutions per minute (rpm). It is necessary therefore to convert the rpm of a drill bit, sized in fractions of an inch, to a lip cutting speed in fpm. A simplified formula for doing this is:

$$CS = \frac{D \times \text{rpm}}{4}$$

where CS = cutting speed, in fpm
D = drill size (diameter), in
rpm = speed of drill

The feed of a drill bit refers to the advancement of the cutting lips into the work and is given in terms of inches per revolution of the bit. A detailed study of feeds and speeds is beyond the scope of this text. A general rule of thumb is that a slow speed and heavy feed are used for hard materials. Softer materials should have a higher speed but lighter feed. Operating a drill at a speed higher than its design speed will usually result in overheating and failure of the drill bit. Table 11-5 gives some cutting speed and feed values for different materials.

Countersinks. A **countersink** is a pointed cutting tool designed to produce a conical-shaped hole in metal or other materials to fit the head of a rivet or screw. The countersunk hole forms the mouth of a previously drilled straight hole. A countersink is usually provided with a straight shank for use in a hand drill, a drill motor, or a bench drill. Two different types of countersinks are shown in Fig. 11-90. When it is desired to drill a hole and countersink in the same operation, a combination drill and countersink is used. This tool is made in small diameters so the body may also serve as the shank to be inserted in a drill chuck. Countersinks are made with a variety of point angles. Technicians should make sure that they use the correct countersink for the screw or rivet to be installed.

Counterbores. The **counterbore,** illustrated in Fig. 11-91, is a tool designed to bore a second hole that is larger than the first and concentric with it. The pilot on the counterbore fits the first hole drilled and keeps the tool concentric with the first hole. The pilot is interchangeable so it may be used with different sizes of tools. One of the purposes for counterboring is to make a recess for a bolt head.

The counterbore is also employed for "spot-facing." Spot-facing is the process of cutting a smooth surface around the edge of a hole to provide a square seat for a bolt head. In this case, the counterbore is permitted to cut just deep enough to make the desired seat.

Reamers

A **reamer** is a cutting tool designed to enlarge a hole, produce an accurately sized hole, or to cut a tapered

TABLE 11-4 Drill Size Chart

Milli-meter	Dec. equiv.	Frac-tional	Num-ber	Milli-meter	Dec. equiv.	Frac-tional	Num-ber	Milli-meter	Dec. equiv.	Frac-tional	Num-ber	Milli-meter	Dec. equiv.	Frac-tional	Num-ber	Milli-meter	Dec. equiv.	Frac-tional
0.10	0.0039			1.75	0.0689				0.1570		22	6.80	0.2677			10.72	0.4219	27/64
0.15	0.0059			—	0.0700		50	4.00	0.1575			6.90	0.2716			11.00	0.4330	
0.20	0.0079			1.80	0.0709			—	0.1590		21	—	0.2720		I	11.11	0.4375	7/16
0.25	0.0098			1.85	0.0728			—	0.1610		20	7.00	0.2756			11.50	0.4528	
0.30	0.0118			—	0.0730		49	4.10	0.1614			—	0.2770		J	11.51	0.4531	29/64
—	0.0135		80	1.90	0.0748			4.20	0.1654			7.10	0.2795			11.91	0.4687	15/32
0.35	0.0138			—	0.0760		48	—	0.1660		19	—	0.2811		K	12.00	0.4724	
—	0.0145		79	1.95	0.0767			4.25	0.1673			7.14	0.2812	9/32	—	12.30	0.4843	31/64
0.39	0.0156	1/64	—	1.98	0.0781	5/64	—	4.30	0.1693			7.20	0.2835			12.50	0.4921	
0.40	0.0157			—	0.0785		47	—	0.1695		18	7.25	0.2854			12.70	0.5000	1/2
—	0.0160		78	2.00	0.0787			4.37	0.1719	11/64	—	7.30	0.2874			13.00	0.5118	
0.45	0.0177			2.05	0.0807			—	0.1730		17	—	0.2900		L	13.10	0.5156	33/64
—	0.0180		77	—	0.0810		46	4.40	0.1732			7.40	0.2913			13.49	0.5312	17/32
0.50	0.0197			—	0.0820		45	—	0.1770		16	7.50	0.2953			13.50	0.5315	
—	0.0200		76	2.10	0.0827			4.50	0.1771			7.54	0.2968	19/64	—	13.89	0.5469	35/64
—	0.0210		75	2.15	0.0846			—	0.1800		15	7.60	0.2992			14.00	0.5512	
0.55	0.0217			—	0.0860		44	4.60	0.1811			—	0.3020		N	14.29	0.5625	9/16
—	0.0225		74	2.20	0.0866			—	0.1820		14	7.70	0.3031			14.50	0.5709	
0.6	0.0236			2.25	0.0885			4.70	0.1850		13	7.75	0.3051			14.68	0.5781	37/64
—	0.0240		73	—	0.0890		43	4.75	0.1870			7.80	0.3071			15.00	0.5906	
—	0.0250		72	2.30	0.0905			4.76	0.1875	3/16	—	7.90	0.3110			15.08	0.5937	19/32
0.65	0.256			2.35	0.0925			4.80	0.1890		12	7.94	0.3125	5/16	—	15.48	0.6094	39/64
—	0.0260		71	—	0.0935		42	—	0.1910		11	8.00	0.3150			15.50	0.6102	
—	0.0280		70	2.35	0.0937	3/32	—	4.90	0.1929			—	0.3160		O	15.88	0.6250	5/8
0.7	0.0276			2.40	0.0945			—	0.1935		10	8.10	0.3189			16.00	0.6299	
—	0.0292		69	—	0.0960		41	—	0.1960		9	8.20	0.3228			16.27	0.6406	41/64
0.75	0.0295			2.45	0.0964			5.00	0.1968			—	0.3230		P	16.50	0.6496	
—	0.0310		68	—	0.0980		40	—	0.1990		8	8.25	0.3248			16.67	0.6562	21/32
0.79	0.0312	1/32	—	2.50	0.0984			5.10	0.2008			8.30	0.3268			17.00	0.6693	
0.80	0.0315			—	0.0995		39	—	0.2010		7	8.33	0.3281	21/64	—	17.06	0.6719	43/64
—	0.0320		67	—	0.1015		38	5.16	0.2031	13/64	—	8.40	0.3307			17.46	0.6875	11/16
—	0.0330		66	2.60	0.1024			—	0.2040		6	—	0.3320		Q	17.50	0.6890	
0.85	0.0335			—	0.1040		37	5.20	0.2047			8.50	0.3346			17.86	0.7031	45/64
—	0.0350		65	2.70	0.1063			—	0.2055		5	8.60	0.3386			18.00	0.7087	
0.90	0.0354			2.75	0.1082			5.25	0.2067			—	0.3390		R	18.26	0.7187	23/32
—	0.0360		64	2.78	0.1094	7/64	—	5.30	0.2086			8.70	0.3425			18.50	0.7283	
—	0.0370		63	—	0.1100		35	—	0.2090		4	8.73	0.3437	11/32	—	18.65	0.7344	47/64
0.95	0.0374			2.80	0.1102			—	0.2130		3	8.75	0.3445			19.00	0.7480	
—	0.0380		62	—	0.1110		34	5.50	0.2165			8.80	0.3465			19.05	0.7500	3/4
—	0.0390		61	—	0.1130		33	5.56	0.2187	7/32	—	—	0.3480		S	19.45	0.7656	49/64
1.00	0.0394			2.90	0.1141			5.60	0.2205			8.90	0.3504			19.50	0.7677	
—	0.0400		60	—	0.1160		32	—	0.2210		2	9.00	0.3543			19.84	0.7812	25/32
—	0.0410		59	3.00	0.1181			5.70	0.2244			—	0.3580		T	20.00	0.7874	
1.05	0.0413			3.10	0.1220			5.75	0.2263			9.10	0.3583			20.24	0.7969	51/64
—	0.0420		58	3.18	0.1250	1/8	—	—	0.2280		1	9.13	0.3594	23/64	—	20.50	0.8071	
—	0.0430		57	3.20	0.1260			5.80	0.2283			9.20	0.3622			20.64	0.8125	13/16
1.10	0.0433			3.25	0.1279			5.90	0.2323			9.25	0.3641			21.00	0.8268	
1.15	0.0452			—	0.1285		30	—	0.2340		A	9.30	0.3661			21.03	0.8281	53/64
—	0.0465		56	3.30	0.1299			5.95	0.2344	15/64	—	—	0.3680		U	21.43	0.8437	27/32
1.19	0.0469	3/64	—	3.40	0.1338			6.00	0.2362			9.40	0.3701			21.50	0.8465	
1.20	0.0472			—	0.1360		29	—	0.2380		B	9.50	0.3740			21.83	0.8594	55/64
1.25	0.0492			3.50	0.1378			6.10	0.2401			9.53	0.3750	3/8	—	22.00	0.8661	
1.30	0.0512			—	0.1405		28	—	0.2420		C	—	0.3770		V	22.23	0.8750	7/8
—	0.0520		55	3.57	0.1406	9/64	—	6.20	0.2441			9.60	0.3780			22.50	0.8858	
1.35	0.0531			3.60	0.1417			6.25	0.2460		D	9.70	0.3819			22.62	0.8906	57/64
—	0.0550		54	—	0.1440		27	6.30	0.2480			9.75	0.3838			23.00	0.9055	
1.40	0.0551			3.70	0.1457			6.35	0.2500	1/4	E	9.80	0.3858			23.02	0.9062	29/32
1.45	0.0570			—	0.1470		26	6.40	0.2520			—	0.3860		W	23.42	0.9219	59/64
1.50	0.0592			3.75	0.1476			6.50	0.2559			9.90	0.3898			23.50	0.9252	
—	0.0595		53	—	0.1495		25	—	0.2570		F	9.92	0.3906	25/64	—	23.81	0.9375	15/16
1.55	0.0610			3.80	0.2496			6.60	0.2598			10.00	0.3937			24.00	0.9449	
1.59	0.0625	1/16	—	—	0.1520		24	—	0.2610		G	—	0.3970		X	24.21	0.9531	61/64
1.60	0.0629			3.90	0.1535			6.70	0.2638			—	0.4040		Y	24.50	0.9646	
—	0.0635		52	—	0.1540		23	6.75	0.2657	17/64	—	10.32	0.4062	13/32	—	24.61	0.9687	31/32
1.65	0.0649			3.97	0.1562	5/32	—	6.75	0.2657			—	0.4130		Z	25.00	0.9843	
1.70	0.0669							—	0.2660		H	10.50	0.4134			25.03	0.9844	63/64
—	0.0670		51													25.40	1.0000	1

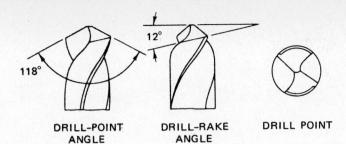

DRILL-POINT ANGLE · DRILL-RAKE ANGLE · DRILL POINT

GENERAL PURPOSE POINT

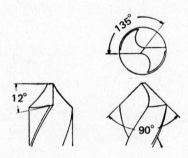

LONG POINT FOR WOOD, BAKELITE, PLASTICS

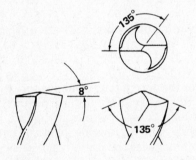

SHORT POINT FOR HARD MATERIAL

FIG. 11-89 Properly sharpened drill points.

PLAIN COUNTERSINK

COUNTERSINK WITH STOP AND PILOT

FIG. 11-90 Countersinks.

FIG. 11-91 A counterbore.

hole. These tools are made in a wide variety of sizes and styles for a multitude of applications. Figure 11-92 illustrates two types of reamers for hand use. The first reamer pictured is a standard straight-fluted reamer, and the second is an adjustable or expansion reamer.

ADJUSTING SCREW

FIG. 11-92 Two types of reamers.

For very accurate work a fixed reamer of a standard size is employed. Standard-sized reamers may be obtained in increments of $\frac{1}{64}$ in. If other sizes are required, special reamers must be ordered from the manufacturer.

Expansion reamers may be used when it is desired to size a hole slightly larger than a standard measurement. Expansion reamers are not as accurate as fixed reamers, but for many purposes they are satisfactory.

TABLE 11-5 Speeds and Feeds

Material	Cutting speed, fpm	Feed per revolution		
		Less than $\frac{1}{8}$	$\frac{1}{8}$ to $\frac{1}{4}$	$\frac{1}{4}$ to $\frac{1}{2}$
Aluminum	200–300	0.0015	0.003	0.006
Bronze	50–100	0.0015	0.003	0.006
Cast iron				
Soft	100–150	0.0025	0.005	0.010
Hard	30–100	0.0010	0.002	0.003
Magnesium	200–400	0.0025	0.005	0.010
CRES	20–100	0.0025	0.005	0.010
Steel				
Low carbon	60–100	0.0015	0.003	0.006
Tool	25–50	0.0025	0.005	0.010
Titanium	15–20	0.0015	0.003	0.006

Reamers must be handled with great care. They should never be placed in a rack or toolbox where they will contact other metal tools. Each reamer should be provided with a case of a soft material or should be placed in a separate space provided in a wooden reamer board. The sharp cutting edges are nicked easily and thus may be rendered useless.

Taps and Dies

Taps and **dies** are used for cutting threads. The tap is employed to cut threads inside a drilled hole, and the die to cut threads on metal rod stock.

Three common taps are shown in Fig. 11-93. These are in order, a **taper tap,** a **plug tap,** and a **bottoming tap.** The taper tap is used more often than the others and is always used to start cutting threads in a hole. If the hole passes through a piece of metal, no other tap is required. The plug tap is used when the hole is blind and it is desired to cut threads nearer the bottom than is possible with a taper tap. If it is required to cut threads all the way to the bottom of a hole, the bottoming tap is used.

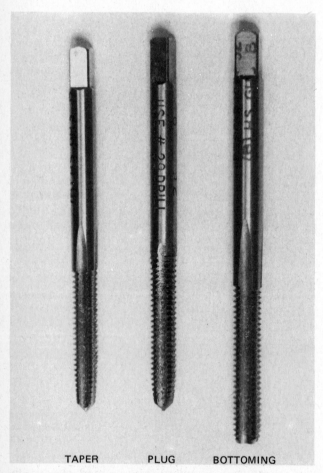

FIG. 11-93   Taps for cutting internal threads.

To hold small taps for cutting threads, the T-handle tap wrench is used. This wrench grips the tap firmly and may be turned with one hand. Care must be taken to hold the handle straight so the tap will be in exact line with the hole to be threaded. It is advisable to apply some light cutting oil to the tap when cutting

threads in steel or any other hard metal. For large taps a larger hand-tap wrench, shown in Fig. 11-94, is used.

FIG. 11-94   A tap wrench.

Figure 11-95 shows the two types of dies and die stock. The die on the left is a solid die. The second die is a split die and can have small adjustments made to it. The solid die has no provision for adjustment.

FIG. 11-95   Dies and die stock.

The threading of a metal rod with a die mounted in a die stock is shown in Fig. 11-96. The rod to be threaded is held securely in a vise, and a light oil is applied to the end. The starting side of the die is placed over the end of the rod and is turned in a clockwise direction. From time to time the direction of the die movement is reversed in order to clear the threads of metal chips which may clog them and cause galling. This same practice is advisable when using a tap to cut threads in a hole.

FIG. 11-96   Threading a metal rod with a die.

● PUNCHES

Punches are made for a variety of purposes, the most common types being the **drift punch,** the **center punch,** and the **pin punch.** The drift punch has a long tapered end with a blunt point. It is normally used for aligning holes for bolts or pins to facilitate installation. The center punch has a 90° conical point and is used to make an indentation in metal to mark the location of holes to be drilled and to make the drill start at the correct point. The **prick punch** resembles the center punch in appearance except it has a sharper point. It is used during layout work to precisely mark center locations. Prick-punch marks should be enlarged with the center punch prior to drilling. The pin punch has a long straight cylindrical end and is used to drive and remove various types of pins from shafts or other parts. Pin punches are sized by the diameter of the punch end. Aviation technicians will need a minimum set of $\frac{3}{32}$-, $\frac{1}{8}$-, and $\frac{5}{32}$-in pin punches. Punches are illustrated in Fig. 11-97.

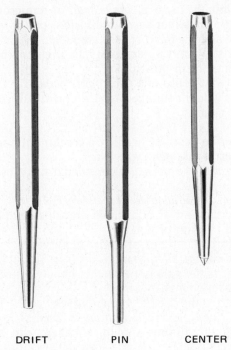

DRIFT   PIN   CENTER

FIG. 11-97   Punches.

Scriber

The **scriber,** illustrated in Fig. 11-98, is used for marking accurate lines on metal for layout purposes. The point of the scriber is kept very sharp and should be removed from the tool and turned "point in" when not in use. This protects the point and also protects the operator from injury. Scribers should not be used on pieces of metal that will be installed on the aircraft. The scratches may cause stress concentrations and lead to metal failure.

FIG. 11-98   A scriber.

● SAFETY EQUIPMENT

The most valuable tool in the toolbox is the one that will prevent personal injury to the individual. Ironi-

cally these pieces of equipment are also among the most inexpensive items that the toolbox will contain. At a bare minimum technicians should have their own personal eye protection, hearing protection, and respirator.

An example of **safety glasses** for eye protection is shown in Fig. 11-99. Some personal glasses have hardened or safety lenses. If it is desired to use personal glasses, it is recommended that clip-on side-shields should be purchased to prevent injury from the side.

FIG. 11-99   Safety glasses.

The aircraft technician works in a noisy environment. The sound of turbine engines, rivet guns, air drills, and numerous electric pumps and motors all contribute to excessive noise pollution. Figure 11-100 shows two kinds of hearing protectors. The **ear plugs** are very inexpensive, easy to store, and adequate for most situations. The **ear-muff**-style protectors will usually do a better job and are felt by many to be more comfortable than plugs.

A **respirator** is shown in Fig. 11-101. The technician has two areas of air pollution to contend with. The first is dust and other solid particles from cutting and sanding. The second area of concern is chemical pollution from solvents, cleaners, finishes, and resins. Certain substances such as polyurethane paint can produce very serious health impairment if proper protective measures are not taken. The respirator shown has two separate "filtration" systems. The first is a paper or **dust filter** that will remove dust particles. The second device contains **activated charcoal** that absorbs **chemical vapors.** The life of the charcoal filter is short when exposed to air. The manufacturer's directions should be followed very closely during use. The two filters can be used individually or together as the situation warrants.

FIG. 11-100   Ear protectors.

FIG. 11-101   A respirator.

262

● **REVIEW QUESTIONS**

1. How would you measure the width of a strip of metal?
2. Why is it a good idea to measure from the 1-in point on a scale rather than from the end of the scale?
3. Give a physical description of an *outside caliper,* an *inside caliper,* and *a set of dividers.*
4. Explain the graduations on a micrometer scale.
5. On a micrometer that measures in fractions of inches, what distance is represented by one turn of the thimble?
6. What is the purpose of a ratchet cap on the micrometer?
7. Explain the importance of careful handling and use of a micrometer.
8. Explain the use of a *vernier scale* on a micrometer.
9. A vernier caliper has four divisions on the vernier scale and the main scale is graduated in $\frac{1}{8}$-in increments. What is the smallest measurement that can be made with this caliper?
10. Explain the use of a *telescoping gage.*
11. How would you take a measurement with a *small hole gage?*
12. What does a *thread gage* measure?
13. Describe the three heads of a combination square.
14. Describe an *open end wrench.*
15. Why are *box wrenches* considered the most suitable wrench?
16. When are *adjustable wrenches* used?
17. Describe how an adjustable wrench should be applied to a nut or bolt.
18. Describe the shape of a *flare-nut wrench.*
19. Describe an *Allen wrench.*
20. What is the value of a ratchet handle in use with a socket wrench?
21. Describe the use of a *crowsfoot wrench.*
22. Describe three types of *torque wrenches.*
23. What precautions should be taken with a *pipe wrench?*
24. Describe three types of *spanner wrenches.*
25. Describe a properly ground tip for a common screwdriver.
26. What damage may occur as the result of using a screwdriver with a worn or incorrect tip?
27. Describe the advantage of *cross-point screwdrivers?*
28. What precaution must be taken when using cross-point screwdrivers?
29. What is the value of an *offset screwdriver?*
30. How may *slip-joint pliers* be used to advantage?
31. What is the principal use of *diagonal-cutting pliers?*
32. What are *Vise-Grip pliers?*
33. What would a technician use *wire twisters* for?
34. What are some of the uses of a *ball peen hammer?*
35. Of what materials are *soft hammers* made?
36. When should a *soft hammer* be used?
37. For what purpose is a *chisel* used?
38. What safety precaution should be observed when using a chisel to chip metal?
39. What is the purpose of "set" in the teeth of a hacksaw blade?

40. Why are hacksaw blades made with different sizes of teeth?
41. What is a general rule with respect to number of teeth (pitch) and thickness of material being cut with a hacksaw?
42. What is the difference between a *single-cut* and *double-cut file?*
43. Describe the different shapes of files.
44. How is the coarseness of file teeth designated?
45. What is the purpose of a *rasp?*
46. What type of file should be used to produce a very smooth surface on aluminum?
47. Explain what is meant by *drawfiling.*
48. Explain how a file should be held and applied to the work.
49. Describe the care of files when they are not in use.
50. Describe a correctly sharpened drill point.
51. What drill-point angle is best for drilling stainless steel?
52. Describe a method for starting a drill to make sure that the hole is in the correct location.
53. Which drill in the number series is the largest; the smallest?
54. List three types of taps and the use of each.
55. What is the purpose of a "split" thread die?
56. What is a *tap drill?*
57. Describe a *countersink.*
58. What is the purpose of a *counterbore?*
59. Under what conditions would you use a *reamer?*
60. Describe the use of an *expansion reamer.*
61. Explain the need for the care and protection of the reamer.
62. What is the difference between a *center punch* and a *prick punch?*

12 AIRCRAFT FLUID LINES AND FITTINGS

● INTRODUCTION

All aircraft depend upon numerous systems of various types to carry on vital functions of operation. Many of these systems utilize fluid-line networks of tubes and fittings made from a variety of materials. Fuel, oxygen, lubricating oil, hydraulic, instrument, fire extinguishing, air conditioning and heating, and water systems all require fluid lines. The malfunction of these systems, due to fluid-line failure, could seriously jeopardize the aircraft's safety.

The pressure of the fluids varies from very low, 5 to 10 psi [34 to 69 KPa], for an instrument system to as much as 5000 psi [34.450 KPa] for a hydraulic system. Low-pressure fluid lines may be made of plastic or rubber hose. High-pressure fluid lines are available in a variety of materials including aluminum alloy, stainless steel, and reinforced rubber hose.

Fluid lines are made of rigid and semirigid tubes or flexible hoses depending upon the application. A tube is defined as a hollow object that is long in relation to its cross section and has a uniform wall thickness. The cross section of a tube may be round, hexagon, octagon, elliptical, or square. Tubes used for fluid lines are usually round in cross section.

A rigid fluid line is one that is not bent to shape or flared. Directional changes and connections are made by the use of threaded fittings. The term **"fitting"** is used for a device used to connect two or more fluid lines. Rigid fluid lines usually have threads cut in their walls.

Semirigid tubes can be bent or formed to shape and have a relatively thin wall thickness in comparison to rigid lines. Various types of fittings are used to make connections between lines.

Hoses are made from rubber or synthetic materials. Depending on the pressure they are designed to carry, a hose may have reinforcing material wrapped around it. Various types of fittings are used to attach hoses to each other or to other components.

Fittings are made from a variety of materials for many different purposes. New designs, some highly specialized, of fluid lines and fittings are being developed continuously. As in other chapters of this text, the emphasis will be placed on basic materials and practices that have wide-spread use. Specialized technical information on a specific system is available from the product's manufacturer and distributer. As with any other type of aircraft maintenance, the technician must ensure that the correct materials are used and the manufacturer's instructions followed.

● TYPES OF FLUID-LINE SYSTEMS

Pipe

Pipe is a rigid fluid line and may be defined as a tube which is made in standardized combinations of outside diameter and wall thickness. It is identified by nominal pipe sizes ($\frac{1}{4}$ in, $\frac{1}{2}$ in, 1 in, etc.) and by ANSI (American National Standards Institute) schedule numbers. Table 12-1 shows how the dimensions of a nominal size of pipe will vary in relation to schedule numbers. The outside diameter of a nominal size is the same for each schedule number. The wall thickness increases with the schedule numbers. This will allow the pressure carried by the pipe to be greater, but it also reduces the inside diameter, or fluid-carrying capability, of the pipe. It should be noted that the pipe's nominal size cannot be obtained by direct measurement.

TABLE 12-1 ANSI Pipe Schedule Dimensions

Nominal size, in	Schedule no.	Outside diameter, in	Wall thickness, in
$\frac{1}{4}$	40	0.540	0.088
$\frac{1}{4}$	80	0.540	0.119
$\frac{1}{2}$	40	0.840	0.109
$\frac{1}{2}$	80	0.840	0.147
$\frac{1}{2}$	160	0.840	0.187
1	40	1.315	0.133
1	80	1.315	0.179
1	160	1.315	0.250

Pipes are joined by fittings utilizing threads cut in the wall of the pipe. In order to provide a fluid-tight seal, the threads are tapered. As two parts are screwed together, the taper will cause a very tight metal-to-metal seal to develop. Figure 12-1 shows how pipe threads taper and form the metal-to-metal seal. Large-scale use of pipe on aircraft is impractical because of weight. However, many components used on aircraft will utilize pipe threads. This requires the use of fittings with pipe thread and special fittings that connect pipe threads to other types of line connecters. Figures 12-2 and 12-3 show fittings used for this purpose. The aircraft maintenance technician should understand the use of pipe threads and be able to identify them.

Tubes

Semirigid fluid lines are usually referred to as tubes or tubing. They can be bent to shape and are usually

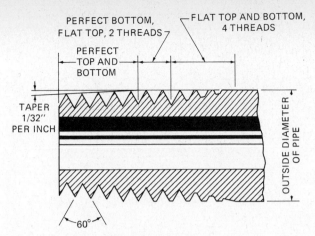

FIG. 12-1   Tapered pipe threads.

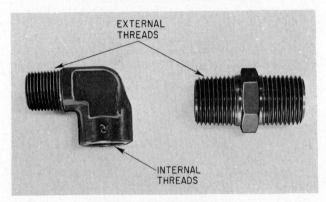

FIG. 12-2   AN fittings with pipe threads.

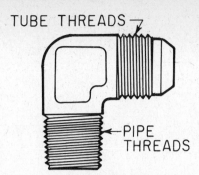

FIG. 12-3   Tube threads and pipe threads.

flared for connecters. Tubes used for fluid lines are sized by the *outside diameter* (OD) in inches and the wall thickness. A typical size tube is $\frac{1}{2}$ in [12.7 mm] OD by 0.035 in [0.89 mm] wall thickness. Tube sizes increase in $\frac{1}{16}$-in increments. Tube fittings use dash

numbers to indicate size in relation to these increments. A -8 fitting is $\frac{8}{16}$ or $\frac{1}{2}$ in. The tube itself may also be referred to as a -6 ($\frac{3}{8}$ in) or -8 ($\frac{1}{2}$ in) tube rather than by its fractional size.

Tubes are made from several metals. Table 12-2 shows some of the metals used and typical applications. Only the materials authorized by the manufacturer, or approved substitutes, may be used on a specific aircraft system.

Flared Tube Connections.   Because of the thin wall thickness of fluid lines, threads cannot be cut in a tube for connections. Special fittings have been designed to allow tubes to be connected to other tubes as well as to the aircraft system components. These fittings may be classified as flared, flareless, swaged, soldered, or brazed. Many of the fittings used for these connections are standard parts and carry AN, AND, or MS specification numbers (see Chap. 10).

Flared fittings require a 37° flare to be formed on the end of the tube. The flare of the tube matches a cone on the fitting. The fitting is also threaded with standard machine-screw threads. A special nut and sleeve are used to pull the flare into contact with the cone and form a fluid-tight metal-to-metal seal (see Fig. 12-4). Small sizes or thin-wall tubing may have a double

TABLE 12-2   Metal Used for Fluid-Line Tubing

Metal	Spec. no.	Application
Aluminum alloy:		
5052	WW-T-787	Used for fuel, oil, instrument, and low-pressure hydraulic lines below 1500 psi where flared fittings are specified.
6061-0	WW-T-789	Used for air ducts where bending is required in diameters up to 6 in. This alloy is weldable.
6061-T6	AMS-4083	Used for hydraulic lines where flareless fittings are specified.
Soft copper	WW-T-799, type N	Used for high-pressure oxygen systems and pressure transmitter lines.
Corrosion-resistant steel (CRES):		
304-$\frac{1}{8}$H	Mil-T-6845	Used for hydraulic lines above 1500 psi and other lines where aluminum alloy is not satisfactory. Temperatures must be below 800° F. This steel is not weldable.
304-1A	Mil-T-8504	Used in place of 304-$\frac{1}{8}$H where greater ductility is necessary and high strength is not required. It is not weldable.
347-1A	WW-T-858	Used in place of 304 steel for welded parts or where operating temperatures are above 800° F.
Titanium	3 AL 2.5V	Used for high-pressure hydraulic systems. Requires swaged fittings.

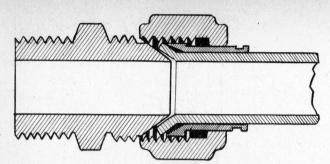

FIG. 12-4 Position of fitting parts for assembly.

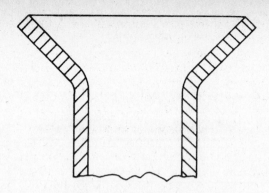

(a) SINGLE-FLARED END

flare as shown in Fig. 12-5. The double flare is used to provide a greater thickness of metal for the seal.

Flared fittings in use today are made to AN or MS standards. Flared fittings made prior to World War II were made to an AC standard. Although AC fittings used a similar flare, the threads are not the same and cannot be used with an AN fitting. AC fittings may still be found on older aircraft. They can be readily recognized as shown in Fig. 12-6. The AN fitting has a space on each side of the threads while the AC has none. AC fittings can be replaced with AN fittings.

The basic components of a flared connection are an AN818 nut, an AN819 sleeve, and one of a number of fittings with a cone to match the flare. Fittings commonly used are classified as tube fittings, universal and bulkhead fittings, and pipe-to-tube (or pipe-to-AN) fittings. Fittings are identified by AN or MS numbers that identify the function of the fitting. Table 12-3 lists some common fittings and their function.

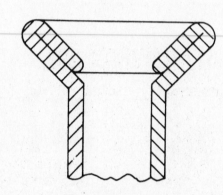

(b) DOUBLE-FLARED END

FIG. 12-5 Single and double flares.

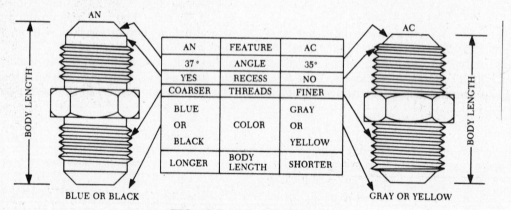

AN	FEATURE	AC
37°	ANGLE	35°
YES	RECESS	NO
COARSER	THREADS	FINER
BLUE OR BLACK	COLOR	GRAY OR YELLOW
LONGER	BODY LENGTH	SHORTER

FIG. 12-6 AN and AC fitting.

TABLE 12-3 Common Fittings and Their Function

Fitting no.	Name	Function
AN818	Coupling nut	Join flared tube to fitting
AN819	Coupling sleeve	Used with AN818 nut
AN815	Union, flared tube	Join two flared tubes
AN821	Elbow, 90°	Join two tubes at 90° angle
AN824	TEE, flared tube	Join three flared tubes
AN827	Cross, flared tube	Join four flared tubes
AN832	Union, universal or bulkhead	Join two tubes through bulkhead or in boss fitting
AN833	Elbow 90° universal or bulkhead	Same as AN821 plus AN832
AN834	TEE, flared tube, universal or bulkhead on tee	Same as AN824 plus AN832
AN816	Nipple, pipe to AN (straight)	Join internal pipe threads to flared tube
AN822	Nipple, pipe to AN (90°)	Join internal pipe threads to flared tube, 90° angle
AN823	Nipple, pipe to AN (45°)	Join internal pipe threads to flared tube, 45° angle

Figure 12-7 shows some common tube fittings used to join two or more tubes together.

It is often necessary to have fuel, oil, or other lines pass through the metal structure of an aircraft. This requires a type of fitting with a long body and provisions for securing the fitting to the bulkhead. Fittings for this purpose are shown in Fig. 12-8 and are called **universal** and **bulkhead fittings.** These same fittings are used where a component with internal screw threads must be connected to a line. Screw threads will not provide a fluid-tight seal. A rubber O-ring seal will be required to prevent fluid leakage and is installed with an AN6289 nut as shown in Fig. 12-9. The use of an angle fitting such as the AN833 (see Fig. 12-8) will allow the fitting to be rotated to any direction before tightening. This provides for proper tube and component alignment (see Fig. 12-10).

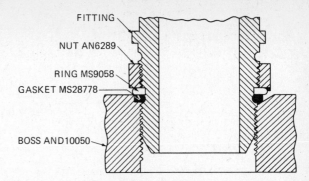

FIG. 12-9   O-ring seal for boss fitting.

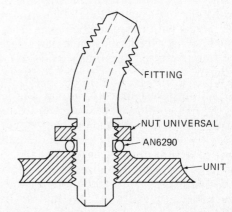

FIG. 12-10   The fitting can be turned to any angle before the nut is tightened.

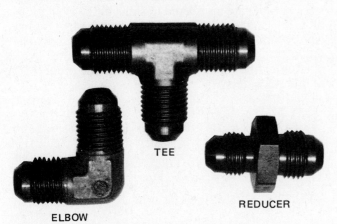

FIG. 12-7   Typical fittings for aircraft tubing.

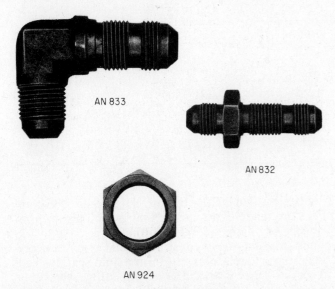

FIG. 12-8   Universal and bulkhead fitting.

FIG. 12-11   Universal fittings.

Another type of fitting, also called a **universal fitting,** is shown in Fig. 12-11. These are used in special installations where a certain degree of flexibility is required. They are used with either stationary or moving units. When used with moving units, such as hydraulic actuating cylinders, they are sometimes called **banjo fittings.** A complete universal assembly includes a bolt (AN775) that has the longitudinal and transverse passages and a fitting that the bolt holds in place. Soft metal "crush" washers (gaskets) are placed under the head of the bolt and between the fitting and the unit to which it is attached. Serrations on the clamping surfaces provide a satisfactory seal of the joints. Universal fittings, or couplings, permit the connection of a section of tubing at any angle of rotation about the bolt center.

Many components on an aircraft use internal pipe threads for connections. This requires the use of a special adaptor called a **pipe-to-AN nipple** as shown in Fig. 12-12. The fluid-tight seal is provided by the pipe threads on one end and the metal-to-metal flare on the other. Various other combinations of pipe and AN fittings are available such as tees with one of the three legs having pipe threads. The use of pipe-to-AN fittings does away with the need for rubber seals and

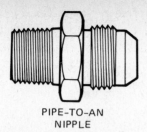

PIPE-TO-AN
NIPPLE

FIG. 12-12 A pipe-to-AN fitting.

nuts. Directional control of the line from the component is not as versatile as with a universal fitting and is primarily a function of tightening the pipe threads.

Flared fittings are usually made of aluminum, steel, or stainless steel with the fitting being the same material as the line. An exception is the use of steel fittings in higher-temperature areas such as in an engine compartment.

Flared Fittings Designation. Fittings are designed by an AN or MS number which indicates the function of the fitting. The fitting is sized by the size of the tube it is used with. The outside diameter of the tube is expressed in sixteenths of an inch. An AN819 fitting for a $\frac{1}{4}$-in tube has a designation of AN819-4 and for a $\frac{3}{4}$-in tube it is AN819-12. The absence of a letter before the dash number indicates that the material is steel. Aluminum-alloy fittings are indicated by a D before the dash. The letter C indicates the fitting is made from corrosion-resistant steel (see Fig. 12-13). Table 12-4 provides additional information on tube fitting sizes.

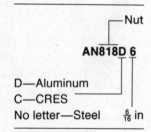

FIG. 12-13 AN fitting designation.

TABLE 12-4 Thread and Wrench Size for AN Fittings

Fitting dash no.	Tube size OD, in	Wrench size, in	Thread
-3	$\frac{3}{16}$	$\frac{1}{2}$	$\frac{3}{8}$-24
-4	$\frac{1}{4}$	$\frac{9}{16}$	$\frac{7}{16}$-20
-5	$\frac{5}{16}$	$\frac{5}{8}$	$\frac{1}{2}$-20
-6	$\frac{3}{8}$	$\frac{11}{16}$	$\frac{9}{16}$-18
-8	$\frac{1}{2}$	$\frac{7}{8}$	$\frac{3}{4}$-16
-10	$\frac{5}{8}$	1	$\frac{7}{8}$-14
-12	$\frac{3}{4}$	$1\frac{1}{4}$	$1\frac{1}{16}$-12

Pipe-to-AN nipples are sized by the tube used on the flared end. Common fittings have a specified relationship between the size of the tube and the pipe thread size as shown in Table 12-5. Other combinations can be obtained when designated as shown.

TABLE 12-5 AN-to-Pipe Nipple Fittings (for AN816, 822, and 823 Fittings)

AN dash no.	Tube size, in	Thread size, in
-2	$\frac{1}{8}$	$\frac{1}{8}$
-3	$\frac{3}{16}$	$\frac{1}{8}$
-4	$\frac{1}{4}$	$\frac{1}{8}$
-4-4	$\frac{1}{4}$	$\frac{1}{4}$
-5	$\frac{5}{16}$	$\frac{1}{8}$
-6	$\frac{3}{8}$	$\frac{1}{4}$
-6-2	$\frac{3}{8}$	$\frac{1}{8}$
-6-6	$\frac{3}{8}$	$\frac{3}{8}$
-8	$\frac{1}{2}$	$\frac{1}{2}$
-10	$\frac{5}{8}$	$\frac{1}{2}$
-12	$\frac{3}{4}$	$\frac{3}{4}$

Fittings designed to connect or adapt two different sizes of tubes are numbered to show the combinations of sizes. Table 12-6 shows the numbering for an AN919 adapter to join two different-size tubes. The manufacturer's specifications or the distributer's catalog should be consulted to determine the proper size of other adapters.

TABLE 12-6 AN 919 Reducers (External Thread)

AN dash no.	Connects
-0	$\frac{3}{16}-\frac{1}{8}$
-1	$\frac{1}{4}-\frac{1}{8}$
-2	$\frac{1}{4}-\frac{3}{16}$
-3	$\frac{5}{16}-\frac{1}{4}$
-4	$\frac{3}{8}-\frac{1}{8}$
-5	$\frac{3}{8}-\frac{3}{16}$
-6	$\frac{3}{8}-\frac{1}{4}$
-7	$\frac{3}{8}-\frac{5}{16}$
-8	$\frac{1}{2}-\frac{1}{8}$
-9	$\frac{1}{2}-\frac{3}{16}$
-10	$\frac{1}{2}-\frac{1}{4}$
-11	$\frac{1}{2}-\frac{5}{16}$
-12	$\frac{1}{2}-\frac{3}{8}$

Flareless Tube Connections. Higher-strength fluid lines are made of a material that is too hard to form satisfactory flares. **Flareless fittings** are made to MS standards and are so constructed that it is not necessary to use flared tubing to form a fluid seal. Instead, a special sleeve is installed with the fitting. The sleeve incorporates a cutting edge that seizes and grooves the tubing as the fitting is tightened, thus producing a fluid-tight seal. The flareless tube fitting therefore eliminates all flaring in connection with tube fittings.

The flareless tube fitting consists of three units; a body, a sleeve, and a nut. There are also special parts for close couplings and reducer couplings. The body of the flareless tube fitting has a counterbored shoulder against which the end of the tube rests. This counterbore has a cone angle of about 24° which, on assembly, causes the cutting edge of the sleeve to cut into the outside of the tubing. Further tightening of the nut forces the sleeve to form the tight seal mentioned above. The nut also engages the bevel of the sleeve, causing it to grip the tube and provide additional support. The pilot edge of the sleeve limits the depth of the tube and prevents any closing of the tube. A cutaway illustration of the MS flareless tube fitting is shown in Fig. 12-14, and a typical fitting is shown in Fig. 12-15. It should be noted that it is easy to identify flareless fittings at the male end because there is no flare cone and no space between the threads and the end of the fitting.

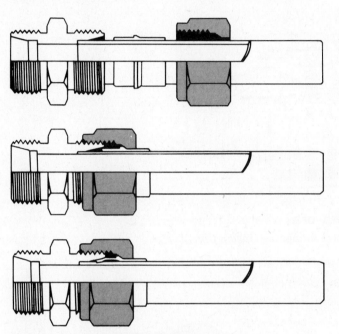

FIG. 12-14   Flareless tube fitting.

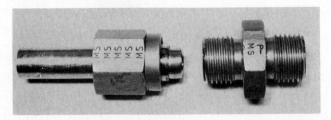

FIG. 12-15   Typical MS fittings.

Swaged Tube Fittings. Many aircraft utilize swaged fittings to join tubes in areas where routine disconnections are not required. Military specifications require that all tubing be permanently joined, either by swaging or welding, except where it is necessary to make disconnections. Swaged fittings are made by Deutsch Metal Components Division from aluminum, stainless steel, and titanium to be used with tubing of the same material. The fittings are attached quickly and easily, either in the aircraft or outside by means of a hydraulically operated, portable swaging tool. A cutaway view of a swaged fitting is shown in Fig. 12-16.

The advantages of the swaged-type tube fittings are several: The original cost is low compared with that of standard AN or MS fittings, the installation takes less time, substantial weight is saved, and repairs can be made in the aircraft without removing complete sections of tubing.

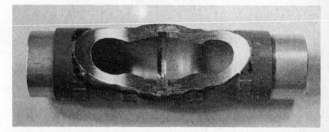

FIG. 12-16   Cutaway view of a swaged splice. *(Deutsch Metal Products Div.)*

Quick-Disconnect Fittings. At various points in aircraft plumbing systems, it is specified that quick-disconnect couplings be installed. The purpose of these couplings is to save time in the removal and replacement of components and at the same time prevent the loss of the fluid in the system. The use of these couplings greatly reduces the cost of maintenance in the systems involved. Typical uses are in fuel, oil, hydraulic, and pneumatic systems.

A typical quick-disconnect fitting is illustrated in Fig. 12-17. This particular assembly is trade-named the Saf-Loc Coupling, 3750 series, and is manufactured by the Aeroquip Corporation. This coupling consists of a male and a female assembly, each incorporating a sealing piston (poppet) that effectively prevents the loss of fluid when the coupling is being connected or disconnected. Fluid loss is prevented at these times by a tubular valve sleeve that remains engaged until the poppet valves are closed. The sleeve is inside the female part of the coupling, and the O-ring seal with which it engages is in the male part.

In the illustration of Fig. 12-17, the coupling is shown in both the disconnected and connected configurations. It can be observed that when the two parts are brought together for coupling, the sleeve fluid seal engages first to prevent fluid loss when the poppet valves engage each other and open. Three check points are used to verify positive connection. These involve sound, visual observation, and touch. A click is heard at the time the coupling is locked, and indicator pins extend from the outer sleeve. These pins can be seen and can be felt by hand.

Ventilator and Hot-Air Ducts

Ventilator and hot-air heater ducts are made from a variety of materials. Among these materials are a lightweight asbestos fabric impregnated with synthetic rubber and having a spiral body wire, plastic-impregnated glass fiber, aluminum, and stainless steel. The type of material used and the attachments employed for connecting sections are determined by the temperatures and pressures under which the duct must operate. For high-pressure and high-temperature air,

269

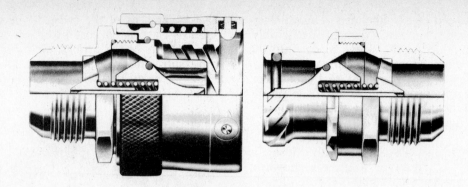

DISCONNECTED—SEALED

INDICATOR PIN ↓

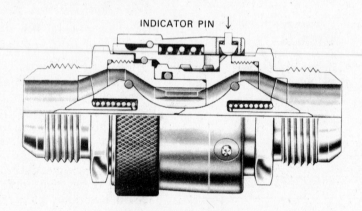

CONNECTED—OPEN

FIG. 12-17 A quick-disconnect fitting. *(Aeroquip Corp.)*

large-diameter, thin-walled stainless steel is used. For medium temperatures and pressures, aluminum tubing may be used.

In the replacement of ducting in a particular aircraft, the technician should use a manufactured part specified by a part number or should construct the part of the same material as that used in the original part. The manufacturer's maintenance manual or the illustrated parts catalog for a particular make and model of aircraft will give the description and a part number of any part required.

Ducting in an aircraft is supported by means of band clamps such as the one illustrated in Fig. 12-18. These clamps are made of aluminum alloy, stainless steel, or titanium, depending upon the requirements of the system. They are quickly removable, durable, and corrosion-resistant.

Joints in ducting are usually connected by means of V-band couplings, flanges, and gaskets. The components of a typical joint are shown in Fig. 12-19. A cross section of a joint as installed is shown in Fig. 12-20.

Hoses and Hose Fittings

Because of the need for flexibility in many areas of aerospace vehicle construction, it is often necessary to employ hose instead of rigid tubing for the transmission of fluids and gases under pressure. Specifications

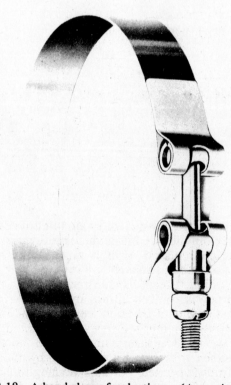

FIG. 12-18 A band clamp for ducting. *(Aeroquip Corp.)*

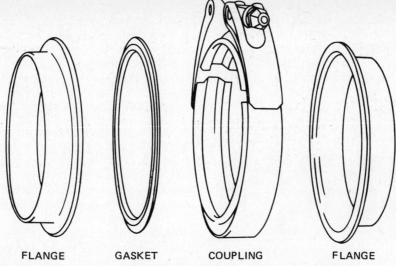

FLANGE   GASKET   COUPLING   FLANGE

FIG. 12-19   Components of a ducting joint.   *(Aeroquip Corp.)*

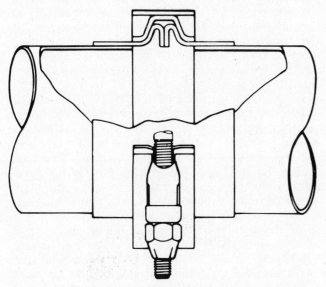

FIG. 12-20   Cross section of a ducting joint.   *(Aeroquip Corp.)*

for hose to be used in aerospace vehicle systems are given in the Air Force–Navy Aeronautical Design Standard (AND) 10340. This specification covers hoses for fuel, oil, hydraulic, pneumatic, and other systems.

Hose and hose assemblies for aircraft plumbing systems are manufactured in four different pressure categories: low-pressure, with a maximum operating pressure of 300 psi [2068 kPa]; medium-pressure (300 to 1500 psi) [2068 to 10 342 kPa]; high-pressure (1500 to 3000 psi) [10 342 to 20 685 kPa]; and extra-high-pressure (3000 to 6000 psi) [20 865 to 41 370 kPa]. It must be noted that the pressure ranges given above are general and that variations will be encountered in specific systems and installations. Hose assemblies are usually pressure-tested at a pressure at least twice the maximum operating pressure. The burst pressure is usually required to be at least 4 times the maximum operating pressure. Thus, a good margin of safety is provided.

The construction of a low-pressure hose is shown in Fig. 12-21. This hose conforms to Specification MIL-H-5593. The inner tube of the hose consists of synthetic rubber with a braided cotton reinforcement. The outer cover is also synthetic rubber. The hose is identified by a linear yellow stripe, called a **lay line,** interspersed with the symbol "LP" and the hose manufacturer's code, size, and date of manufacture with the quarter year and calendar year indicated. The markings are repeated every 6 in [15.24 cm] and provide a means for determining whether the hose is twisted.

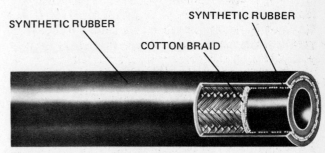

SYNTHETIC RUBBER   SYNTHETIC RUBBER
                  COTTON BRAID

FIG. 12-21   Low-pressure aircraft hose.   *(Aeroquip Corp.)*

Low-pressure hose is used for instrument air or vacuum systems, automatic pilots, and instruments where the maximum pressure for the hose is not exceeded. The hose assembly (with fittings installed) conforms to AN6270, and the fittings conform to MIL-A-38726 and MS27404.

Medium-Pressure Hose. Hose for pressures up to 1500 psi [10 342 kPa] and sometimes higher usually has one or more layers of stainless-steel braid to reinforce the inner tubing. Construction of such hose is shown in Fig. 12-22. The inner tube is seamless synthetic rubber or tetrafluoroethylene (TFE or Teflon). If the hose is of Teflon, it can be used for practically all fluids that may be encountered on an airplane. The Teflon is nonaging, chemically inert, and physically stable and can withstand relatively high temperatures. Specifications for medium-pressure hose are MIL-H-8794 and MIL-H-27267.

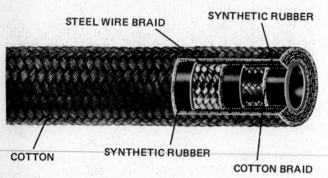

FIG. 12-22 Medium-pressure hose. *(Aeroquip Corp.)*

High-Pressure Hose. The construction of high-pressure hose is shown in Fig. 12-23. The illustration shows a hose consisting of a seamless inner tube made of synthetic rubber and reinforced with high-tensile-strength, carbon-steel wire braid. The smaller sizes have two layers of the wire braid, and the larger sizes (above $\frac{3}{4}$-in [1.905-cm] inner diameters or ID) have a triple-wire-braid reinforcement. This hose complies with Specification MIL-H-8788.

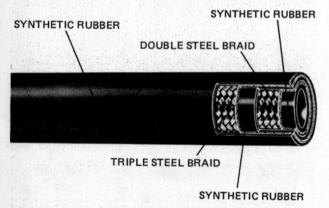

FIG. 12-23 High-pressure hose. *(Aeroquip Corp.)*

Another type of high-pressure hose complies with Specification MIL-H-38360. This hose has a Teflon-type inner tube with reinforcements of stainless-steel wire and braid, the amount of reinforcement depending upon the diameter of the hose. For sizes 4 and 6, two stainless-steel wire braids are applied over the inside tube; for hose size 8, two spiral stainless-steel wire layers and one outer layer of stainless-steel braid are used; and for sizes 10 to 16, four stainless-steel spiral layers are used with an outer layer of stainless-steel braid. Because of the TFE inner tubing, this hose can be used for most of the fluids employed in the aircraft systems.

Extra-High-Pressure Hose. Hose for systems with pressures up to 6000 psi [41 370 kPa] is called extra-high-pressure hose. This hose employs spiral stainless-steel wire layers cushioned with a high-temperature elastomer between layers. An elastomer is a synthetic rubber. This hose is usable for practically all aircraft fluids including phosphate esters. The temperature range is up to 400°F [204°C].

Selection of Hose. Care must be taken in the selection of hose for a particular application. Usually the type of hose will be specified in the maintenance manual; however, there may be instances in which the exact information is not available, and the technician must be able to determine whether the tube is suitable for the system involved.

Hose must be selected on the basis of size, pressure rating, temperature rating, and material. The size is indicated by dash numbers such as 4, 6, 8, 12, and 16. These numbers show the approximate outer diameter (OD) of the inside tube of the hose in sixteenths of an inch. A No. 8 hose has an inner tube with an OD of $\frac{8}{16}$ in, or $\frac{1}{2}$ in [1.27 cm]. The inside diameter will be approximately $\frac{7}{16}$ in [1.11 cm]. It must be pointed out that these diameter indications will not always be exactly in sixteenths of an inch for various makes and types of hose, but a reasonably accurate dimension can be determined by examining the specification chart for the hose in question.

The material of the inner tube of the hose is most important, because some fluids will soften or otherwise damage some materials. When synthetic rubber is used for the inner tubing, it must be of a type that will not be affected by the fluid it carries. Synthetic rubbers are manufactured that are resistant to petroleum products (except aromatics), aromatic petroleum products, and synthetic oils such as phosphate ester fluids. Some hose is marked with the word "Skydrol" to indicate that it is resistant to synthetic, fire-resistant hydraulic fluid. If the inner tubing of the hose is made of tetrafluoroethylene (TFE or Teflon), it can be used for either petroleum products or synthetics. The technician should refer to the maintenance manual for the system involved and use the hose specified by the manufacturer. Usually the hose will carry a manufacturer's part number.

The pressure and temperature ratings for hose are given in the specification chart for the hose. When the hose or hose assembly specified by part number is selected, the pressure and temperature ratings for the hose will be correct for the system.

Hose Identification. Hose with a synthetic rubber covering is usually marked by a stripe, called a lay line, applied lengthwise on the hose, together with identification numbers and codings. If the hose outer surface is metal braid, the identification information is stamped on a metal band secured around the hose.

Hose Fittings. End fittings for aircraft hose assemblies are manufactured for permanent attachment and for reuse. With the permanent-type fitting, if it is necessary to replace the hose, both the hose and the fit-

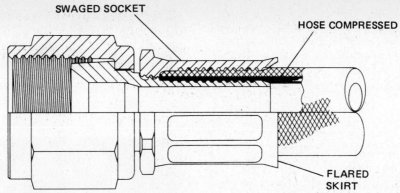

FIG. 12-24   A permanent hose fitting.   *(Resistoflex Corp.)*

tings (complete hose assembly) are discarded. A permanent fitting for a high-pressure hose is shown in Fig. 12-24. During assembly, the socket is swaged circumferentially to compress the hose between the socket wall and the nipple. The serrations inside the socket and outside of the nipple grip the outer and inner surfaces of the hose to prevent slippage. The socket is welded to the nipple at the forward projection and is also provided with an interlock that prevents a blowoff under maximum pressure. The skirt of the socket is flared to allow for the hose to bend at the fitting without creating concentrated stress at the end of the socket.

A fitting for an extra-high-pressure hose is shown in Fig. 12-25. This fitting is designed with an internal locking and sealing member called a **lip seal.** In the illustration, it will be noted that the inner tube of the hose is inside the lip seal and the reinforcing layers are outside the lip seal and inside the socket wall. After the fitting is assembled to the hose, the socket is swaged, causing the circumferential serrations to grip both the inner tubing and the reinforcing layers, inside and outside, this producing a blowoff proof attachment.

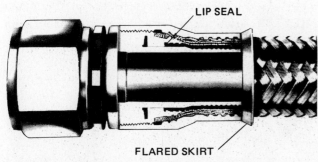

FIG. 12-25   Fitting for extra-high-pressure hose.
(Resistoflex Corp.)

Hose fittings are designed to mate with various tubing fittings and to accommodate different angles of attachment and rotation. They are available to mate with either flared tube fittings (AN and MS types) and with flareless fittings (MS types). Some typical fittings mounted on hose assemblies are shown in Fig. 12-26. These fittings are designed to mate with MS33514

flareless fittings. Hose fittings designed to mate with flared tube fittings incorporate a 37° bevel to match the tube fitting.

The rotation of the fittings on a hose assembly in-

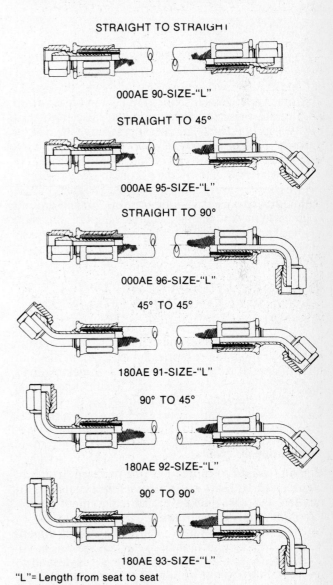

STRAIGHT TO STRAIGHT

000AE 90-SIZE-"L"

STRAIGHT TO 45°

000AE 95-SIZE-"L"

STRAIGHT TO 90°

000AE 96-SIZE-"L"

45° TO 45°

180AE 91-SIZE-"L"

90° TO 45°

180AE 92-SIZE-"L"

90° TO 90°

180AE 93-SIZE-"L"

"L"= Length from seat to seat

FIG. 12-26   Typical hose assemblies to mate with flareless fittings.   *(Resistoflex Corp.)*

dicates the angular distance of one fitting from the other. One fitting is held in a vertical position (straight down), and the rotation angle is measured counterclockwise from the center line of the top fitting to the center line of the bottom fitting. This is illustrated in Fig. 12-27.

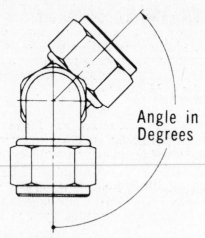

Angle in Degrees

FIG. 12-27   Angular displacement of hose fittings.   *(Resistoflex Corp.)*

Protective Sleeves for Hose. In certain areas within an aircraft, it is advisable to protect hose from heat and wear. For these purposes, protective sleeves of various types have been developed. The illustrations of Fig. 12-28 show some of the types of protective sleeves manufactured by the Aeroquip Corporation.

Fire sleeves are installed on hose in areas where high temperatures exist—for example, on and around engines and in engine compartments. Abrasion sleeves are used where the hose may rub against parts of the aircraft.

● FABRICATION, REPAIR, AND INSTALLATION OF FLUID LINES

Most of the plumbing jobs an aircraft technician will be called upon to do are field and emergency repairs or replacements where fabrication equipment may not be available. It is important that the technician knows basic techniques that can be used without specialized tools or equipment. The techniques covered in this text are basic operations that can be done with tools found at most aviation maintenance facilities.

Preparation of Tubing

When a section of tubing must be replaced, it should be replaced with a tube of the identical material, diameter, and wall thickness. The replacement section should be straight and round.

The ends of the tube are cut to the correct dimension. It is important to make clean, square cuts at 90° to the center line of the tubing. When it is desired to cut aluminum tubing or tubing of any comparatively soft metal, a small tube cutter can be used. A cutter suitable for this purpose is shown in Fig. 12-29. This cutter makes a clean, right-angle cut without leaving

Silicone Coated Asbestos Fire Sleeve
−65°F. to +450°F.

Heat Shrinkable Polyolefin Abrasion Sleeve
−65°F. to +275°F.

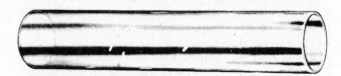

Nylon Spiral Wrap Abrasion Sleeve
−65°F. to +200°F.

FEP100 Teflon Abrasion Sleeve
−65°F. to +400°F.

AEROQUIP AE 138-SIZE

Neoprene Tubing Abrasion Sleeve
−65°F. to +250°F.

FIG. 12-28   Protective sleeves for hose.   *(Aeroquip Corp.)*

FIG. 12-29   A hand-operated tube cutter.

burrs or crushing the tube. A hardened reamer for cleaning the ends of the tubing is often included as part of a cutter. The reamer is used as shown in Fig. 12-30 to smooth the inner edge of the cut where the metal is pressed inward a small amount. If ends of the tube are not properly cleaned and smoothed, the flares will be unsatisfactory, because any nick, cut, or scratch will be enlarged in the flaring operation. A number of tools are made for cleaning and deburring the tube ends.

FIG. 12-30 Dressing the end of cut tubing.

When a section of tubing is replaced in an aircraft, the section to be replaced can be used as a pattern. If this is not possible, a piece of welding rod or stiff wire can be used. The wire is bent as required to conform to the shape of the tube section. This results in a considerable saving of time and material. If the pattern material is marked every few inches before it is bent, it will help to determine the length of tubing needed for the replacement. When the pattern is being bent, special attention must be paid to clearance around obstructions and to the alignment of the ends at the point where they connect to the other parts of the installation.

Short straight sections of tubing between fixed parts of an aircraft are avoided in the designing of an installation because of the danger of excessive strain when the tube expands or contracts as a result of temperature changes. It is therefore the general practice to make installations with bends in the tubing to absorb the changes in length whenever they may occur.

Hand Bending

The wall thickness and the outside diameter govern the minimum permissible bend radius for tubing, but it is advisable to make the bends as large as the installation will permit. It is also desirable to make all bends of the same radius in any one line. As a general rule, at least 5 times the diameter of the tubing should be used as a minimum bend radius, although it is permissible to use a radius as small as 3 times the diameter of the tubing under some conditions. Minimum-bend radii and torque values approved for certificated aircraft aluminum-alloy and stainless-steel tubing installations are given in Table 12-7.

TABLE 12-7 Bend Radii and Torque

Tube OD, in	Torque range for tube nuts, lb/in*		Minimum bend radii, in†	
	Aluminum alloy	Stainless steel	Aluminum alloy	Stainless steel
$\frac{1}{8}$	—	—	$\frac{3}{8}$	
$\frac{3}{16}$	—	30–70	$\frac{7}{16}$	$\frac{21}{32}$
$\frac{1}{4}$	40–65	50–90	$\frac{9}{16}$	$\frac{7}{8}$
$\frac{5}{16}$	60–80	70–120	$\frac{3}{4}$	$1\frac{1}{8}$
$\frac{3}{8}$	75–125	90–150	$\frac{15}{16}$	$1\frac{5}{16}$
$\frac{1}{2}$	150–250	155–250	$1\frac{1}{4}$	$1\frac{3}{4}$
$\frac{5}{8}$	200–350	300–400	$1\frac{1}{2}$	$2\frac{3}{16}$
$\frac{3}{4}$	300–500	430–575	$1\frac{3}{4}$	$2\frac{5}{8}$
1	500–700	550–750	3	$3\frac{1}{4}$
$1\frac{1}{4}$	600–900	—	$3\frac{3}{4}$	$4\frac{3}{8}$

*Pound-inches can be converted into newton-meters by multiplying by a factor of 0.1129.
†Inches can be converted into centimeters by multiplying by a factor of 2.54.

The method for determining the radius of a bend is shown in Fig. 12-31. Observe that the radius of the bend is measured from the center line of the tubing. Correctly bent tubing maintains its circular shape and presents a smooth appearance throughout, without kinks or breaks. Proper and improper bends are shown in Fig. 12-32.

Use of Bending Tools. Soft tubing under $\frac{1}{4}$-in diameter can be bent by hand without a bender. For

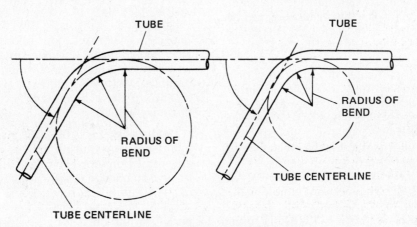

TUBE

TUBE

RADIUS OF BEND

RADIUS OF BEND

TUBE CENTERLINE

TUBE CENTERLINE

FIG. 12-31 Radius of tubing bend.

FIG. 12-32 Good and bad bends.

in the sheave block, and the clip is hooked over the tubing. The slide-bar handle is then rotated so that the groove in the slide bar fits over the tubing. Continued rotation of the slide-bar handle makes the bend with the degree of bend determined by the distance of rotation as indicated by the scale on the sheave block.

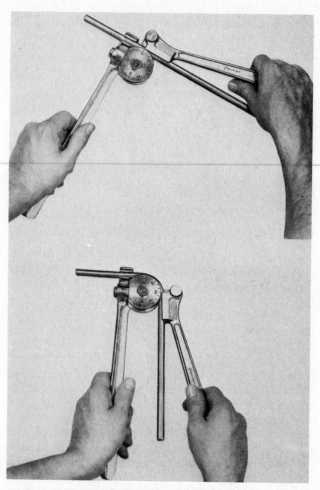

FIG. 12-33 A hand tube bender.

larger sizes specialized tools and equipment will be needed. Bending tools are divided into two types: **hand tube benders,** which require a different bender for each tube OD, and **production benders,** which can be used for different tube sizes by changing the attachments. Production benders may be either manually or power operated.

The choice of the particular model bender to be used depends upon the size and material of the tubing to be bent, the kind of benders available, and the number of bends to be made. If only one or two bends are to be made for a particular job, it is often more economical to use a hand bender rather than take the time to set up a production bender.

A typical hand bender is shown in Fig. 12-33. The bender consists of a sheave block and handle with a clip attached and a slide bar with a handle. To insert a piece of tubing to be bent, the slide bar is turned back away from the sheave block. The tubing is placed

The hand tube bender is a valuable means of fabricating tubing and maintaining the many systems in which tubing plays a vital part. Fuel, lubricating oil, and other liquids and also a variety of gases are carried from supply tanks to various operating mechanisms. In an airplane, wing flaps, cowl flaps, retractable landing gear, dive brakes, power-boosted controls, propeller pitch-changing mechanisms, flight instruments, and various other mechanisms may be operated hydraulically through a plumbing system. For the maintenance of these systems, the hand tube bender is a most useful tool.

Production Tube Bender. Although the aircraft maintenance technician may not be required to operate production tube-bending equipment; it is desirable to have an understanding of such equipment. The production equipment utilizes the same principles as those for the hand bender; that is, the tubing is supported on the outside by suitable blocks designed to accommodate the various sizes of tubes.

The bending mechanism of a typical production bender is shown in Fig. 12-34. The bending mechanism consists of a radius block, clamp block, and sliding block. The clamp block secures the tubing at the point of the bend to the radius block. The clamp block and the sliding block are adjusted for position by means of vise mechanisms.

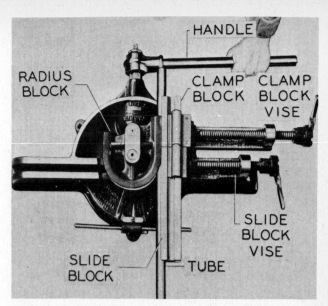

FIG. 12-35   Tubing in the bender ready for bending.

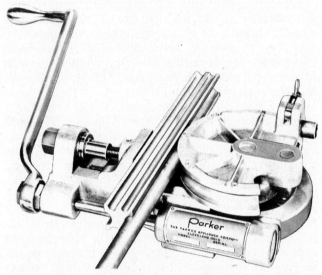

FIG. 12-34   Bending mechanism of a production bender.

When a section of tubing is to be bent, it is placed in the bender as shown in Fig. 12-35. The radius block is chosen for the radius desired and the diameter of the tubing to be bent. The clamp block is grooved to fit the tubing, and when the clamp block vise is tightened, the tubing is held firmly against the radius block. The slide block vise is also tightened to bear against the tubing. When the handle is turned, the radius block and clamp block together rotate, draw the tubing around, and produce the bend. The slide block holds the tubing in shape as the bend is made.

If tubing is to be bent with a radius less than standard, or if the tubing wall is thinner than standard, it is necessary to use a mandrel inside the tubing. The mandrel is mounted on a rod of sufficient length to accommodate the length of tubing to be bent. The rod is secured by an adjustable mount on the bender bench. The rounded end of the mandrel is positioned in the tubing at the point where the sliding block bears against the radius block. As the tubing is drawn around the radius block, the mandrel supports it from the inside as shown in the simplified drawing of Fig. 12-36. The position of the mandrel must be exact; otherwise, the tubing will be stretched and the bend distorted.

The production tube bender is provided with a scale to indicate the degree of bend completed. This makes it possible to produce an accurate bend as required for the installation. In the operation of the tube bender, it

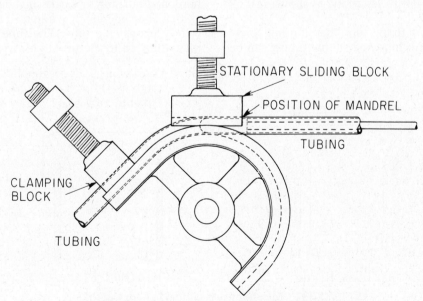

FIG. 12-36   Position of the mandrel for tube bending.

is important to see that all parts are clean and not scratched or nicked, that the radius block, clamp block, and sliding block are of the correct sizes, and that the vises are correctly adjusted. Proper lubrication of tubing and all sliding parts is required.

Flaring

The purpose of a **flare** on the end of a tube is twofold. First, it provides a flange that is gripped between the sleeve or flare nut and the body of a fitting. This prevents the end of the tube from slipping out of the fitting. Second, the flare acts as a gasket between the sleeve and the cone of the fitting, thus providing a tight seal. It is obvious that the flare must be nearly perfect because minute cracks or irregularities would permit leakage at the junction. The flare must be neither too long nor too short. A flare that is too long will bear against the threads of the fitting and may cause damage to both the flare and the threads. A tight seal cannot usually be obtained under such conditions. Maximum and minimum flare lengths are shown in Fig. 12-37. The angle of a tube flare is usually 37°.

Before flaring the end of a tube, it should be remembered that a part of the fitting, in most cases sleeves (if required) and nuts, must be slipped on the tubing before flaring since it is impossible to install them after the flare is formed.

It is not possible to make a satisfactory flare without the aid of a good tool. Several types of flaring tools are available at reasonable cost, but the technician must make sure that the tool selected will produce a suitable flare without damaging the tubing.

A practical hand flaring tool, shown in Fig. 12-38, consists of parallel bars between which are mounted blocks with holes of various sizes drilled between the blocks. The blocks are split at the holes so that they can be separated for the insertion of tubing. The holes between the blocks are drilled with a diameter slightly less than the outside diameter of the tubing so that they will grip the tubing firmly when the clamping screw is tightened. A yoke which carries the flaring cone slides over the entire assembly, as shown in Fig. 12-38.

To produce a flare with this tool, the clamping screw at the end of the tool is loosened so that tubing can be inserted through the correct-sized hole. About $\frac{1}{4}$ in [0.635 cm] of the tubing is extended above the clamping blocks. The clamping screw is then tightened to hold the tubing in place. Next, the yoke with the 37° flaring cone is slid over the tool and positioned so the cone is directly over the end of the tubing. When the flaring-cone screw is turned, the cone is forced into the end of the tubing until the flare is formed.

Difficulties and failures with single flares have been overcome by making a double flare. This requires only slight addition to or alteration of the flaring tools. The use of an adapter will make most of the flaring tools suitable for double flaring. Making a double flare with an adapter is shown in Fig. 12-39. Tubing with a double flare is shown in Fig. 12-40.

The double flare is not required on steel or titanium tubing; however, it should be used on all 6061-T and 5052-0 aluminum tubing with outside diameters of $\frac{1}{8}$ to $\frac{3}{8}$ in [0.317 to 0.952 cm]. The double flare prevents the failures that would result from weak or cut-single flares.

Tube Beading

Even though flexible hose sections in modern installations are almost always made with manufactured hose fittings, there are times when it becomes necessary to join sections of hose to tubing without the use of such fittings. When this is done, it is necessary to bead the tubing if the diameter is more than $\frac{3}{8}$ in [0.952 cm] OD.

A manufactured hand beading tool consists of a body, which carries a hardened bushing for the inner beading rollers, and a guide for a slide block, which carries two rolls that size the outer section of the bead. An adjusting screw whose end is enclosed in the side block is mounted in the body. Stop nuts are placed on this screw for limiting the length of the bead.

To make a bead with the tool, the tubing is placed over the inner roller after being lubricated. The roller must be of the size to accommodate the tube being beaded. The tube end is placed against the face of the hardened bushing to locate the bead in the correct position. The slide block is adjusted to contact the outer rollers against the tube. The tube clamp assembly is then attached to the tube. This assembly is similar to

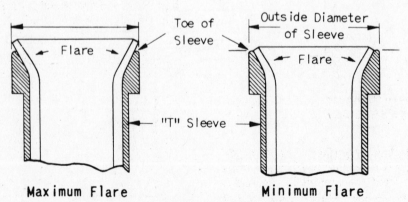

FIG. 12-37 Maximum and minimum flare lengths.

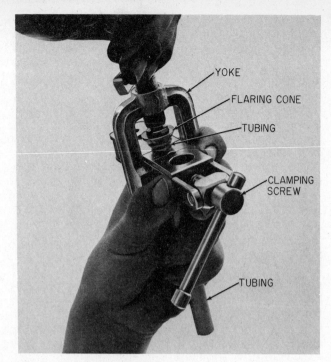

FIG. 12-38　A hand flaring tool.

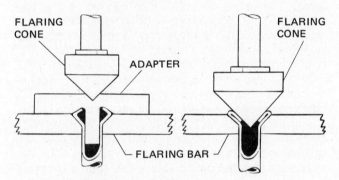

FIG. 12-39　Making a double flare.

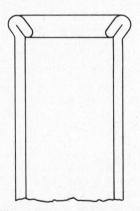

FIG. 12-40　Tubing with a double flare.

the beading tool except that a synthetic rubber block is used instead of rollers on the slide block, and a similar block assembly is rigidly mounted in the body. When the slide block is adjusted against the tube, the friction of the clamp is sufficient to hold the tube while the bead is formed. By holding the clamp assembly in the left hand and turning the beading assembly with the right hand, one can make a bead in a few seconds. This operation is shown in Fig. 12-41. Pressure is applied to the bead-forming roller by means of the hand screw while the tool is being turned.

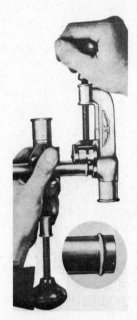

FIG. 12-41　Using a beading tool.

Installation of Flareless Fittings

A flareless fitting consists of a body, sleeve, and nut. When installed, the cutting edge of the sleeve is embedded in the tubing to which it is attached. The best method for installing a flareless fitting is to use a presetting tool to make the installation of the sleeve on the tube.

The presetting tool is a duplicate, dimensionwise, of the body to which the tubing will ultimately be attached. A cross section of such a tool is shown in Fig. 12-42. The bottom of the counterbore correctly positions the tubing so the sleeve will be in the right position. The taper in the counterbore engages the pilot lip on the end of the sleeve to force the cutting edge of the sleeve into the tubing. The coupling nut engages the shoulder on the sleeve to hold the tube firmly in place.

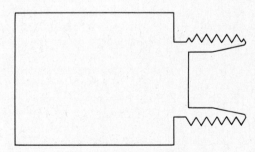

FIG. 12-42　Cross section of a presetting tool.

The following is a typical procedure for installing or presetting a flareless fitting on a tube:

1. See that the end of the tube is properly cut, deburred, and dressed as explained previously.

2. Select a presetting tool of the correct size for the tubing being repaired. Mount the presetting toll in a vise as shown in Fig. 12-43.

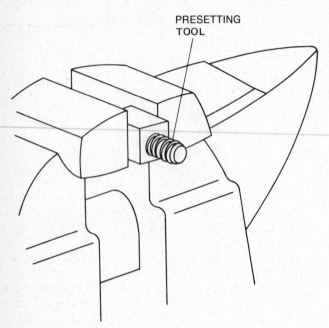

PRESETTING TOOL

FIG. 12-43   Presetting tool mounted in a vise.

3. Select the correct size of sleeve and nut. Slide them onto the end of the tube, the nut first with threads out toward the end of the tube, then the sleeve with the pilot and cutting edge toward the end of the tube.

4. Select the correct lubricant for the type of system in which the tubing will be installed. For example, if the tubing is to be used in a hydraulic system, the lubricant should be the hydraulic fluid used in the system. A petroleum-base oil should be used for fuel system fittings. Lubricate the fitting threads, tool seat, and shoulder sleeve.

5. Insert the tube end into the presetting tool until it is firmly against the bottom of the counterbore. Slowly screw the nut on the tool threads until the tube cannot be turned with the thumb and fingers. At this point, the cutting edge of the sleeve is gripping the tube sufficiently to prevent tube rotation, and the fitting is ready for the final tightening necessary to set the sleeve on the tube.

6. Tighten the nut the number of turns specified in the appropriate table for the size and material of tubing involved. The number of turns will usually be $\frac{5}{6}$, 1, or $1\frac{1}{6}$. The sleeve is now permanently set with the cutting edge seated into the outer surface of the tube. Sleeves should not be removed from tubing and reused under any circumstances.

After the sleeve for a flareless fitting has been seated on the tubing, the nut is backed off and the tube removed from the presetting tool. If the section of tubing is not to be installed at once, the fitting should be capped to prevent the entrance of any foreign matter.

All installations should be carefully inspected and tested before being installed. The interior of the tubing should be checked for metal chips, dirt, or other foreign material.

The inspection procedures for flareless fittings after the presetting procedure is generally as follows:

1. Check the cutting edge and pilot of the sleeve. The cutting edge should be embedded into the tube outside diameter approximately 0.003 to 0.008 in [0.076 to 0.203 mm], depending upon the size and material of the tubing. A lip of material will be raised under the pilot. The pilot of the sleeve should be in contact with or very close to the outside diameter of the tube. The tube projection from the pilot of the sleeve to the end of the tube should conform to the appropriate table.

2. The sleeve should be bowed slightly.

3. The sleeve may rotate on the tube and have a longitudinal movement of not more than $\frac{1}{64}$ in [0.396 mm].

4. The sealing surface of the sleeve which contacts the 24° angle of the fitting body seat should be smooth, free from scores, and showing no longitudinal or circumferential cracks.

5. The minimum internal diameter of the tube at the point where the sleeve cut is made should be checked against the table for the size of tubing used.

6. The tube assembly should be proof-tested at a pressure equal to twice the intended working pressure.

Plumbing Installation

The proper functioning of the many fluid and gas systems in aircraft is assured by the original design and manufacture of the systems; however, continued satisfactory operation depends on the proper maintenance, service, and installation of replacement parts. In the field it is the duty of the technician to inspect, troubleshoot, and repair the systems. Many of the practices and principles for the maintenance of plumbing systems have been covered in this chapter; however, it is important that installation of systems and components be examined from the technician's position.

Installation of Tubing. One of the important steps before tubing is installed is the lubrication of the fittings. Lubrication is not essential to all types of fittings, but it **must** be applied to some, and it is a good practice for others. In the application of a lubricant, it is important that none of the lubricant enter the tubing unless the lubricant is the same material that will be flowing in the system. This applies particularly to hydraulic fluids. In general, no lubricant should be applied to the starting threads, and it should be used sparingly on the rest of the fittings. Figure 12-44 shows

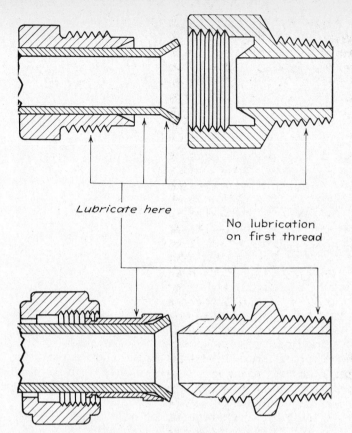

Lubricate here

No lubrication on first thread

FIG. 12-44 Lubrication of tube fittings.

the points of lubrication for typical fittings. The following general rules apply:

Lubricate nuts and fittings on the **outside** of the sleeve and on the male threads of the fittings except for the starting threads.
Lubricate coupling nuts and fittings on the outside of the flare, and lubricate the female threads except for the starting threads.

For fittings on oxygen lines, a petroleum-base or oily lubricant may not be used. For fittings of oxygen systems, a special lubricant must be used. A lubricant conforming to AN-C-86 or MIL-T-5542-B may be used.

Several lubricants are used on hydraulic fittings, including the fluid to be used in the system. Straight threads of brass or steel may be left dry or may be lubricated with the fluid. If the threads are an aluminum alloy, petrolatum (petroleum jelly) may be used.

For pipe threads, the lubricant must be of a type that is not soluble in the fluid being carried in the system. If a petroleum-base lubricant is used on pipe fittings in a fuel system carrying gasoline or jet fuel, the lubricant will be dissolved and a leak will develop. It must be remembered that the lubricant used with a pipe fitting must also serve as a seal and fill the space at the roots of the threads.

Before tubing assemblies are installed, a final inspection should be made. Flares and sleeves must be concentric and free of cracks. The tubing must not be appreciably dented or scratched. Each assembly must be in initial alignment with the fitting to which it is to be attached. **A fitting or assembly must never be forced into position.** A section that must be forced to line up is under initial stress and may fail in operation.

Tubing should be pushed against the fitting snugly and squarely before starting to turn the coupling nut. The tubing should not be drawn up to the fitting by tightening the nut because the flare may be easily turned off. To make sure that a snug fit is effected, all nuts should be started by hand. Tubing installed in an airplane must not be used as a footrest or as a ladder, and lamp cords and other weights should not be suspended from it. The most important of all operations for tubing installation is that of tightening or torquing the nuts. The most common mistake is to overtighten the nuts in order to ensure a leak-free union in a pressure system. Overtightening causes damage to the tubing and fittings and often results in line failure in flight. Correct torque values are given in Table 12-7. Improper tightening effects are illustrated in Fig. 12-45 and are compared with nuts that are properly tightened. These drawings also show that there is an angular difference between the sleeve and the flare on the AN-type fitting.

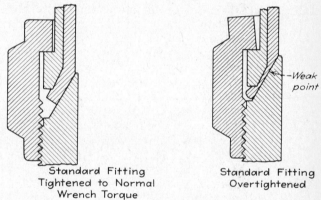

Standard Fitting
Tightened to Normal
Wrench Torque

Weak point

Standard Fitting
Overtightened

FIG. 12-45 Fittings properly and improperly tightened.

To obtain correct torque values when tubing sections are installed and the couplings tightened, it is desirable to use a torque wrench. Special wrenches for fittings are available, and some of these are constructed with snap releases so that they automatically release when the proper torque is attained.

Installation of Hose Fittings. There are a large number of different types and designs of reusable hose fittings for various levels of fluid pressure. The basic principle of the reusable fitting is to clamp the hose between the socket and a nipple with sufficient force to prevent a separation at the maximum pressure for which the hose is designed. In this section, we shall describe the installation procedures for two types of medium-pressure hose. The manufacturers instructions or FAA Advisory Circular 43.13-1A should be consulted for an actual installation.

A reusable fitting that may be used for either low-pressure or medium-pressure hose is shown in Fig.

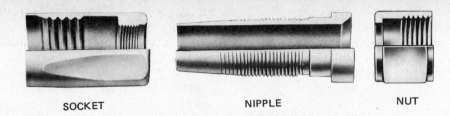

SOCKET   NIPPLE   NUT

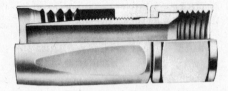

FITTING ASSEMBLED

FIG. 12-46   A reusable hose fitting. *(Aeroquip Corp.)*

12-46. This fitting consists of a nut and nipple that may be made of cadmium-plated steel or anodized aluminum alloy together with a matching socket and conforms to MS24587. The fitting is installed on hose as shown in Fig. 12-47. The procedure is as follows:

Step 1. Place the hose in a suitable holding fixture and cut squarely with a fine-tooth hacksaw or a hose cutter.

Step 2. Place the fitting socket in a vise and screw the hose into it by turning the hose counterclockwise. Turn the hose until it bottoms in the socket.

Step 3. Lubricate the nipple threads and the inside of the hose with a petroleum oil if it is to be used on a fuel, oil, or hydraulic system in which petroleum-base hydraulic fluid will be used. If the fluid in the system is to be a synthetic oil or fluid such as Sky-drol, the lubricant should be compatible.

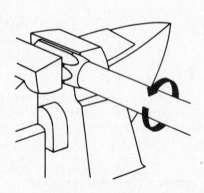

INSERTING HOSE
IN SOCKET

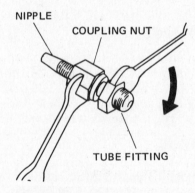

NIPPLE
COUPLING NUT
TUBE FITTING

ASSEMBLING NIPPLE
AND NUT WITH TUBE FITTING

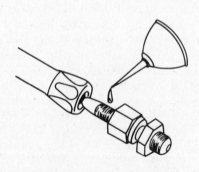

LUBRICATING NIPPLE
THREADS

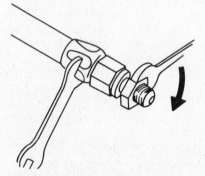

SCREWING THE NIPPLE
INTO THE HOSE AND SOCKET

FIG. 12-47   Installation of a reusable fitting on hose. *(Aeroquip Corp.)*

Step 4. Screw the nipple into the socket using a wrench on the hex of the nipple. Leave a clearance of 0.005 to 0.031 in [0.127 to 0.79 mm] between the nipple hex and the socket. This clearance is to allow freedom of movement for the coupling nut.

Another reusable hose fitting that may be used for low- and medium-pressure applications is the "Little Gem" fitting manufactured by the Aeroquip Corporation. This fitting incorporates a circumferentially machined spur on the nipple that cuts a lip in the tube stock during assembly, and this lip forms a seal that is separated from the reinforcement braid of the tubing. The braid is gripped between the nipple and socket outside the sealing lip. A cutaway view of the fitting on a hose is shown in Fig. 12-48. The procedure for the installation of this hose fitting is illustrated in Fig. 12-49 and is described as follows:

Step 1. Cut the hose squarely to length with a cut-off machine or fine-tooth hacksaw. To minimize wire-braid flare-out, wrap the hose with masking tape and saw through the tape. Remove the tape before the next step.

Step 2. Insert the hose in the socket with a twisting, pushing motion until the hose is in line with the back of the socket threads. Mark the hose position in the socket by placing a line around the hose at the edge of the socket. Use a grease pencil, painted line, or tape.

Step 3. Lubricate the inside of the hose and nipple threads liberally with the appropriate lubricant compatible with the fluid that will be used in the system. Avoid getting oil in the cutting spur of the nipple. Carefully insert the nipple and engage the nipple and socket threads while holding the hose in position with the other hand. Make sure that the

hose does not push out of the socket by observing the position line you placed on the hose.

Step 4. Complete the assembly, using a wrench while holding the socket in a vise. The maximum gap between the nipple hex and the socket is 0.041 in [1.04 mm] for sizes 3, 4, and 5 and 0.031 in [0.79 mm] for size 6 and larger. After completion of assembly, the hose should be checked for push-out by examining the position mark. Upon completion, the hose should be cleaned, inspected, and pressure-tested.

Maintenance Practices for Aircraft Hose

Hose for aircraft fluid systems requires reasonable care and an understanding of the conditions that can cause damage, deterioration, or malfunction. The primary purpose of the hose is to carry a fluid at a required pressure and flow rate to serve the functions of the system involved. The following practices are recommended for the care of hose:

1. Do not use hose assemblies as foot or hand holds.

2. Do not lay hose where it may be stepped upon or run over by a vehicle.

3. Do not lay objects on top of hose assemblies.

4. When loosening or tightening hose fittings, turn the swivel nut only. Do not turn the hexes that form part of the socket or nipple assembly. Hold the socket with a wrench to prevent it from turning.

5. Hold the fitting to which a hose assembly is to be connected to prevent it from turning. Use an end wrench of the correct size.

6. Cover open ends of hose assemblies with caps or plugs until the assemblies are to be installed.

7. Check the hose and fittings for cleanliness, inside and out, before installation.

When one is inspecting hose in aircraft systems, the principal conditions to check for are leaks, wear or damage to outer surface, broken wire strands in metal braid, corrosion of metal braid, evidence of overheating, bulges, twist in the hose alignment, damage or wear of chafe guards, damage or wear of fire sleeves, damage to end fittings, separation of plies, blisters and bulging, cracks in outer cover, and any other indication of damage or deterioration. Any appreciable defect in the condition of the hose or fittings is usually reason for replacement. A leak may be caused by a loose fitting, and this may be corrected by loosening and inspecting the fitting, then tightening it to the proper torque if there is no sign of damage to the fitting. A fitting must not be overtorqued to stop a leak. Leaks or seepage from the hose surface requires replacement of the hose assembly.

If there is more than one broken wire per plait in the covering braid or if there are more than six broken wires per lineal foot, the hose should normally be replaced.

Hose that is reinforced with carbon-steel wire braid is subject to corrosion. This is easily detected by the rust color on the surface. If the corrosion is appreciable, the hose should be replaced. Stainless-steel wire braid often turns a golden yellow to brown color when

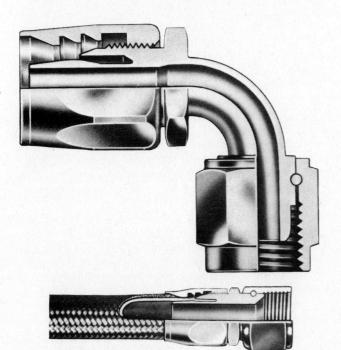

FIG. 12-48 Fitting for hose.

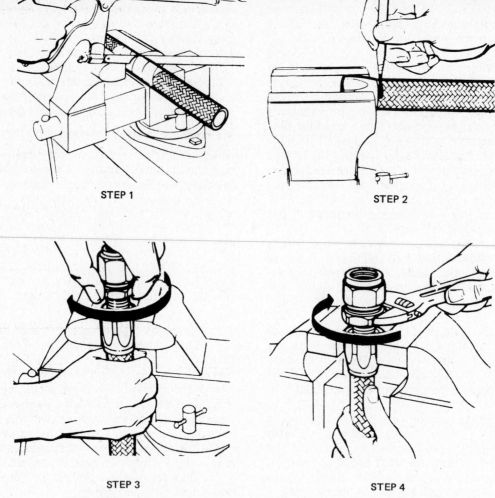

STEP 1

STEP 2

STEP 3

STEP 4

FIG. 12-49   Installation of fitting with lipseal feature.   *(Aeroquip Corp.)*

subjected to heat, and this condition should not be confused with corrosion. If the coloring is extreme, it is possible that the hose has been overheated and may require replacement.

The hose mountings in the aircraft should be inspected for condition of the clamps, bulging of the hose or other damage to the hose at the clamps, condition of the cushioning in the clamps, position of the hose and cushion in the clamp, and security of the clamp screws. The positioning of the cushion material in the clamp must be such that the material does not lodge between the end tabs of the clamp when the clamp is closed.

Hose that is twisted as indicated by the lay line along the hose can be corrected by loosening one of the fittings, straightening the hose, and retorquing the fitting.

Fire sleeves are mounted on hose to protect the hose from excessive heat and flame. If the fire sleeve is worn through, torn, cut, or oil-soaked, the hose assembly should be replaced. The removed hose assembly may be inspected and tested and, if found to be serviceable, may have a new fire sleeve installed and then be returned to service.

End fittings are checked for corrosion, cleanliness, nicks, scratches, cracks, damage to threaded areas, damage to cone-seat sealing surfaces, damage to flanges, and backed-out retaining wires on swivel nuts. The hose assembly should be replaced if any condition found could cause malfunction or deterioration.

In all inspections of hose installations, the technician should consult the applicable manufacturer's manual to assure that specified conditions are met. There are many different types and designs of fittings, and it is essential that the instructions and specifications for the particular type of fitting being inspected are understood.

Installation Practices for Aircraft Hose. Before installation of a section of aircraft hose, the hose should be thoroughly inspected as previously explained. If the hose is straight, the inside can be examined by looking through it toward a light source. If there is an elbow on one end, a flashlight or other light source can be used to illuminate the inside of the elbow, and the interior of the tube can be examined by looking in the opposite end. If it is not possible to look inside the hose, a steel ball slightly smaller than the ID of the hose should be passed through the tube. The ball should roll freely through the tube from one end to the other.

Hose that is preformed to fit certain installations should not be straightened out. Straightening would cause undue stresses, wrinkling inside the hose, and other possible defects. To prevent the straightening of preformed hose, a wire or cord can be attached to each end and pulled taut.

Some installation practices are illustrated by the drawings of Fig. 12-50. The lay line on the hose makes it easy to detect a twist. If the hose does not have a lay line, twist can be avoided by securing one end first while the other end is free. Then connect the other end, taking care to hold the socket hex with a wrench while tightening the coupling nut.

The installation of flexible hose assemblies requires that the hose be of a length that will not be subjected to tension. The hose section should be of sufficient length to provide about 5 to 8 percent slack. The hose should be installed without twisting by keeping the lay line on the hose straight. The coupling nuts for flexible hose assemblies should be torqued to the correct value as specified by the manufacturer. Bends in the hose should not have a radius less than 12 times the ID of the hose for normal installations.

In the few cases where a plain hose is used to provide a flexible joint between two sections of tubing, the ends of the tubing must be beaded. Clamps should not be overtightened because of the danger of damaging the hose. A good practice is to tighten the clamp fingertight plus one-quarter turn. It must be emphasized that plain hose and clamps must not be used where the fluid in the system is under pressure. In this case flexible hose assemblies of an approved type must be employed.

Storage of Aircraft Hose. Synthetic rubber hose and hose assemblies should be stored in a dark, cool, dry area and be protected from circulating air, sunlight, fuel, oil, water, dust, and ozone. Storage life of synthetic-rubber bulk hose normally does not exceed 5 years from the cure date stenciled on the hose, and storage life for hose assemblies does not exceed 4 years. Storage life may vary, and it is necessary to consult the manufacturer's information to determine the exact condition for any particular type of hose.

Inspection and Maintenance of Plumbing Systems

Lines and fittings should be inspected carefully at regular intervals for leaks, damage, loose mountings, cracks, scratches, dents, and other damage. Flexible lines (hose) should be checked for cracks, cuts, abrasions, soft spots, and any other indication of deterioration. Parts with defects should be either replaced or repaired. A damaged metal line should be replaced in its entirety if the damage is extensive. If the damage is localized, it is permissible to cut out the damaged section and insert a new section with approved fittings. Care must be taken that no foreign material enters the line during the repair operation. When soft aluminum tubing utilizing flare fittings is replaced, a double flare should be used on all tubing of $\frac{3}{8}$ in [0.952 cm] OD or smaller.

The following defects are not acceptable for metal lines:

1. Cracked flare.
2. Scratches or nicks greater in depth than 10 percent of the tube wall thickness or in the heel of a bend. Such cracks or nicks may be repaired by burnishing with hand tools.
3. Severe die marks, seams, or splits.
4. A dent of more than 20 percent of the tube diameter or in the heel of a bend.

When it is necessary to replace flexible hose assemblies, the replacement part should be of the same length and type as the original. Usually such hose assemblies can be ordered by part number from the manufacturer. If reusable fittings are employed on the section being removed, the fittings may be attached to a new section of approved hose of the correct length. The installation of the fittings should be performed in

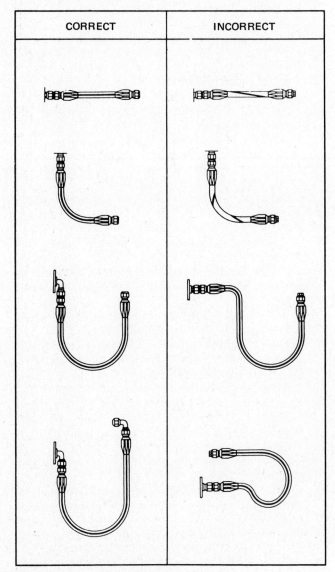

CORRECT	INCORRECT

FIG. 12-50 Correct and incorrect hose installations. *(Stratoflex Inc.)*

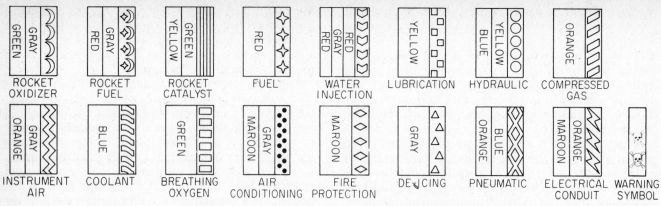

FIG. 12-51   Color-coding bands for plumbing lines.

accordance with the manufacturer's directions and with the proper tools as described previously.

Fuel lines should be of a size adequate to carry not less than double the maximum flow required for take-off power. Fuel lines should not have vertical bends or humps that may trap vapor. There should be a continuous "up" or "down" slope in such lines. The lines must be so supported and installed as to preclude the possibility that structural loads may be imposed upon them.

Where copper lines are installed, it is required that such lines be annealed from time to time to prevent hardening and cracking. This is accomplished by heating the line to a red heat and then quenching in cold water. Aluminum lines should never be annealed except under accurately controlled conditions in a heat-treating oven. Field annealing with a torch is likely to produce an uneven hardness and may also result in intergranular corrosion.

Color Codes for Plumbing Lines

Color codes for aircraft and missile plumbing lines are established by AND10375. These codings replace the plain color system used before August 1949. Because of fading of the colors and the fact that the color perception of some persons is not sufficiently acute, the plain colors were found to be subject to misinterpretation under adverse conditions. For this reason new color coding was established with black-and-white symbols.

Figure 12-51 shows the coding for various systems in an airplane. In practice the coding bands are colored as indicated in the illustration. The symbols are black against a white background. For tubing in which extreme pressures exist or tubing that carries poisonous or corrosive liquids, an additional symbol band is used. This symbol is a black skull and crossbones on a white background.

● REVIEW QUESTIONS

1. Name aircraft systems that require the use of plumbing.
2. What metals and alloys are commonly used for tubing in plumbing systems?
3. Describe the difference between AC-type tube fittings and AN-type tube fittings.
4. What range of numbers apply to AN-type tube fittings; MS-type fittings?
5. Describe a *pipe thread*.
6. What type of fittings are used where a fluid line must pass through a bulkhead?
7. Explain the use of *universal fittings*.
8. How are *pipe-thread dimensions* designated?
9. Why is a sealing compound needed with pipe threads?
10. Explain how a flareless fitting provides a seal against fluid leakage.
11. How may a flareless fitting be identified by inspection?
12. Describe a *swaged tube fitting*.
13. What are the advantages of *swaged fittings*?
14. Describe the construction of a *quick-disconnect fitting* and explain its operation.
15. Of what materials are ventilator and hot-air ducts made?
16. How are large air ducts joined?
17. At what pressure is a hose assembly pressure-tested?
18. Describe the difference in the construction of low-pressure, medium-pressure, and high-pressure hose.
19. What considerations are involved in the selection of replacement hose in a fluid system?
20. What precaution must be taken with respect to the type of fluid to be used in a fluid system?
21. How may hose be identified?
22. Why is the skirt of a hose-fitting socket flared?
23. Explain the uses for protective sleeves for aircraft hose.
24. What type of tool is best for cutting tubing?
25. Why is it poor practice to install short, straight sections of tubing between fixed parts of an aircraft?
26. What is the acceptable procedure for making a pattern for a section of bent tubing that is to be replaced?
27. What is the minimum bend radius for aluminum-alloy tubing of $\frac{1}{2}$ in OD?
28. How may the technician assure that a smooth bend of the correct radius is made in a section of tubing?

29. Describe the use of a *production tube bender.*
30. Why is the positioning of the mandrel critical in a production tube bender?
31. Explain the importance of the flare in a flared tube fitting.
32. Describe the use of a *hand flaring tool.*
33. Explain how a *double flare* is made.
34. On what tubing is double flaring required?
35. Why are *beads* used on some tubing?
36. Describe a procedure for beading a tube by hand.
37. Describe the procedure for presetting a sleeve for a flareless fitting.
38. What lubricant may be used when presetting a flareless-fitting sleeve for use in a hydraulic system?
39. If a section of prepared tubing is not to be installed immediately, what should be done to protect it from contamination?
40. What conditions should the technician check for after having installed flared fittings on a section of tubing?
41. List practices which prevent damage to aircraft hose.
42. What conditions should be checked for during the inspection of aircraft hose?
43. What is the purpose of the lay line on aircraft hose?
44. How may a twist in a section of hose be corrected?
45. If a fire sleeve is worn through, torn, cut, or oil soaked, what should be done?
46. Describe the installation of reusable hose fittings.
47. What is the recommended procedure for cutting aircraft hose?
48. When installing a reusable fitting on high-pressure hose with a synthetic rubber outer ply, how can the amount of the outer ply to be removed be determined?
49. How can the interior of a preformed hose be checked for blisters or collapse of the inner tubing?
50. Describe conditions to be attained in the installation of aircraft hose.
51. Discuss the storage of aircraft hose.
52. How should AN and MS tube fittings be lubricated during assembly?
53. Why should tube fittings not be forced into position?
54. What is the effect of overtightening flared fittings?
55. How much slack should be allowed in the installation of aircraft hose assemblies?
56. What defects are not acceptable in metal tubing?
57. What is the tubing color code for hydraulic, fuel, and oxygen lines?

FEDERAL AVIATION REGULATIONS AND PUBLICATIONS

● INTRODUCTION

When the Federal Aviation Administration (FAA) is mentioned, most people think in terms of its responsibilities in the operation and maintenance of the world's largest and most advanced air traffic control system. Almost half of the agency's work force of approximately 50 000 people is engaged in some phase of air traffic control. No air traffic control system, no matter how automated, can function safely and efficiently unless the people and machines who use the system measure up to certain prescribed standards. The FAA has been charged with the responsibility for establishing and enforcing standards relevant to the training and testing of air personnel and the manufacture and continued airworthiness of aircraft.

There are more than 200 000 civil aircraft in the United States, and the FAA requires that each be certificated as airworthy by the agency. Both the original design and each subsequent aircraft constructed from the design must be approved by FAA inspectors. Even home-built aircraft require FAA certification. All civil aircraft certificated for operation by the Federal Aviation Administration must be maintained in accordance with the requirements of Federal Aviation Regulations (FARs) issued by the FAA. FARs affect aircraft design, operation, maintenance, repair, and alteration. In this chapter, FARs of particular interest to aviation maintenance technicians and publications issued by the FAA that involve aviation maintenance and maintenance personnel will be discussed.

● HISTORY OF THE FEDERAL AVIATION ADMINISTRATION

Twenty-three years after the Wright brothers' flight at Kitty Hawk on December 17, 1903, Congress recognized the potential of a new industry—air transportation—and passed legislation to bring it within federal control. Now an operating arm of the **Department of Transportation,** the **Federal Aviation Administration** traces its ancestry back to the Air Commerce Act of 1926, which led to the establishment of the Aeronautics Branch (later reorganized as the Bureau of Air Commerce) in the Department of Commerce with authority to certificate pilots and aircraft, develop air navigation facilities, promote flying safety, and issue flight information.

Since this date, pilots and aircraft have had to be certificated by the federal government as qualified to fly. In a few short years, radio aids, communications, and ground lights were improved and expanded; airport construction was encouraged; and the first Civil Air Regulations were written.

The government acted just in time. In May 1927, Charles Lindbergh bridged the North Atlantic in $33\frac{1}{2}$ h, generating new interest and enthusiasm for aviation in both Europe and America.

Aviation continued to grow and expand at a very rapid rate in the decade following Lindbergh's flight, creating a need for new machinery to regulate civil flying. The result was the Civil Aeronautics Act of 1938, which established the independent Civil Aeronautics Authority with responsibilities in both safety and economic areas. In 1940, the machinery was readjusted and the powers previously vested in the Civil Aeronautics Authority were assigned to a new **Civil Aeronautics Administration (CAA),** which was placed under an assistant secretary in the Department of Commerce, and a semi-independent Civil Aeronautics Board (CAB), which had administrative ties with the Department of Commerce but reported directly to the Congress, was created to regulate airfares and routes.

The CAA performed very well during World War II but proved unequal to the task of managing the airways in the years after the war because of the tremendous surge of civil air traffic and the introduction of new high-performance aircraft. In 1958, the same year American jets entered commercial service, Congress passed the **Federal Aviation Act,** which created the independent Federal Aviation Agency with broad new authority to regulate civil aviation and provide for the safe efficient utilization of the nation's airspace.

In April 1967, the Federal Aviation Agency became the Federal Aviation Administration and was incorporated into the new Department of Transportation which had been established to give unity and direction to a coordinated national transportation system. The FAA's basic responsibilities remain unchanged, however. While working with other administrations in the Department of Transportation in long-range transportation planning, the FAA continues to concern itself primarily with the promotion and regulation of civil aviation to insure safe and orderly growth.

● AVIATION SAFETY REGULATION

The **Federal Aviation Act of 1958** forms the legal basis for the present system of Federal Aviation Regulations. The provisions of this act are contained in 15 sections referred to as "titles."

Title VI Safety Regulation of Civil Aeronautics is

the section that is of most direct interest to pilots and mechanics from an operational point of view. The first section of Title VI details the duties of the secretary of transportation and thereby the administrator in promoting safety:

> Sec. 601 (a) The Secretary of Transportation is empowered and it shall be his duty to promote safety of flight of civil aircraft in air commerce by prescribing and revising from time to time:
>
> 1. Such minimum standards governing the design, materials, workmanship, construction, and performance of aircraft, engines, and propellers as may be required in the interest of safety.
> 2. Such minimum standards governing appliances as may be required in the interest of safety.
> 3. Reasonable rules and regulations and minimum standards governing, in the interest of safety; (A) the inspection, servicing, and overhaul of aircraft, aircraft engines, propellers, and appliances; (B) the equipment and facilities for such inspection, servicing, and overhaul; and (C) in the discretion of the Secretary of Transportation, the periods for, and the manner in which such inspection, servicing, and overhaul shall be made, including provision for examinations and reports by properly qualified private persons whose examinations or reports the Secretary of Transportation may accept in lieu of those made by its officers and employees.
> 4. Reasonable rules and regulations governing the reserve supply of aircraft, aircraft engines, propellers, appliances, and aircraft fuel and oil required in the interest of safety, including the reserve supply of aircraft fuel and oil which shall be carried in flight.
> 5. Such reasonable rules and regulations, or minimum standards, governing other practices, methods, and procedure as the Secretary of Transportation may find necessary to provide adequately for national security and safety in air commerce."

Most safety standards and regulations are carried out by setting minimum standards and issuing certification that the standards have been met. Six basic certificates have been created in order to enforce and maintain these standards:

1. Airman certificates—pilots, mechanics, air traffic control, repairman, etc.

2. Aircraft certificates—type, production, airworthiness

3. Air carrier operating certificates

4. Air navigation facility rating

5. Air agency rating—schools, repair stations, etc.

6. Airport operating certificates

● ORGANIZATION OF THE FAA

The FAA Act provides for an **administrator, a deputy administrator, and necessary personnel** to carry out the duties and responsibilities as stated in the law.

A large number of personnel and a complex organizational structure are presently involved in regulating aviation. An aircraft technician would be most directly involved with the **Office of Airworthiness (AWS).** The Office of Airworthiness includes the Aircraft Engineering Division, the Aircraft Manufacturing Division, and the Aircraft Maintenance Division. The Aircraft Maintenance Division has a General Aviation Branch and an Air Carrier Branch. The Office of Airworthiness develops rules, standards, and policies to be carried out in the various geographic regions.

Regional Offices

The **FAA regional offices** serve as an extension of the national headquarters and solve the day-to-day problems that arise in the various geographic regions. The regional offices plan the functions that will occur in the region, such as compiling statistics, providing navigational aids and regional personnel, and the administering of examinations and inspections. They are responsible for the standardization of maintenance and engineering practices of the airlines located within the region.[1] There are nine domestic regions as shown in Fig. 13-1 (a list of FAA regional offices is provided in the appendices).

FIG. 13-1 FAA regional offices.

District Offices

The technician's primary contact with the FAA will be with the airworthiness inspectors assigned to the local district offices. There are two types of district offices, the **General Aviation District Office (GADO),** and the **Flight Standards District Office (FSDO).** The FSDO has both an air carrier and a general aviation section while the GADO deals only with general aviation. (See the appendices for a list of district offices.)

---

[1] James R. Rardon, *Understanding the Federal Air Regulations,* Aviation Maintenance Publishers, Inc., 1980, pp. 4–5.

Aeronautical Center

The FAA Mike Monroney Aeronautical Center in Oklahoma City consists of a number of FAA departments. The Airmen Certification Branch has the records of all those that have received airman certificates. One of the largest centers for aviation training is located here, as is the department that develops the various FAA tests. FAA aircraft are maintained and repaired at this installation.

Accident Investigation

The FAA participates with the **National Transportation Safety Board (NTSB)** in the investigation of major aircraft accidents to determine if any immediate action is needed to correct deficiencies and prevent a recurrence. In addition, the agency investigates most nonfatal and many fatal general aviation accidents on behalf of NTSB, although the responsibility of determining probable cause remains with the NTSB. The FAA also investigates accidents to see if any safety rules have been violated.

● FEDERAL AVIATION REGULATIONS

The **Federal Aviation Regulations (FARs)** are published as Chapter 14 of the United States Code of Federal Regulations. The various regulations are the minimum standards which have been set to insure, as much as possible, air safety. The regulations are divided into parts relating to a specialized area such as: FAR Part 25-Airworthiness Standards, Transport Category Aircraft, or FAR Part 147-Aviation Maintenance Technician Schools.

The FAA is continually researching ways of improving and updating a regulation. Once a proposal has been developed for a new regulation or a revision, the public is notified by a Notice of Proposed Rule Making (NPRM). The NPRM states the proposed changes and the logic behind them. The public is invited to comment on the proposal. After public opinion has been gathered, the proposal is either passed, modified, or dropped. FAR Part 11 contains the procedures for regulatory changes.

Of the many parts of the FARs, certain ones are of particular interest to aviation maintenance technicians, repair stations, and others involved in airframe and power plant maintenance and certification. The following parts should be reviewed by the technician:

Part 1. Definitions and Abbreviations
Part 13. Enforcement Procedures
Part 21. Certification Procedures for Products and Parts
Part 23. Airworthiness Standards: Normal-, Utility-, and Acrobatic-Category Airplanes
Part 25. Airworthiness Standards: Transport-Category Airplanes
Part 27. Airworthiness Standards: Normal-Category Rotorcraft
Part 29. Airworthiness Standards: Transport-Category Rotorcraft
Part 33. Airworthiness Standards: Aircraft Engines
Part 35. Airworthiness Standards: Propellers
Part 37. Technical Standard Order Authorizations
Part 39. Airworthiness Directives
Part 43. Maintenance, Preventive Maintenance, Rebuilding, and Alteration
Part 45. Identification and Registration Marking
Part 47. Aircraft Registration
Part 49. Recording of Aircraft Titles and Security Documents
Part 65. Certification: Airmen Other Than Flight Crew Members
Part 91. General Operating and Flight Rules
Part 145. Repair Stations
Part 147. Aviation Maintenance Technician Schools

The principal objective of the FAA is to make flying safe. To accomplish this, FARs are passed to establish standards of aircraft design, performance, and dependability and also to set forth standards of performance for all personnel involved with the operation, flight, and maintenance of aircraft. The FAR parts in the foregoing list are all of interest to the aviation maintenance technician and should become a part of the technician's general knowledge. Parts 23, 37, 39, 43, 65, 91, and 145 deal most specifically with aircraft maintenance and maintenance personnel requirements.

Part 23 of the FAR sets forth standards and requirements for general-aviation aircraft. Aircraft are classified into three categories: normal, utility, and acrobatic categories.

A normal-category aircraft is limited to nonacrobatic operation. It is permitted to execute any maneuver required for normal flight including stalls (except whip stalls), lazy eights, chandelles, and steep turns in which the bank is not more than 60°.

The utility-category aircraft may perform limited acrobatics. In addition to the maneuvers allowed for normal-category aircraft, it may perform spins if it is approved for this maneuver, and it may make turns in which the angle of bank is more than 60°.

Acrobatic aircraft are intended for use without restrictions except for those shown to be necessary as a result of required flight tests.

Part 37 sets forth the requirements for the issuance of Technical Standard Order Authorizations. Technical Standard Orders (TSOs) contain minimum performance and quality-control standards for specified materials, parts, or appliances used on civil aircraft. TSOs have been issued for such items as emergency locator transmitters, anticollision lights, tires, instruments, and seat belts. The performance standards in each TSO are those that the FAA finds necessary to ensure that the article concerned will operate satisfactorily or will accomplish satisfactorily its intended purpose under specified conditions.

Part 39 relates to Airworthiness Directives. These will be discussed later in this chapter.

Part 43 establishes the standards and procedures for the maintenance, preventive maintenance, repair, and alteration on aircraft with a U.S. airworthiness certificate with the exception of aircraft issued an experimental airworthiness certificate. Part 43 establishes who is authorized to perform maintenance, performance rules to be observed, and procedures to be followed in approving the aircraft for return to service.

Appendix A of Part 43 defines major alterations, major repairs, and preventive maintenance.

Appendix B of Part 43 describes the requirements for recording major repairs and alterations using Form 337 or a repair station work order.

Appendix D provides for the scope and detail of 100-h annual inspections. An inspection checklist must include the applicable items listed in Appendix D.

Advisory Circular 43.13-1A was produced to help implement Part 43. It provides detailed descriptions of acceptable procedures for many types of aircraft repairs.

Part 65, Subpart D, establishes the qualifications, experience requirements, and privileges of certificated airframe and power plant mechanics (maintenance technicians). It explains specifically who may approve aircraft for return to service after various types of maintenance, repair, alteration, or inspection. For example, a certificated airframe and power plant mechanic cannot return an aircraft to service after a major repair or alteration unless that mechanic holds an inspection authorization (IA) rating. Part 65 also prescribes similar requirements for air-traffic-control operators, aircraft dispatchers, repair technicians, and parachute riggers.

Part 91 sets forth rules governing the operation of aircraft in the United States. FAR Part 91 is usually thought of as being only of interest to pilots, but Subpart C of Part 91 specifies maintenance requirements for aircraft operated in the United States. It requires that certain inspections be performed at specified intervals to maintain an aircraft in airworthy status. Subpart C also details what aircraft records are required and the time periods that they must be kept.

Part 145 of the FAR establishes the requirements for certificated repair stations. The purpose of Part 145 is to assure that the repair station is adequately equipped, staffed, and organized to perform the types of repairs for which it is certificated. Repair stations are discussed in more detail in Chap. 16 of this text.

Obtaining Federal Aviation Regulations

Advisory Circular AC00-44 sets forth the current publication status of the Federal Aviation Regulations and provides a price list and ordering instructions. Federal Aviation Regulations are sold by the Superintendent of Documents, U.S. Government Printing Office, Washington, D.C. 20402.

FARs having frequent changes are sold on a subscription basis with the revisions sent automatically to the subscribers. Parts which are changed less frequently are sold on a single-sale basis and changes to these parts are sold separately.

● ADVISORY CIRCULARS

The FAA issues **Advisory Circulars (AC)** (see Fig. 13-2) to inform the aviation community, in a systematic way, about material of interest. Unless the Advisory Circular is incorporated into an FAR, its contents are not binding on the public. An Advisory Circular is issued to provide guidance and information in its designated subject area or to show a method acceptable to the FAA for complying with a related FAR.

Advisory Circulars are issued in a numbered-subject system corresponding to the subject areas of the FARs. The Advisory Circular numbers relate to the FAR parts and, when appropriate, to the specific sections of the Federal Aviation Regulations. For example, AC43.13-1A, Acceptable Methods, Techniques, and Practices—Aircraft Inspection and Repair, and AC43.13-2A, Acceptable Methods, Techniques, and Practices—Aircraft Alterations, correspond with FAR

AC NO: 00-34A

DATE: 7/29/74

ADVISORY CIRCULAR

DEPARTMENT OF TRANSPORTATION
FEDERAL AVIATION ADMINISTRATION

SUBJECT: AIRCRAFT GROUND HANDLING AND SERVICING

FIG. 13-2 Advisory Circular.

43.13 which lists the performance rules for repair and alteration of aircraft.

An Advisory Circular Checklist (AC00-2) is issued that lists all of the Advisory Circulars published and gives directions for obtaining them. Most Advisory Circulars are free and may be obtained from the Superintendent of Documents as explained previously.

● AIRWORTHINESS ALERTS

The primary function of the FAA is to ensure that aviation is a safe and dependable form of transportation. One of the means for promoting safety is the **Service Difficulty Program** which is an information system designed to provide assistance to aircraft owners, operators, maintenance organizations, and manufacturers in identifying aircraft problems encountered during service. The program provides for the collection, organization, analysis, and dissemination of aircraft service information so as to improve the service reliability of aviation products.

General aviation service difficulties are generally reported to the FAA by maintenance personnel through the use of a Malfunction or Defect Report. Submission of service difficulty information by the aviation public is voluntary, although certificated repair stations and air taxi and commercial operators are required by the FARs to submit certain specific infor-

mation. Reports of defects should be mailed to the local FAA district office where they are reviewed and then forwarded to the Safety Data Branch in Oklahoma City for processing. The defect reports are retained in a computer bank for a period of 5 years providing a base for detection of trends and failure rates. Analysis of service difficulty information is done primarily by the Safety Data Branch. When trends are detected, they are made available to maintenance personnel through AC43-16, **General Aviation Airworthiness Alerts** (see Fig. 13-3). The intent of the Alerts is to familiarize technicians with problems encountered in the field so that they may pay special attention to these areas when inspecting aircraft of a similar type. Compliance with the recommendations made in Airworthiness Alerts is voluntary.

Airworthiness Alerts are published monthly and are distributed free of charge to all authorized inspectors, repair stations, air taxis, and FAA certified technician schools. Alerts may also be purchased as AC43-16 from the Superintendent of Documents.

● AIRWORTHINESS DIRECTIVES

A primary safety function of the Federal Aviation Administration is to require correction of unsafe conditions found in an aircraft, when such a condition is likely to exist or develop in other aircraft of the same

U.S. Department
of Transportation

**Federal Aviation
Administration**

**General Aviation
Airworthiness
Alerts**

AC No. 43-16

Alert No. 60
July 1983

**Improve Reliability-
Interchange Service
Experience**

FIG. 13-3   Airworthiness Alerts.

design. The unsafe condition may exist because of a design defect, maintenance, or other causes. **Airworthiness Directives (ADs)** are used by the FAA to notify aircraft owners of these unsafe conditions and to require their correction. ADs prescribe the conditions and limitations, including inspections, repair, or alteration, under which the product may continue to be operated.

AD Categories

ADs are Federal Aviation Regulations and are published in the *Federal Register* as amendments to FAR Part 39. Depending on the urgency, ADs are published as follows:

Notice of Proposed Rulemaking (NPRM). An NPRM is issued and published in the *Federal Register* when an unsafe condition is discovered in a product. Interested persons are invited to comment on the NPRM by submitting such written data, views, or arguments as they may desire. The comment period is usually 60 days. Proposals contained in the notice may be changed or withdrawn in light of comments received. When an NPRM is adopted as a final rule, it is published in the *Federal Register*, printed, and distributed by first class mail to the registered owners of the product affected.

Immediate Adopted Rule. ADs of an urgent nature are adopted without prior notice (without first being an NPRM) as "immediately adopted rules." These ADs usually become effective less than 30 days after publication in the *Federal Register* and are distributed by first class mail to the registered owners of the product affected.

Emergency AD. These ADs are issued when immediate corrective action is required. Emergency ADs are distributed to the registered owners of the product affected by telegram or priority mail and are effective upon receipt. Emergency ADs are published in the *Federal Register* as soon as possible after initial distribution.

ADs Issued to Other than Aircraft. ADs may be issued which apply to engines, propellers, or appliances installed on multiple makes or models of aircraft. When the product can be identified as being installed on a specific make or model aircraft, AD distribution is made to the registered owners of those aircraft. However, there are times when a determination cannot be made, and direct distribution to the registered owner is impossible.

Distribution and Publication

As described in the preceding paragraphs, ADs are mailed to registered owners and published in the *Federal Register*. Most of the individuals involved in aircraft maintenance receive their notice of airworthiness directives through a subscription service to the FAA.

Airworthiness Directives are published in two separate volumes:

Volume I relates to small aircraft of 12 500 lb or less maximum certificated takeoff weight.

Volume II relates to large aircraft of more than 12 500 lb maximum certificated takeoff weight.

ADs applicable to engines, propellers, and other appliances are contained in the respective small or large aircraft volume. Each volume consists of two books:

Book 1 contains ADs issued prior to 1971. These ADs remain in effect until otherwise suspended, revoked, or amended. If an AD from Book 1 is amended, it will appear as a supplement to Book 2. Since Book 1 is not revised, the book is bound and its pages are numbered.

Book 2 contains ADs issued during and after 1971. Book 2 is loose leaf to incorporate biweekly revisions.

Each book is divided into four major subjects: aircraft, engines, propellers, and appliances. Within each subject, ADs are listed alphabetically by the manufacturer's or Type Certificate (TC) holder's name.

AD Indexes

The summary of AD notes contains two indexes. One is alphabetical by the manufacturer of the product and the second is a numerical listing of all AD notes from the oldest to the most current.

The alphabetical index divides the ADs into the given subject areas (aircraft, engines, propellers, and appliances). Each model of the manufacturer's product is listed with the applicable AD notes for that model.

A sample of the alphabetical index is shown in Fig. 13-4.

The numerical index provides the name of the manufacturer and, in the event that the AD has been superseded, the new AD to which reference should be made (see Fig. 13-5).

The indexes are updated every 6 months during the 2-year period that the summary is effective. The FAA intends these indexes to be used as a guide only. They should not be relied upon as conclusive evidence of AD applicability.

New and revised ADs are compiled by the FAA every 2 weeks and mailed out in what is called the "Biweekly Listing." AD summaries are published every 2 years. The purchase of this AD summary includes a subscription to the biweekly revisions until the ADs are republished. Because of their importance, Airworthiness Directives may be purchased directly from the FAA in Oklahoma City.

AD Numbers

The AD notes are amendments of FAR Part 39 and therefore carry an amendment number such as 39-3184. This number changes when an AD note is revised, at which time a new amendment number is issued. The AD is normally identified in maintenance records and for other purposes by a six-digit number such as 86-08-03. The 86 identifies the year of issue

Aircraft make and model	TC no.	AD number		Subject	Page no.	Book no.
		48-05-04		Operator limitations placard ..	8	1
		48-07-01		Stabilizer attaching bolts......	8	1
		48-25-02		Welded exhaust muffler	8	1
		48-25-03		Wing drag wire system	8	1
		50-31-01		Fin spar reinforcement	9	1
		51-21-01		Rudder rib flanges	11	1
		61-25-01		Met-Co-Aire landing gear	14	1
		62-24-03		Cabin heat system	15	1
		79-08-03		Electrical system	71	2
140A	5A2	61-25-01		Met-Co-Aire landing gear	14	1
		62-24-03		Cabin heat system	15	1
		79-10-14		Fuel tank venting	73	2
150 Series	3A19	62-22-01		Vacuum pump modification ..	15	1
		67-03-01		Exhaust gas heat exchangers...	17	1
		68-17-04		Stall warning system	19	1
		71-22-02		Cracks in nose gear fork	3	2
		79-08-03		Electrical system	71	2
		79-10-14		Fuel tank venting	73	2
150	3A19	73-23-07		Defective spar attach fittings ..	16	2
		74-06-02		AVCON mufflers.............	18	2
		75-15-08		Engine lubrication	23	2
		77-02-09		Wing flap system	49	2
150A	3A19	75-15-08		Engine lubrication	23	2
150B	3A19	75-15-08		Engine lubrication	23	2
150C	3A19	75-15-08		Engine lubrication	23	2
150D	3A19	75-15-08		Engine lubrication	23	2
		83-17-06		Aileron balance weights	106	2
150E	3A19	75-15-08		Engine lubrication	23	2
		83-17-06		Aileron balance weights	106	2
150F	3A19	72-03-03	R3	Wing flap jack screw	6	2
		75-15-08		Engine lubrication	23	2
		80-11-04		Cracked nutplates	90	2
		83-17-06		Aileron balance weights	106	2
150G	3A19	67-31-04		Removal of glove compartment	18	1
		72-03-03	R3	Wing flap jack screw	6	2
		75-15-08		Engine lubrication	23	2
		80-11-04		Cracked nutplates	90	2
		83-17-06		Aileron balance weights	106	2
150H	3A19	67-31-04		Removal of glove compartment	18	1
		72-03-03	R3	Wing flap jack screw	6	2
		75-15-08		Engine lubrication	23	2
		80-11-04		Cracked nutplates	90	2
		83-17-06		Aileron balance weights	106	2
150J	3A19	72-03-03	R3	Wing flap jack screw	6	2
		75-15-08		Engine lubrication	23	2
		80-11-04		Cracked nutplates	90	2
		83-17-06		Aileron balance weights	106	2

FIG. 13-4 Airworthiness Directive Alphabetical Index.

(1986), the 08 identifies the period as being in the eighth bi-weekly period, which is the fifteenth or sixteenth week. The 03 is a sequential number and simply means that it was the third AD adopted in that period. A revised AD note will usually retain its six-digit number; however, an R will be added after the six-digit number. Number 86-08-03 R_2 would indicate that this AD has been revised twice.

AD Content

The content of AD notes can be divided into several categories such as applicability, compliance time and date, effective dates, and required actions. (Figure 13-6 is a sample AD.)

Applicability of ADs. Each AD contains an applicability statement specifying the product (aircraft, air-

AD no.	Manufacturer	Page no.
82-08-06	Piper . . . . Superseded by 82-27-13	
82-09-01	Teledyne	31
82-09-03	Beech	107
82-09-52	Bell . . . . Superseded by 82-09-53	
82-09-53	Bell	58
82-10-01	Aerospatiale	23
82-10-05	AiResearch, Garrett Superseded by 82-27-07	
82-10-06	Hiller Superseded by 82-15-02	
82-11-01	Robinson	4
82-11-04	Piper	134
82-11-05	Bendix, Equipment	13
82-12-04	Detroit Diesel Allison, Equipment	1
82-12-06	Government Aircraft Factories	7
82-12-07	Aerospatiale	23
82-13-01	Bendix, Equipment	14
82-13-02	Piccard . . . . Superseded by 83-15-03	
82-13-03	Allison	8
82-13-04	Hiller	18
82-13-05	Aerospatiale	24
82-13-06	Schweizer	8
82-14-01	Hughes	33
82-15-01	Hughes	33
82-15-02	Hiller	19
82-15-03	Hughes	34
82-15-06	Empresa Brasileira Superseded by 82-20-02	
82-15-07	Robinson	5
82-16-05	Piper	BW
82-16-06	Bell	58
82-16-07	Hiller	20
82-16-08	Helic	1
82-16-09	Enstrom	11
82-16-11	Government Aircraft Factories	8
82-16-12	Bell	59
82-17-01	Hughes	34
82-17-02	Rolladen Schneider	1
82-17-03	Sikorsky	17

FIG. 13-5 Airworthiness Numerical Index.

81-04-07 R1 *PIPER:* Amendment 39-4044 as amended by Amendment 39-4272. Applies to Model PA-38-112, Serial Nos. 38-78A0001 thru 38-80A0198, certificated in all categories.

To avoid possible hazards in flight associated with cracks in the fin forward spar, P/N 77601-03, and the fuselage bulkhead assembly, P/N 77553-02, accomplish the following:

a. Within the next 25 hours in service from the effective date of this AD or upon the attainment of 300 hours in service, whichever is later, unless accomplished within the previous 75 hours in service and thereafter, at intervals not to exceed 100 hours in service from the last inspection, accomplish the following:

(1) Inspect the forward surface of the fin forward spar web in the area of the vertical edges of the fin forward spar attachment fitting P/N 77553-05 for cracks using a dye penetrant method or equivalent. Remove two forward fin attachment bolts and displace fin spar $\frac{1}{8}''$ laterally in each direction to increase visibility of spar area adjacent to edge of attachment fitting. Remove any scuff marks on spar by sanding prior to applying dye penetrant.

(2) Inspect the fuselage bulkhead assembly P/N 77553-02, at fuselage station 221.42, in the area of the fin forward spar attachment fitting, P/N 77553-05, for cracks using a dye penetrant method or equivalent. Access to aft side of bulkhead may be obtained by removing rudder and adjacent access door and to front side by removing the luggage compartment rear partition. When using luggage compartment, provide a stand to support the aft fuselage and, in order to assure that no associated damage will occur during the inspection, provide a support board for the mechanic.

b. Replace fin spars with cracks exceeding one-half inch and bulkheads with cracks exceeding three-quarter inch prior to further flight with undamaged parts of the same part number, an equivalent part, or accomplish a repair which must be approved by the Chief, Engineering and Manufacturing Branch, FAA, Eastern Region. Parts which have cracks less than these values must be replaced or repaired within 25 hours in service. A ferry flight may be authorized under FAR 21.197 to a place of repair with prior approval of the Chief, Engineering and Manufacturing Branch, FAA, Eastern Region.

c. Inspect replacement and repaired parts in accordance with the AD unless otherwise authorized by the Chief, Engineering and Manufacturing Branch, FAA, Eastern Region.

d. Equivalent inspections and parts must be approved by the Chief, Engineering and Manufacturing Branch, FAA, Eastern Region.

e. Upon submission of substantiating data by an owner or operator through an FAA Maintenance Inspector, the Chief, Engineering and Manufacturing Branch, FAA, Eastern Region may adjust the compliance times specified in this AD.

Amendment 39-4044 was effective February 19, 1981.

This amendment 39-4272 is effective December 9, 1981.

FIG. 13-6 Sample Airworthiness Directive.

craft engine, propeller, or appliance) to which it applies. The applicability statement is usually identified in the first part of the AD. Applicability is given by product model and, as applicable, serial number and category of certification. In some ADs with multiple parts the applicability statements may be made at different places, depending upon the required action. It is very important for the mechanic to research carefully the applicability statements to see if part or all of the AD note applies.

Compliance Time or Date. Compliance requirements specified in ADs are established for safety reasons and may be stated in numerous ways. Some ADs are of such a serious nature they require compliance before further flight. In some instances these ADs authorize flight, providing a ferry permit is obtained, but without such authorization in the ADs, further flight is prohibited. Other ADs express compliance time in terms of a specific number of hours of operation, for example, "compliance required within the next 50-hours time in service after the effective date of this AD." Compliance times may also be expressed in operational terms such as "within the next 10 landings after the effective date of this AD."

For turbine engines, compliance times are often expressed in terms of cycles. A cycle normally consists of an engine start, takeoff operation, landing, and engine shutdown. When a direct relationship between airworthiness and calendar time is identified, compliance time may be expressed as a calendar date. Another aspect of compliance times to be emphasized is that not all ADs have a one-time compliance. Repetitive inspections at specified intervals after initial compliance may be required. Repetitive inspections are used in lieu of a permanent fix because of costs or until a permanent fix is developed.

Effective Dates. The effective date of the AD is usually published toward the end of the note. In the case of a revision, the effective date of the revision, as well as the date of the original amendment, is given. Care should be taken to ensure that when a revision becomes effective, it does not require additional action on a previously received AD note.

Required Action. The required action may take the form of inspection, replacement of parts, modification of design, or changes in operating procedures. The required action could be totally described in the AD note, or reference may be made to the manufacturer's service instructions which will give the details of the required action. Notification to the FAA upon compliance or other special record entries in the aircraft maintenance records may be required. The action may call for recurring inspections or other action until certain parts are replaced or modified.

Compliance Responsibility

Responsibility for AD compliance always lies with the registered owner or operator of the aircraft. This responsibility is usually met by having appropriately rated maintenance technicians accomplish the maintenance required by the AD.

Maintenance technicians may also have direct responsibility for AD compliance, aside from the times when AD compliance is the specific work contracted for by the owner or operator. When a 100-h annual or progressive inspection or an inspection required under Part 123 or 125 is accomplished, FAR Section 43.15(a) requires the person performing the inspection to perform it so as to determine that *all* applicable airworthiness requirements are met, which includes compliance with ADs.

Maintenance technicians should be aware that even though an inspection of the complete aircraft is not made, if the inspection conducted is a progressive inspection or an inspection required by Part 123 or 125, determination of AD compliance for those portions of the aircraft inspected is required.

In conducting an AD search, the technician must first obtain the make, model, and serial number of the aircraft, engine, propeller, and appliances. The four-subject indexes may be used as an aid in performing the AD search. In referring to Fig. 13-4 it should be noted that there are 11 possible ADs that could apply to a Cessna 150G Airframe. The 11 ADs include six listed under 150 Series and five listed under 150G. After noting the possible ADs from the index, it is then necessary to refer to the specific AD to check model and serial number applicability.

When performing an AD search, it is also necessary to check the biweekly listing since these are not included in the index.

AD Record Entries

FAR Part 91.173 requires that the aircraft records contain the current status of applicable airworthiness directives including, for each, the method of compliance, the AD number, and the revision date. If the AD involves recurring action, the time and date when the next action is due is required. An Airworthiness Directive Summary such as shown in Fig. 13-7 is recommended as a record of AD compliance.

● TYPE CERTIFICATE DATA SHEETS

The FAA prescribes minimum standards governing the design, materials, workmanship, construction, and performance of aircraft, aircraft engines, and propellers. When an aircraft or component meets or exceeds these requirements, it is issued a **Type Certificate.** The Type Certificate consists of the type design, the operating limitations, the **Type Certificate Data Sheet,** and the applicable regulations with which the FAA requires compliance.

Type Certificate Data Sheets (TCDSs) (see Fig. 13-8) were originated and first published in January 1958. FAR 21.41 indicates they are part of the Type Certificate. A Type Certificate Data Sheet is evidence that the product has been type certificated. Data sheets contain information on the design specifications of the aircraft. In addition, they show the applicable regulations under which the aircraft was certificated. The technician makes extensive use of the TCDS during an inspection to see that the aircraft does conform to its certificated design.

Make _____ Model _____ S/N _____ N _____

Date of Manufacture _____ Tach Time _____ Total Time _____

Date _____ ADs OK'd thru _____ Name and A&P No. _____

AD number	AD rev. date	Subject	Compliance		Method of compliance	One-time	Reoccuring	Next compliance due date/time	Authorized signature and number
			Date	Total time					
74-20-05		EXHAUST MUFFLER	3-2-82	1471.0	BY INSP PER PAR. (A)(1)		x	1571.0	A&P 2033266
79-12-02		CONTROL CABLES	3-2-82	1471.0	BY VISUAL INSP PER PAR (B)		x	1671.0	A&P 2033266
81-14-07		LIFT STRUT FORKS	4-6-81	1380.2	BY REPLACEMENT PER S.B. 121	x			A&P 2033266
82-05-01R₁	08-02-81	INSPECT MAIN SPAR	—	—	N/A BY SERIAL NUMBER			—	—

FIG. 13-7 Airworthiness Directive Summary.

VOLUME I

SINGLE-ENGINE AIRPLANES

TYPE CERTIFICATE DATA SHEETS AND SPECIFICATIONS

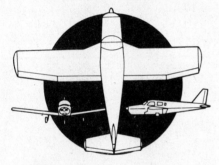

DEPARTMENT OF TRANSPORTATION
FEDERAL AVIATION ADMINISTRATION

FIG. 13-8 Type Certificate Data Sheets.

Aircraft type-certificated under the old civil air regulations have **Aircraft Specifications** rather than Type Certificate Data Sheets. Specifications originated during implementation of the Air Commerce Act of 1926. Specifications are FAA record-keeping documents issued for both type-certificated and non-type-certificated products which have been found eligible for U.S. airworthiness certifications. Specifications are basically the same as TCDSs with one major difference. Aircraft Specifications have the required and optional equipment listed for each model of aircraft. To find this information on aircraft with a Type Certificate Data Sheet would require the equipment list for that particular aircraft. Although they are no longer issued, specifications remain in effect and will be further amended. Specifications covering type-certificated

products may be converted to Type Certificate Data Sheets at the option of the type certificate holder. However, to do so the type certificate holder must provide an equipment list.

The Aircraft, Aircraft Engine, and Propeller Type Certificate Data Sheets and Specifications are issued in six volumes, each of which may be purchased separately in paper copy form or as a complete library on microfiche. Information concerning the current prices and how to order may be obtained from: U.S. Dept. of Transportation, Publications Section, Washington, D.C. The volumes are:

Volume I, Single-Engine Airplanes. Contains documents for all single-engine, fixed-wing airplanes regardless of takeoff weight.

Volume II, Small Multiengine Airplanes. Contains data sheets for multiengine, fixed-wing airplanes of 12 500 lb or less maximum takeoff weight.

Volume III, Large Multiengine Airplanes. Contains data sheets for multiengine, fixed-wing airplanes of more than 12 500 lb maximum takeoff weight.

Volume IV, Rotorcraft, Gliders, and Balloons. Contains data sheets for all rotorcraft, gliders, and piloted balloons.

Volume V, Aircraft Engines and Propellers. Contains data sheets for engines and propellers of all types and models.

Volume VI, Aircraft Listing and Aircraft Engine and Propeller Listing. Contains information pertaining to older aircraft models of which 50 or less are shown on the FAA Aircraft Registry. It also includes engine and propeller information for which approvals have expired or for which the manufacturer no longer holds a production certificate.

The Type Certificate Data Sheets and Specifications are listed in the master alphabetical index under the same name of the current type certificate holder. The index is cross referenced to assist in locating a specific data sheet or specification that has been transferred to a person other than the original manufacturer (see Fig. 13-9). A revised master index is issued annually in January and incorporates all changes made during the preceding calendar year.

Supplements containing newly issued data sheets and revisions to existing ones are issued monthly. For a revised data sheet specification, the revision number appears directly below the data sheet or specification number. The revised material is indicated by either a bracket or a solid vertical bar along the left-hand margin.

Content

The content of the various Type Certificate Data Sheets and Specifications varies but, for the purposes of this book, will be broken down into the following subparts (see Fig. 13-10):

1. Identification
2. Data for a specific model
3. Data pertinent to all models
4. Notes

Type Certificate Data Sheets are issued for aircraft, aircraft engines, and propellers. Since the information for each of these three products is different, they will be covered separately.

Aircraft Data Sheets

Identification. The upper right-hand corner of the Type Certificate Data Sheet or Specification contains a box with the data sheet number, the latest revision number and date, and the various aircraft models cov-

Type certificate holder name, location, aircraft model no.	Controlling region*	Approval no.	Data sheet spec. listing page no.	Certification basis	Volume no.
LOCKHEED-GEORGIA COMPANY A Div. of Lockheed Aircraft Corp. Marietta, GA	CE-A				
Jetstar 1329-23A, -23D, -23E, 1329-25		2A15	2A15-10	CAR 4b	III
382, 382B, 382E, 382F, 382G		A1SO	A1SO-11	CAR 1 (CAR 9a)	III
300-50A-01 (USAF C-141A)		A2SO	A2SO-3	CAR 1 (CAR 4b)	III
282-44A-05 (C-130B)		A5SO	A5SO	FAR 21.25 (a) (2)	III
LONGREN AIRCRAFT COMPANY (See Mitchell)					
LUSCOMBE AIRCRAFT CORPORATION Atlanta, GA (See Temco)	CE-A				
(Larsen Luscombe) 8, 8A, 8B, 8C, 8D, 8E, 8F, T-8F		A-694	A-694-22	Part 4a	I
LUSCOMBE AIRPLANE CORPORATION Dallas, TX (See Luscombe Aircraft Corp., Temco)	SW				
Phamtom 1, 1S		ATC 552	Page 266 (124)	—	VI
4		TC 687	Page 267 (124)	—	VI
MACCHI (See Aeronautica Macchi & Aerfer)					
MAEL AIRCRAFT CORPORATION Portage, WI	CE-C				
(Burrs) BA-42		A6SO	A6SO-2	FAR 21 (FAR 23)	II

The page number in parenthesis in the "Data Sheet/Spec/Listing Page No." column identifies the page number of the original publication. This page number is sometimes referenced in the TCDS and on the Application for Airworthiness Certification.

*Region in which the technical data are filed. Addresses on last page of index.

FIG. 13-9 Type Certificate Master Index.

DEPARTMENT OF TRANSPORTATION
FEDERAL AVIATION ADMINISTRATION

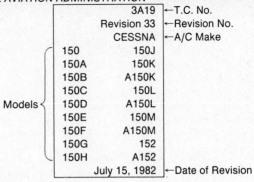

	3A19	←T.C. No.
	Revision 33	←Revision No.
	CESSNA	←A/C Make
150	150J	
150A	150K	
150B	A150K	
150C	150L	
Models 150D	A150L	
150E	150M	
150F	A150M	
150G	152	
150H	A152	
	July 15, 1982	←Date of Revision

TYPE CERTIFICATE DATA SHEET NO. 3A19

This data sheet which is part of Type Certificate No. 3A19 prescribes conditions and limitations under which the product for which the type certificate was issued meets the airworthiness requirements of the Federal Aviation Regulations.

Type Certificate Holder   Cessna Aircraft Company
Pawnee Division   (Does not have to be the manufacturer)
Wichita, Kansas 67201
Class

1 ⎰ -Model 150, 2 PCLM (Utility Category), Approved July 10, 1958

First Models ⎱ Model 150A, 2 PCLM (Utility Category), Approved June 14, 1960
Model 150B, 2 PCLM (Utility Category), Approved June 20, 1961 ←Date Approved
Model 150C, 2 PCLM (Utility Category), Approved June 15, 1962

Engine	Continental 0-200-A
*Fuel	80/87 Min. grade aviation gasoline
*Engine Limits	For all operations, 2750 rpm (100 hp.)
Propeller and Propeller Limits	1. Sensenich M69CK 24 lb. (−32) Diameter: not over 69 in., not under 67.5 in. Static rpm at maximum permissible throttle setting: not over 2470, not under 2320 No additional tolerance permitted 2. McCauley 1A100/MCM 21 lb. (−32) Diameter: not over 69 in., not under 67.5 in. Static rpm at maximum permissible throttle setting: not over 2475, not under 2375 No additional tolerance permitted

*Airspeed limits (CAS)		
Never exceed	157 mph	(136 knots)
Maximum structural cruising	120 mph	(104 knots)
Maneuvering	106 mph	(92 knots)
Flaps extended	85 mph	(74 knots)

C.G. Range	(+33.4) to (+36.0) at 1500 lb. (+32.2) to (+36.0) at 1250 lb. or less Straight line variation between points given
Empty Wt. C.G. Range	None
Leveling Means	Top Edge of fuselage splice plate
*Maximum Weight	1500 lb.
No. of Seats	2 at (+39); (for child's optional jump seat refer to Equipment List)
Maximum Baggage	80 lb. at (+65)

FIG. 13-10 Sample Type Certificate Data Sheet.

Fuel Capacity		26 gal. (22.5 gal. usable, two 13 gal. tanks in wings at +42) See NOTE 1 for data on system fuel	

Oil Capacity		6 qt. (−13.5; unusable 2 qt.) See NOTE 1 for data on system oil	

Control Surface MOVEMENTS	Wing Flaps		
		Retracted	0°
		1st Notch	10°
		2nd Notch	20°
		3rd Notch	30°
		4th Notch	40°
	Ailerons	Up 20°	Down 15°
	Elevator	Up 25°	Down 15°
	Elevator tab	Up 10°	Down 20°
	Rudder	Right 16°	Left 16°

Serial Nos. Eligible		
	Model 150:	617, 17001 thru 17999, 59001 thru 59018
	Model 150A:	628, 15059019 thru 15059350
	Model 150B:	15059351 thru 15059700
	Model 150C:	15059701 thru 15060087

Data Pertinent to All Models

Datum	Fuselage station 0.0 front face of firewall
Certification Basis	Part 3 of the Civil Air Regulations dated May 15, 1956, as amended by 3-4. In addition, effective S/N 15282032 and on for 152 and S/N 681, A1520809 and on for A152, FAR 23.1559 effective March 1, 1978. FAR 36 dated December 1, 1969, plus Amendments 36-1 thru 36-5 for 152 and A152 only.

Application for Type Certificate dated August 13, 1956.

Type Certificate No. 3A19 issued July 10, 1958, obtained by the manufacturer under delegation option procedures.

Equivalent Safety Items	S/N 15077006 thru 15079405
	S/N 15279406 and on
	S/N A1500610 thru A1500734
	S/N 681, A1500433, A1520735 and on

Airspeed Indicator	CAR 3.757 (See NOTE 4)
Operating Limitations	CAR 3.778(a)

Production Basis	Production Certificate No. 4. Delegation Option Manufacturer No. CE-1 authorized to issue airworthiness certificates under delegation option provisions of Part 21 of the Federal Aviation Regulations.
Equipment:	The basic required equipment as prescribed in the applicable airworthiness regulations (see Certification Basis) must be installed in the aircraft for certification. This equipment must include a current Airplane Flight Manual effective S/N 15282032 and on, S/N 681, and S/N A1520809 and on. In addition, the following item of equipment is required:

1. Stall warning indicator, audible, Cessna Dwg. 0511062 (Model 150 thru 150E)

2. Stall warning indicator, audible, Cessna Dwg. 0413029 (Model 150F thru 150 M, 1977 Model) (A150K thru A150M, 1977 Model) (152 and on, A152 and on)

NOTE 1. Current weight and balance report together with list of equipment included in certificated empty weight and loading instructions when necessary must be provided for each aircraft at time of original certification.

Serial Nos. 17001 thru 17999, 59001 thru 59018, 15059019 thru 15077005 and A1500001 thru A1500609

The certificated empty weight and corresponding center of gravity location must include unusable fuel of 21 lb. at (+40) for landplanes or 27 lb. at (+40) for seaplanes and an undrainable oil of (0) lb. at (−13.5) for both landplane and seaplane.

FIG. 13-10 *(Continued)*

NOTE 2. The following information must be displayed in the form of composite or individual placards.

    A. In full view of the pilot:
        (1) "This airplane must be operated in compliance with the operating limitations stated in the form of placards, markings and manuals."

        (2) (a) Model 150A, 150B and 150C
              "Acrobatic maneuvers are limited to the following":

Maneuver	Entry Speed
Chandelle	106 m.p.h. (92 knots)
Steep turns	106 m.p.h. (92 knots)
Lazy eights	106 m.p.h. (92 knots)
Stalls (except whip)	Use slow deceleration
Spins	Use slow deceleration

FIG. 13-10 (*Continued*)

ered under that type certificate. The Type Certificate Data Sheet number is the same as the type certificate number. The name and address of the current type certificate holder is also given.

Data for a Specific Model. A Type Certificate may have more than one model of an aircraft approved under it due to changes in the type design. In most cases the manufacturer will designate the changes with a new model number such as Cessna 150E. In a few cases, the manufacturer may stay with the basic model designation and indicate the changes with serial numbers. All of the information which pertains specifically to that model will be presented; then the next change or model of aircraft will be covered. The aircraft will first be identified by model, type certificate category(s), and date of approval.

Many aircraft specifications and some Type Certificate Data Sheets carry coded information to describe the general characteristics of the product. These may be found in the model caption line or a separate line entry titled "Type" or "Designation." Aircraft codes (Designations) are as follows for the example, 2 PO-CLM:

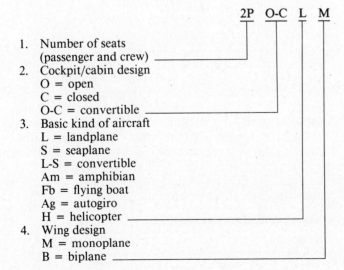

1. Number of seats
   (passenger and crew)
2. Cockpit/cabin design
   O = open
   C = closed
   O-C = convertible
3. Basic kind of aircraft
   L = landplane
   S = seaplane
   L-S = convertible
   Am = amphibian
   Fb = flying boat
   Ag = autogiro
   H = helicopter
4. Wing design
   M = monoplane
   B = biplane

The other information contained for a particular model will include:

Engine(s)
Fuel grade and capacity
Engine operational limits
Propeller and propeller limits
Airspeed limits
No. of seats and location
Baggage capacity
Oil capacity
Control surface movement
Center of gravity weight information

This section also contains a list of serial numbers produced for each model, which is very important when determining conformity with Type Certificate Data Sheets.

Data Pertinent to All Models. This section contains the location of the datum and the leveling means. It also includes the certification basis, production basis, and required equipment. The certification basis lists the exact regulations and special conditions, if any, that apply to each of the models. The production basis will list whether or not a production certificate was used and other pertinent information.

On aircraft that are still listed on an Aircraft Specification, the equipment will be listed and numbered to correspond to the required equipment listed for specific models. The list also shows the weight of the equipment and the arm for weight-and-balance purposes. Optional equipment is also included in this list.

A Type Certificate Data Sheet contains the statement: "The basic required equipment as prescribed in applicable airworthiness regulations (see certification basis) must be installed in the aircraft for certification." Any special equipment not covered by the above statement will be listed. The required equipment is listed in the aircraft's equipment list.

Notes. This section contains information that may apply to all models or certain aircraft within a given model. The information listed is very important and should not be ignored. Note No. 1 in all cases reads, "Current weight and balance data report together with the list of equipment included in certificated empty weight and loading instructions when necessary must be provided for each aircraft at the time of original certification." The note on an Aircraft Specification is slightly different in wording but not in meaning or intent. Unique weight-and-balance information for this aircraft is also located in this note.

Note No. 2 states that "The following placards must be displayed in front of and in clear view of the pilot." Placards, if any are listed, must be present if the aircraft is to conform to its type certificate. Other notes contain information about equipment installed in specific models, special operating procedures for various models, modifications, conversion from one model to another, etc.

Aircraft Engine Data Sheets

Identification. This section is similar to that found on the Aircraft Type Certificate Data Sheets. Engine TCDSs also utilize coded information in identifying engines. Engine Codes (Type) are as follows for the example, 4LIA (sometimes 4LAI):

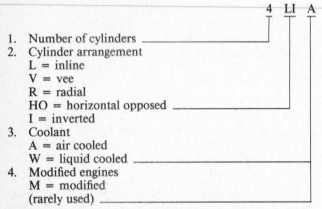

1. Number of cylinders
2. Cylinder arrangement
    L = inline
    V = vee
    R = radial
    HO = horizontal opposed
    I = inverted
3. Coolant
    A = air cooled
    W = liquid cooled
4. Modified engines
    M = modified
    (rarely used)

Data Pertinent to Specific Models. Data in this section includes the following for the various models of engines:

Type	Compression ratio
Power rating (maximum continuous and takeoff)	Oil sump capacity
Shaft (Type)	Weight
Fuel grade	Propeller
Oil grade versus air temperature (type)	Carburetion ignition
Bore and stroke	Magnetos ignition
Displacement	
Timing	

Data Pertinent to All Models. This section contains information on the certification and production basis.

Notes. The notes contain information on engine temperatures, fuel and oil pressure limits, spark plugs approved for use, and accessory drives. There is also information on the difference between the various models, approved optional equipment, procedures for conversion to other models, and special operating procedures as applicable.

Propellers

The propeller Type Certificate Data Sheets contain identification information that is similar to the aircraft and engine sheets. They also include the type of propeller, the engine shaft it fits, the hub and blade material, the number of blades, and the hub models eligible for this certificate. A listing of the eligible blades, by model number, follows with the power ratings with which they can be used, the diameter limits, and the weights (approximate) of the complete propellers with the various blades installed.

Certification and product basis are also listed. The notes explain the hub and blade model designations and special provisions for pitch control, operation, and interchangeability of parts.[2]

● SUPPLEMENTAL TYPE CERTIFICATES

Part 21 of the Federal Aviation Regulations provides for the issuance of **Supplemental Type Certificates (STCs)** to persons who make major changes on a previously approved type design when the change is not so extensive as to require a new Type Certificate. An applicant for an STC must show that the altered product meets all applicable airworthiness requirements.

The holder of an STC may:

1. In the case of aircraft, obtain airworthiness certificates
2. In the case of other products, obtain approval for installation on certificated aircraft
3. Obtain a production certificate for the change in the type design that was approved by the supplemental type certificate

A summary has been published that lists those STCs in which the holders have indicated that the design, parts, or kits will be made available to the public. Figure 13-11 shows an example from the summary of STCs. The publication contains three sections identified as: (1) Aircraft, (2) Engines, and (3) Propellers. These sections, arranged alphabetically by manufacturer's name, contain information on STCs approved for a specific make and model type-certificated product. Some of the STCs are listed as applying to a series of aircraft, engines, or propellers.

● MANUFACTURERS' PUBLICATIONS

Over an aircraft's service life virtually all of its components are involved in some form of inspection, preventive maintenance, overhaul, repair, or replacement. The technical information to carry out these activities must be available in clear, concise language to the technicians. This information may be in the form of maintenance, overhaul, and parts manuals as well as service letters or bulletins. The proper use of these publications by the technician is a must in maintaining today's complex aircraft.

Service Bulletins (Letters)

Service bulletins, or letters, are the medium utilized by manufacturers to communicate with owners and op-

---

[2]James R. Rardon, *Understanding the Federal Air Regulations,* Aviation Maintenance Publishers, Inc., 1980, pp. 36–39.

Aircraft make model & T.C. no.	STC no.	Description	RG	STC code	STC holder
PA-28-140, -150, -151, -160, -161; T.C. 2A13	SA793CE	Installation of Lycoming O-360-A1A engine and Hartzell HC-C2YK-1B/7666A-2 or HC-C2YK-1BF/F7666A-2 propeller. Amended 10/20/83.	CE	86195	Robert L. and Barbara V. Williams Box 654 Udall, KS 67146
PA-28-140, -150, -151; T.C. 2A13	SA802GL	Modify airplane to fly on unleaded automotive gasoline, 87 minumum antiknock index. Issued 7/30/84.	CE-W	64274	Petersen Aviation, Inc. Route 1, Box 18 Minden, NE 68959
PA-28-140, -150, -160, -180, -235, 28-S-160, S-180, 28R-180, -200; T.C. 2A13	SA819EA	Alteration to replace standard anti-collision light with Grimes Series 550 single anti-collision red or white strobe kit, P/N 34-0001-1 or 34-0001-2. Amended 3/30/78.	GL	39500	Grimes Manufacturing Company 515 North Russell Street Urbana, OH 43078
PA-28-140, -150, -160, -180, -235, -S-160, -S-180, -R-180; T.C. 2A13	SA822SW	Mitchell automatic flight system AK217 consisting of Century III autopilot with optional automatic pitch trim, aileron stabilizer, and radio coupler.	SW	60000	Mitchell Industries, Inc. P.O. Box 610 Municipal Airport Mineral Wells, TX 76067
PA-28-140, -150, -160, -180, -28R-180, 28R-200; T.C. 2A13	SA880WE	Installation of new design fiberglass wing tip. Amended 1/4/83.	NM	58500	Met-Co Aire P.O. Box 2216 Fullerton, CA
PA-28-140, -150, -160, -180, PA-28R-180, -200, PA-34-200; T.C. 2A13, A7SO	SA979EA	Installation of L/R wing tip transparent fairings and sensor mounting bracketry. Issued 4/5/74.	EA	68700	Rock Avionic Systems, Inc. 412 Avenue of the Americas New York, NY 10011
PA-28-140, -28-150, -28-160, -28-180, -28R-180, -28R-200; T.C. 2A13	SA1072CE	Installation of wing leading edge cuffs and droop tips, dorsal fin and vertical stabilizer vortex generator. Amended 11/6/84.	CE-W	42560	Horton Stol-Craft, Inc. Wellington Municipal Airport Wellington, KS 67152

FIG. 13-11 Summary of STCs index.

erators. Service bulletins are issued for such items as notification of design defects, possible modifications, or a change in approved maintenance practices. Service bulletins are usually mailed to registered owners and obtained by maintenance personnel on a subscription basis.

Maintenance Manual

The aircraft maintenance manual contains the information necessary to enable the technician to check, service, troubleshoot, and repair the airplane and its associated systems. It includes information necessary for the technician to perform maintenance or make minor repairs to components while they are installed on the airplane. The maintenance manual does not contain information relative to work normally performed on components when they are removed from the aircraft. For instance, the airplane maintenance manual provides the instructions for timing a magneto to the engine, but information on overhauling the magneto is located in the component maintenance manual.

Overhaul Manual

The manufacturer's overhaul manual contains detailed step-by-step instructions covering work normally performed on a unit away from the aircraft. The manual which includes the approved and recommended overhaul procedures is written to provide a technician with the information necessary to overhaul and test the part, thus making it available for reinstallation on an aircraft.

Structural Repair Manual

The structural repair manual contains descriptive information for the identification and repair of the aircraft's primary and secondary structure. The manual assists in defining damage that has a significant effect on the strength or life of the aircraft and that which

does not. For significant damage, the structural repair manual provides the technical data for repairing the aircraft. On some smaller aircraft, structural repair information is included in the aircraft's maintenance manual.

Parts Catalog

An illustrated parts catalog is designed to assist maintenance personnel in the ordering, storing, and issuing of aircraft parts. In addition, a parts catalog is often helpful in locating and identifying specific aircraft components with which the technician is unfamiliar. When ordering parts, it is important that the technician pay careful attention to the applicable serial numbers and codes.

● AIR TRANSPORT ASSOCIATION SPECIFICATION NO. 100

The Air Transport Association of America (A.T.A.) has established a specification to establish a standard for the presentation of technical data by aircraft, engine, or component manufacturers that is required for their respective products. This specification is identified as **A.T.A. No. 100**.

It is the primary intent of A.T.A. No. 100 to clarify the general requirements of the airline industry with reference to the coverage, preparation, and organization of technical data. The standards which provide for a uniform method of arranging technical materials is an effort to simplify the technician's problem in locating instructions and parts. The Standard Numbering System, which utilizes a three-part number, is designed to provide numbers for the identification of all systems, subsystems, and subjects. A standard number is composed of three elements which consist of two digits each, as shown for example in Table 13-1. A complete table of the A.T.A. No. 100 system, subsys-

TABLE 13-1 A.T.A. Standard Numbering System

First element, chapter (system)		Second element, section (subsystem)		Third element, subject (unit)	Coverage
26 (System) Fire protection	—	00	—	00	Material which is applicable to the system as a whole.
26	—	20 (subsystem) Extinguishing	—	00	Material which is applicable to the subsystem as a whole.
26	—	22 (subsubsystem) Engine fire extinguishing	—	00	Material which is applicable to the subsubsystem as a whole. This number (digit) is assigned by the manufacturer.
26	—	22	—	03 (Unit) Bottles	Material which is applicable to a specific unit of the subsystem. Both digits are assigned by the manufacturer.

tem, and titles is included in the appendices. The A.T.A. has also developed Specification No. 102, "Specifications for Computer Software Manual," which provides a standard for the presentation of digital computer software documentation.

● GAMA SPECIFICATION NO. 2

GAMA specification No. 2 was developed by the General Aviation Manufacturers Association (GAMA) for use in preparing manufacturer's maintenance data in a standardized format. General aviation manufacturers may prepare maintenance data in conformance with either Air Transport Association of America Specification No. 100 or with GAMA Specification No. 2, which closely follows and is compatible with A.T.A. Specification No. 100. This GAMA specification provides for the preparation and issuance of less complex manuals for less complex airplanes than those operated by the large scheduled airlines, but it still achieves industry-wide standardization of nomenclature and format.

GAMA Specification No. 2 has the provision for simple handbooks for simple airplanes and much more comprehensive handbooks for complex airplanes. Although much flexibility is built into Specification No. 2 that will allow it to be adopted to a variety of aircraft, its utilization will result in a high degree of standardization by providing for uniformity of arrangement of systems and components.

● MICROFICHE

Microfiche is fast becoming the industry standard for producing and storing technical information. Microfiche, often referred to as aerofiche by manufacturers, is a piece of photographic film, measuring 4 by 6 in which is capable of storing up to 288 pages of information. This tremendous reduction in size makes it possible to now store a roomful of information on a table top within one's reach. Microfiche is less expensive to produce than traditional hard copy manuals and is also much easier to revise as technical information is updated. A disadvantage of microfiche is that to read it requires the use of a microfiche reader (see Fig. 13-12). Machines called reader printers are capable of printing out pages from the fiche. Technical information is available on microfiche directly from the government printing office and manufacturers or may be purchased from independent publishers who reproduce technical information and add features which make it easier to use.

● MANUFACTURERS' OPERATING PUBLICATIONS

Approved Flight Manual

The **FAA Approved Flight Manual** contains all the pertinent information essential to proper operation of the airplane. An Approved Flight Manual listed on an aircraft's Type Certificate Data Sheet is considered a required piece of equipment that must be kept on

FIG. 13-12   Microfiche reader.

board the airplane. The manual contains an airplane's standard operating procedures and techniques as well as emergency procedures and weight-and-balance data.

Minimum Equipment List

If some deviations from the type-certificated configuration and equipment required by the FAR operating rules were not permitted, an aircraft could not be flown unless all such equipment were operating. Experience has proven that the operation of every system or component installed on an aircraft is not necessary when the remaining operating instruments and equipment provide for continued safe operation. The Federal Aviation Regulations permit the publication of a **Minimum Equipment List (MEL)** which provides for operation of an aircraft with certain items or components inoperative, provided the FAA finds an acceptable level of safety is maintained by a transfer of the function to another operating component or by reference to other instruments or components providing the required information.

Briefly, the MEL does not include obviously required items such as wings, rudders, flaps, engines, landing gear, etc. Also, the list may not include items which do not affect the airworthiness of the aircraft, such as galley equipment, entertainment systems, passenger convenience items, etc.

Unless otherwise specified in the remarks column, the FAA does not define where or when an inoperative item is to be repaired or replaced. The failure of instruments or items of equipment in addition to those allowed to be inoperative by the MEL causes the aircraft to be unairworthy. The MEL is not intended to provide for continued operation of the aircraft for an indefinite period with inoperative items. The basic purpose of the MEL is to permit the operation of an aircraft with inoperative equipment within the framework of a controlled and sound program of repairs and parts replacement. It is important that the owner or operator make repairs at the first airport, since additional malfunctions may require the aircraft to be taken out of service. Obtaining an MEL is accomplished by applying to the local Flight Standards District Office.

Operation Specifications

Operation Specifications are issued to supplement air carrier and air taxi rules by listing additional privileges and limitations that are not specifically covered by the regulations. Operation Specifications are divided into different subparts such as: the kinds of operations authorized, airport limitations, weight-and-balance data, and maintenance procedures.

Operation Specifications for maintenance cover items such as the aircraft's inspection program, overhaul time limits, component replacement times, and contract maintenance.

Operation Specifications are prepared by the applicant and submitted to the FAA Flight Standards District Office for approval. When approved, the provisions of the operation specifications are as legally binding as the regulations themselves.

● REVIEW QUESTIONS

1. When was the FAA given responsibility for safety legislation and enforcement?
2. To whom does the administrator of the FAA report?
3. When working in the field, the technician's primary contact with the FAA would be with whom?
4. Who has primary responsibility for determining the causes of aircraft accidents?
5. Why does the FAA participate in aircraft accident investigation?
6. What is an *NPRM?*
7. Which FAR deals with the certification of mechanics?
8. What is the purpose of an *Advisory Circular?*
9. How may an Advisory Circular be made mandatory?
10. What is the purpose of *Airworthiness Alerts?*
11. Alerts are published as what Advisory Circular?
12. When does the FAA issue an *Airworthiness Directive?*
13. Airworthiness Directives are published as amendments to which FAR?
14. How are *AD Notes* indexed?
15. How often is the *AD Summary* published?
16. What are the subject areas into which ADs are divided?
17. How often are newly issued and revised AD Notes mailed out?
18. What do the digits "06" signify in the AD number 86-06-07?
19. When is a technician responsible for performing an AD search?
20. What items need to be included in an AD record entry?
21. What is the difference in content between a TCDS and Aircraft Specifications?
22. TCDSs are published in how many volumes?
23. Technical information about older aircraft models of which no more than 50 remain in service would be found in which volume?
24. What does the designation code 4P-CSM mean?
25. To what will Note 2 on a TCDS always refer?
26. What medium is utilized by manufacturers to inform aircraft owners and operators of design defects?
27. To which manufacturer's publication would you refer in locating information on troubleshooting an aircraft system?
28. What publication provides a list of the aircraft equipment that must be operable in order to initiate a flight?
29. What is the purpose of the A.T.A. 100 code?
30. What are the advantages of utilizing microfiche?

14 GROUND HANDLING AND SAFETY

INTRODUCTION

Most aviation technicians devote a portion of their time to ground-handling and taxiing aircraft. The complexity of today's aircraft and accompanying ground support equipment creates a very expensive and dangerous work environment. The purpose of this chapter is to give the technician information on the procedures, techniques, and safety precautions involved in handling and servicing aircraft on the ground. Because of the many different types of aircraft and aircraft operations, it is possible to provide only general information and standard practices. For specific operations applying to a particular aircraft, the technician should consult the aircraft's service manual.

GENERAL SAFETY PRECAUTIONS

By their very nature and design, airplanes are dangerous to work around. The dangerous possibilities are further increased by the wide variety of machines, tools, and materials required to support and maintain aircraft. Developing and following a set of personal work safety habits is in a technician's own self-interest. Personal safety begins with the proper use of eye and ear protection as well as proper dress for the job being performed. Technicians should only operate equipment with which they are familiar and can operate safely. Hand tools should be kept in proper working order. Technicians should know the location of first aid and emergency equipment.

Good housekeeping in hangars, shops, and on the flight line is essential to safe and efficient maintenance. The highest standards of orderly work arrangements and cleanliness should be observed while maintaining an aircraft. When the maintenance task is complete, remove and properly store maintenance stands, hoses, electrical cords, hoists, crates, boxes, and anything else used to perform the work.

Pedestrian lanes and fire lanes should be marked and utilized as a safety measure to prevent accidents and keep pedestrian traffic out of work areas.

Power cords and air hoses should be straightened, coiled, and properly stored when not in use. Oil, grease, and other substances spilled on hangar or shop floors should be immediately cleaned or covered with an absorbent material to prevent fire or personal injury. Drip pans should be placed beneath engines and engine parts wherever dripping exists. **Under no circumstances should oil or cleaning fluid be emptied into floor drains.** Fumes from this type "disposal" may be ignited and cause severe property damage. Gasoline spills on the hangar floor should be flushed away with water. Sweeping these fuel spills with a dry broom could cause static electricity that could ignite the fuel.

Application of aircraft finishes should be accomplished in a controlled environment (paint room) whenever possible. Never do this type work in the presence of lights that are not explosion proof or near an open flame. Other work should not be done on an aircraft while it is being painted.

Welding should only be performed in designated areas. Any part to be welded should be removed from the aircraft, if possible. Repair would then be accomplished in the welding shop under a controlled environment. A welding shop should be equipped with proper tables, ventilation, tool storage, and fire prevention and extinguishing equipment.

COMPRESSED GAS SAFETY

Compressed gases are frequently used in the maintenance and servicing of aircraft. The utilization of compressed gases requires a special set of safety measures. **The following rules for use of compressed gases apply:**

1. Cylinders of compressed gases should be handled as high-energy sources and therefore as potential explosives.

2. Always use safety glasses when handling and using compressed gases.

3. Never use a cylinder that cannot be positively identified.

4. When storing or moving a cylinder, have the cap securely in place to protect the valve stem.

5. When large cylinders are moved, they should be strapped to a properly designed wheeled cart to insure stability.

6. Use the appropriate regulator on each gas cylinder. Adaptors or homemade modifications can be dangerous.

7. Under no condition should high-pressure gases be directed at a person.

8. Compressed gas or compressed air should not be used to blow away dust or dirt, since the resultant flying particles are dangerous.

9. Rapid release of a compressed gas will cause an unsecured gas hose to whip dangerously and also may build up a static charge which could ignite a combustible gas.

10. Oil or grease on an oxygen cylinder can cause an explosion.

● FIRE SAFETY

Fire is one of the greatest discoveries. For all its many advantages, however, fire is capable of producing disaster in a matter of seconds. Year-in and year-out fires continue to take their toll even though we have the technological knowledge and capability to prevent and retard fires.

Nature and Classification of Fires

Fire results from the chemical reaction that occurs when oxygen combines rapidly with fuel to produce heat and light (see Fig. 14-1). The essentials of this process are: fuel—a combustible gas, liquid, or solid; oxygen—sufficient in volume to support the process of combustion and usually supplied from surrounding air; and heat—sufficient in volume and intensity to raise the temperature of fuel to its ignition or kindling point.

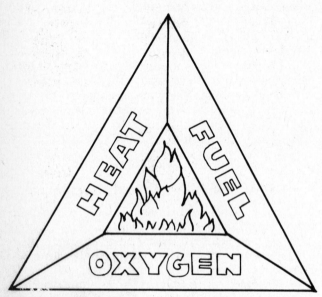

FIG. 14-1 The fire triangle consists of: (1) heat, (2) fuel, (3) oxygen.

There are four classes of fires, each determined by what is burning. The most common is that which occurs in ordinary combustible materials: paper, wood, textiles, and rubbish. It is designated as a class A fire. Fires in combustible liquids, such as gasoline, alcohol, oil, grease, and oil-base paint, form a second category known as class B fires. The third, class C fires, are those occurring in live electrical equipment such as fuse boxes, switches, appliances, motors, or generators. Class D fires are those of high intensity that may occur in certain metals such as magnesium, sodium, potassium, titanium, and zirconium. The greatest hazard exists when these metals are in a molten state or in finely divided forms of dust, chips, turnings, or shavings.

Spontaneous Ignition

Aviation technicians need to be particularly aware of **spontaneous ignition** caused by the lubricants and solvents that are used in maintaining aircraft. Certain materials such as rags soaked with oil or solvents are capable of generating sufficient heat to cause combustion. These rags should be disposed of in airtight cans.

Principles of Extinguishing Fires

Based on the principle of the fire triangle (see Fig. 14-1), there have long been three ways to extinguish fires: **(1) cooling the fuel below its kindling point, (2) excluding the oxygen supply, and (3) separating the fuel from the oxygen.** These methods led to the development of different types of extinguishers for different types of fires.

Types of Fire versus Extinguishing Agents

Class A fires respond best to water or water-type extinguishers which cool the fuel below combustion temperatures. Class B and C extinguishers are effective but not equal to the wetting/cooling action of the class A extinguisher.

Class B fires respond to carbon dioxide (CO_2), halogenated hydrocarbons (halons), and dry chemicals, all of which displace the oxygen in the air thereby making combustion impossible. Foam is effective, especially when used in large quantities. Water is ineffective on class B fires and will cause the fire to spread.

Class C fires involving electrical wiring, equipment, or current respond best to carbon dioxide (CO_2), which displaces the oxygen in the atmosphere making combustion improbable. The CO_2 extinguisher must be equipped with a nonmetallic horn to be approved for use on electrical fires. Two reasons for this must be considered:

1. The discharge of CO_2 through a metal horn can generate static electricity. The static discharge could reignite the fire.
2. The metal horn, if in contact with the electric current, would transmit that current to the extinguisher operator.

Halogenated hydrocarbons are very effective on class C fires. The vapor reacts chemically with the flame to extinguish the fire. Dry chemicals are effective but have the disadvantage of contaminating the local area with powder. Also, if used on wet and energized electrical equipment, they may aggravate current leakage.

Water or foam are not acceptable agents for use on electrical equipment, as they also may aggravate current leakage.

Class D fires respond to application of dry powder, which prevents oxidation and the resulting flame. Application may be from an extinguisher, a scoop, or a shovel. Special techniques are needed in combating fires involving metal. Manufacturers' recommendations should be followed at all times. Areas which could be subjected to metal fires should have the proper protective equipment installed. Under **no** conditions use water on a metal fire. It will cause the fire to burn more violently and can cause explosions.

Identification of Fire Extinguishers

Applying the incorrect material on a specific fire can do more harm than good and may actually be dangerous. It is important that fire extinguishers be well marked for quick identification under emergency conditions. In the excitement of a fire, it is very easy to grab the wrong type of extinguisher. Before using any extinguisher, read the labels. Usually fire extinguishers are marked with decals, paint codes, or similar methods. The National Fire Protection Association recommends the following markings for various extinguishers:

1. Extinguishers suitable for use on class A fires should be identified by a triangle containing the letter A. If the triangle is colored, it should be colored green.

2. Extinguishers suitable for use on class B fires should be identified by a square containing the letter B. If the square is colored, it should be colored red.

3. Class C fire extinguishers should be identified by a circle containing the letter C. If colored, the circle should be colored blue.

4. Extinguishers suitable for fires involving metals should be identified by a five-point star that contains the letter D. If the star is colored, it should be yellow.

Extinguishers suitable for more than one class of fire may be identified by multiple symbols. Figure 14-2 shows typical fire extinguisher markings in use today. The Underwriters' Laboratories (UL) is a nationally recognized testing agency established and supported by manufacturers and insurance companies. Throughout the years, the UL label has become a symbol of quality and efficiency in fire extinguishing equipment. One of the functions of UL is that of testing and listing all types of electrical and fire-fighting equipment. Therefore, all such equipment should have the UL label.

All fire extinguishers, whether in hangars, airplanes, or elsewhere, should be checked frequently to be sure they are in working order. This check should determine if the extinguishers are full and whether they have been tampered with. Proper maintenance, including checking for needed repairs, recharging, or replacement, should be carried out annually.

● **FLIGHT LINE SAFETY**

The most obvious source of accidents on the flight line is that of propellers or rotors. The propeller or rotor is difficult to see when in operation. Even personnel familiar with the danger of a turning propeller or rotor sometimes forget about its presence. Propeller- and rotor-to-person accidents differ from other aircraft accidents in that they usually result in fatal or serious injury. This is because of the fact that a propeller or rotor rotating under power, even at slow idling speed, has sufficient force to inflict serious injury. It should be remembered that a rotating propeller or rotor is extremely dangerous and should be treated with all due caution. Some manufacturers of propeller and rotor blades use paint schemes to increase the visibility of the blades. Technicians should give strong consideration to maintaining the visibility paint scheme of the original manufacturer.

Persons directly involved with aircraft service are most vulnerable to injuries by propellers or rotors. Working around aircraft places these people in the

1. WATER

2. CARBON DIOXIDE, DRY CHEMICAL BROMOCHLORODIFLUOROMETHANE AND BROMOTRIFLUOROMETHANE

3. MULTIPURPOSE DRY CHEMICAL

4. MULTIPURPOSE DRY CHEMICAL (INSUFFICIENT AGENT FOR "A" RATING)

METALS

5. DRY POWDER

FIG. 14-2 Typical extinguisher markings.

most likely position for possible propeller or rotor accidents. **Aircraft service personnel should develop the following safety habits:**

1. Treat all propellers as though the ignition switches are ON.

2. Chock airplane wheels before working around aircraft.

3. Use wheel chocks and parking brakes before starting engines.

4. Attach pull ropes to pull chocks from wheels close to a rotating propeller or rotor blades.

5. After an engine run and before the engine is shut down, perform an ignition switch test to detect a faulty ignition switch.

6. Before moving a propeller or connecting an external power source to an aircraft, be sure that the aircraft is chocked, ignition switches are in the OFF position, throttle is closed, mixture is in IDLE CUT-OFF position, and all equipment and personnel are clear of the propeller or rotor. Faulty diodes in aircraft electrical systems have caused starters to engage when external power was applied regardless of the switch position.

7. Remember, when removing an external power source from an aircraft, keep the equipment and yourself clear of the propeller or rotor.

8. Always stand clear of rotor and propeller blade paths, especially when moving the propeller. Particular caution should be practiced around warm engines.

Ground support personnel that are in the vicinity of aircraft that are being run up need to wear proper eye and ear protection. Ground personnel must also exercise extreme caution in their movements about the ramp; a great number of very serious accidents have occurred involving personnel in the area of turbojet engine air inlets. The turbojet engine intake and exhaust hazard areas are illustrated in Fig. 14-3. Care

should also be taken to ensure that the run-up area is clear of all items such as nuts, bolts, rocks, rags, or other loose debris.

● TOWING AIRCRAFT

Particular care must be exercised when pulling or pushing an aircraft. Persons performing towing operations should be thoroughly familiar with the procedures that apply to the type of aircraft to be moved. When towing aircraft, the proper tow bar must be used. The wrong type of tow bar, or makeshift equipment, can cause damage to the aircraft.

As illustrated in Fig. 14-4, the tow bar is usually made with fittings for attachment to the ends of the axles or to the tow fittings on the landing-gear strut or axle. Large aircraft are towed by means of specially designed towing vehicles. Figure 14-5 shows a Boeing 747 airplane being towed by such a vehicle. The aircraft is attached to the towing vehicle by a tow bar designed for the purpose. In this type of towing, the operator of the towing vehicle must be in constant

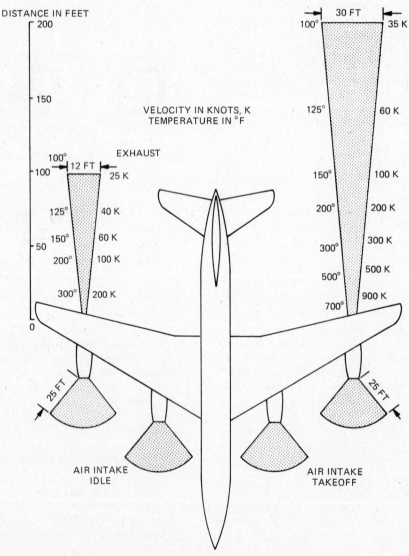

FIG. 14-3   Engine intake and exhaust hazard areas.

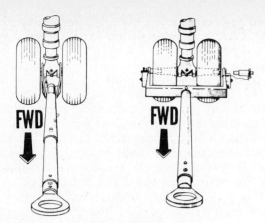

FIG. 14-4 Tow bar attachment.

FIG. 14-5 Towing a very large airplane.

communication with the operator of the aircraft in the cockpit. The operator of the aircraft is also in radio contact with ground traffic control in the tower.

During the towing of aircraft, care must be taken to make sure that the following precautions are followed; otherwise damage to the aircraft may result:

1. One should never tow an aircraft in congested areas without guidemen to assist in determining that there is adequate clearance.

2. When a small aircraft is being moved manually, care must be taken to apply hand pressure to the aircraft at only those areas that have structural strength sufficient to withstand the pressure without damage.

3. When towing, do not turn the nose gear beyond its steering radius in either direction, as this will result in damage to the nose gear and steering mechanism.

4. Always make sure that the airplane control locks are removed before towing the airplane. Serious damage to the steering linkage can result if the airplane is towed while the control locks are installed.

5. Prior to movement of any aircraft, all landing-gear struts and tires should be properly inflated and brake pressure built up when applicable.

6. The tow vehicle operator should avoid sudden starts and stops. Although the tug will control the steering of the airplane, someone should be positioned in the pilot's seat to operate the brakes in case of an emergency.

7. Clearance must be obtained from the airport control tower, either by appropriate radio frequency or by prior arrangement through other means, before moving aircraft across runways or taxiways.

● TAXIING AND STARTING

The taxiing of an airplane can be a relatively simple matter, or it can be a very complex and critical operation, depending upon the size and type of aircraft being taxied. Before any airplane can be taxied, it is, of course, necessary to start the engine or engines. Here again, the operation can be simple or complex, depending upon the size and type of engine.

The following procedures are typical of those used to start reciprocating engines. There are, however, wide variations in the procedures for the many reciprocating engines. **No attempt should be made to use the methods presented here for actually starting an engine.** Instead, always refer to the procedures contained in the applicable manufacturer's instructions.

Starting Small Engines

The starting of small aircraft engines is not difficult, but it should be done carefully to avoid damage and to provide for the safety of personnel. Certain general procedures apply and should be understood by the technician.

Three conditions are necessary for the starting of any internal-combustion engine: (1) the presence of fuel in the combustion chamber, (2) a source of ignition for the fuel, and (3) a method for rotating the engine to start the sequence of operating events. The methods by which these conditions are satisfied depend to some extent upon the type of engine and the design of fuel, ignition, and starting systems.

For a light-aircraft engine equipped with a float-type carburetor, the starting procedure may be as follows:

1. See that the engine is supplied with oil and that an adequate supply of fuel is available in the fuel tank or tanks.

2. With the ignition switch off, rotate the engine at least three complete revolutions by hand or with the starter to see that the engine is clear of obstructions. When the engine is rotated by hand, the person turning the propeller must remember to stand well out of the range of the propeller in case the engine should fire. In some cases the engine can be "hot" even though the ignition switch is in the OFF position.

3. Prime the engine with the priming pump, using approximately three strokes, provided that the engine is cold. The fuel valve must be ON.

4. Place the mixture control in the FULL RICH position.

5. Open the throttle to about one-eighth of the FULL OPEN position.

6. Turn the ignition switch on and press the starter button, or turn the starter switch key to the start position.

If prolonged cranking is necessary, allow the starter motor to cool at frequent intervals, since excessive heat may damage the armature.

7. After the engine starts, adjust the throttle for approximately 800 to 1000 rpm [84 to 105 rad/s] and check the oil-pressure gage for pressure. If oil pressure does not show within about 30 s, shut the engine down and check for the trouble. With some small engines, after a long period of storage, the oil pump will lose its prime and it is necessary to reprime the pump by removing the oil screen and pouring oil into the opening.

When starting a light engine with a pressure-type carburetor or a fuel-injection system, it is not usually necessary to use a primer. In these cases, fuel will be supplied as soon as the mixture control is moved from the IDLE CUT-OFF position toward the FULL RICH position. The fuel booster pump must be turned on to provide fuel pressure. The mixture control should be moved to the FULL RICH position after the engine is started by means of the starter. Another procedure recommended for starting certain engines with fuel injection is to start the engine on the priming fuel only and then move the mixture control to the FULL RICH position as soon as the engine starts.

Starting Large Reciprocating Engines

Large radial engines installed on such aircraft as the DC-3, DC-6, Constellation, and other large aircraft should be started according to the manufacturer's instructions. The steps in starting are similar to those used for light-aircraft engines, but additional precautions are necessary. First, a fire guard should be placed to the rear and outboard of the engine being started in case the engine backfires and fire burns in the induction system of the engine. The fire guard should have an adequate supply of carbon dioxide gas in suitable fire extinguisher bottles in order to immediately direct the gas into the induction system of the engine. The engine should be kept turning so that the fire will be drawn into the cylinders. Very often, it is not even necessary to use the extinguisher because the air rushing into the engine carries the fire with it and, as the engine starts, the fire cannot continue to burn in the induction system.

Before attempting to start a large reciprocating engine, the engine should be rotated several complete revolutions to eliminate the possibility of **liquid lock** caused by oil in the lower cylinders. If the engine stops suddenly while being rotated by hand or with the starter, it is apparent that oil has collected in a lower cylinder, and before the engine can be started, the oil must be removed. This is best accomplished by removing a spark plug from the cylinder. It is not recommended that the engine rotation be reversed to clear the oil. After the oil is drained from the cylinder, the spark plug can be replaced and the engine can be started.

For large reciprocating engines, priming is usually accomplished by means of a fuel-pressure pump and an electrically operated priming valve. The fuel is carried from the primer to a spider (distributing fitting) and then to the top cylinders of the engine. This applies to a radial engine, either single or twin row. In a nine-cylinder radial engine, the top five cylinders of the engine receive priming. Priming is accomplished by pressing the priming switch while the fuel booster pump is turned on.

Large reciprocating engines may have direct-cranking starters similar to those used on light-aircraft engines but much more powerful ones, or they may have inertia starters in which the cranking energy is stored in a rapidly rotating flywheel. With the inertia starter, the flywheel must be energized by means of an electric motor or by hand crank until enough energy is stored to turn the engine for several revolutions. The ENGAGE switch is then turned on to connect the flywheel-reduction gearing to the crankshaft through the starter jaws. A plate clutch, located between the flywheel and the starter jaws, allows slippage to avoid damage due to inertial shock when the starter is first engaged.

With the engine properly primed, the throttle set, and the ignition switch ON, the engine should start very soon after it is rotated by the starter. The throttle is then adjusted for proper warmup speed.

Hand Cranking

Hand cranking a starter-equipped engine with a low battery or defective starter, although convenient, can expose personnel to a possible accident. For safety reasons, the replacement of the faulty starter and the use of a ground power source should be considered rather than hand cranking. **Only experienced persons should do the hand cranking and a reliable person should be in the cockpit.** Hand cranking with the cockpit unoccupied has resulted in many serious accidents.

If the aircraft has no self-starter, the engine must be started by swinging the propeller. The person who is turning the propeller calls, **"fuel on, switch off, throttle closed, brakes on."** The person operating the engine will check these items and repeat the phrase. The switch and throttle must not be touched again until the person swinging the prop calls **"contact."** The operator will repeat "contact" and then turn on the switch. Never turn on the switch and then call "contact."

When swinging the prop, a few simple precautions will help to avoid accidents. When touching a propeller, always assume that the ignition is ON. The switches which control the magnetos operate on the principle of short-circuiting the current to turn the ignition off. If the switch is faulty, it can be in the OFF position and still permit current to flow in the magneto primary circuit.

Be sure the ground is firm. Slippery grass, mud, grease, or loose gravel can lead to a fall into or under the propeller. Never allow any portion of your body to get in the way of the propeller. This applies even though the engine is not being cranked.

Stand close enough to the propeller to be able to step away as it is pulled down. Stepping away after cranking is a safeguard in case the brakes fail. Do not stand in a position that requires leaning toward the propeller to reach it. This throws the body off balance and could cause you to fall into the blades when the engine starts. In swinging the prop, always move the blade downward by pushing with the palms of the hand. Do not grip the blade with the fingers curled over the edge, since "kickback" may break them or draw your body into the blade path.

Starting Turbine Engines

Before starting a gas-turbine engine, it is important to see that the areas in front of and to the rear of the engines are clear. The airplane should be resting on clean concrete or asphalt so that no loose material can be drawn into the inlet of the engines. The area to the rear of a gas-turbine engine is exposed to intense heat and high-velocity gases when the engine is running. Materials or objects that can be damaged by heat should be cleared from the exhaust area.

With a gas-turbine engine, either a turboprop or a pure jet, care must be taken to follow the manufacturer's instructions for starting. Since these engines have several different types of fuel-control units, the operator must know the procedures to be followed for the particular engine. Basically, the engine must be rotated to approximately 10 percent of full speed, ignition must be provided, and the proper amount of fuel must be delivered to the fuel nozzles. A manual start for some types of jet engines can be accomplished as follows:

1. Turn on the master switch.
2. Turn on the fuel boost pump.
3. Turn on the starter switch and hold it until the engine speed as shown on the percent-of-power gage is approximately 10 percent.
4. Turn on the ignition switch.
5. Move the power-control lever (throttle) slowly forward while watching the exhaust gas temperature (EGT) gage. As soon as the fuel ignites, the EGT gage will show a rapid rise in temperature.
6. Stop moving the power-control lever until the temperature stabilizes and then slowly move it forward, being careful to see that the EGT does not exceed the proper limit (approximately 600°C). The engine should accelerate smoothly as the power lever is moved forward.
7. If the engine does not accelerate properly or if the EGT exceeds the safe limit, the power-control lever should be retarded to cut off the fuel and stop the engine.

When, during the starting of a gas-turbine engine, the EGT exceeds the prescribed safe limit, the engine is said to have a **hot start.** As explained above, when this occurs, the engine should be immediately shut down. Hot starts are usually caused by an excess of fuel entering the combustion chamber. With an automatic fuel-control unit, a hot start would not occur unless the unit were malfunctioning.

If the engine fails to accelerate properly, it is said to have a **false start,** also called a **hung** start. This type of start is caused by attempting to ignite the fuel before the engine has been accelerated sufficiently by the starter. The automatic systems used on modern engines prevent this type of problem unless the fuel-control unit is malfunctioning.

With an automatic system on a modern jet aircraft, the starting procedure is comparatively simple. With the electric power turned on in the airplane and the ground air starter unit connected to the aircraft, the pilot or flight engineer needs merely to initiate the starting sequence with the START switch and the power-control lever. The starting sequence will then take place under control of the fuel-control unit and associated equipment. After one engine is started on a multiengine aircraft by means of the ground starter unit, the other engines can be started by using bleed air from the running engine.

A schematic diagram of the starting system for the McDonnell Douglas DC-10 airliner is shown in Fig. 14-6. The engine is equipped with a pneumatic starter mounted on the accessory section of the engine. Air to operate the starter can be supplied either from the auxiliary power unit (APU) in the aircraft or from a ground power unit. The ground starting unit is essentially a small gas-turbine engine or an engine-driven compressor from which an adequate supply of air is drawn to operate the engine starter.

To start the engine on the DC-10 airplane, it is necessary merely to set the throttle (power-control lever) in the proper position, see that air is available (either through the APU or the ground-air supply unit), have electric power turned on, correctly position the engine ignition switch, have the fuel boost pumps turned on, then press the starter switch. The starter switch is equipped with a solenoid that holds it in until the engine rpm, N_2 section, attains 45 percent speed. The sequence of starting occurs automatically, and after the speed of 45 percent is reached, the START switch disconnects.

Taxiing Aircraft

The taxiing of small aircraft is not difficult, but it does require watchfulness and care. Before starting to taxi, the pilot or technician observes carefully to see that no other aircraft, persons, vehicles, or obstructions are likely to be in or near the taxi route. A small aircraft's direction is controlled during taxiing with the use of rudder pedal steering, brakes, and in the case of twin-engine aircraft, differential engine power. A large aircraft such as the Douglas DC-9 will control its movements during taxiing by the use of selective engine thrust; the nose-wheel steering system, which is hydraulically controlled through its full range of 164° (82° to either side of center); a steering wheel located on the captain's left console; the rudder pedals which can be used for 17° or less steering control to either side of the neutral position; and the brakes.

Manufacturers often supply detailed instructions for taxiing, towing, and handling of aircraft. Typical of such instructions are the following for the Cessna Model 421 airplane: Before attempting to taxi the aircraft, ground personnel should be checked out by qualified persons. When it is determined that the propeller blast area is clear, apply power and start the taxi roll and perform the following checks:

1. Taxi forward a few feet and apply the brakes to determine their effectiveness.

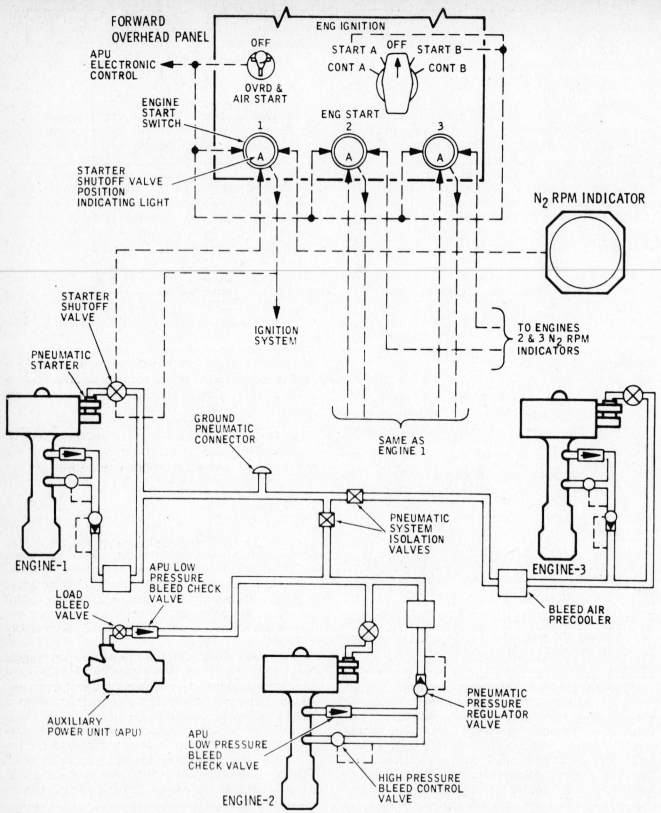

FIG. 14-6   Starting system for a DC-10 airliner.   *(McDonnell Douglas Corporation)*

2. While taxiing, make slight turns to determine the effectiveness of the nose-gear steering.

3. Check the operation of the turn-and-bank indicator and the directional gyro.

4. Check for sluggish instruments during taxiing.

5. In cold weather, make sure all instruments have warmed sufficiently for normal operation.

6. Minimum turning distance must be strictly observed when the aircraft is being taxied close to buildings or other stationary objects (see Fig. 14-7).

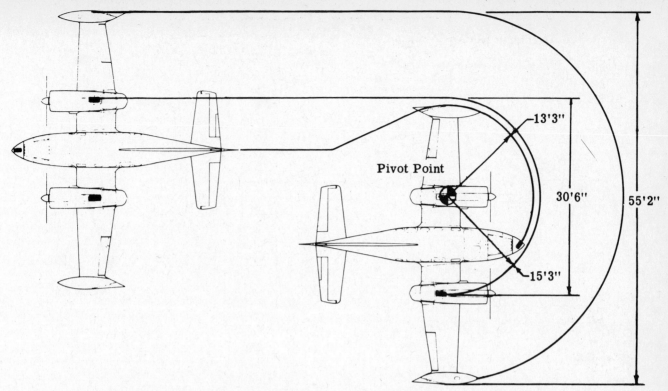

FIG. 14-7　Minimum turning distance for a light twin-engine airplane. *(Cessna Aircraft Co.)*

7. Do not operate the engine at high rpm when taxiing over ground containing loose stones, gravel, or any loose material that may cause damage to the propeller blades.

When a large aircraft is being taxied in the vicinity of other aircraft or near terminal buildings, one or more linemen to watch for proper clearance should be standing on the ground forward and to the right and/ or left of the aircraft so as to be clearly visible to the operator of the aircraft. The lineman should use standard FAA hand signals to indicate to the operator of the aircraft what action is to be taken. Some of the standard signals are shown in Fig. 14-8. It will be noted that these signals are easily understood because the motions represent the action to be taken by the operator of the aircraft. On a field with an operating tower, taxiing of aircraft must be accomplished in accordance with directions from the ground-control operator in the airport tower. The operator of the aircraft must keep the radio tuned to the ground-control frequency at all times while taxiing. In the case where the aircraft is not radio equipped or in the event of a radio failure, the use of lights are used to regulate taxiing. The following signal lights from the tower are used when necessary:

Steady red	Stop.
Flashing red	The aircraft should immediately taxi clear of the runway it is on.
Steady green	OK to taxi.
Alternating red and green	OK to taxi but with extreme caution.
Flashing white	Return to starting point.

● TYING DOWN OR MOORING AIRCRAFT

Small aircraft should be tied down after each flight to preclude damage from sudden wind gusts. Aircraft should be headed, as nearly as possible, into the wind, depending on the locations of the fixed, parking area tie-down points.

When parking the airplane, be sure that it is sufficiently protected from adverse weather conditions and that it presents no danger to other aircraft. When parking the airplane for any length of time or overnight, it should be moored securely.

Airport tie-down areas are usually equipped with three-point rings, hooks, or other devices embedded in concrete for the purpose of attaching tie-down ropes, chains, or cables. Location of tie downs are usually indicated by some means such as white or yellow paint markings.

Tie-down devices are often left attached to the fitting in the concrete or other pavement, and it is necessary merely to attach the rope, chain, or cable to the aircraft after parking. If a rope is used, it should be attached to the aircraft with a nonslip knot such as bowline or square knot. Examples of these knots are shown in Fig. 14-9. It is recommended that the tie-down rope or cable be attached in such a manner that

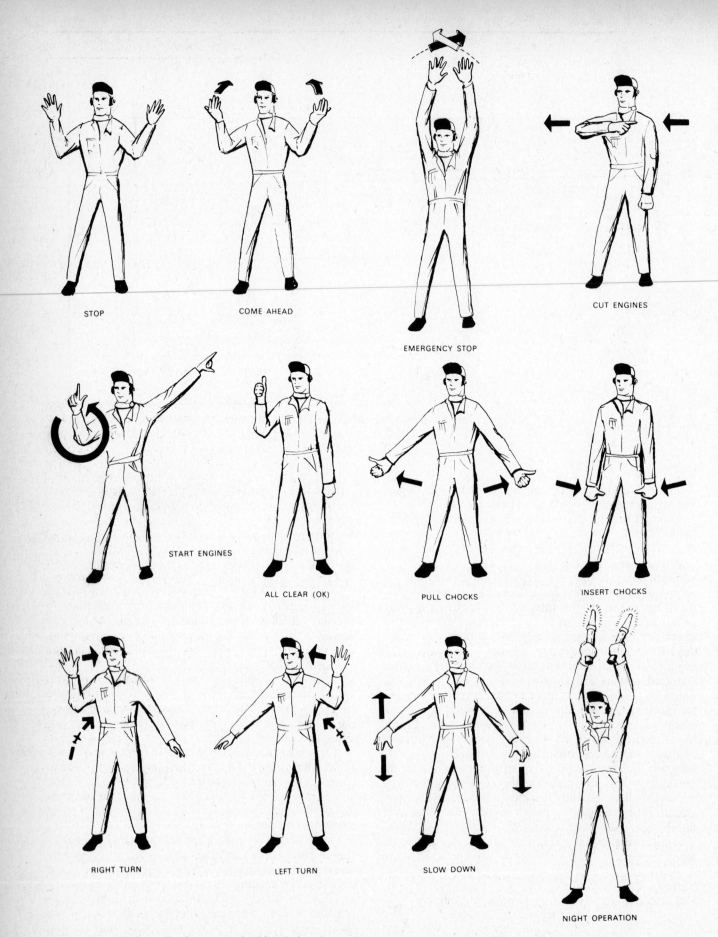

STOP

COME AHEAD

EMERGENCY STOP

CUT ENGINES

START ENGINES

ALL CLEAR (OK)

PULL CHOCKS

INSERT CHOCKS

RIGHT TURN

LEFT TURN

SLOW DOWN

NIGHT OPERATION

FIG. 14-8 Some standard hand signals required for directing aircraft during taxiing or towing.

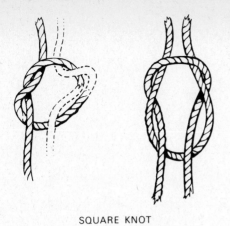

SQUARE KNOT

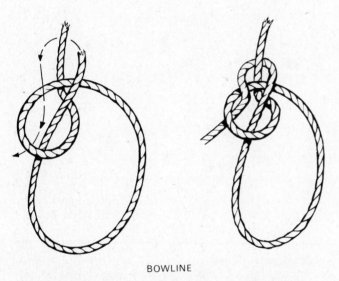

BOWLINE

FIG. 14-9 Knots that may be used for aircraft tie-down.

			Minimum tensile strength, lb			
			Dacron		Yellow polypropylene	
Size, in	Manila	Nylon	Twist	Braid	Twist	Braid
$\frac{3}{16}$	—	960	850	730	800	600
$\frac{1}{4}$	600	1 500	1 440	980	1 300	1 100
$\frac{5}{16}$	1 000	2 400	2 200	1 650	1 900	1 375
$\frac{3}{8}$	1 350	3 400	3 120	2 300	2 750	2 025
$\frac{7}{16}$	1 750	4 800	4 500	2 900	—	—
$\frac{1}{2}$	2 650	6 200	5 500	3 800	4 200	3 800
$\frac{5}{8}$	4 400	10 000	—	—	—	—
$\frac{3}{4}$	5 400	—	—	—	—	—
1	9 000	—	—	—	—	—

FIG. 14-10 Comparison of common tie-down ropes.

it is at an angle of approximately 45° with the ground if possible.

Tie-down ropes may be made of Manila hemp, cotton, nylon, Dacron, or other synthetic material. The synthetic ropes are less likely to deteriorate than hemp or cotton. Hemp has the disadvantage of shrinking when it becomes damp, and, for this reason, care must be taken to see that the rope is not pulled tight when the aircraft is tied down. A slack of 2 or 3 percent of the rope length normally should be sufficient to prevent aircraft damage due to shrinking of the rope. The strength of various types of commonly used tie-down rope are compared in Fig. 14-10.

When an aircraft is being secured in a tie-down area, **control locks** and **gust locks** should be installed to prevent the controls and surfaces from being damaged by wind. Wind can cause control surfaces to move violently from one extreme to the other, and this often results in damage. A twin-engine airplane with gust locks installed is shown in Fig. 14-11. On some smaller aircraft the aileron and elevator controls may be secured by using the front seat belts. Care must be taken to see that all such locks are removed before attempting to taxi or fly the airplane.

When an aircraft is operated from an airport where tie-down facilities are not installed, steel tie-down stakes should be carried in the aircraft. One of the most effective types of tie-down stakes has the appearance of a large corkscrew with a loop at the top to which the tie-down rope is attached. The stakes are screwed into the ground and are normally adequate to hold the aircraft secure under all conditions of wind except hurricane or tornado. If the ground is saturated with water, the ability of the stakes to hold will be impaired. Tie-down kits containing stakes and ropes or chains are available from aircraft-supply companies.

It is recommended that airplanes be headed into the wind when parked; however, since tie-down spaces are usually laid out by the airport management, there may be no choice as to the direction in which the aircraft will be headed when it is tied down.

All parked aircraft, whether tied down or not, should have chocks placed against the front and rear of the main wheels to prevent the aircraft from rolling. Parking brakes should be set except in cases where the brakes are overheated or during cold weather when accumulated moisture may freeze a brake. When parking a helicopter, the ground handling wheels should be retracted thus allowing the helicopter to rest on the skid type landing gear. The main and tail rotor blades should be secured if a helicopter is parked in an area subjected to turbulence created by jet, prop, or rotor blast from other aircraft.

● JACKING AND HOISTING AIRCRAFT

From time to time in performing maintenance and inspection procedures it becomes necessary to jack or hoist the aircraft. In the jacking or hoisting of aircraft, it is essential that the operation be performed according to the manufacturer's instructions. The points where hoisting or jacking fittings are attached to the aircraft are suitably designed to withstand the stresses imposed when the aircraft is hoisted or jacked. If the

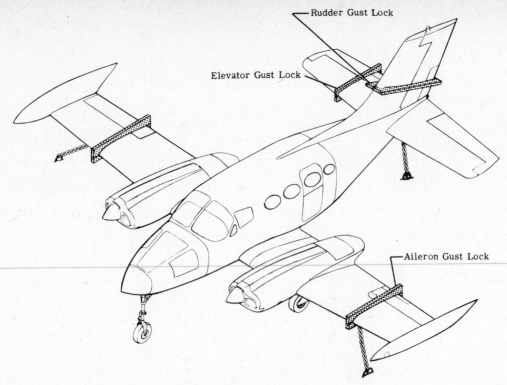

FIG. 14-11   Gust locks on an airplane.   *(Cessna Aircraft Co.)*

aircraft is lifted or jacked at a point other than those designed for the operation, severe damage will often be caused. The locations of jacking points for a light twin-engine aircraft are shown in Fig. 14-12.

Jacking Aircraft

Since jacking procedures and safety precautions vary for different types of aircraft, only general jacking procedures and precautions are discussed here. Consult the applicable aircraft manufacturer's maintenance instructions for specific jacking procedures. In jacking one wheel only, a small jack can be used at the landing-gear jack pad. This procedure is employed when tires are being changed and when brakes are being repaired or serviced. When only one set of wheels has to be raised, a low single-base jack is used, as shown in Fig. 14-13. Before the wheel is raised, the remaining wheels must be chocked fore and aft to prevent movement of the aircraft. The wheel should be raised only high enough to clear the floor.

For jacking the complete aircraft, wide-base jacks such as the large tripod types shown in Fig. 14-14 should be used because of the greater stability they will afford. The size and configuration of the aircraft will dictate the type and number of jacks needed to raise it. Many small aircraft are raised by using a jack under each wing spar and a weighted tail stand as shown in Fig. 14-15. If this method is utilized, be sure to consult the manufacturer's recommendations on the amount of weight needed.

Many aircraft have jack pads located at the jack points. Others have removable jack pads that are inserted into receptacles prior to jacking (see Fig. 14-16).

The correct jack pad should be used in all cases. The function of the jack pad is to ensure that the aircraft load is properly distributed at the jack point and to provide a convex bearing surface to mate with the concave jack stem.

Typical instructions for jacking procedures and precautions follow.

Raising the Aircraft

1. Jacking of aircraft should be done in a hangar, if possible, to avoid the effects of wind. (Head airplane into wind if it is to be jacked out-of-doors.)

2. See that the areas under and around the airplane are clear of obstructions.

3. If any other work is in progress on the aircraft, ascertain if any critical panels have been removed. On some aircraft the stress panels or plates must be in place when the aircraft is jacked to avoid structural damage.

4. The jacks should be placed directly under the center of each jack point and extended until they touch the jack points (most accidents during jacking are the result of misaligned jacks).

5. The legs of the jacks should be checked to see that they will not interfere with the operations to be performed after the aircraft is jacked, such as retracting the landing gear.

6. Operate all jacks evenly so that the airplane remains as nearly level as possible.

7. Keep the amount of lift to an absolute minimum and always within safe limits of the jack.

8. Set locking devices on the jacks to prevent accidental lowering of airplane.

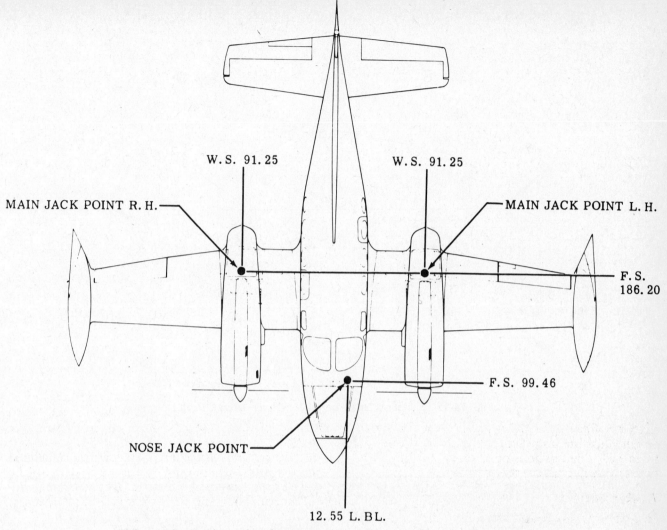

FIG. 14-12 Jacking points for a light twin-engine aircraft. *(Cessna Aircraft Co.)*

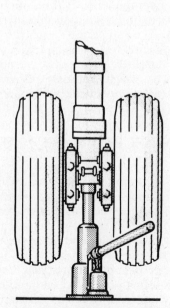

FIG. 14-13 Jacking a single set of wheels.

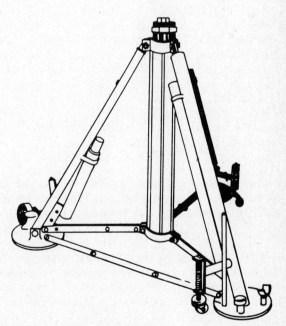

FIG. 14-14 Typical tripod jack.

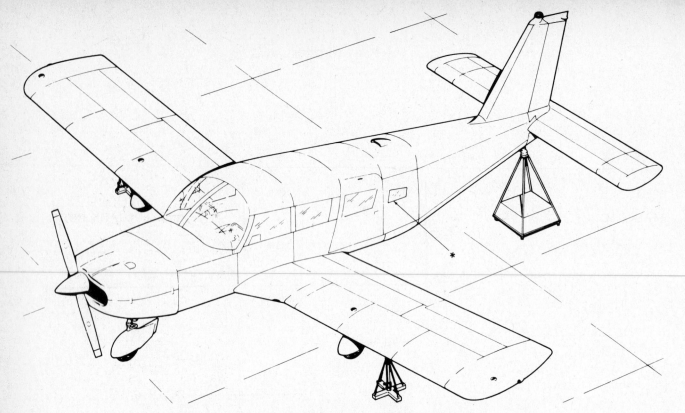

FIG. 14-15   Jacking a small airplane.   *(Piper Aircraft Corp.)*

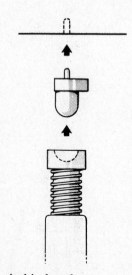

FIG. 14-16   Typical jack pad.

9. Climbing on the aircraft should be held to an absolute minimum, and no violent movements should be made by persons who are required to go aboard.

Lowering the Aircraft

1. Before lowering the aircraft, make certain that all working stands, equipment, and persons are clear of the aircraft, that the landing gear is down and locked, and that all ground locking devices are properly installed.

2. Release the mechanical locks, then slowly and carefully release the pressure, lowering the jacks evenly. *Caution:* As the aircraft is lowered, watch that the oleo struts do not bind up.

3. As soon as possible remove jacks out from under the aircraft.

Hoisting

It is often necessary to hoist airplanes and helicopters in order to perform certain service and maintenance operations. When hoisting the entire airplane or any of the airplane components, it is recommended that hoisting slings, manufactured specifically for the airplane, be used. These slings are designed to lift the airplane or components from the approximate center of gravity. Most fuselage hoist slings are adjustable to allow for different weight and center-of-gravity variations. Figure 14-17 illustrates a helicopter and components being hoisted.

● GROUND-SUPPORT EQUIPMENT

Ground-support equipment is needed primarily for the operation of aircraft on the ground when the aircraft engines are not operating. In some cases, small ground-support units such as battery carts, preoil units, and test units are used with light aircraft, but this is not usually a regular and ongoing procedure as with large aircraft. Ground-support units for large aircraft include electrical power supplies, air conditioning and/or air supply units, hydraulic test units, and various service units. The units used on a regular basis are those supplying electric power and air for starting engines, ventilating, heating, and cooling.

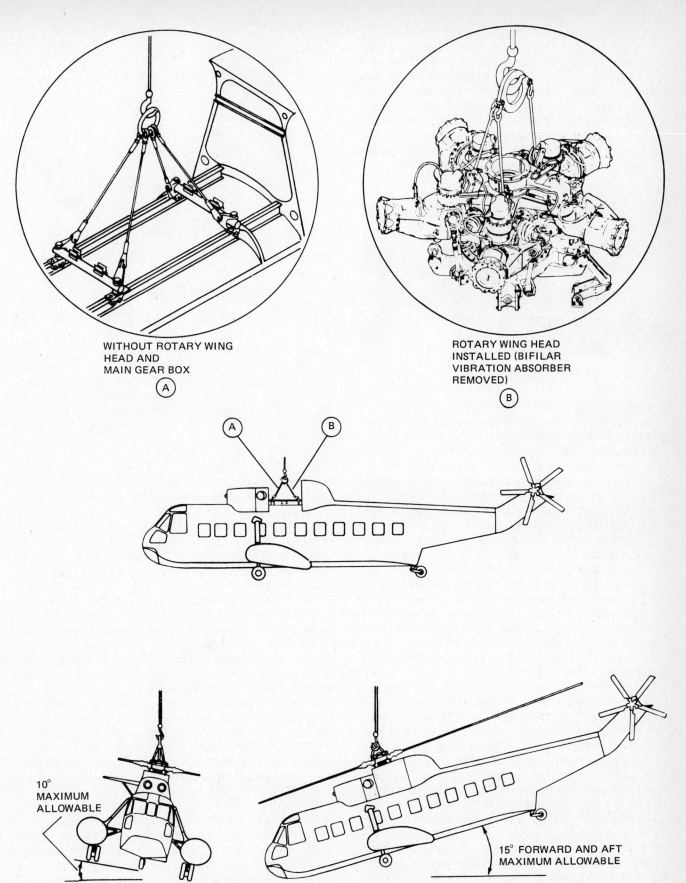

WITHOUT ROTARY WING
HEAD AND
MAIN GEAR BOX
(A)

ROTARY WING HEAD
INSTALLED (BIFILAR
VIBRATION ABSORBER
REMOVED)
(B)

10°
MAXIMUM
ALLOWABLE

15° FORWARD AND AFT
MAXIMUM ALLOWABLE

FIG. 14-17   How a large helicopter is hoisted.   (Sikorsky Aircraft, Division of United Technologies)

Electrical Power Supplies

Electric power for light aircraft is usually supplied by means of a **battery cart.** Such carts contain one or more storage batteries connected to supply 12, 24, or 28.5 volts (V) dc and having sufficient capacity to provide several engine starts before recharging (see Fig. 14-18). Battery carts should be kept clean and fully charged. Since they are not used on a regular basis, they are often overlooked and not properly cared for. It is good practice to schedule a regular time for checking and servicing ground-power batteries.

FIG. 14-19   Ground power unit.   *(Hobart Brothers Co.)*

FIG. 14-18   Battery power cart.

When a battery cart is to be used, the technician must ascertain the voltage and capacity of the unit and the system voltage of the aircraft being serviced. A 12-V power supply cannot service a 24-V system, and a 24-V power supply will cause severe damage if connected to a 12-V system. In all cases, the master switch of the aircraft should be OFF when the battery cart is being connected or disconnected to an aircraft. A specially designed receptacle is usually installed in the aircraft system to make it impossible to connect the power supply with the wrong polarity. The plug on the power supply cable cannot be inserted into such an aircraft receptacle unless the polarity is correct.

Mobile electrical power units consist of engine-driven generators mounted on a trailer or truck such as the power supply pictured in Fig. 14-19. They are designed to supply the correct type of power (ac or dc) and the correct voltage and capacity for the aircraft that they are to service.

All large aircraft, such as commercial airliners, require electric power when in flight and when on the ground preparing for flight. It is therefore necessary to have electric power available at the gate where an airliner loads and unloads passengers. Most large airliners are equipped with auxiliary power units (APUs), which are small gas-turbine engines that drive generators (alternators) and also supply bleed air for engine starting and air conditioning. These units are expensive to operate, and so ground power supplies are used as much as possible.

A typical electrical power supply for ground service of airliners consists of a diesel-engine-driven or turbine-engine-driven alternator that produces 400 Hz 115/200 V ac. The power unit is plugged into the aircraft as soon as it is parked at the gate and assumes the electrical load of the aircraft when the engines are shut down.

Mobile ground-power units, even though very widely used for many years, are now being replaced by **fixed power supplies** where possible. The fixed power supply is less costly, eliminates the air pollution and noise caused by mobile units, and reduces the clutter of equipment around the aircraft. Until recently, a separate fixed power supply has been used at each airline terminal; however, some airports are now equipped with a single, central power station, and electric power is delivered through cables to the various terminals on the airport. This has been made possible by transmitting the power from the central station at a high voltage (4160 V) to the terminal, then reducing it by means of transformers to the correct voltage for the aircraft (115/200 V). If 400-Hz power is transmitted any appreciable distance at 115/200 V, the line losses are excessive. At 4160 V, 400-Hz power can be transmitted 2 or 3 mi [3 to 5 km] without serious power loss.

Fixed electric power is delivered to the aircraft by means of a multiple cable. This cable is either mounted on the side of the passenger bridge or stored in a pit adjacent to the aircraft parking space. When an aircraft arrives at the gate, the ground service personnel can quickly and easily connect the power cable and supply the aircraft with electric power.

It must be noted that many airports do not yet have fixed electric power available and it is therefore necessary to use mobile units or the APU. Regardless of what type of power supply is employed, the technician must follow carefully the procedures specified for the airplane that is being serviced.

Ground Air Supply

In addition to electric power, an airliner requires an adequate air supply. Low-pressure—35 to 45 psi [241 to 310 kPa]—high-volume air is required for starting engines, ventilation, heating, and cooling.

Air supply units are made in a variety of different types. Typical of these are turbine-engine compressors, diesel-engine-driven compressors, and electric-motor-driven compressors. The compressors may be axial flow, centrifugal, or screw type. The compressor and driving unit may be mounted on a trailer that requires towing, or it may be on a self-propelled vehicle such as a small truck. A unit of this type is shown in Fig. 14-20.

FIG. 14-20   A mobile air supply unit.   *(Atlas Copco Inc.)*

The air needed for servicing an airliner must be warm and oil free. Some types of compressors require that oil separators be installed in the outlet line to eliminate oil from the air. The screw-type compressor does not require an oil separator because no oil is used to lubricate the screw elements. The screws are precision-machined and timed so there is no contact between them as they rotate.

A cutaway view of a screw compressor is shown in Fig. 14-21. It consists of two screw helical rotors that mate with a very small clearance that does not allow an appreciable amount of air leakage. The male rotor has four lobes and turns 50 percent faster than the female rotor. The effect is to continuously compress the air as it moves from the inlet to the outlet. This is because the air space between the rotor lobes decreases continuously from the inlet to the outlet.

Air is drawn into the compressor through the inlet

FIG. 14-21   Cutaway view of a screw compressor.   *(Atlas Copco Inc.)*

port into the space between the lobes. As the rotors revolve further, the air inlet is sealed and the compression of the air begins. The rotary motion produces a smooth compression that continues until each groove in turn reaches the beginning of the outlet port. The air is then forced smoothly out of the compressor, and the outlet end is sealed again ready for the next cycle. No special inlet and outlet valves are needed, because both inlet and outlet ports are automatically covered and uncovered by the ends of the rotors.

The most effective and economical air supply for large aircraft is the stationary type. A system of this type is limited to new airports or those that are being extensively remodeled. This is because the air piping must be installed under the runways and aprons. The initial cost of an installation is higher than for the use of mobile units; however, the savings in operating costs quickly make up for the difference.

With a fixed air supply, there is one location for all compressors and the compressors are driven by efficient electric motors. Since electric-power costs are substantially lower than fuel costs, the economic advantages are considerable. The compressors are large screw-type units capable of starting several turbine engines at once. Maintenance costs are very low because of the trouble-free electric motors and screw-type compressors.

The fixed air system eliminates the need for ground units adjacent to the airplanes. This eliminates air and noise pollution and reduces traffic clutter around the airplanes.

● FUELING

Perhaps the most common service operation on an aircraft is the filling of fuel tanks. Improper fueling procedures have caused many aircraft accidents and in-flight incidents. Fueling personnel should be familiar with the fuel requirements for the models and types of aircraft they are servicing. The following paragraphs contain a description of problems that may be encountered in fueling aircraft and recommended procedures for combating these problems.

Fuel Contamination

Fuel is contaminated when it contains any material that was not provided under the fuel specification. This material generally consists of water, rust, sand, dust, microbial growth, and certain additives that are not compatible with the fuel, fuel system materials, and engines.

Water Contamination.   All aviation fuels absorb moisture from the air and contain water in both suspended particle and liquid form. The amount of suspended particles varies with the temperature of the fuel. Whenever the temperature of the fuel is decreased, some of the suspended particles are drawn out of the solution and slowly fall to the bottom of the tank. Whenever the temperature of the fuel increases, water is drawn from the atmosphere to maintain a saturated solution. Changes in fuel temperature, therefore, result in a continuous accumulation of water. During freezing temperatures, this water may turn to ice, restricting or stopping fuel flow.

There is no way of preventing the accumulation of water formed through condensation in fuel tanks. The accumulation is certain, and the rate of accumulation will vary; therefore it is recommended that storage tanks, fuel truck tanks, and aircraft fuel tanks be checked daily for the presence of water. Any water discovered should be removed immediately. Adequate settling time is necessary for accurate testing. The minimum settling time for aviation gasoline is 15 min per foot-depth of fuel and 60 min per foot-depth of turbine fuel. Testing storage tanks and fuel trucks may be done by attaching water-detecting paste or litmus paper to the bottom of the tank dip stick. The procedure is to push the dip stick to the bottom of the tank and hold it there for 30 s. When the stick is removed, the detecting paste or litmus paper will have changed color if water is present.

Microorganism Contamination. Many types of microbes have been found in unleaded fuels, particularly in the turbine engine fuels. The microbes, which may come from the atmosphere or storage tanks, live at the interface between the fuel and liquid water in the tank. These microorganisms of bacterial and fungi rapidly multiply and cause serious corrosion in tanks and may clog filters, screens, and fuel-metering equipment. The growth and corrosion are particularly serious in the presence of other forms of contamination.

Other Contaminations. Pipelines, storage tanks, fuel trucks, and drum containers tend to produce rust that can be carried in the fuel in small-size particles. Turbine fuels tend to dislodge rust and scale and carry it in suspension. Fuel may also be contaminated with dust and sand which enters through openings in tanks and from the use of fuel-handling equipment that is not clean.

Contamination Control

The presence of any contamination in fuel systems is dangerous. Laboratory and field tests have demonstrated that when water is introduced into the gasoline tank, it immediately settles to the bottom. Fuel tanks are constructed with sumps to trap this water. Most fuel dispensing equipment is equipped with a high degree of filtration to remove the liquid, dust, and rust particles from the fuel. It is advisable when fueling from drums to use a 5-micron (μm) filter or, as a last resort, a chamois-skin filter and filter funnel.

It is practically impossible to keep water from forming in airplane tanks; therefore it becomes necessary to regularly drain the airplane's fuel sumps in order to remove all water from the system. It may be necessary to gently rock the wings of some aircraft while draining sumps to completely drain all the water. On certain tailwheel-type aircraft, raising the tail to level flight attitude may result in additional flow of water to the gascolator or main fuel strainer. If left undrained, the water accumulates and will pass through the fuel line to the engine and may cause the engine to stop

operating. The elimination of contaminants from aviation fuel may not be entirely possible, but it can be controlled by the application of good housekeeping habits.

Aviation Gasoline Grades and Color Codes

Technicians and refueling personnel should be familiar with aviation gasoline (avgas) grades and respective color codes in order to assure proper servicing of engines. Three grades of avgas are now produced for civil use: grades 80, 100LL (low lead), and 100. These grades replace 80/87, 91/96, 100/130, and 115/145 avgas.

The lead quantity or concentration of lead in aviation gasoline is expressed in terms of milliliters ($\frac{1}{1000}$ of a liter) per gallon of avgas. The Standard Specification for Aviation Gasolines, Specification D 910-75, developed by the American Society for Testing and Materials, established that grade 80 should be red in color and contain 0.5 milliliter (mL) maximum of tetraethyl lead per gallon. Grade 100LL is blue in color and contains a maximum of 2.0 mL/gal. Grade 100 is green in color and contains a maximum of 3.0 mL/gal.

Grades 100LL and 100 represent two aviation gasolines which are identical in antiknock quality but differ in maximum lead content and color. The color identifies the difference for those engines which have a low tolerance to lead.

Limited availability of grade 80 in some geographical areas of the country has forced owners and operators to use the next higher grade of avgas. Specific use of higher grades is dependent on the applicable manufacturer's recommendations. Continuous use of higher-lead fuels in low-compression engines designed for low-lead fuels can cause erosion or necking of the exhaust valve stems and spark plug lead fouling.

Turbine Fuels

There are two types of jet fuel in common use today: (1) kerosene-grade turbine fuel, now named Jet A, and (2) a blend of gasoline and kerosene fractions, designated Jet B. There is a third type, called Jet A-1, which is made for operation at extremely low temperatures.

There is very little physical difference between Jet A (JP-5) fuel and commercial kerosene. Jet A was developed as a heavy kerosene having a higher flash point and lower freezing point than most kerosenes. It has a very low vapor pressure, so there is little loss of fuel from evaporation or boil-off at higher altitudes. It contains more heat energy per gallon than does Jet B (JP-4).

Jet B is similar to Jet A. It is a blend of gasoline and kerosene fractions. Most commercial turbine engines will operate on either Jet A or Jet B fuel. However, the difference in the specific gravity of the fuels may require fuel control adjustments. Therefore, the fuels cannot always be considered interchangeable. Both Jet A and Jet B fuels are blends of heavy distillates and tend to absorb water. The specific gravity of jet fuels, especially kerosene, is closer to water than is aviation gasoline; thus, any water introduced into the fuel, either through refueling or condensation, will take an appreciable time to settle out. At high altitudes, where low temperatures are encountered, water droplets combine with the fuel to form a frozen substance referred to as "gel." The mass of gel, or "icing," that may be generated from moisture held in suspension in jet fuel can be much greater than in gasoline.

Additives

Certain turbine-engine-powered aircraft require the use of fuel containing **anti-icing additives.** Therefore, fuel personnel must know whether or not the fuels they dispense contain additives. When anti-icing additives are to be added to the fuel, the manufacturer's instructions (usually printed on the container) should be followed to assure proper mixture. Anti-icing additive content in excess of 0.15 percent by volume of fuel is not recommended as higher concentration can cause the aircraft fuel capacitance system to give erroneous indications.

Certain oil companies, in developing products to cope with aircraft fuel icing problems, found that their products also checked "bug" growth. These products, known as "biocides," are usually referred to as additives. However, some additives may not be compatible with the fuel or the materials in the fuel system and may be harmful to other parts of the engine with which they come in contact. Additives that have not been approved by the manufacturer and FAA should not be used.

Fuel Tank Markings

Federal Aviation Regulations Part 23, Section 23.1557(c) (1), requires that aircraft fuel filler openings be marked to show the word "FUEL" and the minimum fuel grade or designation for the engines. Typical fuel tank markings are illustrated in Fig. 14-22.

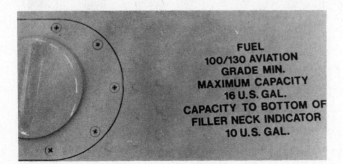

FIG. 14-22 Typical fuel tank markings.

It is equally important that tank vehicles be conspicuously marked to show the type of fuel carried. It is suggested that the marking be of a color in sharp contrast to that of the vehicle and in lettering at least 12 in tall. This marking should be on each side and on the rear of the tank vehicles. Additionally, it is suggested that the tank vehicle hose lines be marked by labels next to the nozzle and every 6 ft.

Fueling Precautions and Procedures

Fuel-dispensing vehicles and stationary facilities should be equipped with appropriate fire extinguishers, fire blankets, static grounding cables, explosion-proof flashlights, and ladders.

Fueling vehicles should be positioned as distant from the aircraft as permitted by the length of the fuel-dispensing hose. Mobile units should be parked parallel to or heading away from the aircraft wing leading edge so it may be moved away quickly in the event of an emergency.

Fueling personnel should first check with the flight crew to determine the type and grade of fuel required, including additives, for the aircraft. In the absence of the flight crew, fueling personnel should check the placard located near the aircraft fuel tank filler port or the aircraft flight manual to determine the type and grade of fuel required. Check to make sure that:

1. No electrical or radio equipment in the aircraft is energized or being maintained while fuel is being dispensed into the aircraft except those switches that may require energizing to operate fuel-selector valves and quantity gage systems.

2. Qualified personnel should be stationed at the aircraft fuel control panel during pressure fueling operations.

3. Fueling personnel should not carry objects in the breast pockets of their clothing when servicing aircraft or filling fuel service vehicles because loose objects may fall into fuel tanks.

4. Matches or lighters should never be carried during fueling operations.

5. Because of the high lead content, direct fuel contact with skin or the wearing of fuel-saturated clothing should be avoided. Skin irritation or blisters may result from direct contact with fuel.

6. Immediate medical attention should be sought if fuel enters the eyes.

7. In the event of fuel spillage, discontinue fueling operations until the spill can be removed, using proper safety precautions.

The following is a typical sequence that should be followed by fueling crew:

1. Connect a grounding cable from the fueling vehicle to a satisfactory ground. Grounding posts usually consist of pipes or rods driven far enough into the ground to result in a zero potential.

2. Connect a grounding cable from ground to the aircraft (on landing gear axle or other unpainted surface). Do not attach ground cables to the propeller or radio antenna.

3. Connect a grounding cable from the fueling vehicle to the aircraft. The fueling vehicle may be equipped with a T or Y cable permitting ground attachment first and the grounding of the aircraft with the other end.

4. Connect a grounding cable from the fuel nozzle to the aircraft before removing the aircraft tank cap. This bond is most essential and needs to be maintained throughout the fueling operation and until the fuel cap is replaced.

Overwing Fueling

When using the **overwing method,** the fuel-filler hose should be draped over the wing leading edge (as is demonstrated in Fig. 14-23). Never lay the fuel-filler hose over the wing trailing edge because aircraft structural damage may result. A simple rubber shower mat may be used to provide protection for wing leading edges during fuel operation. Step ladders or padded upright ladders may be used to provide easy access to high-wing and large aircraft. Avoid standing on wing surfaces and never stand on wing struts. Hold the fuel nozzle firmly while it is inserted in the fuel tank filler neck and never block the nozzle lever in the open position. Be sure that fuel filler caps are replaced and securely latched when fueling is completed.

FIG. 14-24 Pressure fueling

FIG. 14-23 Overwing fueling.

FIG. 14-25 Pressure fueling control panel.

Pressure Fueling

Most large late-model aircraft are fueled from a **single point** under the wing or fuselage as demonstrated in Fig. 14-24. The proper fuel distribution to the various tanks is regulated by the pilot at an on-board control panel as is pictured in Fig. 14-25. The fueling method referred to as **pressure fueling** greatly reduces the time required to fuel large aircraft. Other advantages in the pressure fueling process are that it eliminates aircraft skin damage and hazards to personnel as well as reduces the chances for fuel contamination. Pressure fueling also reduces the chance of static electricity igniting fuel vapors; thus the need for the static ground wire is eliminated.

● REVIEW QUESTIONS

1. What method is used to remove spilled gasoline from a hangar floor?
2. What are some safety precautions that should be followed when moving compressed gas cylinders?
3. What three ingredients are essential in causing a fire?
4. List the different classifications of fire.
5. Why are oil-soaked rags a fire hazard?
6. What type of fire extinguisher agent is recommended for a class B fire?
7. How may a class C fire extinguisher be identified?

8. List five precautions that should be followed in towing aircraft.
9. What three conditions are necessary for the starting of an internal-combustion engine?
10. Why should a reciprocating engine be rotated by hand or with the starter before it is started?
11. What check should be made immediately after starting a reciprocating engine?
12. What is a liquid lock and how is it detected and eliminated?
13. Describe the precautions that should be taken with respect to the surrounding area before starting a gas-turbine engine.
14. Describe briefly the manual starting of a gas-turbine engine.
15. What is the first indication that the fuel in a gas-turbine engine has ignited?
16. What is a *hot start* and what action should be taken when such a start occurs?
17. What is a *hung start?*
18. What precautions should be taken in taxiing small aircraft?
19. How are aircraft steered during taxiing?
20. When a taxiing aircraft must be signaled from the control tower, what light colors and signals are used?
21. By what means are aircraft tied down when parked?
22. What precaution must be observed in using hemp rope to tie down an airplane?
23. What is the purpose of *gust locks?*
24. Under what conditions is it advisable to leave the parking brakes off on a parked aircraft?
25. Why is it important to use especially designed jacking points when jacking aircraft?
26. What type of jack should be used for jacking aircraft?
27. What precautions must be observed in jacking aircraft?
28. Describe types of ground power units used for aircraft.
29. What are the advantages of fixed electrical power supplies over mobile units?
30. What type of electric power is required for large aircraft?
31. For what purposes is a *ground air supply* required on large aircraft?
32. How is air provided for ground service?
33. What are three types of fuel contamination?
34. How may water be detected in storage tanks?
35. What color is 100LL avgas?
36. What is done to prevent jet fuel from *gelling?*
37. What markings should be found adjacent to a fuel filler opening on an aircraft?
38. What are some general precautions that should be followed when fueling an aircraft?
39. What are two methods of fueling aircraft?
40. What are the advantages of *pressure fueling?*

15 AIRCRAFT INSPECTION AND SERVICING

● INTRODUCTION

Airplanes are designed and built to provide many years of service. For the airplane to remain airworthy and safe to operate, it should be operated in accordance with the recommendations of the manufacturer and cared for with sound inspection and maintenance practices. The Federal Aviation Regulations (FARs) require the inspection of all civil aircraft at specific intervals to make sure that the aircraft's condition is equal to its original or properly altered condition with regard to aerodynamic function, structural strength, and resistance to vibration.

Aircraft inspection may range from a casual "walk around" to a detailed inspection involving complete disassembly and the use of complex inspection aids. This chapter will discuss aircraft inspection requirements and practices as well as review activities such as servicing and lubrication that generally accompany inspections.

● REQUIRED AIRCRAFT INSPECTIONS

In establishing an aircraft's inspection requirements it is necessary to consider the aircraft's size and type as well as the purpose for which it is utilized and its operating environment.

Some aircraft must be inspected each 100 h of time in service while others must be inspected only once every 12 calender months.

The inspection requirements for aircraft in various types of operation are stated in FAR 91.169. Small aircraft usually fall under the requirements of **annual, 100-h,** and **progressive inspections.** Large aircraft (over 12 500 lb) and turbine-powered multiengine airplanes (turbojet and turboprop) fall under the jurisdiction of a different set of inspection programs.

The Annual and 100-h Inspection

The **annual** and **100-h inspections** are designed to provide a complete inspection of aircraft performed at specified intervals. The inspection will determine the condition of the aircraft and the maintenance required to return the aircraft to an acceptable condition of airworthiness.

For aircraft operating under FAR Part 91 the maximum interval between annual inspections is 12 **calendar months,** meaning it will again become due for inspection upon the last day of the same month, 12 months later. In addition to an annual inspection, aircraft operated commercially are also required to have a 100-h inspection. The procedures and scope of the

inspections are set forth in **Appendix D of FAR Part 43** and should be followed in detail (see Fig. 15-1). The regulations speak of 100-h and annual inspections as being of identical scope; the only difference between the two is the persons authorized to perform them. Certificated airframe and power plant maintenance technicians are authorized to perform a 100-h inspection. A certificated airframe and power plant maintenance technician holding an **inspection authorization** (IA) issued by the FAA may perform the annual inspection. An annual inspection may be substituted for a 100-h inspection. The 100-h time limitation may be exceeded by not more than 10 h, if necessary, to reach a place where the inspection can be performed. The excess time, however, is included in computing the next 100 h of time in service. As an example, an aircraft that flew 105 h between inspections would have only 95 h until the next inspection is due. However, the reverse does not apply. For example, an aircraft that has been inspected after only 90 h does not have 110 h before the next inspection. There is no provision for exceeding an annual inspection. To move an aircraft that is "out of annual" would require obtaining a special flight permit from the local FAA FSDO office. FAR 43.15 provides a list of rules for persons performing inspections. One of the rules is that a **checklist** must be used by a person performing an annual or 100-h inspection.

The technician (see Fig. 15-1) may use the checklist in FAR 43, Appendix D, the manufacturer's inspection checklist, or a checklist designed by the technician that includes the items listed in Appendix D to check the condition of the entire aircraft. In most instances, it is preferable to use the manufacturer's checklist since it was written specifically to include the procedures and details necessary to adequately inspect that particular make and model of aircraft (see Fig. 15-2). The checklist will usually be found in the aircraft maintenance manual.

Progressive Inspections

The **progressive inspection** system has been designed to schedule inspections of aircraft on a predetermined basis. The purpose of the program is to allow maximum utilization of the aircraft, to reduce maintenance inspection cost, and to maintain a maximum standard of continuous airworthiness. This system is particularly adaptable to larger multiengine aircraft and aircraft operated by companies and corporations where high utilization is demanded. A progressive inspection satisfies the complete airplane inspection requirements of both the 100-h and annual inspections. The

(a) Each person performing an annual or 100-hour inspection shall, before that inspection, remove or open all necessary inspection plates, access doors, fairing, and cowling. He shall thoroughly clean the aircraft and aircraft engine.

(b) Each person performing an annual or 100-hour inspection shall inspect (where applicable) the following components of the fuselage and hull group:

(1) Fabric and skin—for deterioration, distortion, other evidence of failure, and defective or insecure attachment of fittings.

(2) Systems and components—for improper installation, apparent defects, and unsatisfactory operation.

(3) Envelope, gas bags, ballast tanks, and related parts—for poor condition.

(c) Each person performing an annual or 100-hour inspection shall inspect (where applicable) the following components of the cabin and cockpit group:

(1) Generally—for uncleanliness and loose equipment that might foul the controls.

(2) Seats and safety belts—for poor condition and apparent defects.

(3) Windows and windshields—for deterioration and breakage.

(4) Instruments—for poor condition, mounting, marking, and (where practicable) for improper operation.

(5) Flight and engine controls—for improper installation and improper operation.

(6) Batteries—for improper installation and improper charge.

(7) All systems—for improper installation, poor general condition, apparent and obvious defects, and insecurity of attachment.

(d) Each person performing an annual or 100-hour inspection shall inspect (where applicable) components of the engine and nacelle group as follows:

(1) Engine section—for visual evidence of excessive oil, fuel, or hydraulic leaks, and sources of such leaks.

(2) Studs and nuts—for improper torquing and obvious defects.

(3) Internal engine—for cylinder compression and for metal particles or foreign matter on screens and sump drain plugs. If there is weak cylinder compression, for improper internal condition and improper internal tolerances.

(4) Engine mount—for cracks, looseness of mounting, and looseness of engine to mount.

(5) Flexible vibration dampeners—for poor condition and deterioration.

(6) Engine controls—for defects, improper travel, and improper safetying.

(7) Lines, hoses, and clamps—for leaks, improper condition, and looseness.

(8) Exhaust stacks—for cracks, defects, and improper attachment.

(9) Accessories—for apparent defects in security of mounting.

(10) All systems—for improper installation, poor general condition, defects, and insecure attachment.

(11) Cowling—for cracks, and defects.

(e) Each person performing an annual or 100-hour inspection shall inspect (where applicable) the following components of the landing gear group:

(1) All units—for poor condition and insecurity of attachment.

(2) Shock absorbing devices—for improper oleo fluid level.

(3) Linkage, trusses, and members—for undue or excessive wear, fatigue, and distortion.

(4) Retracting and locking mechanism—for improper operation.

(5) Hydraulic lines—for leakage.

(6) Electrical system—for chafing and improper operation of switches.

(7) Wheels—for cracks, defects, and condition of bearings.

(8) Tires—for wear and cuts.

(9) Brakes—for improper adjustment.

(10) Floats and skis—for insecure attachment and obvious or apparent defects.

(f) Each person performing an annual or 100-hour inspection shall inspect (where applicable) all components of the wing and center section assembly for poor general condition, fabric or skin deterioration, distortion, evidence of failure, and insecurity of attachment.

(g) Each person performing an annual or 100-hour inspection shall inspect (where applicable) all components and systems that make up the complete empennage assembly for poor general condition, fabric or skin deterioration, distortion, evidence of failure, insecure attachment, improper component installation, and improper component operation.

(h) Each person performing an annual or 100-hour inspection shall inspect (where applicable) the following components of the propeller group:

(1) Propeller assembly—for cracks, nicks, binds, and oil leakage.

(2) Bolts—for improper torquing and lack of safetying.

(3) Anti-icing devices—for improper operations and obvious defects.

(4) Control mechanisms—for improper operation, insecure mounting, and restricted travel.

(i) Each person performing an annual or 100-hour inspection shall inspect (where applicable) the following components of the radio group:

(1) Radio and electronic equipment—for improper installation and insecure mounting.

(2) Wiring and conduits—for improper routing, insecure mounting, and obvious defects.

(3) Bonding and shielding—for improper installation and poor condition.

(4) Antenna including trailing antenna—for poor condition, insecure mounting, and improper operation.

(j) Each person performing an annual or 100-hour inspection shall inspect (where applicable) each installed miscellaneous item that is not otherwise covered by this listing for improper installation and improper operation.

FIG. 15-1   FAR 43 Appendix D.

instructions and schedule for a progressive inspection must be approved by a representative of the Local District Office of the FAA having jurisdiction over the area in which the applicant for the progressive inspection is located. Approval for such an inspection system requires that a person holding an Inspection Authorization (IA) supervise the inspection procedures and that an inspection-procedures manual be availa-ble and readily understandable to the pilot and maintenance personnel.

The frequency and detail of the progressive inspection must provide for the complete inspection of the aircraft within each 12 calendar months and be consistent with the manufacturer's recommendations, field service experience, and the kind of operation in which the aircraft is engaged. The progressive inspec-

NOTE

Perform inspection or operation at each of the inspection intervals as indicated by a circle (O).

A. PROPELLER GROUP

Nature of Inspection	50	100	500	1000
1. Inspect spinner and back plate for cracks		○	○	○
2. Inspect blades for nicks and cracks		○	○	○
3. Check for grease and oil leaks		○	○	○
4. Lubricate propeller per Lubrication Chart		○	○	○
5. Check spinner mounting brackets for cracks		○	○	○
6. Check propeller mounting bolts and safety (Check torque if safety is broken)		○	○	○
7. Inspect hub parts for cracks and corrosion			○	○
8. Rotate blades of constant speed propeller and check for tightness in hub pilot tube			○	○
9. Remove constant speed propeller; remove sludge from propeller and crankshaft				○
10. Inspect complete propeller and spinner assembly for security, chafing, cracks, deterioration, wear and correct installation		○	○	○
11. Check propeller air pressure (at least once a month)	○			
12. Overhaul propeller			○	○

B. ENGINE GROUP

CAUTION: Ground Magneto Primary Circuit before working on engine.

Nature of Inspection	50	100	500	1000
1. Remove engine cowl		○	○	○
2. Clean and check cowling for cracks, distortion and loose or missing fasteners		○	○	○
3. Drain oil sump (See Note 2)	○	○	○	○
4. Clean suction oil strainer at oil change (Check strainer for foreign particles)	○	○	○	○
5. Clean pressure oil strainer or change full flow (cartridge type) oil filter element (Check strainer or element for foreign particles)	○	○	○	○
6. Check oil temperature sender unit for leaks and security		○	○	○
7. Check oil lines and fitting for leaks, security, chafing, dents and cracks (See Note 4)		○	○	○
8. Clean and check oil radiator cooling fins		○	○	○
9. Remove and flush oil radiators				○
10. Fill engine with oil per information on cowl or Lubrication Chart	○	○	○	○
11. Clean engine		○	○	○

CAUTION: Use caution not to contaminate vacuum pump with cleaning fluid. Refer to Lycoming Service Letter 1221A.

Nature of Inspection	50	100	500	1000
12. Check condition of spark plugs (Clean and adjust gap as required; adjust per Lycoming Service Instruction No. 1042.)		○	○	○
13. Check cylinder compression (Refer to AC 43.13-1A)		○	○	○
14. Check ignition harness and insulators (High tension leakage and continuity)		○	○	○

Nature of Inspection	50	100	500	1000
15. Check magneto points for proper clearance (Maintain clearance at .018 ± .006)		○	○	○
16. Check magneto for oil seal leakage		○	○	○
17. Check breaker felts for proper lubrication		○	○	○
18. Check distributor block for cracks, burned areas or corrosion and height of contact springs			○	○
19. Check magnetos to engine timing		○	○	○
20. Overhaul or replace magnetos (See Note 3)				○
21. Remove air filters and tap gently to remove dirt particles (Replace as required)	○	○	○	○
22. Clean fuel injector inlet line strainer (Clean injector nozzles as required) (Clean with acetone only)	○	○	○	○
23. Check condition of injector alternate air doors and boxes		○	○	○
24. Remove induction air box valve and inspect for evidence of excessive wear or cracks. Replace defective parts (See Note 7)			○	○
25. Inspect fuel injector attachments for loose hardware (See Note 8)			○	○
26. Check intake seals for leaks and clamps for tightness			○	○
27. Inspect all air inlet duct hoses (Replace as required)		○	○	○
28. Inspect condition of flexible fuel lines			○	○
29. Replace flexible fuel lines (See Note 3)				○
30. Check fuel system for leaks		○	○	○
31. Check fuel pumps for operation (Engine driven and electric)			○	○
32. Overhaul or replace fuel pumps (Engine driven and electric) (See Note 3)				○
33. Check vacuum pumps and lines			○	○
34. Overhaul or replace vacuum pumps (See Note 3)				○
35. Check throttle, alternate air, mixture and propeller governor controls for travel and operating condition		○	○	○

NOTE: Visually inspect the exhaust system per Piper Service Bulletin No. 373A at each 25 hours of operation. (See Note 10.)

Nature of Inspection	50	100	500	1000
36. Inspect exhaust stacks, connections and gaskets for cracks and loose mounting (Replace gaskets as required)	○	○	○	○
37. Inspect muffler, heat exchange, baffles and "augmentor" tube (See Note 6)	○	○	○	○
38. Check breather tubes for obstructions and security		○	○	○
39. Check crankcase for cracks, leaks and security of seam bolts		○	○	○
40. Check engine mounts for cracks and loose mountings		○	○	○
41. Check engine baffles for cracks and loose mounting		○	○	○
42. Check rubber engine mount bushings for deterioration (Replace as required)			○	○
43. Check fire wall seals			○	○
44. Check condition and tension of alternator drive belt	○	○	○	○
45. Check condition of alternator and starter		○	○	○
46. Check fluid in brake reservoir (Fill as required)	○	○	○	○
47. Inspect all lines, air ducts, electrical leads and engine attachments for security, proper routing, chafing, cracks, deterioration and correct installation		○	○	○
48. Lubricate all controls		○	○	○
49. Overhaul or replace propeller governor (See Note 3)				○
50. Complete overhaul of engine or replace with factory rebuilt (See Note 3)				○
51. Reinstall engine cowl	○	○	○	○

NOTES:

1. Both the annual and 100 hour inspections are complete inspections of the airplane, identical in scope, while both the **500 and 1000** hour inspections are extensions of the annual or 100 hour inspection, which require a more detailed examination of the airplane, and overhaul or replacement of some major components. Inspection must be accomplished by persons authorized by the FAA.

2. Intervals between oil changes can be increased as much as 100% on engines equipped with full flow (cartridge type) oil filters - provided the element is replaced each 50 hours of operation.

3. Replace or overhaul as required or at engine overhaul. (For engine overhaul, refer to Lycoming Service Instructions No. 1009.)

4. Replace flexible oil lines as required, but no later than 1000 hours of service.

5. Refer to Piper Service Letter No. 597 for flap control cable attachment bolt use.

6. Refer to Piper Service Bulletin No. 373A for exhaust system inspection.

7. Refer to Piper Service Bulletin No. 358.

8. Torque all attachment nuts to 135 to 150 inch-pounds. Seat "Pal" nuts finger tight against plain nuts and then tighten an additional 1/3 to 1/2 turn.

9. Inspect rudder trim tab for "free play;" must not exceed .125 inches. Refer to Service Manual for procedure, Section V. Refer to Piper Service Bulletin No. 390A.

10. Compliance with Piper Service Letter No. 673 eliminates repetitive inspection requirements of Piper Service Bulletin No. 373A and FAA Airworthiness Directive No. 73-14-2.

11. Piper Service Letter No. 704 should be complied with.

FIG. 15-2 Inspection schedule for a light airplane. *(Piper Aircraft Corp.)*

C. CABIN GROUP

Nature of Inspection	50	100	500	1000
1. Inspect cabin entrance, doors and windows for damage and operation		O	O	O
2. Check upholstery for tears		O	O	O
3. Check seats, seat belts, security brackets and bolts		O	O	O
4. Check trim operation		O	O	O
5. Check operation and condition of rudder pedals		O	O	O
6. Check parking brake handle and toe brakes for operation and cylinder leaks		O	O	O
7. Check control wheels, column, pulleys and cables		O	O	O
8. Check flap control cable attachment bolt per Note 5		O	O	O
9. Check landing, navigation, cabin and instrument lights		O	O	O
10. Check instruments, lines and attachments		O	O	O
11. Check manifold pressure gauge filters (2); replace as required		O	O	O
12. Check gyro operated instruments and electric turn and bank (Overhaul or replace as required)			O	O
13. Replace filters on gyro horizon and directional gyro or replace central air filter			O	O
14. Clean or replace vacuum regulator filter	O			
15. Check altimeter (Calibrate altimeter system in accordance with FAR 91.170, if appropriate)			O	O
16. Perform pitot-static tests, if appropriate (Refer to FAR 91.170)			O	O
17. Check operation of fuel selector valves		O	O	O
18. Check operation of fuel drains (See Note 11)		O	O	O
19. Check condition of heater controls and ducts		O	O	O
20. Check condition and operation of air vents		O	O	O

D. FUSELAGE AND EMPENNAGE GROUP

Nature of Inspection	50	100	500	1000
1. Remove inspection plates and panels		O	O	O
2. Check baggage doors, latches and hinges		O	O	O
3. Check battery, box and cables (Check at least every 30 days. Flush box as required and fill battery per instructions on box)	O	O	O	O
4. Check electronic installation		O	O	O
5. Check bulkheads and stringers for damage		O	O	O
6. Check antenna mounts and electric wiring		O	O	O
7. Check hydraulic pump fluid level (Fill as required)	O	O	O	O
8. Check hydraulic pump lines for damage and leaks		O	O	O
9. Check fuel lines, valves and gauges for damage and operation		O	O	O
10. Check security of all lines		O	O	O
11. Check vertical fin and rudder surfaces for damage		O	O	O
12. Check rudder hinges, horn and attachments for damage and operation		O	O	O
13. Check vertical fin attachments		O	O	O
14. Check ELT installation and condition of battery and antenna		O	O	O
15. Check rudder, tab hinge bolts for excess wear (Replace as required) (See Note 9)		O	O	O
16. Check rudder trim mechanism (See Note 9)		O	O	O
17. Check stabilator surfaces for damage		O	O	O
18. Check stabilator, tab hinges, horn and attachments for damage and operation		O	O	O

(CABIN GROUP, continued)

Nature of Inspection	50	100	500	1000
19. Check stabilator attachments		O	O	O
20. Check stabilator and tab hinge bolts and bearings for excess wear (Replace as required)		O	O	O
21. Check stabilator trim mechanism (Check tab free play) (Refer to Section V)		O	O	O
22. Check aileron, rudder, stabilator, stabilator trim cables, turnbuckles, guides and pulleys for safety, damage and operation		O	O	O
23. Inspect all control cables, air ducts, electrical leads and attaching parts for security, routing, chafing, deterioration, wear and correct installation		O	O	O
24. Clean and lubricate stabilator trim drum screw		O	O	O
25. Clean and lubricate all exterior needle bearings		O	O	O
26. Lubricate per Lubrication Chart	O	O	O	O
27. Check rotating beacon for security and operation		O	O	O
28. Check condition and security of Autopilot bridle cable and clamps		O	O	O
29. Reinstall inspection plates and panels		O	O	O

E. WING GROUP

Nature of Inspection	50	100	500	1000
1. Remove inspection plates and fairings		O	O	O
2. Check surfaces and tips for damage, loose rivets, and condition of walk-way		O	O	O
3. Check aileron hinges and attachments		O	O	O
4. Check aileron cables, pulleys and bellcranks for damage and operation		O	O	O
5. Check flaps and attachments for damage and operation		O	O	O
6. Check condition of bolts used with hinges (Replace as required)		O	O	O
7. Lubricate per Lubrication Chart		O	O	O
8. Check wing attachment bolts and brackets		O	O	O
9. Inspect all control cables, air ducts, electrical leads, lines and attaching parts for security, routing, chafing, deterioration, wear and correct installation	O	O	O	O
10. Check fuel tanks and lines for leaks and water		O	O	O
11. Remove, drain and clean fuel gascolator bowls (Drain and clean at least every 90 days)	O	O	O	O
12. Fuel tanks marked for capacity			O	O
13. Fuel tanks marked for minimum octane rating			O	O
14. Check fuel tank vents (Refer to Piper Service Bulletin No. 382)		O	O	O
15. Reinstall inspection plates and fairings		O	O	O

NOTES:

1. Both the annual and 100 hour inspections are complete inspections of the airplane, identical in scope, while both the **500 and 1000** hour inspections are extensions of the annual or 100 hour inspection, which require a more detailed examination of the airplane, and overhaul or replacement of some major components. Inspection must be accomplished by persons authorized by the FAA.
2. Intervals between oil changes can be increased as much as 100% on engines equipped with full flow (cartridge type) oil filters – provided the element is replaced each 50 hours of operation.
3. Replace or overhaul as required or at engine overhaul. (For engine overhaul, refer to Lycoming Service Instructions No. 1009.)
4. Replace flexible oil lines as required, but no later than 1000 hours of service.
5. Refer to Piper Service Letter No. 597 for flap control cable attachment bolt use.
6. Refer to Piper Service Bulletin No. 373A for exhaust system inspection.
7. Refer to Piper Service Bulletin No. 358.
8. Torque all attachment nuts to 135 to 150 inch-pounds. Seat "Pal" nuts finger tight against plain nuts and then tighten an additional 1/3 to 1/2 turn.
9. Inspect rudder trim tab for "free play," must not exceed .125 inches. Refer to Service Manual for procedure, Section V. Refer to Piper Service Bulletin No. 390A.
10. Compliance with Piper Service Letter No. 673 eliminates repetitive inspection requirements of Piper Service Bulletin No. 373A and FAA Airworthiness Directive No. 73-14-2.
11. Piper Service Letter No. 704 should be complied with.

FIG. 15-2 *(Continued)*

F. LANDING GEAR GROUP

Nature of Inspection	50	100	500	1000
1. Check oleo struts for proper extension (Check for proper fluid level as required)	O		O	O
2. Check nose gear steering control and travel		O	O	O
3. Check wheel alignment		O	O	O
4. Put airplane on jacks (Refer to Section II)		O	O	O
5. Check tires for cuts, uneven or excessive wear and slippage		O	O	O
6. Remove wheels, clean, check and repack bearings		O	O	O
7. Check wheels for cracks, corrosion and broken bolts		O	O	O
8. Check tire pressure (N-31 psi/M-50 psi)	O	O	O	O
9. Check brake lining and disc		O	O	O
10. Check brake backing plates		O	O	O
11. Check brake lines and retaining clamps		O	O	O
12. Check condition of center spring		O	O	O
13. Check gear forks for damage		O	O	O
14. Check oleo struts for fluid leaks and scoring		O	O	O
15. Check gear struts, attachments, torque links, retraction links and bolts for condition and security		O	O	O
16. Check downlocks for operation and adjustment			O	O
17. Check torque link bolts and bushings (Rebush as required)			O	O
18. Check drag end side brace link bolts (Replace as required)			O	O
19. Check gear doors and attachments		O	O	O
20. Check gear warning horn and light for operation		O	O	O
21. Check hydraulic fluid level in pump reservoir		O	O	O
22. Retract gear - check operation		O	O	O
23. Retract gear - check doors for clearance and operation		O	O	O
24. Check operation of squat switch		O	O	O
25. Check downlock switches, up-switches and electrical leads for security	O	O	O	O
26. Lubricate per Lubrication Chart		O	O	O
27. Remove airplane from jacks		O	O	O

G. OPERATIONAL INSPECTION

Nature of Inspection	50	100	500	1000
1. Check fuel pumps, fuel tank selector and crossfeed operation	O	O	O	O
2. Check fuel quantity and pressure or flow gauges	O	O	O	O
3. Check oil pressure and temperatures	O	O	O	O
4. Check alternator output	O	O	O	O
5. Check manifold pressure	O	O	O	O
6. Check alternate air	O	O	O	O
7. Check parking brake and toe brakes	O	O	O	O
8. Check vacuum gauge	O	O	O	O
9. Check gyros for noise and roughness	O	O	O	O
10. Check cabin heater operation	O	O	O	O
11. Check magneto switch operation	O	O	O	O
12. Check magneto RPM variation	O	O	O	O
13. Check throttle and mixture operation	O	O	O	O
14. Check propeller smoothness	O	O	O	O
15. Check constant speed propeller action	O	O	O	O
16. Check engine idle	O	O	O	O
17. Check electronic equipment operation	O	O	O	O
18. Check operation of controls	O	O	O	O
19. Check operation of flaps	O	O	O	O

H. GENERAL

Nature of Inspection	50	100	500	1000
1. Aircraft conforms to FAA Specifications	O	O	O	O
2. All FAA Airworthiness Directives complied with	O	O	O	O
3. All Manufacturers Service Letters and Bulletins complied with	O	O	O	O
4. Check for proper Flight Manual	O	O	O	O
5. Aircraft papers in proper order	O	O	O	O

NOTES:

1. Both the annual and 100 hour inspections are complete inspections of the airplane, identical in scope, while both the 500 and 1000 hour inspections are extensions of the annual or 100 hour inspection, which require a more detailed examination of the airplane, and overhaul or replacement of some major components. Inspection must be accomplished by persons authorized by the FAA.

2. Intervals between oil changes can be increased as much as 100% on engines equipped with full flow (cartridge type) oil filters - provided the element is replaced each 50 hours of operation.

3. Replace or overhaul as required or at engine overhaul. (For engine overhaul, refer to Lycoming Service Instructions No. 1009.)

4. Replace flexible oil lines as required, but no later than 1000 hours of service.

5. Refer to Piper Service Letter No. 597 for flap control cable attachment bolt use.

6. Refer to Piper Service Bulletin No. 373A for exhaust system inspection.

7. Refer to Piper Service Bulletin No. 358.

8. Torque all attachment nuts to 135 to 150 inch-pounds. Seat "Pal" nuts finger tight against plain nuts and then tighten an additional 1/3 to 1/2 turn.

9. Inspect rudder trim tab for "free play." must not exceed .125 inches. Refer to Service Manual for procedure, Section V. Refer to Piper Service Bulletin No. 390A.

10. Compliance with Piper Service Letter No. 673 eliminates repetitive inspection requirements of Piper Service Bulletin No. 373A and FAA Airworthiness Directive No. 73-14-2.

11. Piper Service Letter No. 704 should be complied with.

FIG. 15-2 (Continued)

tion schedule must ensure that the aircraft, at all times, shall be airworthy and shall conform to all applicable Aircraft Specifications, Type Certificate Data Sheets, Airworthiness Directives, and other approved data such as service bulletins and service letters issued by the manufacturer. If the progressive inspection is discontinued, the owner or operator shall immediately notify the local General Aviation District Office of the FAA in writing. After the discontinuance, the first annual inspection is due within 12 calendar months after the last complete inspection under the progressive inspection schedule. The 100-h inspection required by FAR Part 91 is due within 100 h of operation after that same complete inspection.

A typical progressive inspection schedule is shown in Fig. 15-3. Under this program, the airplane is inspected and maintained in four operations, called **"events,"** scheduled at 50-h intervals. The events are arranged so that a 200-h flying **cycle** results in a complete aircraft inspection.

An Event Inspection is a group of several predetermined location inspections, both **"routine"** and **"detailed"** as shown in Fig. 15-3. A Routine Inspection consists of a visual examination or check of the aircraft and its components and systems insofar as is practicable without disassembly. The Detailed Inspections consist of a thorough examination of the aircraft and its components and systems with such disassembly as is necessary. Figure 15-4 shows typical routine and detailed inspections for a propeller.

An event may be conducted and recorded in the event-record by an airframe and power plant mechanic. When the four "events" are complete and recorded, an entry should be made in the cycle record by the IA who is supervising the inspection.

With a schedule such as is shown in Fig. 15-3, unless an aircraft is operated a minimum of 200 h a year, it would not be beneficial to place it on a progressive inspection since one of the requirements of a progressive schedule is that it be completed at least every 12 calendar months. The decreased maintenance cost and increased utilization stem from the fact that the inspection is divided up into smaller segments than a 100-h inspection and also that some parts of the airplane such as the fuselage, wings, and empennage receive a detailed inspection only every 200 h.

● LARGE AND TURBINE-POWERED MULTIENGINE AIRPLANES

Inspection Programs

Large aircraft (over 12 500 lb) and **turbine-powered multiengine aircraft** are excluded from using 100-h or annual and progressive inspections. Due to the complexity of the aircraft, the inspection programs for these aircraft tend to be more specific than 100-h or annual inspections. FAR Section 91.169 provides four options to the owner or operator in the selection of an inspection program:

Option 1. A continuous airworthiness inspection program that is a part of a continuous airworthiness maintenance program currently in use by a person holding a certificate issued under FAR Part 121.

Option 2. An approved aircraft inspection program currently in use by a person holding an air taxi certificate under FAR Part 135.

Option 3. A current inspection program recommended by the manufacturer.

Event Inspection #1

To be performed at the 50–250–450–650–850 Flying Hour Intervals.
Consist of:
1. Engine Routine   4. Cabin Detail
2. Propeller Routine   5. Operational Inspection
3. Fuselage Detail

Event Inspection #2

To be performed at the 100–300–500–700–900 Flying Hour Intervals.
Consist of:
1. Engine Detail   4. Landing Gear Routine
2. Empennage Detail   5. Operational Inspection
3. Propeller—Routine

Event Inspection #3

To be performed at the 150–350–550–750–950 Flying Hour Intervals.
Consist of:
1. Engine Routine   4. Fuselage—Routine
2. Propeller Detail   5. Landing Gear—Detailed
3. Wings Detailed   6. Operational Inspection

Event Inspection #4

To be performed at the 200–400–600–800–1000 Flying Hour Cycles.
Consist of:
1. Engine Detail   3. Landing Gear—Routine
2. Propeller Routine   4. Operational Inspection

FIG. 15-3 Sample Progressive Inspection Schedule.

Routine propeller inspection	Detailed propeller inspection
1. Check spinner attachments. 2. Inspect propeller for condition	1. Remove and inspect spinner and back plate. 2. Inspect blades for nicks and cracks. 3. Check for grease and oil leaks. 4. Lubricate per lubrication chart. 5. Check spinner mounting brackets for cracks. 6. Check propeller mounting bolts for safety. 7. Inspect hub parts for cracks and corrosion. 8. Check blades for tightness in hub pilot tube. 9. Check propeller air pressure. Refer to Pressure Temperature Chart. 10. Check condition of propeller deicer system.

FIG. 15-4   Sample routine and detailed propeller inspections.

Option 4. Any other inspection program established by the registered owner or operator of the airplane and approved by the Administrator under FAR Section 91.169(g).

In the first three options listed, the provision "current" is mentioned. This current requirement is intended to prevent use of obsolete programs.

Continuous Airworthiness Inspection Programs

Air carriers operating under FAR Part 121 are required to have a **continuous airworthiness maintenance program.** A continuous airworthiness maintenance program is a compilation of the individual maintenance and inspection functions utilized by an operator to fulfill its total maintenance needs. Continuous airworthiness maintenance programs are included in the maintenance section of an air carrier's "Operations Specifications" approved by the Federal Aviation Administration. These specifications prescribe the scope of the program, including limitations, and they reference manuals and other technical data as supplements to these specifications. Following are the basic elements of continuous airworthiness maintenance programs:

1. Aircraft inspection. This element deals with the routine inspections, servicing, and tests performed on the aircraft at prescribed intervals. It includes detailed instructions and standards (or references thereto) by work forms, job cards, etc., which also serve to control the activity and to record and account for the tasks that comprise this element.

2. Scheduled maintenance. This element concerns maintenance tasks performed at prescribed intervals. Some are accomplished concurrently with inspection tasks that are part of the inspection element and may be included on the same form. Other tasks are accomplished independently. The scheduled tasks include replacement of life-limited items, components requiring replacement for periodic overhaul, special inspections such as x-rays, checks, or tests for on-condition items, lubrications, etc. Special work forms can be provided for accomplishing these tasks or they can be specified by a work order or some other document. In any case, instructions and standards for accomplishing each task should be provided to ensure its proper accomplishment and that it is recorded and signed for.

3. Unscheduled maintenance. This element provides instructions and standards for the accomplishment of maintenance tasks generated by the inspection and scheduled maintenance elements, pilot reports, failure analyses, or other indications of a need for maintenance. Procedures for reporting, recording, and processing inspection findings, operational malfunctions, or abnormal operations such as hard landings are an essential part of this element. A continuous aircraft logbook can serve this purpose for occurrences and resultant corrective action between scheduled inspections. Inspection discrepancy forms are usually used for processing unscheduled maintenance tasks in conjunction with scheduled inspections. Instructions and standards for unscheduled maintenance are normally provided by the operator's technical manuals. The procedures to be followed in using these manuals and for recording and certifying unscheduled maintenance are included in the operator's procedural manual.

Approved Aircraft Inspection Program

The **Approved Aircraft Inspection Program** (AAIP) concept was first developed for the benefit of FAR Part 135 air taxi operators who requested regulatory authority to develop and utilize inspection programs more suitable to aircraft in their operating environments than the conventional 100-h or annual inspections required by Part 91.

The AAIP allows each operator to develop a program which is tailored to meet their particular needs in satisfying aircraft inspection requirements. It provides for the operator to adjust the intervals between individual inspection tasks in accordance with the needs of the aircraft rather than repeat all tasks at each 100-h increment. It also allows them to develop procedures and standards for the accomplishment of those tasks. Along with these benefits is the responsibility to achieve an acceptable level of equivalency to the conventional Part 91 inspection requirements. The AAIP serves as the operator's specification for each segment of the program. This is in contrast to the 100-h or annual inspection wherein the performing mechanic or repair station determines, in accordance with Appendix D of Part 43, what work is required. Under the AAIP, the operator is responsible for the program content and standards, and the performing mechanic or repair station is responsible for accomplishment of the inspection as specified by worksheets and other criteria designated by the program.

An approved aircraft inspection program should encompass the total aircraft including all installed equipment such as communications and navigational gear, cargo provisions, etc. It should include a schedule of the individual tasks or groups of tasks that comprise the program and their frequency of accomplishment.

Manufacturer's Recommended Inspection Program

One of the more popular ways to satisfy the inspection requirements of FAR 91.169 is by the adoption of an aircraft manufacturer's program. Under this arrangement, the aircraft manufacturer's program including methods, techniques, practices, and standards for its accomplishment, along with inspection intervals, is adopted in its entirety.

The aircraft manufacturer's program in most cases contains the frequency and the extent of maintenance necessary for the aircraft, engine, propeller, and rotors. It may also include the frequency of overhauls and the life limit of components requiring replacement.

Operator-Developed Inspection Program

An operator-developed inspection program is developed and published by the operator. It must include methods, techniques, practices, and standards necessary for proper accomplishment of the program. These programs are usually not developed from scratch but instead are continuous or are manufacturer's programs that have been modified to suit the operator's particular need. If the program is, in effect, a manufacturer's maintenance program with variations such as a higher engine overhaul period, that variation categorizes it as an operator-developed program, not an adoption of a manufacturer's program. An operator-developed program bears no prior FAA approval.

Significant variation from the manufacturer's recommendations have to be fully substantiated by the applicant and the program being approved must provide an alternative action which will ensure an equivalent level of safety. This includes such things as man-

ufacturer-recommended special inspections and special structural inspections.

Walk-Around Inspections

To keep an aircraft in proper operating condition and to locate defects that arise between major inspections, manufacturers will recommend various types of **walk-around** and **preflight** inspections. Frequent minor inspections of airliners are conducted by flight engineers or maintenance personnel. The inspections are usually conducted at every stop at which time permits. The inspector carries a flashlight and checks the interior of the tailpipe for the condition of the turbines and thrust reversers. The inspector further checks tires, looks for fluid leaks, and examines the fuselage and control surfaces for wrinkles and any other condition that indicates deterioration or damage.

Daily and Preflight Inspections

Before the first flight of an airplane each day, a daily inspection should be performed. This inspection will usually include the following checks:

1. Check the fuel tanks for quantity by removing the fuel caps and observing the level of fuel in the tanks. A dip stick is sometimes necessary.

2. Check the quantity of oil in the oil tank by means of a dip stick or sight gage.

3. Drain a small amount of fuel from each of the fuel drains and strainers. This is to ensure that sediment and water are removed from the tanks.

4. Check the inflation of all tires.

5. Check the extension of shock struts to ensure that they are properly inflated.

6. Check the engine compartment for loose wires and fittings. Visually inspect spark-plug leads, nuts, air ducts, exhaust pipes, controls, and accessories. Check for oil leaks.

7. Examine the propeller blades for nicks, cuts, and evidence of any other damage.

8. With the ignition switch off, turn the propeller by hand at least two revolutions and note any unusual noises or other indication of malfunction. Stand clear of the propeller's plane of rotation.

9. Examine cowling and inspection plates for proper fastening.

10. Inspect hinges and control attachments for all control surfaces. Test each control for freedom of movement.

11. Visually inspect the exterior of the aircraft for damage, loose parts, or any other unsatisfactory condition. Check for fluid leaks.

12. Inspect the windshield, windows, and doors for condition. Check the door or doors for proper latching and locking.

13. Inspect the interior of the cabin and cockpit, including seat belts, seats, and loose parts on the floor.

14. Test the operation of all controls from within the cockpit including operation of the brakes.

15. After starting the engine, check the operation of all instruments, radio, and propeller (if the propeller is a constant-speed type). See that engine oil pressure is indicated within 30 s after the engine starts.

16. Test the operation of the landing, navigation, cabin, and instrument lights.

If the aircraft is flown several times during the day, it is not necessary to make all the inspections listed above; however, a general check of the aircraft, including the quantity of fuel and oil, should be made before each flight.

Special Inspection

In addition to the regularly scheduled inspections, many manufacturers provide for special inspections that are to be performed in the event that the aircraft is subjected to stresses outside of its normal operating environment. These inspections may be for such events as lightning strikes, sudden engine stoppage, severe wind gust loads, or extreme hard landings. Since these inspections tend to be very specific in nature, it is necessary to refer to the manufacturer's maintenance manual for performance details.

Altimeter and Transponder Inspections

In order to operate an airplane under Instrument Flight Rules in controlled airspace it must have the altimeters and the static system inspected in accordance with FAR Part 43 Appendix E every 24 calendar months. This test is usually performed by an appropriately rated repair station; however, airframe technicians may perform the test and inspection on the static pressure system.

ATC transponders are required to have an inspection every 24 calendar months in accordance with FAR 43 Appendix F, if they are to be operated. Although altimeter and transponder inspections are primarily the pilot's responsibility, technicians can provide valuable assistance by pointing out overdue inspections when they are reviewing the maintenance records.

● CONDUCTING A 100-H OR ANNUAL INSPECTION

Although specific inspection practices and procedures will vary depending on the size of aircraft and type of inspection being conducted, the basic fundamentals followed in conducting an inspection do not change.

Inspection Preparation

The inspection process commences with the owner requesting that an inspection be performed on the airplane. At this point a work order should be filled out itemizing just what work is to be done. A firm understanding should be reached about the cost of the inspection as well as discussing what is included in this cost, e.g., servicing, lubrication, airworthiness directive compliance, etc., and the approximate time period planned for completing the inspection. In order to gain a better understanding of the history and present condition of the aircraft, it will be necessary to obtain the airplane's maintenance records for thorough study and review.

An important part of the preinspection process is researching airworthiness directives and service bulletins. The technician must determine whether all applicable airworthiness directives on the aircraft, power plant, propeller, instruments, and appliances have actually been accomplished.

If the maintenance records indicate compliance with an AD, the technician should make a reasonable attempt to verify this. The reason for this is that it is not uncommon for a component to have an AD complied with and properly recorded and then later be replaced by another on which the AD had not yet been accomplished.

The FAA General Aviation Airworthiness Alerts (AC43-15) are also an important source of service experience. These alerts are selected service difficulties reported to the FAA on Malfunction or Defect Reports. It makes sense to utilize the experience other people have had on similar products. These publications will help in not overlooking problem areas.

Prior to beginning the inspection, the checklist to be utilized should be located along with discrepency forms and the appropriate maintenance manual and Type Certificate Data Sheet.

The tools needed to perform the inspection should be readied. Inspection tools can be many and varied ranging from a pocket-sized magnifying glass to a complete x-ray machine. The principal tools for most inspectors are a flashlight and an inspection mirror; however, additional items such as a magneto timing light, compression tester, and jacks are among other tools usually required.

Opening and Cleaning

FAR 43, Appendix D, which lists the scope and detail of items to be included in a 100-h or annual inspection begins by saying:

> Each person performing an annual or 100-h inspection shall, before that inspection, remove or open all necessary inspection plates, access doors, fairing, and cowling. He shall thoroughly clean the aircraft and aircraft engine.

New technicians often have difficulty in knowing which inspection plates and panels must be removed. Many manufacturers provide assistance in the form of an inspection panel diagram such as is illustrated in Fig. 15-5. When opening inspection plates and cowlings, the technician should take note of any oil or other foreign material accumulation which may offer evidence of fluid leakage or other abnormal condition that should be corrected.

An engine and accessories wash-down should be accomplished prior to each 100-h inspection to remove oil, grease, salt corrosion, or other residue that might conceal component defects during inspection. Precautions, such as wearing rubber gloves, an apron or coveralls, and a face shield or goggles, should be taken when working with cleaning agents. Use the least toxic of available cleaning agents that will satisfactorily accomplish the work. These cleaning agents include: (1) Stoddard solvent, (2) a water-base alkaline detergent

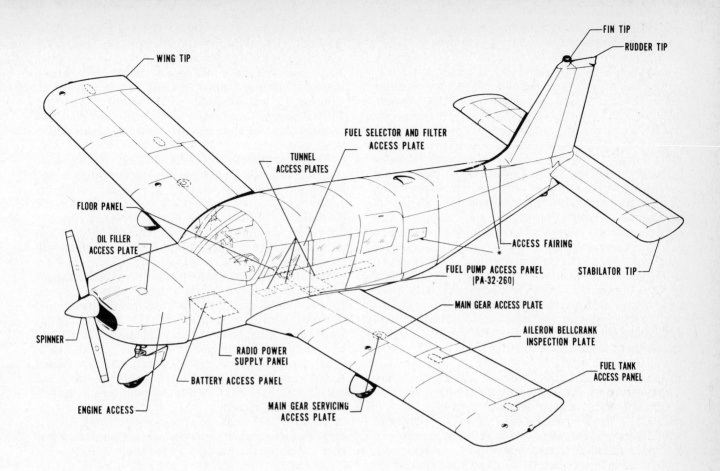

FIG. 15-5   Access plates and panels.

cleaner mixed 1 part cleaner, 2 to 3 parts water, and 8 to 12 parts Stoddard solvent, or (3) a solvent-base emulsion cleaner mixed 1 part cleaner and 3 parts Stoddard solvent.

WARNING: *Do not use gasoline or other highly flammable substance for wash-down.*

Perform all cleaning operations in well-ventilated work areas, and ensure that adequate fire-fighting and safety equipment is available. Compressed air, used for cleaning agent application or drying, should be regulated to the lowest practical pressure. Use of a stiff bristle fiber brush, rather than a steel brush, is recommended if cleaning agents do not remove excess grease and grime during spraying.

Before cleaning the engine compartment, place a strip of tape on the magneto vents to prevent any solvent from entering these units.

Place a large pan under the engine to catch waste. With the engine cowling removed, spray or brush the engine with solvent or a mixture of solvent and degreaser. In order to remove especially heavy dirt and grease deposits, it may be necessary to brush areas that were sprayed.

WARNING: *Do not spray solvent into the alternator, vacuum pump, starter, or air intakes.*

Allow the solvent to remain on the engine from 5 to 10 min. Then rinse the engine clean with additional solvent and allow it to dry. Cleaning agents should never be left on engine components for an extended period of time. Failure to remove them may cause damage to components such as neoprene seals and silicone fire sleeves and could cause additional corrosion. Completely dry the engine and accessories using clean, dry compressed air. If desired, the engine cowling may be washed with the same solvent.

Remove the protective tape from the magnetos and lubricate the controls, bearing surfaces, etc., in accordance with the Lubrication Chart.

Other parts of an airplane that often need cleaning prior to inspection are the landing gear and the underside of the aircraft. Most compounds used for removing oil, grease, and surface dirt from these areas are emulsifying agents. These compounds, when mixed with petroleum solvents, emulsify the oil, grease, and dirt. The emulsion is then removed by rinsing with water or by spraying with a petroleum solvent. Openings such as air scoops should be covered prior to cleaning.

Inspecting the Airplane

The most important part of any inspection is the visual examination given to the airplane to determine its airworthiness. All the related steps in the inspection process, such as service and repair, are dependent on a thorough visual exam. On a 100-h or annual inspection this step cannot be supervised but instead must be performed by the person who is returning the aircraft to service. As required by FAR 43.15, a checklist must be utilized and should be carefully followed. While performing the visual inspection, the inspector should avoid getting sidetracked on related service and repair problems but instead should make a written list of all discrepancies found on the aircraft. A **discrepancy report** such as is shown in Fig. 15-6 should provide a location for the discrepancy writeup, the corrective action taken, the technician making the repair, and the inspector returning the aircraft to service.

The principal purpose in performing an inspection is to determine if the aircraft is airworthy. In order for an airplane to be declared airworthy it must meet two criteria: It must be in condition for safe operation, and it must conform to its Type Certificate Data Sheet. Safe operation refers to the condition of the aircraft with relation to wear and deterioration. Some general inspection guidelines that may be followed in making this determination are inspection of:

Metal parts for: Security of attachment, cracks, metal distortion, broken spotwelds, corrosion, condition of paint, and any other apparent damage
Movable parts for: Lubrication, servicing, security of attachment, binding, excessive wear, safetying, proper operation, proper adjustment, correct travel, cracked fittings, security of hinges, defective bearings, cleanliness, corrosion, deformation, sealing, and tension
Bolts in critical areas for: Proper installation, correct safetying, and correct torque values when visual inspection indicates the need for a torque check

Wiring for: Security, chafing, burning, defective insulation, loose or broken terminals, heat deterioration, and corroded terminals
Fluid lines and hoses for: Leaks, cracks, dents, kinks, chafing, proper radius, security, corrosion, deterioration, obstruction, and foreign matter
Filters, screens, and fluids for: Cleanliness, contamination, and/or replacement at specified intervals

The manufacturer's checklist and maintenance manual will provide assistance in determining which items and tolerances need to be checked during the inspection.

Conformity to its Type Certificate Data Sheet is considered attained when the required and proper components are installed and they are consistent with the drawings, specifications, and other data that is a part of the type certificate. In addition to providing for required equipment, the data sheet will provide such information as static rpm, control surface travel, and if a flight manual is required. It also identifies limitations which must be displayed in the form of markings and placards. Any deviation from the type design is considered a major alteration and should be provided for in an FAA Form 337. An aircraft that does not conform to its Type Certificate Data Sheet is considered to be unairworthy.

An additional item included on most checklists is a review of the **aircraft paperwork**. The **aircraft registration** and **airworthiness certificate** should be located on board the aircraft. If these certificates are not available, the aircraft should not be reported as unairworthy. Instead, the owner should be informed that the documents must be in the aircraft, with the airworthiness certificate displayed as required in FAR 91.27, when the aircraft is operated.

PAGE _____ DISCREPANCY FORM

A/C MAKE/MODEL _____ S/N _____ N- _____

TYPE INSP. _____ TACH TIME _____ DATE _____

ITEM NO.	DISCREPANCY	AUTH.	CORRECTIVE ACTION	MECH.	INSP.
1	Cabin door seal is loose	*gl*	Reglued seal with #1300L	*K*	*MK*
2	Copilot's seat has rip in back	N/A	No action	—	—
3	Left magneto drop 250 rpm	*gl*	Installed new points Performed internal and external timing	*K*	*MK*
4	Left flap chaffing inboard side	*gl*	Adjusted flap clearance Checked travel	*K*	*MK*
5	Left aileron outboard hinge nutplate cracked	*gl*	Replaced nutplate #PN13762-5	*K*	*MK*

FIG. 15-6 Sample Discrepancy Form.

Other documents often needed but not a part of the airworthiness requirements might be state registration, FCC radio station licenses, etc. The owner or operator is responsible for the proper display of these documents. However, the technician will be performing an appreciated service by informing the operator of any deficiencies in the display and carriage of these documents.

At the conclusion of the visual inspection, a meeting with the airplane's owner is usually in order to discuss the discrepancies located during the inspection. Discrepancies need to be divided into two categories: those that affect the airworthiness of the airplane and those that do not. Those items not affecting the airworthiness of the airplane may have corrective action delayed to a later date. For those discrepancies that affect airworthiness, the owner has the option of having them repaired by the maintenance technician performing the inspection or using the procedures specified in FAR 43.11 and having the aircraft declared unairworthy. This will permit an owner to assume responsibility for having the discrepancies corrected prior to operating the aircraft. The discrepancies can be cleared by an A&P technician unless they consist of major repairs or major alterations. If the repairs are preventive maintenance, they could be cleared by the owner or pilot.

There is no stigma attached to the aircraft because it is reported "unairworthy." In effect, the report says the aircraft is airworthy with the exception of the items on the discrepancy list. When those listed items are corrected, the aircraft is eligible to be operated. The owner may want the aircraft flown to another location to have repairs completed in which case the owner should be advised that a **Special Airworthiness Certificate,** FAA Form 8130-7 (formerly referred to as a ferry permit), is necessary. A Special Airworthiness Certificate may be obtained at the local FAA district office.

● LUBRICATION AND SERVICING

The visual inspection is required to determine the current condition of the aircraft and its components. The repair of discrepancies is required to bring the aircraft back up to airworthy standards. In an effort to keep the aircraft in airworthy condition, the manufacturer may recommend that certain services be performed at various operating intervals. While servicing and lubrication is often conveniently accomplished during an inspection, it should not be considered a part of the inspection itself.

Lubrication

As is true of any item of machinery where moving parts bear against one another, lubrication is required at many locations in an aircraft. The type of lubrication for each point is determined by the type of bearings, bearing loads, frequency and speed of movement, temperatures at the bearing, and the materials that bear against one another. Lubricants used for aircraft may be ordinary lubricating oil such as that used in the engine, lightweight lubricating oil, various

weights of greases, high-pressure (HP) grease, low-temperature grease, high-temperature grease, graphite, and other special lubricants.

The frequency of lubrication for each lubrication point in an aircraft is specified by the manufacturer together with the type of lubricant, the method of application, and special instructions. Information regarding lubrication is provided in the manufacturer's maintenance manual for each model of aircraft.

Commercial airlines develop their own schedules and procedures for lubrication of their aircraft. Lubrication is usually accomplished along with the progressive maintenance, inspections, and other service operations.

Lubrication information for light airplanes is often presented in the form of charts and tables. Figure 15-7 is a chart for a light-twin airplane. Information and instructions for use of the chart are shown in Fig. 15-8. Parts nomenclature is given in Fig. 15-9.

Some general guidelines that should be followed in the application of lubricants are as follows:

● Cleanliness is essential to good lubrication. Lubricants and dispensing equipment must be kept clean. Use only one lubricant in a grease gun or oil can.
● Store lubricants in a protected area. Containers should be closed at all times when not in use.
● Wipe grease fittings, oil holes, etc., with clean, dry cloths before lubricating.
● When lubricating bearings which are vented, force grease into fittings until all old grease is extruded, unless otherwise noted.
● When lubricating sealed bearings, use extreme care not to dislodge the seals.
● Work moving parts, if practical, to assure thorough lubrication.
● After any lubrication, clean surplus lubricant from all but actual working surfaces.

● SERVICING AIRCRAFT

Servicing aircraft requires great care and attention to detail, regardless of the type of aircraft being serviced or the particular service being performed. In this section we shall discuss general principles involved in engine, aircraft, and system servicing. The servicing of particular aircraft and components is usually described in great detail in the manufacturers' maintenance and service manuals. Instructions given in these manuals should be followed for satisfactory and safe performance of aircraft.

Engine Oil Service

Oil quantity for small aircraft should be checked daily or before each flight. The dip stick is marked to show the maximum quantity of oil permitted in the engine sump or tank. If the dip stick shows that the oil is near or below the required minimum level, oil must be added. In no case should the oil level be permitted to rise above the maximum line. The oil sump or tank must have sufficient **foaming space** to allow for expansion of the oil and the development of foam.

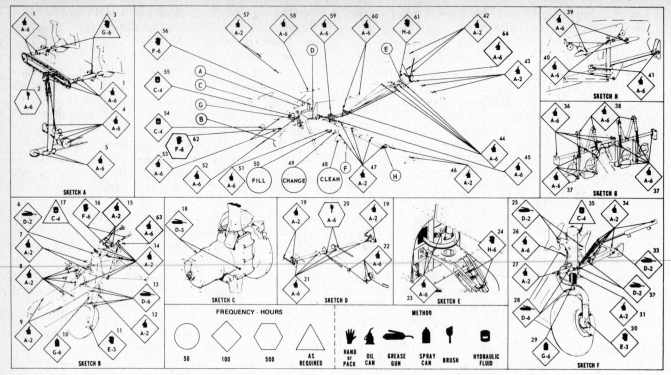

FIG. 15-7   Lubrication chart for a light twin-engine airplane.   *(Piper Aircraft Corp.)*

Aircraft manufacturers usually give specific instructions regarding the engine oil service. Typical of such instructions are those provided by the Cessna Aircraft Company for the Model 421 airplane:

Check oil level before each flight. Capacity for each engine oil sump is 13 U.S. quarts, which include 1 quart for oil filter. (Do not take off on less than 9 quarts.) When preflight shows less than 9 quarts, service with aviation grade engine oil, SAE 30 below 40°F, SAE 50 above 40°F. The aircraft is delivered from the factory with straight mineral oil; therefore it will be necessary during the break-in period or first 50 hours of operation to add straight mineral oil. Multiviscosity oil with a range of SAE 10W30 is recommended for improved starting and turbocharger controller operation in cold weather. Detergent or dispersant oil conforming to Continental Motors Specification MHS-24A must be used after the first 50 hours of new or overhauled engine operation.

For many reciprocating engines that use petroleum oils, it is recommended that the oil be changed after every 50 h of operation. At the same time, the oil screens should be cleaned and filters should be replaced. If an aircraft is not operated frequently, the oil should be changed at least every 4 months, even though the aircraft has not operated for 50 h. Some manufacturers recommend changing the oil at intervals of not more than 90 days. If the oil on the dip stick appears dirty when the oil quantity is being checked, the oil should be changed, the screens cleaned, and the filter replaced.

Gas-turbine engines generally utilize high-temperature synthetic lubricants conforming to MIL-L-7808 or MIL-L-23699. These lubricants are now commonly tested by chemical means to determine their quality. Oil changes are made only if the chemical test indicates that the oil is no longer satisfactory for lubrication.

When changing or adding oil to a gas-turbine-engine system, the technician must be certain that the correct type and grade of lubricant are used. In handling synthetic lubricants, care must be taken to prevent the lubricant from spilling or coming into contact with the skin. If the lubricant should come into contact with the hands or any other part of the body, it should be wiped off immediately and the affected area should be washed with soap and water.

Oil changes for jet aircraft are governed by the approved service and maintenance procedures established by the airline operating the aircraft. The technicians performing service operations will be thoroughly informed regarding the procedures to be employed. These procedures are developed through cooperative planning with the manufacturers of the engines and aircraft.

Servicing Brake and Hydraulic Systems

Brake and hydraulic systems must be serviced in accordance with manufacturers' instructions. As long as these systems operate satisfactorily, service is not normally required at intervals of less than 100 h of operation. When it is necessary to add fluid, it is most important that the correct fluid be added. If the wrong fluid is used, it will be necessary to drain and flush the system and probably change all the seals in compo-

TYPE OF LUBRICANT

IDENTIFICATION LETTER	SPECIFICATION	LUBRICANT
A	MIL-L-7870	LUBRICATING OIL, GENERAL PURPOSE, LOW TEMPERATURE
B	MIL-L-6082 7	LUBRICATING OIL, AIRCRAFT RECIPROCATING ENGINE (PISTON) GRADE AS SPECIFIED
SAE 50 ABOVE 60°F AIR TEMP.
SAE 40 30° TO 90° AIR TEMP.
SAE 30 0° TO 70°F AIR TEMP.
SAE 20 BELOW 10°F AIR TEMP.
C	MIL-H-5606	HYDRAULIC FLUID, PETROLEUM BASE (OR UNIVIS —40 OR MOBIL AERO-HF)
D	MIL-G-23827	GREASE, AIRCRAFT AND INSTRUMENT, GREASE AND ACTUATOR SCREW
E	(NONE)	TEXACO MARFAX ALL PURPOSE GREASE OR MOBIL MOBIL GREASε 77 (OR MOBIL EP2 GREASE)
F	MIL-H-7711	GREASE—LUBRICATION, GENERAL PURPOSE, AIRCRAFT
G		FLUOROCARBON RELEASE AGENT DRY LUBRICANT #MS-122 (PURCH)
H		AERO LUBRIPLATE (PURCH) FISKE BROS. REFINING CO.

SPECIAL INSTRUCTIONS

1. AIR FILTER—TO CLEAN FILTER, TAP GENTLY TO REMOVE DIRT PARTICLES. DO NOT BLOW OUT WITH COMPRESSED AIR OR USE OIL. REPLACE FILTER IF PUNCTURED OR DAMAGED.
2. BEARING AND BUSHINGS—CLEAN EXTERIOR WITH A DRY TYPE SOLVENT BEFORE LUBRICATING.
3. WHEEL BEARINGS—DISASSEMBLE AND CLEAN WITH A DRY TYPE SOLVENT. ASCERTAIN THAT GREASE IS PACKED BETWEEN THE BEARING ROLLER AND CONE. DO NOT PACK GREASE IN WHEEL HOUSING.
4. OLEO STRUTS, HYDRAULIC PUMP RESERVOIR AND BRAKE RESERVOIR—FILL PER IN- STRUCTIONS ON UNIT OR CONTAINER, OR REFER TO SERVICE MANUAL.
5. PROPELLER—REMOVE ONE OF THE TWO GREASE FITTINGS FOR EACH BLADE. APPLY GREASE THROUGH FITTING UNTIL FRESH GREASE APPEARS AT HOLE OF REMOVED FITTING.
6. LUBRICATION POINTS—WIPE ALL LUBRICATION POINTS CLEAN OF OLD GREASE, OIL, DIRT, ETC. BEFORE LUBRICATING.

NOTES

1. PILOT AND PASSENGER SEATS—LUBRICATE TRACK ROLLERS AND STOP PINS AS REQUIRED (TYPE OF LUBRICANT: "A")
2. WHEEL BEARINGS REQUIRE CLEANING AND REPACKING AFTER EXPOSURE TO AN ABNORMAL QUANTITY OF WATER.
3. SEE LYCOMING SERVICE INSTRUCTIONS NO. 1014 FOR USE OF DETERGENT OIL.
4. FUEL SYSTEM—SERVICE REGULARLY—FUEL PUMP STRAINER INJECTOR SCREEN—FILTER BOWL—QUICK DRAIN UNIT.
5. BATTERY—FLUID LEVEL & CONDITION CHECK EVERY 25 HOURS
6. MIL-C-6529 TYPE 2/OIL IS THE OIL THE ENGINE IS SERVICED WITH AT INSTALLATION. THE ENGINE MUST OPERATE ON THIS OIL 25 HOURS MINIMUM, 50 HOURS MAXIMUM. (SEE LYCOMING SERVICE LETTER NO. L121A)
7. THIS CHART IS FOR INITIAL FACTORY LUBRICATION AND SERVICE.

CAUTIONS

1. DO NOT USE HYDRAULIC FLUID WITH A CASTOR OIL OR ESTER BASE.
2. DO NOT OVER-LUBRICATE COCKPIT CONTROLS.
3. DO NOT APPLY LUBRICANT TO RUBBER PARTS.
4. DO NOT LUBRICATE CABLES—THIS CAUSES SLIPPAGE.

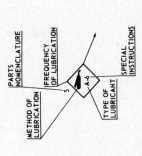

FIG. 15-8 Information and instructions for lubrication chart.

1. CONTROL COLUMN FLEX JOINT, SPROCKET AND O-RING
2. AILERON AND STABILATOR CONTROL CHAIN
3. O-RING CONTROL SHAFT BUSHING
4. TEE BAR PIVOT POINTS, AILERON AND STABILATOR CONTROL PULLEYS
5. STABILATOR CONTROL ROD AND IDLER PULLEY
6. NOSE GEAR STRUT HOUSING
7. NOSE GEAR PIVOT POINT AND HYDRAULIC CYLINDER ROD END
8. NOSE GEAR DOOR RETRACTION MECHANISM
9. NOSE GEAR DOOR HINGES
10. EXPOSED OLEO STRUT (NOSE)
11. NOSE WHEEL BEARINGS
12. NOSE GEAR TORQUE LINK ASSEMBLY
13. NOSE GEAR TORQUE LINK ASSEMBLY AND STRUT HOUSING
14. NOSE GEAR PIVOT POINT, DRAG LINK ASSEMBLY, DOWNLOCK AND CYLINDER ASSEMBLY, STEERING ROLLER AND CENTERING SPRING PIVOT POINTS
15. STEERING BELLCRANK PIVOT POINTS AND ROD ENDS
16. NOSE GEAR ROLLER TRACK AND BUNGEE
17. NOSE GEAR OLEO STRUT FILLER POINT
18. PROPELLER ASSEMBLY
19. FLAP CONTROL ROD END BEARINGS
20. FLAP RETURN AND TENSION CHAIN
21. FLAP TORQUE TUBE BEARING BLOCK
22. FLAP HANDLE PIVOT POINT LOCK MECHANISM AND CABLE PULLEY
23. RUDDER SECTOR AND STABILATOR TRIM PIVOT POINTS
24. STABILATOR TRIM SCREW
25. MAIN GEAR PIVOT POINTS
26. MAIN GEAR DOOR HINGE
27. MAIN GEAR TORQUE LINKS
28. MAIN GEAR TORQUE LINKS
29. EXPOSED OLEO STRUT (MAIN)
30. MAIN GEAR WHEEL BEARINGS
31. MAIN GEAR DOOR CONTROL ROD ENDS
32. MAIN GEAR SIDE BRACE LINK ASSEMBLY
33. UPPER SIDE BRACE SWIVEL FITTING
34. MAIN GEAR DOWNLOCK ASSEMBLY RETRACTION FITTING AND CYLINDER ATTACHMENT POINTS
35. OLEO STRUT FILLER POINT (MAIN GEAR)
36. RUDDER TUBE CONNECTIONS, TUBE CABLE ENDS AND STEERING ROD ENDS
37. BRAKE ROD ENDS
38. TOE BRAKE ATTACHMENTS
39. AILERON BELLCRANK CABLE ENDS
40. AILERON BELLCRANK PIVOT POINTS
41. AILERON CONTROL ROD END BEARINGS
42. RUDDER HINGE AND TAB HINGE BEARINGS
43. STABILATOR TRIM TAB HINGE PINS
44. CONTROL CABLE PULLEYS
45. TRIM CONTROL WHEELS - STABILATOR AND RUDDER
46. AILERON HINGE PINS
47. FLAP HINGE BEARINGS AND ALTERNATE AIR DOORS
48. INDUCTION AIR FILTERS
49. CARTRIDGE TYPE OIL FILTERS
50. ENGINE OIL SUMP (8 QTS. CAPACITY)
51. GOVERNOR CONTROLS
52. CONTROL QUADRANT CONTROLS
53. FORWARD BAGGAGE DOOR HINGE AND LATCH PINS
54. HYDRAULIC PUMP RESERVOIR
55. BRAKE RESERVOIR
56. PILOT AND COPILOT SEAT ADJUSTMENT
57. AILERON HINGE PINS
58. MAIN DOOR HINGES AND LATCH MECHANISM
59. CONTROL CABLE PULLEYS
60. BAGGAGE AND REAR DOOR HINGES AND LATCH MECHANISM
61. RUDDER AND STABILATOR TRIM SCREWS
62. LATCH MECHANISM
63. LINK BUSHING
64. ARM BUSHING

FIG. 15-9   Parts nomenclature of lubrication chart.

nents of the system. Petroleum solvents can be used to flush a system using petroleum fluids. Many aircraft of today, particularly large aircraft, use a synthetic, fire-resistant hydraulic fluid. One such fluid is Monsanto Skydrol 500, which is purple in color, and another is Skydrol 7000, which is green. If these fluids or any of a similar type are put into a system designed for petroleum fluid MIL-H-5606A (red), considerable damage can result. In such cases it is necessary to flush the system and replace seals that may be damaged.

Oxygen-System Service

Oxygen supplies should be replaced whenever the oxygen-pressure gage (or gages) show that the supply is low. Oxygen bottles (tanks) should be serviced only with breathing oxygen. In removing and replacing units of an oxygen system, oils, greases, or other petroleum products must not be used for lubrication of parts or fittings. Only an approved oxygen-system lubricant should be used. A suitable lubricant for oxygen system fittings is military specification MIL-T-5542-B or an equivalent.

Oxygen systems that utilize compressed gaseous oxygen are replenished by means of an oxygen-service unit containing breathing oxygen. Service is accomplished by connecting the ground-service unit to the remote fill line of the system and opening the valves required to permit oxygen to flow into the system.

Care must be taken to observe the oxygen pressure and to avoid overfilling the system.

WARNING: *Avoid making sparks and keep all burning cigarettes or fire away from the vicinity of the airplane. Inspect the filler connection for cleanliness before attaching it to the filler valve. Make sure that your hands, tools, and clothing are clean, particularly of grease and oil, because these contaminants will ignite upon contact with pure oxygen under pressure. As a further precaution against fire, open and close all oxygen valves slowly during filling.*

Some airliners are equipped with chemical oxygen generators instead of supplying the system with gaseous oxygen from a tank. These systems utilize a sodium chlorate core, which, when fired, generates pure oxygen that is fed through a filter to the oxygen outlets. Service of these units is accomplished by replacing the oxygen generator.

Servicing Batteries

In servicing batteries, it is most important to know whether a battery is a lead-acid type or an alkaline type, such as the nickel-cadmium battery. Service procedures for the different types vary considerably. The service manual or instructions for the battery should be consulted unless the technician is thoroughly familiar with the procedure.

Lead-Acid Batteries. Whenever checking the battery, ascertain that all connections are clean and tight and the fluid level is above the baffle plates. If it is necessary to add fluid, use distilled water. Do not overfill the battery. When the cells are overfilled, water and acid may spill on the lower portions of the fuselage. A hydrometer check should be performed to determine the percentage of charge present in the battery.

Corrosion on the battery terminals and connections may be neutralized by applying a solution of baking soda and water mixed to the consistency of thin cream. Do not allow any of this soda solution to enter the battery. Repeat this application until all bubbling action has ceased before washing the battery and box with clean water.

Nickel-Cadmium Batteries. The following list of battery servicing precautions and checks is meant to be a general guide. For specific details and complete procedures, refer to the battery manufacturer's instructions.

- After each 100 h of operation or every 30 days, whichever occurs first, check the electrolyte level and clean the battery and filler vent plugs.
- Periodically check that the cell vents are clean and open. Plugged vents may cause excessive internal cell pressure and cause leaks.
- Never remove a cell from the battery case unless a replacement is immediately available; otherwise, the remaining cells may swell making replacement of the removed cells difficult. Loosen the vents before cell replacement to eliminate the possibility of cells swelling from internal gas pressure.
- Check the torque of the terminal screws securing the cross links connecting the cells together (according to manufacturer's specifications).
- Check that no carbon deposit has built up on the cross links or between them and the battery case. If such deposits are present, clean the affected areas.
- When there is any indication of oil in the battery, remove all cells from the case and check all rubber parts for deterioration. Remove the oil and replace all damaged rubber parts. To remove the oil, use soap and water only.
- Keep nickel-cadmium and lead-acid batteries stored separately to prevent mutual contamination. Unless kept in closed storage containers, nickel-cadmium electrolyte (potassium hydroxide) will absorb enough carbon dioxide from the air to render it ineffective.

Tires

Maintaining proper tire inflation will minimize tread wear and aid in preventing tire rupture caused from running over sharp stones and ruts. When inflating tires, visually inspect them for cracks and breaks. Reverse the tires on the wheels, if necessary, to produce even wear. All tires and wheels are balanced before installation and the relationship of tire, tube, and wheel should be maintained upon reinstallation. Out-of-balance wheels can cause extreme vibration in the landing gear during takeoff and landing. In service, tire carcasses grow slightly due to shock loads in landing. Normally, this growth is balanced by tread wear so there is no increase in tire diameter.

Other Service Items

In addition to the services described in the foregoing paragraphs, it is necessary to service other items and systems from time to time as specified in the maintenance manual. Among these are oleo struts, shimmy dampers, bungee cylinders, instrument filters, fuel screens, heaters, etc.

The service of these items is described in the maintenance manual for the aircraft and may or may not be established on a regular basis. For example, oleo struts need service only when inspection shows that they have lost air or fluid. Otherwise, they may need attention only at a time of major reconditioning or overhaul.

● OPERATIONAL INSPECTION

Before an aircraft may be approved for return to service after an annual or 100-h inspection, FAR 43.15 requires that the engines be operated

To determine satisfactory performance, in accordance with the manufacturer's recommendations, of—
  (i) Power output (static and idle r.p.m.);
  (ii) Magnetos;
  (iii) Fuel and oil pressure; and
  (iv) Cylinder and oil temperature.

In addition to FAR 43.15, most manufacturer's inspection checklists include an operational inspection which provides for the operational testing of a wide variety of equipment ranging from parking brakes to radios.

After an operational check is performed, the engine should be given a brief visual inspection for oil leaks resulting from the inspection and servicing performed.

The conclusion of the operational inspection is also an excellent time to give the aircraft a final walk-around visual inspection, checking to see that all the inspection panels that were removed to perform the visual inspection have been correctly reinstalled.

● CLEANING AIRCRAFT AND PARTS

A most important factor in the maintenance of an aircraft is cleanliness. Although cleaning is not actually a part of the inspection process, it is an important step in preparing an aircraft for redelivery to the owner as well as being a good preventative maintenance practice. A film of dirt on the surface of a metal or a painted aircraft will attract moisture and chemical pollutants from the air and permit these materials to react chemically with the surface of the aircraft. The result is oxidation, corrosion, and general deterioration of the surface. Cleanliness is particularly important in areas near the ocean where the air is salty or in areas where air pollution is comparatively severe. Regular cleaning and waxing of exterior surfaces of an aircraft will effectively reduce corrosion and other forms of deterioration. It is desirable to keep an airplane hangared when possible. Since this cannot always be done, the owner or operator of an aircraft should try to keep it as clean as possible.

In this section we do not intend to describe all the methods for cleaning aircraft and parts; however, some general principles will be discussed, with particular emphasis on the dangers involved in the use of improper cleaning methods and unsuitable materials.

Exterior Painted Surfaces

When an airplane is covered with dust, sand, or other types of dirt, no attempt should be made to wipe the surface clean with a dry cloth, no matter how soft the material is. Wiping a dirty surface with dry material will scratch the metal or paint on the surface because of the abrasiveness of the fine particles of sand or dust. Scratching the surface will destroy the smooth finish and will increase the rate of deterioration, particularly if the pure aluminum coating on clad material is scratched through to the base metal.

Generally, the painted surfaces can be kept bright by washing with water and mild soap, followed by a rinse with water and drying with cloths or a chamois. Harsh or abrasive soaps or detergents which cause corrosion or scratches should never be used. Remove stubborn oil and grease with a cloth moistened with Stoddard solvent. Many of the new-type paints being used on aircraft, such as urethane, do not require waxing to keep the painted surfaces bright. However, if desired, the airplane may be waxed with a good automotive wax. Soft cleaning cloths or a chamois should be used to prevent scratches when cleaning or polishing. A heavier coating of wax on the leading surfaces will reduce the abrasion problems in these areas.

Cleaning Plastic Windows

In the cleaning of plastic windshields, it is especially important to avoid wiping a dirty surface with a dry cloth. The first step in cleaning the aircraft and windshield is to flush the surface with clean water, using the bare hand to dislodge any dirt or abrasives. This will prevent the possibility of scratching the surface during the washing procedure. Wash thoroughly with a mild soap solution, taking care that the water is free from all possible abrasives. A soft cloth, sponge, or chamois may be used to apply the soap solution. Remove oil and grease with a cloth moistened with kerosene, naptha, or methanol.

Never use gasoline, benzine, alcohol, acetone, anti-ice fluid, lacquer thinner, or glass cleaner to clean the plastic. These materials will attack the plastic and may cause it to craze.

Waxing with a good commercial wax will finish the cleaning job. A thin, even coat of wax, polished out by hand with clean, soft, flannel cloths, will fill in minor scratches and help prevent further scratching.

Cleaning Tires

Tires should be cleaned with an emulsion cleaner or with soap or detergent and water. Petroleum solvents cause deterioration of natural rubber and therefore should not be applied to rubber tires. Small oil or grease spots can be cleaned off with a cloth dampened with petroleum solvent provided that the spot is wiped dry immediately.

Cleaning DeIce Boots

To prolong the life of deice boots, they should be washed and serviced on a regular basis. The boots should be kept clean and free from oil, grease, and other solvents which cause rubber to swell and deteriorate. Clean the boots with mild soap and water, then rinse thoroughly with clean water.

Isopropyl alcohol can be used to remove grime which cannot be removed using soap. If isopropyl alcohol is used for cleaning, wash the area with mild soap and water, then rinse thoroughly with clean water.

Interior Cleaning

To remove dust and loose dirt from the upholstery and carpet, clean the interior regularly with a vacuum cleaner.

Oily spots may be cleaned with household spot removers, used sparingly. Before using any solvent, read the instructions on the container and test it on an obscure place on the fabric to be cleaned. Never saturate the fabric with a volatile solvent because it may damage the padding and backing materials.

Soiled upholstery and carpet may be cleaned with foam-type detergent, used according to the manufacturer's instructions. To minimize wetting the fabric, keep the foam as dry as possible and remove it with a vacuum cleaner.

If the airplane is equipped with leather seating, cleaning of the seats is accomplished using a soft cloth or sponge dipped in mild soap suds. The soap suds, used sparingly, will remove traces of dirt and grease. The soap should be removed with a clean damp cloth.

The plastic trim, headliner, instrument panel, and control knobs need only be wiped off with a damp cloth. Oil and grease on the control wheel and control knobs can be removed with a cloth moistened with Stoddard solvent. Volatile solvents, such as were mentioned in the paragraphs on the care of the windshield, must never be used since they soften and craze plastic.

● MAINTENANCE RECORDS

In the process of returning an aircraft to service after completing an inspection or performing maintenance, technicians will encounter a number of official forms including maintenance records relating to their work. A properly completed maintenance record provides the information needed by the owner or operator and maintenance personnel to determine when scheduled maintenance is to be performed. Aircraft maintenance recordkeeping is a responsibility shared by the owner and maintenance technician, with the ultimate responsibility assigned to the owner by FAR Part 91, Section 91.165.

A properly executed set of maintenance records will save the owner money since maintenance personnel will spend less time in research to establish the status of the item to be worked on. Good records are also invaluable to maintenance personnel in troubleshooting.

Maintenance Record Format

Maintenance records may be kept in any format which provides record continuity and includes the required information. There is no requirement that the records be bound; however, bound records or records which have some system of page control normally have greater credibility. Many owners and operators, however, have found it advantageous to keep separate records for the airframe, engine, and propeller, particu-larly on multiengine aircraft even though this is not required. Engines, propellers, and appliances are often changed from one aircraft to another. The use of individual records facilitates transfer of the record with the item when ownership changes. The important thing is to have a record system that will provide for storing and retrieving the necessary information.

FAR 91.173 sets forth the requirements on retaining aircraft records. As illustrated in Fig. 15-10, records are divided into basically two groups: those items that must be retained until the work is superseded by other work or for a period of 1 yr after the work was performed, and those records which are considered permanent and are kept for the life of the aircraft.

Inspection Record Entries

FAR 43.11 contains the requirements for inspection entries. When a technician approves or disapproves an aircraft for return to service after an inspection, an entry must be made including:

1. The type of inspection and a brief description of the extent of the inspection.
2. The date of the inspection and aircraft total time in service.
3. The signature, the certificate number, and kind of certificate held by the person approving or disapproving returning the aircraft to service.
4. If the aircraft is found to be airworthy and approved for return to service, the following or a similarly worded statement: "I certify that this aircraft has been inspected in accordance with (insert type) inspection and was determined to be in airworthy condition."

If an aircraft is disapproved for return to service because during the course of the inspection it was found to be unairworthy, a signed and dated list of the discrepancies must be provided to the aircraft owner.

In addition to the above information for progressive or continuous type inspections where only part of the inspection is conducted at a time, the record entry must identify which part of the inspection was accom-

---

Records that shall be retained until work is repeated or superseded by other work or 1 year after work is performed.

Records of: ___ Maintenance
___ Alterations
___ 100-hour
___ Annual
___ Progressive
___ Other approved inspection programs

Which shall include:

___ Description of work performed.
___ Date.
___ Certificate number and signature of person *approving for return* to service.

Records that shall be retained and transferred with the aircraft.
___ T.T. in service: Airframe
Engine
Propeller
___ Current status of life-limited parts of airframe, engine, propeller, rotor, appliance.
___ TSOH of all items requiring OH.
___ Identification of current inspection program and times since last inspection.
___ Current status of AD notes and recurring action.
___ Copies of FAA Form 337 for each major alteration of the airframe, engine, propeller, appliance.

FIG. 15-10 Record keeping requirements.

DEPARTMENT OF TRANSPORTATION
FEDERAL AVIATION ADMINISTRATION

MAJOR REPAIR AND ALTERATION
(Airframe, Powerplant, Propeller, or Appliance)

FOR FAA USE ONLY

OFFICE IDENTIFICATION

INSTRUCTIONS: Print or type all entries.   See FAR 43.9, FAR 43 Appendix B, and AC 43.9–1 (or subsequent revision thereof) for instructions and disposition of this form.

1. AIRCRAFT	MAKE		MODEL	
	SERIAL NO.		NATIONALITY AND REGISTRATION MARK	
2. OWNER	NAME (As shown on registration certificate)		ADDRESS (As shown on registration certificate)	

3. FOR FAA USE ONLY

4. UNIT IDENTIFICATION

UNIT	MAKE	MODEL	SERIAL NO.	5. TYPE REPAIR	5. TYPE ALTER-ATION
AIRFRAME	◆◆◆◆◆◆◆◆◆◆◆◆◆ (As described in item 1 above) ◆◆◆◆◆◆◆◆◆◆◆◆◆				
POWERPLANT					
PROPELLER					
APPLIANCE	TYPE				
	MANUFACTURER				

6. CONFORMITY STATEMENT

A. AGENCY'S NAME AND ADDRESS	B. KIND OF AGENCY	C. CERTIFICATE NO.
	U.S. CERTIFICATED MECHANIC	
	FOREIGN CERTIFICATED MECHANIC	
	CERTIFICATED REPAIR STATION	
	MANUFACTURER	

D. I certify that the repair and/or alteration made to the unit(s) identified in item 4 above and described on the reverse or attachments hereto have been made in accordance with the requirements of Part 43 of the U.S. Federal Aviation Regulations and that the information furnished herein is true and correct to the best of my knowledge.

DATE	SIGNATURE OF AUTHORIZED INDIVIDUAl.

7. APPROVAL FOR RETURN TO SERVICE

Pursuant to the authority given persons specified below, the unit identified in item 4 was inspected in the manner prescribed by the Administrator of the Federal Aviation Administration and is ☐ APPROVED ☐ REJECTED

BY	FAA FLT. STANDARDS INSPECTOR	MANUFACTURER	INSPECTION AUTHORIZATION	OTHER (Specify)
	FAA DESIGNEE	REPAIR STATION	CANADIAN DEPARTMENT OF TRANSPORT INSPECTOR OF AIRCRAFT	

DATE OF APPROVAL OR REJECTION	CERTIFICATE OR DESIGNATION NO.	SIGNATURE OF AUTHORIZED INDIVIDUAL

FAA Form 337 (7–67)   ☆ U.S. Government Printing Office 1977—772-646/141   (8320)

FIG. 15-11   FAA Form 337.

plished. If the owner maintains separate records for the airframe, power plants, and propellers, the entry for the inspection must be entered in each.

Maintenance Record Entries

Maintenance which is performed on aircraft must be properly recorded in the maintenance records. This includes discrepancies corrected during the course of an inspection. FAR 43.9 governs the recording of aircraft maintenance and states that any mechanic who maintains, rebuilds, or alters an aircraft must make an entry containing:

1. A description of the work or some reference to data acceptable to the FAA
2. The date the work was completed
3. The mechanic's name
4. If it is approved for return to service, the signature and certificate number of the approving mechanic.

Major Repair and Alteration Form FAA 337

Major repairs and alterations, in addition to being recorded in the maintenance record according to FAR 43.9, must also be recorded on FAA **Form 337.** FAA Form 337 serves two purposes; one is to provide owners with a record of major repairs and major alterations to their aircraft and the other is to provide the FAA with a copy for their records. A copy of FAA Form 337 is shown in Fig. 15-11. The maintenance technician who performed or supervised the major repair or alteration prepares the original FAA Form 337 in duplicate.

An airframe and power plant technician who holds an inspection authorization rating will then further process the form. The IA will inspect the major alteration or repair to see that it conforms to FAA approved data, review the Form 337 for completeness and accuracy, and then complete item 7, "Approval for Return to Service."

After Form 337 is completed, the original copy should be given to the owner with the duplicate copy forwarded to the local FAA District Office within 48 h. It is generally a good idea for the technician performing the work and the IA signing the Form 337 to retain copies for their files.

MALFUNCTION OR DEFECT REPORT

The FAA requests the cooperation of maintenance technicians in reporting discrepancies located during inspections. The reporting of malfunctions and defects found in aircraft and engines is an important step in detecting and eliminating unsafe conditions. All persons concerned with the maintenance, repair, inspection, or operation of aircraft are encouraged to make such reports on the **Malfunction or Defect Report** (M or D report) form and forward the form to the FAA. A copy of the Malfunction or Defect Report is shown in Fig. 15-12.

M or D Reports provide the FAA and industry with a very essential service record of mechanical difficulties encountered in aircraft operations. Such reports contribute to the correction of conditions or situations which otherwise would continue to prove costly and/or could cause a serious accident or incident.

INSPECTION REMINDER

Whenever a maintenance technician completes an inspection of an aircraft, he or she should prepare an Inspection Reminder, FAA Form 8320-2, and affix it to the aircraft inside the cockpit in a location where it cannot be overlooked. This provides a constant re-

1. REGISTRATION NO. N–	DEPARTMENT OF TRANSPORTATION FEDERAL AVIATION ADMINISTRATION MALFUNCTION OR DEFECT REPORT			Form Approved Budget Bureau No. 04–R0003	8. DATE SUB.	FOR FAA USE ONLY CONTROL NO.			
	A. MAKE	B. MODEL	C. SERIAL NO.	7A. COMMENTS *(Describe the malfunction or defect and the circumstances under which it occurred. State probable cause and recommendations to prevent recurrence.)*					
2. AIRCRAFT									
3. POWERPLANT									
4. PROPELLER									
5. APPLIANCE/COMPONENT *(assy. that includes part)*									
A. NAME	B. MAKE	C MODEL	D. SERIAL NO.						
6. SPECIFIC PART *(of component)* CAUSING TROUBLE									
A. NAME	B. NUMBER	C. PART/DEFECT LOCATION		Continue on reverse					
				SUBMITTED BY					
FAA USE	E. PART TT	F. PART TSO	G. PART CONDITION	B.	C.	D.	E.	F.	G. H. I.
D. ATA CODE				REP. STA.	OPER.	MECH.	AIR TAXI	MFG.	FAA / OTHER

FAA FORM 8330–2 (9–70) SUPERSEDES PREVIOUS EDITIONS

FIG. 15-12 A Malfunction or Defect Report Form.

minder to owners and operators that an inspection will be due on a certain date or time. Inspection Reminder forms are available at all FAA General Aviation District Offices. A copy of an Inspection Reminder form is shown in Fig. 15-13.

DEPARTMENT OF TRANSPORTATION
FEDERAL AVIATION ADMINISTRATION

INSPECTION REMINDER

The next inspection of this aircraft is required by Federal Aviation Regulation

Section:_____

Hours in
Service:_____

OR

Date due: _____

FAA FORM 8600-1 (4-78)
FORMERLY FAA FORM 8320-2

FIG. 15-13   An Inspection Reminder Form.

● REVIEW QUESTIONS

1. What is the difference between an annual inspection and a 100-h inspection? How are they alike?
2. What condition requires the performance of 100-h inspections on aircraft?
3. Why is a checklist required when making an annual or 100-h inspection?
4. Who may perform a 100-h inspection?
5. List a typical grouping of items that must be inspected during an annual inspection or during a 100-h inspection.
6. What is a *progressive inspection?*
7. If a progressive inspection program is authorized for an aircraft, does the aircraft have to undergo an annual inspection?
8. From what sources may a mechanic select an inspection checklist?
9. What inspection program options are available to the operators of turbine-powered multiengine airplanes?
10. What is a *walk around?*
11. Describe a preflight inspection.
12. What condition requires the performance of an altimeter and static system inspection?
13. What steps should be taken in preparation for conducting a visual inspection?
14. What is the purpose of a discrepancy list?
15. What should be done if the Airworthiness Certificate cannot be located during the inspection?
16. What information does a *lubrication chart* provide?
17. What precautions should be taken when servicing an oxygen system?
18. What items are required to be checked during the performance of an operational inspection?
19. What should be used to clean aircraft windows?
20. Why should oily solvents be kept off of deice boots?
21. Which records are considered permanent aircraft records?
22. Which records are considered temporary aircraft records?
23. What format is required for aircraft maintenance records?
24. What items must be included in an inspection record entry?
25. What items must be included in a maintenance record entry?
26. Under what conditions is the preparation of a Major Repair and Alteration Form, FAA Form 337, required?
27. Give the number and distribution of copies required when Form 337 is prepared.
28. What persons or agencies may approve an FAA Form 337?
29. What is the purpose of the *Malfunction or Defect Report?*
30. Under what conditions should a Malfunction or Defect Report be prepared?
31. What is the value of an Inspection Reminder form to the owner of an aircraft?

16 MAINTENANCE SHOP REQUIREMENTS AND PRACTICES

The aviation **repair station** is to the aviation industry what the garage or automobile repair service is to the automotive industry. However, the aviation repair station, because of the need for airworthiness in all aircraft, parts, and accessories, must be under strict direction and control in order to meet the standards required for safe aircraft operation.

The function of the repair station is to make repairs on aircraft, engines, propellers, and appliances and to provide routine maintenance and inspection services. The certificated repair station must perform these functions in accordance with the rules, regulations, and specifications set forth in its application for certification and its **inspection-procedures manual.** In addition, the repair station is obligated to abide by the regulations set forth in Federal Aviation Regulations Part 145.

The Federal Aviation Administration has the responsibility for establishing standards of aircraft and power plant repair and provides for the maintenance of these standards by certificating repair stations and repair-station personnel, inspecting repair facilities and equipment, and examining the methods and procedures employed by certificated repair stations.

● ORGANIZATION OF A REPAIR STATION

Depending upon its type and size, a repair station may be organized in many ways. To be certificated by the Federal Aviation Administration, a repair station must meet rigid standards and adhere to certain principles. One of the most important of these principles is complete separation of the inspection function from other operating divisions of the station. The purpose of this is to make it possible for the inspection personnel to be independent in their judgments regarding the airworthiness of any repaired item. Inspection divisions should therefore be responsible to the administration only.

A suggested organization for a reasonably large repair station is shown by Fig. 16-1. This could be varied in many ways; however, it provides a guide to the principal operating divisions and subdivisions. It will be noted that the repair station is organized in three major divisions and that the inspection division is separate from the others.

Divisions and Functions

The production division of a repair station must operate efficiently and effectively in order to make the business a success. This division receives aircraft, engines, accessories and components for repair or maintenance. At the time the unit is received, a cost estimate is usually made and a **work order** is prepared. The work order includes a complete description of the item with serial numbers, model number, make, type, etc. The owner's name and address and the date the unit is received are recorded. The owner provides information regarding the repair and maintenance to be performed and offers all possible assistance in determining the cause of any discrepancies or malfunctions. The **service manager** records all required repair items on the work order; if additional repair or maintenance are found necessary after the initial **receiving inspection,** the service manager makes provision for these. The owner is given an estimate of the costs involved and informed that any additional work found necessary will involve additional cost. To avoid misunderstandings, the owner must be given complete information on all known and potential costs for the job. During the receiving inspection, a member of the inspection department should be present to note the existing conditions and prepare a checklist for future reference. This list will be used in the inspections as the aircraft or component continues through the repair processes to final assembly. Receiving inspection is performed, primarily, to record all conditions existing and to list any discrepancies noted in components, accessories, etc. If a particular instrument or other component is missing, the owner should be notified.

The next step, or steps, in the processing of an aircraft repair is the required disassembly. This is often done by components or units, especially in cases where an airplane is not to be overhauled completely. For example, one wheel and brake assembly can be disassembled, inspected, repaired, reassembled, and inspected for airworthiness before another section or component of the airplane is begun. Parts are inspected during disassembly to determine the need for repair or replacement. The sequence of events is governed by type and size of aircraft, nature of repair operation required, personnel and facilities available, and the schedule most likely to return the aircraft to service in the shortest possible time.

The function of the inspection department should be carefully noted. Figure 16-1 shows connecting lines from the inspection functions to the repair functions to indicate points at which the inspections are made. Receiving inspection is twofold. First, the inspection division examines the aircraft and components at the time they are received for repair. Second, the inspection division examines new parts and materials when they are received from vendors. This is to assure that all materials received meet airworthiness requirements. It is not necessary to inspect standard manufactured parts which are received from a dealer or distributor in properly identified, sealed packages. Care must be taken, however, to make sure that a part scheduled for a particular work order bears the correct part number.

After a part or subassembly has been repaired, the inspection division examines the repair for quality of workmanship and conformance with approved practice and specifications. This inspection is usually performed before final assembly.

Assembly inspection and final assembly inspection may be separate or identical, depending upon the unit inspected. The guiding principle is that every repair or assembly operation must be checked by an official inspector of the repair station. Units that cannot be inspected after final assembly must be inspected before.

Completed units or aircraft must be given an operational test where an overhaul or repair procedure could affect the operation of the aircraft or component. An aircraft may be given a flight test, and a component may be operated either on the aircraft or on a test stand.

The staff services division provides necessary supporting operations for the entire company. In Fig. 16-1, sections are shown which supply services as indicated. The organization of these sections can and

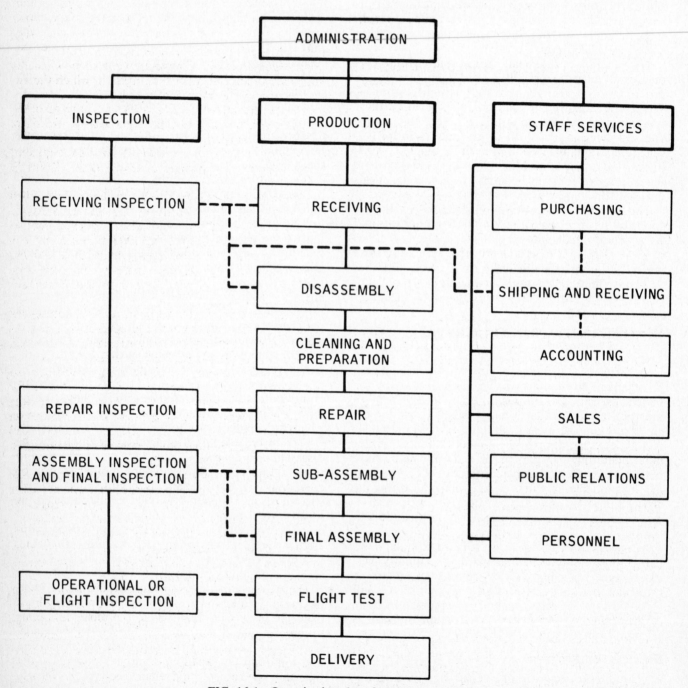

FIG. 16-1 Organization chart for a repair station.

does vary considerably; however, all the services listed must be performed in one way or another.

The sales division of a company is often separate from other divisions and is responsible directly to the administration. If the organization sells both products and services, the sales division is likely to be separate from, but will work closely with, the public relations division.

Certification Requirements

As mentioned previously, the requirements or regulations governing a certificated repair station are set forth in FAR Part 145. These regulations cover both **domestic** and **foreign repair stations** and state that no person may set up and operate as a certificated repair station without, or in violation of, a repair-station certificate. Furthermore, an applicant for a certificate cannot advertise as a certificated repair station until the certificate is issued.

The regulation (145.11) covering the application for and issue of a repair-station certificate is as follows:

An application for a repair station certificate and rating, or for an additional rating, is made on a form and in a manner prescribed by the Administrator (of the FAA) and submitted with duplicate copies of

1. Its inspection procedures manual
2. A list of the maintenance functions to be performed for it, under contract, by another agency (as approved)
3. In the case of an applicant for a propeller rating (class 2) or any accessory rating (class 1, 2, or 3), a list, by type or make, as applicable, of the propeller or accessory for which he seeks approval

An applicant who meets the requirements of this Part is entitled to a repair-station certificate with appropriate ratings prescribing such operations, specifications, and limitations as are necessary in the interests of safety.

All the documents submitted to the Federal Aviation Administration in support of the application for a repair-station certificate are examined and evaluated by representatives of the FAA. Facilities and equipment of the repair station are inspected to assure compliance with the documents submitted and ascertain that the repair station is adequately equipped to perform the services for which it is being certificated. The employment and experience records of key personnel are checked to see that those responsible for the repair and final approval of aircraft and components for return to service are qualified for their responsibilities. This includes verification of the FAA certification of individual employees of the organization.

Continuation of Certificate

To retain approval as a certificated repair station, an organization must comply with further requirements set forth in FAR Part 145. These regulations include application for approval whenever certain changes are made or contemplated, proper display of the repair-station certificate, submission to periodic inspections of the station by FAA representatives, and conformance with rules regarding the use of the station certificate number in advertising.

The certificate of a domestic repair station remains in effect until it is surrendered, suspended, or revoked. Suspension or revocation of a certificate will occur only when the repair station violates the terms of its certification or otherwise fails to meet its obligations and responsibilities.

Repair-Station Ratings

A repair station may be certificated for one or more of the following ratings: (1) airframe, (2) power plant, (3) propeller, (4) radio, (5) instrument, (6) accessory, and (7) limited.

Airframe ratings are classified in four categories. Class 1 is approved for small aircraft of composite construction, and class 2 is for large aircraft of composite construction. **Composite construction** usually includes all aircraft other than those in which the airplane structure is entirely of metal.

The class 3 airframe rating is for small aircraft of all-metal construction, and class 4 is for large aircraft of all-metal construction.

A **large aircraft** is one which has an approved maximum takeoff weight of more than 12 500 lb [5670 kg].

Power plant ratings for a repair station are classified in three categories. A repair station having a class 1 rating is approved for the overhaul of reciprocating engines rated at 400 hp [298.3 kW] or less. Class 2 is for reciprocating engines having more than 400 hp, and class 3 is for gas-turbine engines. The term "gas-turbine" includes turbojet, turboprop, and turboshaft engines.

Propeller ratings are classified in two cateogries: class 1 for all fixed-pitch and ground-adjustable propellers of wood, metal, or composite construction and class 2 for all other propellers, by make.

Radio ratings are issued for three categories of equipment. Briefly, these are as follows: class 1 for general communications radio; class 2 for navigational radio equipment, except that which operates on the radar, or pulsed-radio-frequency, principle; and class 3 for radar and all other equipment operating on the pulsed-radio-frequency principle.

Instrument ratings are issued in four categories. Class 1 is for mechanically operated instruments, class 2 for electrical instruments, class 3 for gyroscopic instruments, and class 4 for electronic instruments.

Accessory ratings cover three classes of equipment. Class 1 is for mechanical accessories, class 2 for electrical accessories, and class 3 for electronic or electronically controlled accessories.

Limited ratings are issued to repair stations which limit their repair functions to particular makes and models of aircraft, engines, propellers, instruments, radio equipment, landing gear, or other components; to special services such as nondestructive testing, inspection, and processing; and to any other purpose for which the FAA finds the applicant's request appropriate.

Requirements for Repair Stations

Housing. A repair station must have adequate housing that provides protected space for any and all operations that the repair station is certificated to perform. Such housing must include space to store at least one of the largest type of aircraft that it is certificated to repair, alter, or maintain.

The housing must provide adequate space for all types of work that the repair station is certificated to perform, space to store and protect tools and materials, space that is separate from the shop to store and protect parts so there will be no danger of damage or deterioration to such parts, and a separate shop space where machine tools are located and where bench-work is done. All areas must have adequate ventilation, lighting, and temperature control.

Work areas for different types of operations must be so arranged or partitioned that operations will not interfere with one another or cause contamination from sawdust, metal particles, paint spray, paint fumes, cleaning fumes, oil spray, smoke, or any other contaminating materials. Workers must be protected so their efficiency will not be impaired and the quality of their work affected. Goggles, face shields, rubber gloves, rubber or plastic aprons, respirators, and other safety items must be available as needed. Spray painting should be done in an enclosed, fireproof paint booth if possible. The painting booth should be equipped with high-volume exhaust fans to draw out the paint fumes and paint particles. The air being exhausted should be drawn through a water curtain or screen to remove the suspended paint particles.

Facilities and Equipment. A repair station must have on hand all the facilities required to perform the work for which it is certificated. Among items that are required are hand tools, power tools, metal-working equipment, welding equipment, woodworking equipment, work stands, test stands, jacks, platform scales, benches, parts racks, support stands, inspection equipment, parts trays, cleaning equipment, painting equipment, machine tools, and special tools required for working on particular types of aircraft, power plants, and accessories. A repair-station operator who has made provision for certain types of work to be performed at another approved facility need not have the special equipment required for such work. Before a repair station is certificated, it will be inspected by a representative of the FAA to assure that it is properly housed and equipped.

Repair-Station Personnel. A certificated repair station must provide adequate personnel who can perform, supervise, and inspect the work for which it is rated. The management of the repair station must carefully consider the qualifications and abilities of the employees, both certificated and uncertificated, and in the case of uncertificated employees, shall determine their abilities by means of practical tests or employment records. The management of a repair station is primarily responsible for the satisfactory work of the employees.

The number of repair-station employees will vary according to the type and volume of work being per-formed. In any case, the repair station must have enough properly qualified employees to keep up with the volume of work in process and may not reduce the number of employees below that necessary to efficiently produce airworthy work.

The repair-station management must determine the abilities of its supervisors and must provide an adequate number of supervisors for all phases of its activities. The FAA may determine the ability of any supervisors by inspecting their employment and experience records or by personal tests.

Repair stations must keep accurate personnel records on all supervisory employees. These records must show the work history of the employees in the field of aviation maintenance, the types of work performed, positions held, certificates and ratings held, and a current record of their positions and duties. Each record must be updated each time there is a change in any of the data.

Inspection Systems. A certificated repair station must have an inspection system separate from the production department. Thus the supervision of the production department will have no influence over the decisions made in the inspection department.

Each inspector must be thoroughly qualified in the areas the inspection encompasses. If instruments or equipment are used in making various types of inspections, the inspector must be fully qualified in the use of the instruments or equipment. The inspector must have available and understand current specifications involving inspection tolerances, limitations, and procedures established by the manufacturer of the product being inspected and with other forms of inspection information such as FAA Airworthiness Directives and manufacturer's service letters and bulletins.

The repair-station management must have an inspection procedures manual that explains the internal inspection system of the repair station in a manner easily understood by any employee of the station. This manual must be kept up to date and be immediately available to anyone requiring it.

Types of Maintenance, Alterations, and Repairs

Maintenance, alterations, and repairs are classified as to their effect on the airworthiness of the product involved and their effect on the flight characteristics and performance of the aircraft, power plant, or other product.

Maintenance systems are generally divided into **preventive maintenance,** discussed below; **periodic maintenance,** in which certain maintenance operations and overhauls are carried out on the basis of how many hours the aircraft has flown; and **progressive maintenance,** in which the aircraft is continually being repaired on a section-by-section and system-by-system basis so that the aircraft is not ever out of service except for a relatively short time.

Alterations and repairs are classified as major or minor, depending upon the effect they may have on the airframe, power plant, propeller, or appliance. If the alteration or repair of a product is major, the product can be returned to service only by a technician holding an IA rating, a certificated repair station rated for the product, an FAA inspector, or the manufacturer of the product.

Preventive Maintenance. **Preventive maintenance** is defined as simple or minor preservation operations and the replacement of small standard parts not involving complex assembly operations. Operations classed as preventive maintenance are as follows:

Removal, installation, and repair of landing-gear tires

Replacing elastic shock-absorber cords on landing gear

Servicing landing-gear shock struts by adding oil, air, or both

Servicing landing-gear wheel bearings, such as cleaning and greasing

Replacing defective safety wiring or cotter keys

Lubrication not requiring disassembly other than removal of nonstructural items such as cover plates, cowlings, and fairings

Making simple fabric patches not requiring rib stitching or the removal of structural parts or control surfaces

Replenishing hydraulic fluid in the hydraulic reservoir

Refinishing decorative coating of fuselage, wings, tail group surfaces (excluding balanced control surfaces), fairings, cowling, landing gear, cabin, or cockpit interior when removal or disassembly of any primary structure or operating system is not required

Applying preservative or protective material to components where no disassembly of any primary structure or operating system is involved and where such coating is not prohibited or is not contrary to good practices

Repairing upholstery and decorative furnishings or the cabin or cockpit interior when the repairing does not require disassembly of any primary structure or operating system or affect the primary structure of the aircraft

Making small simple repairs to fairings, nonstructural cover plates, cowlings, and small patches and reinforcements not changing the contour so as to interfere with the proper airflow

Replacing side windows where that work does not interfere with the structure of any operating system such as controls, electrical equipment, etc.

Replacing safety belts

Replacing seats or seat parts with replacement parts approved for the aircraft, not involving disassembly of any primary structure or operating system

Troubleshooting and repairing broken circuits in landing light wiring circuits

Replacing bulbs, reflectors, and lenses of position and landing lights

Replacing wheels and skis where no weight-and-balance computation is involved

Replacing any cowling not requiring removal of the propeller or disconnecting of flight controls

Replacing or cleaning spark plugs and setting of spark-plug gap clearance

Replacing any nose connections except hydraulic connections

Replacing prefabricated fuel lines

Cleaning fuel and oil strainers

Replacing batteries and checking fluid level and specific gravity

Removing and installing glider wings and tail surfaces that are specifically designed for quick removal and installation and when such removal and installation can be accomplished by the pilot

The holder of a pilot certificate issued under FAR Part 61, may perform preventive maintenance on any aircraft owned or operated by the holder that is not used in air carrier service. Preventive maintenance may also be performed by certificated mechanics, repair stations, repair technicians, air carriers, and others authorized by the FAA. Persons who plan to perform preventive maintenance must ascertain that the operation falls within this category and that they are authorized to perform the work.

Classification of Alterations. A **major alteration** is an alteration not listed in the aircraft, aircraft engine, or propeller specifications (1) that might appreciably affect weight, balance, structural strength, performance, power plant operation, flight characteristics, or other factors of airworthiness or (2) that is not done according to accepted practices or cannot be done by elementary operations.

Alterations of the following parts and alterations of the following types, when not listed in the aircraft specifications issued by the FAA, are **airframe major alterations:**

Wings

Tail surfaces

Fuselage

Engine mounts

Control system

Landing gear

Hull or floats

Elements of an airframe, including spars, ribs, fittings, shock absorbers, bracing, cowlings, fairings, and balance weights

Hydraulic and electrical actuating systems or components

Rotor blades

Changes to the empty weight or empty balance which result in an increase in the maximum certificated weight or CG limits of the aircraft

Changes in the basic design of the fuel, oil, cooling, cabin pressurization, electrical, hydraulic, deicing, or exhaust systems

Changes to the wing or to fixed or movable control surfaces which affect flutter and vibration characteristics

The following alterations of a power plant, when not listed in the Engine Specification issued by the FAA, are **power plant major alterations:**

Conversion of an aircraft engine from one approved model to another, involving any changes in compression ratio, propeller reduction gear, impeller gear ratios, or the substitution of major engine parts which requires extensive rework and testing of the engine

Changes to the engine by replacing aircraft engine structural parts with parts not supplied by the original manufacturer or parts not specifically approved by the FAA administrator

Installation of an accessory which is not approved for the engine

Removal of accessories that are listed as required equipment on the Aircraft or Engine Specification

Installation of structural parts other than the type of parts approved for the installation

Conversions of any sort for the purpose of using fuel of a rating or grade other than that listed in the engine specifications

The following alterations of a propeller, when not authorized in the Propeller Specification issued by the FAA, are classed as **propeller major alterations:**

Changes in blade design

Changes in hub design

Changes in the governor or control design

Installation of a propeller governor or feathering system

Installation of a propeller deicing system

Installation of parts not approved for the propeller

Appliance major alterations are those alterations of the basic design not made in accordance with recommendations of the appliance manufacturer or in accordance with an FAA Airworthiness Directive. In addition, changes in the basic design of radio communication and navigation equipment approved under type certification or a Technical Standard Order (TSO) that have an effect on frequency stability, noise level, sensitivity, selectivity, distortion, spurious radiation, automatic-volume-control (AVC) characteristics, or ability to meet environmental test conditions and other changes that have an effect on the performance of the equipment are also major alterations.

Minor alterations are alterations other than major alterations.

Classifications of Repairs. Repairs are classified as either **major** or **minor** depending upon the type and effect of the repair. A major repair is one which, if improperly done, might appreciably affect the weight, balance, structural strength, performance, power plant operation, flight characteristics, or other qualities affecting airworthiness or one which is not done according to accepted practices or cannot be done by elementary operations.

Repairs to the following parts of an airframe and repairs of the following types, involving the strengthening, reinforcing, splicing, and manufacturing of primary structural members, or their replacement (when replacement is by fabrication such as riveting or welding), are **airframe major repairs:**

Box beams

Monocoque or semimonocoque wings or control surfaces

Wing stringers or chord members

Spars

Spar flanges

Members of truss-type beams

Thin sheet webs of beams

Keel and chine members of boat hulls or floats

Corrugated-sheet compression members which act as flange material of wings or tail surfaces

Wing main ribs and compression members

Wing- or tail-surface brace struts

Engine mounts

Fuselage longerons

Members of the side truss, horizontal truss, or bulkheads

Main seat support braces and brackets

Landing-gear brace struts

Axles

Wheels

Skis and ski pedestals

Parts of the control system such as control columns, pedals, shafts, brackets, or horns

Repairs involving the substitution of material

Repair of damaged areas in metal or plywood stressed covering exceeding 6 in [15.24 cm] in any direction

Repair of portions of skin sheets by making additional seams

Splicing of skin sheets

Repair of three or more adjacent wing- or control-surface ribs or the leading edge of wings and control surfaces between such adjacent ribs

Repair of fabric covering involving an area greater than that required to repair two adjacent ribs

Replacement of fabric on fabric-covered parts such as wings, fuselages, stabilizers, and control surfaces

Repairing of removable or integral fuel tanks and oil tanks, including rebottoming the tanks

Repairs of the following parts of an engine and repairs of the following types are **power plant major repairs:**

Separation or disassembly of a crankcase or crankshaft of a reciprocating engine equipped with an integral supercharger.

Separation or disassembly of a crankcase or crankshaft of a reciprocating engine equipped with other than spur-type propeller reduction gearing

Special repairs to structural engine parts by welding, plating, metalizing, or other methods

Repairs of the following types to a propeller are **propeller major repairs:**

Any repairs to or straightening of steel blades

Repairing or machining of steel hubs

Shortening of blades

Retipping of wood propellers

Replacement of outer laminations on fixed-pitch wood propellers

Repairing elongated bolt holes in the hub of fixed-pitch wood propellers

Inlay work on wood blades

Repairs to composition blades

Replacement of tip fabric

Replacement of plastic covering

Repair of propeller governors
Overhaul of controllable-pitch propellers
Repairs to deep dents, cuts, scars, nicks, etc., and straightening of aluminum blades
Repair or replacement of internal elements of blades

Repair and maintenance work of the following types are classed as **appliance major repairs:**

Calibration and repair of instruments
Calibration of radio equipment
Rewinding the field coil of an electrical accessory
Complete disassembly of complex hydraulic power valves
Overhaul of pressure-type carburetors and pressure-type fuel, oil, and hydraulic pumps

Records

Every operating aircraft is required by FAR to have certain records available, and it is the responsibility of a repair station to supply and maintain records when inspections, repairs, or alterations are performed on airframes, power plants, propellers, helicopter rotors, or appliances. It is the responsibility of the owner or operator of an aircraft to see that all required records are kept up to date.

Maintenance Release. Upon completion of maintenance or alterations, the repair station or individual approving an aircraft for return to service must prepare a **maintenance release.** The maintenance release must contain the following information and be delivered to the owner or operator of the aircraft:

Identity of the aircraft, airframe, engine, propeller, helicopter rotor, or appliance
For an aircraft: the make, model, serial number, nationality and registration marks, and location of the repaired area
For an airframe, aircraft engine, propeller, helicopter rotor, or appliance: the manufacturer's name, name of the part, model, and serial number (if any)

In addition to the information listed above, the maintenance release should contain a statement worded as follows or in a similar manner:

The aircraft, airframe, aircraft engine, propeller, helicopter rotor, or appliance identified above was repaired in accordance with current regulations of the Federal Aviation Administration and is approved for return to service.

Pertinent details of the repair are on file at this repair station under Order No _____ Date _____

Signed _____
             (Signature of authorized representative)

for _____ _____
    (Repair station name) (certificate number)

_____
                  (Address)

● SHOP MANAGEMENT RESPONSIBILITIES

When first employed by a repair station, a certificated technician may not be expected to know much about shop management or management responsibilities. However, as experience is gained, a greater understanding of management will be expected, not only as it relates to the shop or repair station but also as it relates to the technician's own performance and activities. It is particularly important that the technician learn the fundamentals of management and the responsibilities of management if supervisory status and its resultant benefits are a goal.

Application of Federal Aviation Regulations

Since a certificated repair station must operate in accordance with FARs and all its activities are under the scrutiny and inspection of the FAA, the management of the repair station must make certain that the requirements of FARs are met. As explained previously, this applies to organization, equipment, housing, facilities, personnel, and methods of operation. The aviation maintenance technician employed by a certificated repair station should make every effort to assist management in meeting the requirements of FAR.

Assignment of Personnel

The effectiveness and success of a repair station depend to a large extent upon the proper assignment of personnel. Those who are responsible for the management of the repair station must carefully evaluate the capabilities of the employees, both supervisory and nonsupervisory, from the viewpoint of technical skills and knowledge and also in relation to personality and emotional stability. When this is effectively done, it is possible to assign positions that will most completely utilize the qualities of the employees. It is important that no employee be placed in a situation involving so much pressure that it affects the employee's performance.

Maintenance of Records

Management must assure that the proper records necessary to operate the repair station efficiently and within the requirements of the law are prepared and maintained. Business records such as accounts are usually handled by qualified office personnel; however, the aircraft maintenance records must be produced by properly certificated technicians and supervisors. Records maintained under the requirements of FAR must always be available for inspection by FAA inspectors.

Purchasing

The purchasing and handling of aircraft and engine parts, standard parts, approved materials, and other items require meticulous attention to detail. In a large organization, the line technician will not be required to purchase parts, but in small repair stations the technician must often place orders for parts and materials.

Aircraft parts, engine parts, propeller parts, etc., are

usually selected from the **illustrated parts catalog** provided by the manufacturer. The part is specified by manufacturer's part number and by name, and the order is placed with a dealer that represents the manufacturer. In some cases, it is necessary to place an order directly with a manufacturer.

Standard parts (AN, MS, MIL, etc.) are specified in the illustrated parts catalog and may be designated either by a manufacturer's part number or by a specification number. In either case, the purchaser orders by the number given in the catalog as specified for the aviation product involved.

The primary responsibility of the repair station and its employees is to assure that the correct part is ordered and installed on the product being repaired. This assurance is accomplished by using the illustrated parts catalog for the make and model of aircraft, engine, propeller, or appliance being repaired.

Approved supplies and materials are ordered from supply firms that specialize in distributing the product involved. Suppliers for aviation products are careful to deliver the supplies and materials indicated by a particular standard or specification number.

The care and storage of parts have been mentioned previously. The primary objectives are to prevent damage or deterioration and to organize the parts, supplies, and materials in such a manner that they can be requisitioned and delivered to the work station expeditiously. FAR Part 145 requires that the repair station have space and facilities for the proper storage and care of parts, supplies, and materials.

Inventory Control

In the operation of any organization where a stock of parts and materials must be maintained, effective inventory control is essential to good management and efficiency. A variety of systems have been developed for inventory control including manually operated card systems, automatic card systems, and computer systems.

Inventory control means simply that information is periodically or continuously maintained regarding the numbers of parts and hardware items and the quantities of materials that are in stock. This makes it possible for the purchasing agent to order parts and materials that are in short supply before they run out, thus avoiding the delays that may occur if something is needed for a repair job and it is not available.

A continuous manual inventory control system involves the use of card files in which there is a separate card for each item in stock. The card shows the number of items of each type in stock, and when an item is withdrawn for use, the number on the card is reduced. When new stock is received, the number of items received is entered on the card. Personnel can quickly and easily determine what items are in stock.

Larger organizations usually employ a computer system for inventory control. In such a system, information regarding items withdrawn from or placed into stock are fed to the computer. Anyone desiring to know whether a certain part or material is in stock keys the computer control with the part number or other identification code, and the information desired is immediately printed out or displayed on a screen.

Job Estimating

Although the certificated aviation maintenance technician may not be involved in job estimating when first hired, particularly in large organizations, many technicians in smaller organizations do become involved after a reasonable time of employment. Job estimates are the responsibility of the service manager, shop supervisor, or other person with equivalent authority. These individuals often delegate part of the work to trusted assistants.

A job estimate is a cost analysis of a particular repair or service job to be performed on an aircraft, engine, or other part. The first step is to determine exactly what repairs or services are to be performed. When this is established, the cost of parts and materials is computed together with the cost of labor based on the number of worker-hours estimated to do the job.

Many operators utilize a flat-rate system for computing labor charges for standard services such as 100-h or annual inspections for particular aircraft. Repair-station operators who are also dealers for aircraft of a particular make know with reasonable accuracy the time it takes to perform a certain job on a particular make and model of aircraft.

After a cost estimate is completed as accurately as possible, it is discussed with the owner or operator to clarify exactly what is to be done and what the cost will be. The owner or operator should also be informed that any additional work that is found to be necessary will cost an additional amount. If additional work must indeed be done, the owner or operator should be contacted and approve the work before it is begun.

● LEGAL RESPONSIBILITIES

Liability of the Technician

Any organization that provides services for the public assumes responsibility for performing the services in a manner that will not result in loss for those served. The employees involved in the service also share in the responsibility. It is not unusual, therefore, for individuals to be named with companies when damage suits are filed.

If a technician should approve an aircraft for return to service and subsequently a failure should occur that causes loss to the owner or operator, it is likely that the technician would be named as a codefendant in a damage suit. For this reason, liability insurance is a must to protect both the employees and the repair station.

Bailment

The term **bailment** is defined as a delivery of personal property by a bailor to a bailee for specific purposes under an express or implied agreement of the parties that when those purposes are accomplished the property will be returned to the bailor, kept until he reclaims it, or disposed of in accordance with the agreement. An aircraft, engine, or other item that is turned over to the care of a repair station or to an operator for hire is therefore *bailed* from the owner to the other party. Under the terms of a bailment, the bailee has responsibility for the proper care and use of the bailed property.

Mechanics' Liens

When a technician or repair station has performed repair or service work on an aircraft or other item and the owner or operator fails to pay the costs involved in the repair or service, the technician or repair station has a legal claim against the item upon which work was performed. To obtain payment for services, it is sometimes necessary to file a **mechanic's lien** against the property involved. A repair station will usually have legal counsel that takes care of such matters.

● THE REPAIR STATION, TECHNICIAN, AND ETHICS

Successful technicians and repair stations and other businesses observe certain ethical principles in the conduct of their affairs. This is particularly true in the field of aviation maintenance because human lives depend upon the skill, workmanship, and integrity of those responsible for the maintenance of aircraft and components.

Ethics deals with right and wrong, or moral duty and obligation. A **code of ethics** is a set of moral principles by which a person guides his or her conduct, performance, and relations with other human beings.

Responsibilities of Aviation

The role of general and commercial aviation in the socioeconomical-political structure of the world is far greater than the average person realizes. Millions of persons fly throughout the world because of the safe and rapid transportation that aviation provides. National leaders are whisked from continent to continent to meet with one another in order to deal with political and economic crises that affect all nations. Military and civil aircraft are mobilized to provide the swift transportation needed to supply relief when disasters strike in any part of the world. Aircraft are responsible for a large segment of the national defense system and thereby deter would-be aggressors. All the foregoing responsibilities of aviation plus many others are possible because of skilled maintenance technicians and maintenance organizations. Certified maintenance technicians can therefore take pride in their service to humanity.

Responsibility of Technicians

Understanding their critical role in the field of aviation, certificated technicians function with a high degree of integrity. They are especially careful to assure that any repair or service for which they are responsible is accomplished according to high standards of quality and that the end result is a completely airworthy product. They will not allow themselves to be influenced by pressure from supervision or management to approve a repair job that does not meet required standards and specifications. They will not return an aircraft or other product to service unless they are cer-

tain that it is completely airworthy. They take pride in their craftsmanship, their job, and the fact that they are an essential factor in aviation.

Customer Relations

An essential factor in any successful business is good customer relations. If customers are to continue to do business with a repair station, they must be pleased with the service provided and the prices charged. The customers must also feel that the employees of the repair station are personally concerned with their needs and desires.

The technician, supervisor, or manager who exercises a high degree of integrity and friendliness in dealing with a customer is a prime factor in good customer relations. In estimating the cost of a repair job, the ethical technician will be careful to list only required work and will not try to "sell" the customer unnecessary services. Such a technician will try to save the customer money rather than exploit possibilities to add unnecessary work and charges.

Important factors in good customer relations are appearance, conduct, sincerity, and dependability. When customers are pleased with the persons with whom they do business and know that they can trust those persons, they will most likely become steady customers.

Neatness, cleanliness, efficiency, and orderliness in a repair station are prime requisites for good customer relations. The majority of aircraft owners are proud of their equipment and want to know that it will be given a maximum of care and consideration while it is in the shop for service or repairs. The appearance and general atmosphere that are noted by the aircraft owner will either create confidence or instill doubts.

Employer-Technician Relations

The successful operation of a repair station requires good relations between the employees and the employer. Such relations produce good morale, and good morale results in increased satisfaction and higher productivity on the part of each person in the organization.

Technicians who are dedicated to their craft understand that the employer must make a reasonable profit in order to stay in business. Such technicians will, therefore, carry out their responsibilities in such a manner that their work is profitable for the employer. They will avoid waste of time and materials and will not expect the employer to condone sluggardly performance.

Responsible technicians will not attempt to conceal mistakes; neither will they attempt to perform work for which they are not qualified in order to impress their employer. Mistakes will be acknowledged and corrected: If technicians are not well qualified to do a particular job, they will seek help from their supervisor or a qualified fellow technician.

If technicians note conditions in the repair station that can be improved to enhance the quality of work or increase productivity, they may make tactful suggestions to their superiors, explaining why and how they think that an improvement can be made. Successful employers will appreciate constructive suggestions and will usually reward those who make them. The reward may be in the form of a pay increase, promotion, bonus, or some other benefit.

● SUMMARY OF REPAIR-STATION REQUIREMENTS

The operation of a certificated repair station must be carried out in accordance with the provisions or the operation and inspection methods approved for the repair station by the FAA. The tools, equipment, and test apparatus must be adequate to complete the repair work in accordance with accepted industry practices. If special equipment is required by the manufacturer of an item, the repair station should use this equipment or the equivalent.

The repair work performed should be accomplished in a manner which will restore the item being repaired to a condition at least equal to its original condition with respect to airworthiness and function.

The repair station must have an adequate inspection system to assure that all items are inspected independently. The final inspectors must be persons other than those who made the repair.

During the process of overhaul, annual inspection, major repair, or major alteration, the repair station must make certain that all pertinent manufacturer's bulletins, FAA airworthiness directives, and discrepancy reports are given consideration and complied with as indicated. The Aircraft Specification, Type Certificate Data Sheet, and equipment lists should also be examined to ascertain that the aircraft or other item has been kept up as required.

The repair station must prepare the required records and log-book entries to show the work performed, and must maintain records as required by Federal Aviation Regulations.

● REVIEW QUESTIONS

1. What is a *certificated repair station?*
2. Describe the organization of a repair station.
3. Why should the inspection department of a repair station be separate from the repair department?
4. Explain the purpose of a *work order.*
5. What are the purposes of *receiving inspections?*
6. Discuss *assembly inspections* and *final inspections.*
7. Describe the responsibility of the Federal Aviation Administration in approving a repair station for certification.
8. What are repair-station *ratings?*
9. Define a *large aircraft.*
10. Describe the requirements for a certificated repair station.
11. What is the purpose of the *inspection procedures manual?*
12. Who is authorized to return an aircraft to service after a major alteration or repair?
13. Explain *preventive maintenance.*
14. Define a *major alteration;* a *major repair.*
15. What is a *minor alteration?*
16. If the metal or plywood covering of a wing is damaged, how would you determine whether the repair is minor or major?
17. If the tip of a metal wing is damaged to the extent that the last four ribs need repair, is the repair minor or major?
18. List *appliance major repairs.*
19. Describe a *maintenance release.*
20. List the responsibilities of the management of a repair station.
21. Discuss the importance of the repair-station records.
22. From what document should aircraft parts be ordered?
23. How are aircraft parts identified?
24. What is the value of *inventory control?*
25. What are important factors in making a *job estimate?*
26. Discuss the liability of a technician with respect to aircraft repair work.
27. Define *bailment.*
28. What is a *mechanic's lien?*
29. What is meant by *ethics?*
30. Discuss the responsibilities of aviation.
31. How does the technician's role relate to the responsibilities of aviation?
32. Why is integrity an important characteristic of the aviation maintenance technician?
33. What condition must exist with respect to an aircraft before the technician returns it to service?
34. Describe factors that contribute to good *customer relations.*
35. Why is good customer relations important in the operation of a repair station?
36. What factors are important in the relationship between technicians and their employer?

APPENDIX

● **ATA SPEC. 100—SYSTEMS**

Sys. Sub Title

5 TIME LIMITS/MAINTENANCE CHECKS

00 General
10 Time Limits
20 Scheduled Maintenance Checks
30 Reserved
40 Reserved
50 Unscheduled Maintenance Checks

6 DIMENSIONS AND AREAS

7 LIFTING AND SHORING

00 General
10 Jacking
20 Shoring

8 LEVELING AND WEIGHING

00 General
10 Weight and Balance Computer

9 TOWING AND TAXIING

00 General
10 Towing
20 Taxiing

10 PARKING AND MOORING

00 General
10 Parking/Storage
20 Mooring
30 Return to Service

11 REQUIRED PLACARDS

00 General
10 Exterior Color Schemes and Markings
20 Exterior Placards and Markings
30 Interior Placards and Markings

12 SERVICING

00 General
10 Replenishing
20 Scheduled Servicing
30 Unscheduled Servicing

20 STANDARD PRACTICES AIRFRAME

21 AIR CONDITIONING

00 General
10 Compression
20 Distribution
30 Pressurization Control
40 Heating
50 Cooling
60 Temperature Control
70 Moisture/Air Contaminate Control

22 AUTO FLIGHT

00 General
10 Autopilot
20 Speed-Attitude Correction
30 Auto Throttle
40 System Monitor

23 COMMUNICATIONS

00 General
10 High Frequency (HF)
20 VHF/UHF
30 Passenger Address and Entertainment
40 Interphone
50 Audio Integrating
60 Static Discharging
70 Audio and Video Monitoring

24 ELECTRIC POWER

00 General
10 Generator Drive
20 AC Generation
30 DC Generation
40 External Power
50 Electrical Load Distribution

25 EQUIPMENT FURNISHINGS

00 General
10 Flight Compartment
20 Passenger Compartment
30 Buffet/Galley
40 Lavatories
50 Cargo Compartments/AG Spray Apparatus
60 Emergency
70 Accessory Compartments

26 FIRE PROTECTION

00 General
10 Detection
20 Extinguishing
30 Explosion Suppression

27 FLIGHT CONTROLS

00 General
10 Aileron and Tab
20 Rudder/Ruddervator and Tab
30 Elevator and Tab
40 Horizontal Stabilizers/Stabilator
50 Flaps

60 Spoiler, Drag Devices and Variable Aerodynamic Fairings
70 Gust Lock and Dampener
80 Lift Augmenting

28 FUEL

00 General
10 Storage
20 Distribution/Drain Valves
30 Dump
40 Indicating

29 HYDRAULIC POWER

00 General
10 Main
20 Auxiliary
30 Indicating

30 ICE AND RAIN PROTECTION

00 General
10 Airfoil
20 Air Intakes
30 Pilot and Static
40 Windows and Windshields
50 Antennas and Radomes
60 Propellers/Rotors
70 Water Lines
80 Detection

31 INDICATING/RECORDING SYSTEMS

00 General
10 Unassigned
20 Unassigned
30 Recorders
40 Central Computers
50 Central Warning System

32 LANDING GEAR

00 General
10 Main Gear
20 Nose Gear/Tail Gear
30 Extension and Retraction, Level Switch
40 Wheels and Brakes
50 Steering
60 Position, Warning and Ground Safety Switch
70 Supplementary Gear/Skis/Floats

33 LIGHTS

00 General
10 Flight Compartment and Annunciator Panels
20 Passenger Compartments
30 Cargo and Service Compartments
40 Exterior Lighting
50 Emergency Lighting

34 NAVIGATION

00 General
10 Flight Environment Data
20 Attitude and Direction
30 Landing and Taxiing Aids
40 Independent Position Determining
50 Dependent Position Determining
60 Position Computing

35 OXYGEN

00 General
10 Crew
20 Passenger
30 Portable

36 PNEUMATIC

00 General
10 Distribution
20 Indicating

37 VACUUM/PRESSURE

00 General
10 Distribution
20 Indicating

38 WATER/WASTE

00 General
10 Potable
20 Wash
30 Waste Disposal
40 Air Supply

39 ELECTRICAL/ELECTRONIC PANELS AND MULTIPURPOSE COMPONENTS

00 General
10 Instrument and Control Panels
20 Electrical and Electronic Equipment Racks
30 Electrical and Electronic Junction Boxes
40 Multipurpose Electronic Components
50 Integrated Circuits
60 Printed Circuit Card Assemblies

49 AIRBORNE AUXILIARY POWER

00 General
10 Power Plant
20 Engine
30 Engine Fuel and Control
40 Ignition/Starting
50 Air
60 Engine Controls
70 Indicating
80 Exhaust
90 Oil

51 STRUCTURES

00 General

52 DOORS

00 General
10 Passenger/Crew
20 Emergency Exit
30 Cargo
40 Service
50 Fixed Interior
60 Entrance Stairs
70 Door Warning
80 Landing Gear

53 FUSELAGE

00 General
10 Main Frame
20 Auxiliary Structure

30 Plates/Skin
40 Attach Fittings
50 Aerodynamic Fairings

54 NACELLES/PYLONS

00 General
10 Main Frame
20 Auxiliary Structure
30 Plates/Skin
40 Attach Fittings
50 Fillets/Fairings

55 STABILIZERS

00 General
10 Horizontal Stabilizers/Stabilator
20 Elevator/Elevon
30 Vertical Stabilizer
40 Rudder/Ruddervator
50 Attach Fittings

56 WINDOWS

00 General
10 Flight Compartment
20 Cabin
30 Door
40 Inspection and Observation

57 WINGS

00 General
10 Main Frame
20 Auxiliary Structure
30 Plates/Skin
40 Attach Fittings
50 Flight Surfaces

61 PROPELLERS

00 General
10 Propeller Assembly
20 Controlling
30 Braking
40 Indicating

65 ROTORS

00 General
10 Main Rotor
20 Antitorque Rotor Assembly
30 Accessory Driving
40 Controlling
50 Braking
60 Indicating

71 POWER PLANT

00 General
10 Cowling
20 Mounts
30 Fireseals and Shrouds
40 Attach Fittings
50 Electrical Harness
60 Engine Air Intakes
70 Engine Drains

72 (T) TURBINE/TURBOPROP

00 General
10 Reduction Gear and Shaft Section

20 Air Inlet Section
30 Compressor Section
40 Combustion Section
50 Turbine Section
60 Accessory Drives
70 By-pass Section

73 ENGINE FUEL AND CONTROL

00 General
10 Distribution
20 Controlling/Governing
30 Indicating

74 IGNITION

00 General
10 Electrical Power Supply
20 Distribution
30 Switching

75 BLEED AIR

00 General
10 Engine Anti-Icing
20 Accessory Cooling
30 Compressor Control
40 Indicating

76 ENGINE CONTROLS

00 General
10 Power Control
20 Emergency Shutdown

77 ENGINE INDICATING

00 General
10 Power
20 Temperature
30 Analyzers

78 ENGINE EXHAUST

00 General
10 Collector/Nozzle
20 Noise Suppressor
30 Thrust Reverser
40 Supplementary Air

79 ENGINE OIL

00 General
10 Storage (Dry Sump)
20 Distribution
30 Indicating

80 STARTING

00 General
10 Cranking

81 TURBINES (RECIPROCATING ENG)

00 General
10 Power Recovery
20 Turbo-Supercharger

82 WATER INJECTION

00 General
10 Storage
20 Distribution

30 Dumping and Purging
40 Indicating

83 REMOTE GEAR BOXES (ENG DR)

00 General
10 Drive Shaft Section
20 Gearbox Section

84 PROPULSION AUGMENTATION

00 General
10 Jack Assist Takeoff

91 CHARTS

● **SYSTÈME INTERNATIONALE D'UNITÉS (INTERNATONAL-METRIC SYSTEM)**

Numerical Prefixes

Exponential expression	Decimal equivalent	Prefix	Symbol
10^9	1 000 000 000	giga	G
10^6	1 000 000	mega	M
10^3	1 000	kilo	k
10^2	100	hecto	h
10	10	deka	da
10^{-1}	0.1	deci	d
10^{-2}	0.01	centi	c
10^{-3}	0.001	milli	m
10^{-6}	0.000 001	micro	μ

Common Units of the SI Metric System

Parameter measured	Unit	Description
Area	square meter (m^2)	10.76 ft^2
	square centimeter (cm^2)	0.1550 in^2
Density	kilograms per cubic meter (kg/m^3)	0.0624 lb/ft^3
Distance	meter (m)	39.37 in
	kilometer (km)	0.6214 mi
	centimeter (cm)	0.3937 in
Energy (work)	erg	1 dyn of force acting through 1 cm
	joule (J)	10^7 erg
		0.7377 ft·lb
Force	dyne (dyn)	1 dyn will accelerate 1 g 1 cm/s^2
	newton (N)	0.2248 lb
		100 000 dyn
Power	watt (W)	1 J/s
		0.001 34 hp
	kilowatt (kW)	1.340 48 hp
Pressure	pascal (Pa)	1 N/m^2
		0.000 145 psi
	kilopascal (kPa)	0.145 032 psi
	megapascal (MPa)	1000 kPa
	centimeter of mercury (cmHg)	0.1934 psi
Torque	newton-meter (N·m)	A force of 1 N with an arm of 1 m
		0.7375 lb·ft
		8.85 lb·in
Velocity	kilometers per hour (km/h)	0.6214 mph
		0.9113 ft/s
	meters per second (m/s)	3.281 ft/s
		2.237 mph
Volume	cubic centimeters (cm^3)	0.061 02 in^3
	liter (L)	1000 cm^3
	cubic meter (m^3)	1 000 000 cm^3
		35.31 ft^3
Weight (mass)	gram (g)	Weight (mass) of 1 cm^3 of water at its greatest density
		0.035 27 oz avdp
	kilogram (kg)	1000 g
		2.205 lb avdp

Standards

	English system	SI metric system
Gravity (g)	32.174 ft/s²	9.806 65 m/s²
Absolute zero	−459.7°F	−273.1°C
Horsepower	550 ft·lb/s	76.04 kg·m/s
		746W
π (pi)	3.141 59	
Density (ρ_0)	0.002 37 slug/ft³	
	1.221 kg/m³	

STANDARD ATMOSPHERE AT SEA LEVEL:

Temperature	59°F	15°C
Absolute temperature	518.4°R	288.2°K
Specific weight (gρ_0)	0.076 51 lb/ft³	1.2255 kg/m³
Pressure (P_0)	29.92 inHg	760 mmHg
	2 116 lb/ft³	10 332 kg/m²
		101.33 kPa

STANDARD VALUES AT ALTITUDE:

Isothermal level (Z_i)	35 332 ft	10 769 m
Isothermal temperature (t_i)	−69.7°F	−56.5°C
Temperature gradient (a)	0.003 566°F/ft	0.0065°C/m

Laws of Gases

BOYLE'S LAW: When the temperature of a given mass of gas remains constant, the volume of the gas varies inversely at its pressure.

$$\frac{V_1}{V_2} = \frac{P_2}{P_1}$$

CHARLES' LAW: When the pressure of a given mass of gas is kept constant, the volume of the gas is approximately proportional to its absolute temperature.

$$\frac{V_1}{V_2} = \frac{T_1}{T_2}$$

GENERAL LAW

$$\frac{P_1 V_1}{T_1} = \frac{P_2 V_2}{T_2}$$

Standard Atmosphere

Altitude, ft	Temperature t		Pressure P			Density	
	°F	°C	inHg	cmHg	kPa	ρ	ρ/ρ_0
0	59.0	15.0	29.920	76.00	101.33	0.002 378	1.0000
1 000	55.4	13.0	38.860	73.30	97.74	0.002 309	0.9710
2 000	51.9	11.0	27.820	70.66	94.22	0.002 242	0.9428
3 000	48.3	9.1	26.810	68.10	90.80	0.002 176	0.9151
4 000	44.7	7.1	25.840	65.63	87.51	0.002 112	0.8881
5 000	41.2	5.1	24.890	63.22	84.29	0.002 049	0.8616
6 000	37.6	3.1	23.980	60.91	81.21	0.001 988	0.8358
7 000	34.0	1.1	23.090	58.65	78.20	0.001 928	0.8106
8 000	30.5	−0.8	22.220	56.44	75.25	0.001 869	0.7859
9 000	26.9	−2.8	21.380	54.31	72.41	0.001 812	0.7619
10 000	23.3	−4.8	20.580	52.27	69.70	0.001 756	0.7384
11 000	19.8	−6.8	19.790	50.27	67.02	0.001 702	0.7154
12 000	16.2	−8.8	19.030	48.34	64.45	0.001 648	0.6931
13 000	12.6	−10.8	18.290	46.46	61.94	0.001 596	0.6712
14 000	9.1	−12.7	17.570	44.63	59.50	0.001 545	0.6499
15 000	5.5	−14.7	16.880	42.88	57.17	0.001 496	0.6291

Standard Atmosphere (*Continued*)

Altitude, ft	Temperature t °F	Temperature t °C	Pressure P inHg	Pressure P cmHg	Pressure P kPa	Density ρ	Density ρ/ρ_0
16 000	1.9	−16.7	16.210	41.47	54.90	0.001 448	0.6088
17 000	−1.6	−18.7	15.560	39.52	52.70	0.001 401	0.5891
18 000	−5.2	−20.7	14.940	37.95	50.60	0.001 355	0.5698
19 000	−8.8	−22.6	14.330	36.40	48.87	0.001 311	0.5509
20 000	−12.3	−24.6	13.750	34.93	46.57	0.001 267	0.5327
21 000	−15.9	−26.6	13.180	33.48	44.64	0.001 225	0.5148
22 000	−19.5	−28.6	12.630	32.08	42.77	0.001 183	0.4974
23 000	−23.0	−30.6	12.100	30.73	40.98	0.001 143	0.4805
24 000	−26.6	−32.5	11.590	29.44	39.25	0.001 103	0.4640
25 000	−30.2	−34.5	11.100	28.19	37.59	0.001 065	0.4480
26 000	−33.7	−36.5	10.620	26.97	35.97	0.001 028	0.4323
27 000	−37.3	−38.5	10.160	25.81	34.41	0.000 992	0.4171
28 000	−40.9	−40.5	9.720	24.69	32.92	0.000 957	0.4023
29 000	−44.4	−42.5	9.293	23.60	31.47	0.000 922	0.3879
30 000	−48.0	−44.4	8.880	22.56	30.07	0.000 889	0.3740
31 000	−51.6	−46.4	8.483	21.55	28.73	0.000 857	0.3603
32 000	−55.1	−48.4	8.101	20.58	27.44	0.000 826	0.3472
33 000	−58.7	−50.4	7.732	19.64	26.19	0.000 795	0.3343
34 000	−62.2	−52.4	7.377	18.74	24.98	0.000 765	0.3218
35 000	−65.8	−54.3	7.036	17.87	23.83	0.000 736	0.3098
36 000	−69.4	−56.3	6.708	17.04	22.72	0.000 704	0.2962
37 000	−69.7	−56.5	6.395	16.24	21.62	0.000 671	0.2824
38 000	−69.7	−56.5	6.096	15.48	20.65	0.000 640	0.2692
39 000	−69.7	−56.5	5.812	14.76	19.68	0.000 610	0.2566
40 000	−69.7	−56.5	5.541	14.07	18.77	0.000 582	0.2447
41 000	−69.7	−56.5	5.283	13.42	17.89	0.000 554	0.2332
42 000	−69.7	−56.5	5.036	12.79	17.06	0.000 529	0.2224
43 000	−69.7	−56.5	4.802	12.20	16.26	0.000 504	0.2120
44 000	−69.7	−56.5	4.578	11.63	15.50	0.000 481	0.2021
45 000	−69.7	−56.5	4.364	11.08	14.78	0.000 459	0.1926
46 000	−69.7	−56.5	4.160	10.57	14.09	0.000 437	0.1837
47 000	−69.7	−56.5	3.966	10.07	13.43	0.000 417	0.1751
48 000	−69.7	−56.5	3.781	9.60	12.81	0.000 397	0.1669
49 000	−69.7	−56.5	3.604	9.15	12.21	0.000 379	0.1591
50 000	−69.7	−56.5	3.436	8.73	11.64	0.000 361	0.1517

Greek Alphabet

Name	Capital	Lowercase	Use	Name	Capital	Lowercase	Use
Alpha	A	α	Angles	Xi	Ξ	ξ	
Beta	B	β		Omicron	O	o	
Gamma	Γ	γ	Ratio of specific heats	Pi	Π	π	Ratio of circumference to diameter (3.1416)
Delta	Δ	δ	Relative absolute pressure				
Epsilon	E	ϵ	Expansion ratio, surface emissivity	Rho	P	ρ	Mass density
				Sigma	Σ	σ	Capital: sign of summation; lowercase: relative density
Zeta	Z	ζ					
Eta	H	η	Coefficient of kinematic viscosity, efficiency				
				Tau	T	τ	
Theta	Θ	θ		Upsilon	Υ	υ	Kinematic viscosity
Iota	I	ι		Phi	Φ	ϕ	Capital: relative viscosity; lowercase: phase
Kappa	K	κ					
Lambda	Λ	λ	Capital: wing sweep angle; lowercase: taper rate, wavelength	Chi	X	χ	
				Psi	Ψ	ψ	
				Omega	Ω	ω	Capital: ohms; lowercase: specific weight
Mu	M	μ	Absolute viscosity				
Nu	N	ν					

Multiply	By	To obtain	Multiply	By	To obtain
Atmospheres (atm)	76.0	cmHg at 0°C	ft^3	2.832×10^4	cm^3
	29.92	inHg at 0°C		1728	in^3
	33.90	ftH_2O at 4°C		3.704×10^{-2}	yd^3
	1.033	kg/cm^2		7.481	U.S. gal
	14.696	psi		28.32	L
	101.33	kPa		2.83×10^{-2}	m^3
	2116.0	psf	ft^3/min	4.719×10^{-1}	L/s
	1.0133	bar, hectopieze		2.832×10^{-2}	m^3/min
Bars (bar)	75.01	cmHg at 0°C	ft^3H_2O	62.428	lb
	29.53	inHg at 0°C	ftH_2O at 4°C	2.950×10^{-2}	atm
	14.50	psi		4.335×10^{-1}	psi
	100.0	kPa		62.43	psf
Barns (b)	10^{-24}	cm^2 (nuclear cross section)		3.048×10^2	kg/m^2
				2.999	kPa
British thermal units (Btu)	778.26	ft·lb		8.826×10^{-1}	inHg at 0°C
	3.930×10^{-4}	hp·h		2.240	cmHg at 0°C
	2.931×10^{-4}	kW·h	ft/min	1.136×10^{-2}	mph
	2.520×10^{-1}	kg·cal		1.829×10^{-2}	km/h
	1.076×10^2	kg·m		5.080×10^{-1}	cm/s
	1055	J	ft/s	6.818×10^{-1}	mph
Btu/s	1055	W		1.097	km/h
Centimeters (cm)	0.3937	in		30.48	cm/s
	3.281×10^{-2}	ft		5.925×10^{-1}	kn
cmHg	5.354	inH_2O at 4°C	ft·lb	1.383×10^{-1}	kg·m
	4.460×10^{-1}	ftH_2O at 4°C		1.356	N·m *or* J
	1.934×10^{-1}	psi		1.285×10^{-3}	Btu
	27.85	psf		3.776×10^{-7}	kW·h
	135.95	kg/m^2	ft·lb/min	3.030×10^{-5}	hp
	1.333	kPa		4.06×10^{-5}	kW
cm/s	3.281×10^{-2}	ft/s	ft·lb/s	1.818×10^{-3}	hp
	2.237×10^{-2}	mph		1.356×10^{-3}	kW
	3.60×10^{-2}	km/h	fluidounce (fluid oz)	8	drams
cm^2	1.550×10^{-1}	in^2		29.6	cm^3
	1.076×10^{-3}	ft^2	gallon (gal), Imperial	277.4	in^3
	10^4	m		1.201	U.S. gal
cm^3	10^{-3}	L		4.546	L
	6.102×10^{-2}	in^3	gal, U.S., dry	268.8	in^3
	2.642×10^{-4}	U.S. gal		1.556×10^{-1}	ft^3
centipoise (cP)	6.72×10^{-4}	lb/s·ft		1.164	U.S. gal, liquid
	3.60	kg/h·m		4.405	L
circular mils (cmil)	7.854×10^{-7}	in^2	gal, U.S., liquid	231.0	in^3
	5.067×10^{-4}	mm^2		1.337×10^{-1}	ft^3
	7.854×10^{-1}	mil^2		3.785	L
curies (Ci)	3.7×10^{10}	disintegrations/s		8.327×10^{-1}	imperial gal
				128.0	fluid oz
degrees (arc) (°)	1.745×10^{-2}	rad	gigagram (Gg)	10^6	kg
dynes (dyn)	1.020×10^{-3}	g	grains	6.480×10^{-2}	g
	2.248×10^{-6}	lb	grams (g)	15.43	grains
	10^5	N		3.527×10^{-2}	oz avdp
	7.233×10^{-5}	poundals		2.205×10^{-3}	lb avdp
electron volts (eV)	1.602×10^{-12}	ergs		1000	mg
ergs (*or* dyn·cm)	9.478×10^{-11}	Btu		10^{-3}	kg
	6.2×10^{11}	eV		980.67	dyn
	7.376×10^{-8}	ft·lb	g-cal	3.969×10^{-3}	Btu
	1.020×10^{-3}	g·cm	g of U^{235} fissioned	23 000	kW·h heat generated
	10^{-7}	Joule			
	2.388×10^{-11}	kg·cal	g/cm	0.1	kg/m
feet (ft)	3.048×10^{-1}	m		6.721×10^{-2}	lb/ft
	3.333×10^{-1}	yd		5.601×10^{-3}	lb/in
	1.894×10^{-4}	mi	g/cm^3	1000	kg/m^3
	1.646×10^{-4}	nm		62.43	lb/ft^3
ft^2	929.0	cm^2	hectare	10^4	m^2
	144.0	in^2		2.471	acre
	9.294×10^{-2}	m^2			
	1.111×10^{-1}	yd^2			
	2.296×10^{-5}	acres			

Conversion Factors (*Continued*)

Multiply	By	To obtain	Multiply	By	To obtain
hectopieze	29.53	inHg	km/h	9.113×10^{-1}	ft/s
	75.01	cmHg		5.396×10^{-1}	kn
horsepower (hp)	33 000	ft·lb/min		6.214×10^{-1}	mph
	550	ft·lb/s		2.778×10^{-1}	m/s
	76.04	m·kg/s	kPa	1000	N/m²
	1.014	metric hp		14.503×10^{-2}	psi
	7.457×10^{-1}	kW	kilowatts (kW)	9.480×10^{-1}	Btu/s
	745.7	W		7.376×10^{2}	ft·lb/s
	7.068×10^{-1}	Btu·s		1.341	hp
hp, metric	75.0	m·kg/s		2.839×10^{-1}	kg·cal/s
	9.863×10^{-1}	hp	kW·h heat generated	4.35×10^{-5}	g U^{235} fissioned
	7.355×10^{-1}	kW			
	6.971×10^{-1}	Btu·s	knots (kn)	1.0	nmi/h
hp·h	2.545×10^{3}	Btu		1.688	ft/s
	1.98×10^{6}	ft·lb		1.151	mph
	2.737×10^{5}	m·kg		1.852	km/h
inch (in)	2.54	cm		5.148×10^{-1}	m/s
	83.33×10^{-3}	ft	liters (L)	10^{2}	cm³
inHg at 0°C	40.66	inAcBr₄		61.03	in³
	3.342×10^{-2}	atm		3.532×10^{-2}	ft³
	13.60	inH₂O at 4°C		2.642×10^{-1}	U.S. gal
	1.133	ftH₂O		2.200×10^{-1}	imperial gal
	4.912×10^{-1}	psi		1.057	qt
	70.73	psf	megagrams (Mg)	10^{3}	kg
	3.386	kPa	megapascals (MPa)	10^{3}	kPa
	3.453×10^{2}	kg/m²	meters (m)	39.37	in
inH₂O at 4°C	2.99	AcBr₄		3.281	ft
	7.355×10^{-2}	inHg at 0°C		1.094	yd
	1.868×10^{-1}	cmHg at 0°C		6.214×10^{-4}	mi
	3.613×10^{-2}	psi	m/s	3.281	ft/s
	2.49×10^{-1}	kPa		2.237	mph
	25.40	kg/m²		3.600	km/h
in²	6.452	cm²	m²	10.76	ft²
	6.94×10^{-3}	ft²		1.196	yd²
	6.452×10^{-4}	m²	m³	35.31	ft³
in³	16.39	cm³		264.17	gal (U.S.)
	1.639×10^{-2}	L		61023	in³
	4.329×10^{-3}	U.S. gal		1.308	yd³
	1.639×10^{-5}	m³	microamperes (μA)	6.24×10^{12}	unit charges/s
	1.732×10^{-2}	qt	microns (μm)	3.937×10^{5}	in
joules (J)	9.480×10^{-4}	Btu		10^{-6}	m
	7.375×10^{-1}	ft·lb	miles (mi)	5280	ft
	2.389×10^{-4}	kg·cal		1.609	km
	1.020×10^{-1}	kg·m		8.690×10^{-1}	nmi
	2.778×10^{-4}	Watt·h	mph	1.467	ft/s
	3.725×10^{-7}	hp·h		4.470×10^{-1}	m/s
	10^{7}	ergs		1.609	km/h
kilograms (kg)	2.205	lb		8.690×10^{-1}	kn
	9.808	N	millibars (mbar)	2.953×10^{-2}	inHg at 0°C
	35.27	oz		100.0×10^{-3}	kPa
	10^{3}	g	nautical miles (nmi)	6076.1	ft
kg·cal	3.9685	Btu		1.852	km
	3087	ft·lb		1.151	mi
	4.269×10^{2}	kg·m		1852	m
	4.1859×10^{3}	J	newtons (N)	10^{5}	dyn
kg·m	7.233	ft·lb		2.248×10^{-1}	lb
	9.809	J	ounces (oz) avdp	6.250×10^{-2}	lb avdp
kg/m³	62.43×10^{-3}	lb/ft³		28.35	g
	10^{-3}	g/cm³		4.375×10^{2}	grains
kg/cm²	14.22	psi	oz, fluid	29.57	cm³
	2.048×10^{3}	psf		1.805	in³
	28.96	inHg at 0°C	pascals (Pa)	1.0	N/m²
	32.8	ftH₂O at 4°C		14.503×10^{-5}	psi
	98.077	kPa		29.5247×10^{-5}	inHg at 0°C
kilometers (km)	3.281×10^{3}	ft		10^{-1}	bar
	6.214×10^{-1}	mi		100	mbar
	5.400×10^{-1}	nmi			
	10^{5}	cm			

Multiply	By	To obtain	Multiply	By	To obtain
pounds (lb)	453.6	g	stokes	10^{-4}	m^2/s
	7000	grains	stones	14	lb
	16.0	oz avdp		6.35	kg
	4.448	N	tons	2×10^3	lb
	3.108×10^{-2}	slugs		907.2	kg
lb/ft^3	16.02	kg/m^3	tons, metric	10^3	kg
lb/in^3	1728	lb/ft^3		2.205×10^3	lb
	27.68	gm/cm^3	unit charges/s	1.6×10^{-13}	μA
psi	2.036	inHg at 0°C	watts (W)	9.481×10^{-4}	Btu/s
	2.307	ftH_2O at 4°C		1.340×10^{-3}	hp
	6.805×10^{-2}	atm	yards (yd)	3.0	ft
	7.031×10^2	kg/m^2		36.0	in
	6.895	kPa		9.144×10^{-1}	m
radians (rad)	57.30	° (arc)	yd^2	9.0	ft^2
rad/s	57.30	°/s		1.296×10^3	in^2
	15.92×10^{-2}	rev/s		8.361×10^{-1}	m^2
	9.549	rpm	yd^3	27.0	ft^3
revolutions (rev)	6.283	rad		2.022×10^2	gal (U.S.)
rpm	1.047×10^{-1}	rad/s		7.646×10^{-1}	m^3
slugs	14.59	kg			
	32.18	lb			

Trigonometric Functions: Degrees, Radians

Deg	Rad	Sin	Cos	Tan	Cot	Sec	Csc		
0.0	0.0000	0.000 00	1.0000	0.000 00	—	1.000	—	1.5708	90.0
.1	0.0017	0.001 75	1.0000	0.001 75	573.0	1.000	573.0	1.5691	.9
.2	0.0035	0.003 49	1.0000	0.003 49	286.5	1.000	286.5	1.5673	.8
.3	0.0052	0.005 24	1.0000	0.005 24	191.0	1.000	191.0	1.5656	.7
.4	0.0070	0.006 98	1.0000	0.006 98	143.2	1.000	143.2	1.5638	.6
.5	0.0087	0.008 73	1.0000	0.008 73	114.6	1.000	114.6	1.5621	.5
.6	0.0105	0.010 47	0.9999	0.010 47	95.49	1.000	95.49	1.5603	.4
.7	0.0122	0.012 22	0.9999	0.012 22	81.85	1.000	81.85	1.5586	.3
.8	0.0139	0.013 96	0.9999	0.013 96	71.62	1.000	71.62	1.5568	.2
.9	0.0157	0.015 71	0.9999	0.015 71	63.66	1.000	63.67	1.5551	.1
1.0	0.0175	0.017 45	0.9998	0.017 46	57.29	1.000	57.30	1.5533	89.0
.1	0.0192	0.019 20	0.9998	0.019 20	52.08	1.000	52.09	1.5516	.9
.2	0.0209	0.020 94	0.9998	0.020 95	47.74	1.000	7.75	1.5499	.8
.3	0.0227	0.022 69	0.9997	0.022 69	44.07	1.000	44.08	1.5481	.7
.4	0.0244	0.024 43	0.9997	0.024 44	40.92	1.000	40.93	1.5464	.6
.5	0.0262	0.026 18	0.9997	0.026 19	38.19	1.000	38.20	1.5446	.5
.6	0.0279	0.027 92	0.9996	0.027 93	35.80	1.000	35.82	1.5429	.4
.7	0.0297	0.029 67	0.9996	0.029 68	33.69	1.000	33.71	1.5411	.3
.8	0.0314	0.031 41	0.9995	0.031 43	31.82	1.000	31.84	1.5394	.2
.9	0.0332	0.033 16	0.9995	0.033 17	30.14	1.001	30.16	1.5376	.1
2.0	0.0349	0.034 90	0.9994	0.034 92	28.64	1.001	28.65	1.5359	88.0
.1	0.0367	0.036 64	0.9993	0.036 67	27.27	1.001	27.29	1.5341	.9
.2	0.0384	0.038 39	0.9993	0.038 42	26.03	1.001	26.05	1.5324	.8
.3	0.0401	0.040 13	0.9992	0.040 16	24.90	1.001	24.92	1.5307	.7
.4	0.0419	0.041 88	0.9991	0.041 91	23.86	1.001	23.88	1.5289	.6
.5	0.0436	0.043 62	0.9990	0.043 66	22.90	1.001	22.93	1.5272	.5
.6	0.0454	0.045 36	0.9990	0.045 41	22.02	1.001	22.04	1.5254	.4
.7	0.0471	0.047 11	0.9989	0.047 16	21.20	1.001	21.23	1.5237	.3
.8	0.0489	0.048 85	0.9988	0.048 91	20.45	1.001	20.47	1.5219	.2
.9	0.0506	0.050 59	0.9987	0.050 66	19.74	1.001	19.77	1.5202	.1
		Cos	Sin	Cot	Tan	Csc	Sec	Rad	Deg

Deg	Rad	Sin	Cos	Tan	Cot	Sec	Csc		
3.0	0.0524	0.052 34	0.9986	0.052 41	19.08	1.001	19.11	1.5184	87.0
.1	0.0541	0.054 08	0.9985	0.054 16	18.46	1.001	18.49	1.5167	.9
.2	0.0559	0.055 82	0.9984	0.055 91	17.89	1.002	17.91	1.5149	.8
.3	0.0576	0.057 56	0.9983	0.057 66	17.34	1.002	17.37	1.5132	.7
.4	0.0593	0.059 31	0.9982	0.059 41	16.83	1.002	16.86	1.5115	.6
.5	0.0611	0.061 05	0.9981	0.061 16	16.35	1.002	16.38	1.5097	.5
.6	0.0628	0.062 79	0.9980	0.062 91	15.90	1.002	15.93	1.5080	.4
.7	0.0645	0.064 53	0.9979	0.064 67	15.46	1.002	15.50	1.5062	.3
.8	0.0663	0.066 27	0.9978	0.066 42	15.06	1.002	15.09	1.5045	.2
.9	0.0681	0.068 02	0.9977	0.068 17	14.67	1.002	14.70	1.5027	.1
4.0	0.0698	0.069 76	0.9976	0.069 93	14.30	1.002	14.34	1.5010	86.0
.1	0.0716	0.071 50	0.9974	0.071 68	13.95	1.003	13.99	1.4992	.9
.2	0.0733	0.073 24	0.9973	0.073 44	13.62	1.003	13.65	1.4975	.8
.3	0.0750	0.074 98	0.9972	0.075 19	13.30	1.003	13.34	1.4957	.7
.4	0.0768	0.076 72	0.9971	0.076 95	13.00	1.003	13.03	1.4940	.6
.5	0.0785	0.078 46	0.9969	0.078 70	12.71	1.003	12.75	1.4923	.5
.6	0.0803	0.080 20	0.9968	0.080 46	12.43	1.003	12.47	1.4905	.4
.7	0.0820	0.081 94	0.9966	0.082 21	12.16	1.003	12.20	1.4888	.3
.8	0.0838	0.083 68	0.9965	0.083 97	11.91	1.004	11.95	1.4870	.2
.9	0.0855	0.085 42	0.9963	0.085 73	11.66	1.004	11.71	1.4853	.1
5.0	0.0873	0.087 16	0.9962	0.087 49	11.43	1.004	11.47	1.4835	85.0
.1	0.0890	0.088 89	0.9960	0.089 25	11.20	1.004	11.25	1.4818	.9
.2	0.0908	0.090 63	0.9959	0.091 01	10.99	1.004	11.03	1.4800	.8
.3	0.0925	0.092 37	0.9957	0.092 77	10.78	1.004	10.83	1.4783	.7
.4	0.0942	0.094 11	0.9956	0.094 53	10.58	1.004	10.63	1.4765	.6
.5	0.0960	0.095 85	0.9954	0.096 29	10.39	1.005	10.43	1.4748	.5
.6	0.0977	0.097 58	0.9952	0.098 05	10.20	1.005	10.25	1.4731	.4
.7	0.0995	0.099 32	0.9951	0.099 81	10.02	1.005	10.07	1.4713	.3
.8	0.1012	0.101 1	0.9949	0.101 6	9.845	1.005	9.896	1.4696	.2
.9	0.1030	0.102 8	0.9947	0.103 3	9.677	1.005	9.728	1.4678	.1
6.0	0.1047	0.104 5	0.9945	0.105 1	9.514	1.006	9.567	1.4661	84.0
6.0	0.1047	0.104 5	0.9945	0.105 1	9.514	1.006	9.567	1.4661	84.0
.1	0.1065	0.106 3	0.9943	0.106 9	9.357	1.006	9.411	1.4643	.9
.2	0.1082	0.108 0	0.9942	0.108 6	9.205	1.006	9.259	1.4626	.8
.3	0.1100	0.109 7	0.9940	0.110 4	9.058	1.006	9.113	1.4608	.7
.4	0.1117	0.111 5	0.9938	0.112 2	8.915	1.006	8.971	1.4591	.6
.5	0.1134	0.113 2	0.9936	0.113 9	8.777	1.006	8.834	1.4574	.5
.6	0.1152	0.114 9	0.9934	0.115 7	8.643	1.007	8.700	1.4556	.4
.7	0.1169	0.116 7	0.9932	0.117 5	8.513	1.007	8.571	1.4539	.3
.8	0.1187	0.118 4	0.9930	0.119 2	8.386	1.007	8.446	1.4521	.2
.9	0.1204	0.120 1	0.9928	0.121 0	8.264	1.007	8.324	1.4504	.1
7.0	0.1222	0.121 9	0.9925	0.122 8	8.144	1.008	8.206	1.4486	83.0
.1	0.1239	0.123 6	0.9923	0.124 6	8.028	1.008	8.091	1.4469	.9
.2	0.1257	0.125 3	0.9921	0.126 3	7.916	1.008	7.979	1.4451	.8
.3	0.1274	0.127 1	0.9919	0.128 1	7.806	1.008	7.870	1.4434	.7
.4	0.1292	0.128 8	0.9917	0.129 9	7.700	1.008	7.764	1.4416	.6
.5	0.1309	0.130 5	0.9914	0.131 7	7.596	1.009	7.661	1.4399	.5
.6	0.1326	0.132 3	0.9912	0.133 4	7.495	1.009	7.561	1.4382	.4
.7	0.1344	0.134 0	0.9910	0.135 2	7.396	1.009	7.463	1.4364	.3
.8	0.1361	0.135 7	0.9907	0.137 0	7.300	1.009	7.368	1.4347	.2
.9	0.1379	0.137 4	0.9905	0.138 8	7.207	1.010	7.276	1.4329	.1
8.0	0.1396	0.139 2	0.9903	0.140 5	7.115	1.010	7.185	1.4312	82.0
.1	0.1414	0.140 9	0.9900	0.142 3	7.026	1.010	7.097	1.4294	.9
.2	0.1431	0.142 6	0.9898	0.144 1	6.940	1.010	7.011	1.4277	.8
.3	0.1449	0.144 4	0.9895	0.145 9	6.855	1.011	6.927	1.4259	.7
.4	0.1466	0.146 1	0.9893	0.147 7	6.772	1.011	6.845	1.4242	.6
.5	0.1484	0.147 8	0.9890	0.149 5	6.691	1.011	6.765	1.4224	.5
.6	0.1501	0.149 5	0.9888	0.151 2	6.612	1.011	6.687	1.4207	.4
.7	0.1518	0.151 3	0.9885	0.153 0	6.535	1.012	6.611	1.4190	.3
.8	0.1536	0.153 0	0.9882	0.154 8	6.460	1.012	6.537	1.4172	.2
.9	0.1553	0.154 7	0.9880	0.156 6	6.386	1.012	6.464	1.4155	.1
		Cos	Sin	Cot	Tan	Csc	Sec	Rad	Deg

Deg	Rad	Sin	Cos	Tan	Cot	Sec	Csc		
9.0	0.1571	0.156 4	0.9877	0.158 4	6.314	1.012	6.392	1.4137	81.0
.1	0.1588	0.158 2	0.9874	0.160 2	6.243	1.013	6.323	1.4120	.9
.2	0.1606	0.159 9	0.9871	0.162 0	6.174	1.013	6.255	1.4102	.8
.3	0.1623	0.161 6	0.9869	0.163 8	6.107	1.013	6.188	1.4085	.7
.4	0.1641	0.163 3	0.9866	0.165 5	6.041	1.014	6.123	1.4067	.6
.5	0.1658	0.165 0	0.9863	0.167 3	5.976	1.014	6.059	1.4050	.5
.6	0.1676	0.166 8	0.9860	0.169 1	5.912	1.014	5.996	1.4032	.4
.7	0.1693	0.168 5	0.9857	0.170 9	5.850	1.015	5.935	1.4015	.3
.8	0.1710	0.170 2	0.9854	0.172 7	5.789	1.015	5.875	1.3998	.2
.9	0.1728	0.171 9	0.9851	0.174 5	5.730	1.015	5.816	1.3980	.1
10.0	0.1745	0.173 6	0.9848	0.176 3	5.671	1.015	5.759	1.3963	80.0
.1	0.1763	0.175 4	0.9845	0.178 1	5.614	1.016	5.702	1.3945	.9
.2	0.1780	0.177 1	0.9842	0.179 9	5.558	1.016	5.647	1.3928	.8
.3	0.1798	0.178 8	0.9839	0.181 7	5.503	1.016	5.593	1.3910	.7
.4	0.1815	0.180 5	0.9836	0.183 5	5.449	1.017	5.540	1.3893	.6
.5	0.1833	0.182 2	0.9833	0.185 3	5.396	1.017	5.487	1.3875	.5
.6	0.1850	0.184 0	0.9829	0.187 1	5.343	1.017	5.436	1.3858	.4
.7	0.1868	0.185 7	0.9826	0.189 0	5.292	1.018	5.386	1.3840	.3
.8	0.1885	0.187 4	0.9823	0.190 8	5.242	1.018	5.337	1.3823	.2
.9	0.1902	0.189 1	0.9820	0.192 6	5.193	1.018	5.288	1.3806	.1
11.0	0.1920	0.190 8	0.9816	0.194 4	5.145	1.019	5.241	1.3788	79.0
.1	0.1937	0.192 5	0.9813	0.196 2	5.097	1.019	5.194	1.3771	.9
.2	0.1955	0.194 2	0.9810	0.198 0	5.050	1.019	5.148	1.3753	.8
.3	0.1972	0.195 9	0.9806	0.199 8	5.005	1.020	5.103	1.3736	.7
.4	0.1990	0.197 7	0.9803	0.201 6	4.959	1.020	5.059	1.3718	.6
.5	0.2007	0.199 4	0.9799	0.203 5	4.915	1.020	5.016	1.3701	.5
.6	0.2025	0.201 1	0.9796	0.205 3	4.872	1.021	4.973	1.3683	.4
.7	0.2042	0.202 8	0.9792	0.207 1	4.829	1.021	4.931	1.3666	.3
.8	0.2059	0.204 5	0.9789	0.208 9	4.787	1.022	4.890	1.3648	.2
.9	0.2077	0.206 2	0.9785	0.210 7	4.745	1.022	4.850	1.3641	.1
12.0	0.2094	0.207 9	0.9781	0.212 6	4.705	1.022	4.810	1.3614	78.0
1	0.2112	0.209 6	0.9778	0.214 4	4.665	1.023	4.771	1.3596	.9
.2	0.2129	0.211 3	0.9774	0.216 2	4.625	1.023	4.732	1.3579	.8
.3	0.2147	0.213 0	0.9770	0.218 0	4.586	1.023	4.694	1.3561	.7
.4	0.2164	0.214 7	0.9767	0.219 9	4.548	1.024	4.657	1.3544	.6
.5	0.2182	0.216 4	0.9763	0.221 7	4.511	1.024	4.620	1.3526	.5
.6	0.2199	0.218 1	0.9759	0.223 5	4.474	1.025	4.584	1.3509	.4
.7	0.2217	0.219 8	0.9755	0.225 4	4.437	1.025	4.549	1.3491	.3
.8	0.2234	0.221 5	0.9751	0.227 2	4.402	1.025	4.514	1.3474	.2
.9	0.2251	0.223 3	0.9748	0.229 0	4.366	1.026	4.479	1.3456	.1
13.0	0.2269	0.225 0	0.9744	0.230 9	4.331	1.026	4.445	1.3439	77.0
.1	0.2286	0.226 7	0.9740	0.232 7	4.297	1.027	4.412	1.3422	.9
.2	0.2304	0.228 4	0.9736	0.234 5	4.264	1.027	4.379	1.3404	.8
.3	0.2321	0.230 0	0.9732	0.236 4	4.230	1.028	4.347	1.3387	.7
.4	0.2339	0.231 7	0.9728	0.238 2	4.198	1.028	4.315	1.3369	.6
.5	0.2356	0.233 4	0.9724	0.240 1	4.165	1.028	4.284	1.3352	.5
.6	0.2374	0.235 1	0.9720	0.241 9	4.134	1.029	4.253	1.3334	.4
.7	0.2391	0.236 8	0.9715	0.243 8	4.102	1.029	4.222	1.3317	.3
.8	0.2409	0.238 5	0.9711	0.245 6	4.071	1.030	4.192	1.3299	.2
.9	0.2426	0.240 2	0.9707	0.247 5	4.041	1.030	4.163	1.3282	.1
14.0	0.2443	0.241 9	0.9703	0.249 3	4.011	1.031	4.134	1.3265	76.0
.1	0.2461	0.243 6	0.9699	0.251 2	3.981	1.031	4.105	1.3247	.9
.2	0.2478	0.245 3	0.9694	0.253 0	3.952	1.032	4.077	1.3230	.8
.3	0.2496	0.247 0	0.9690	0.254 9	3.923	1.032	4.049	1.3212	.7
.4	0.2513	0.248 7	0.9686	0.256 8	3.895	1.032	4.021	1.3195	.6
.5	0.2531	0.250 4	0.9681	0.258 6	3.867	1.033	3.994	1.3177	.5
.6	0.2548	0.252 1	0.9677	0.260 5	3.839	1.033	3.967	1.3160	.4
.7	0.2566	0.253 8	0.9673	0.262 3	3.812	1.034	3.941	1.3142	.3
.8	0.2583	0.255 4	0.9668	0.264 2	3.785	1.034	3.915	1.3125	.2
.9	0.2600	0.257 1	0.9664	0.266 1	3.758	1.035	3.889	1.3107	.1
		Cos	Sin	Cot	Tan	Csc	Sec	Rad	Deg

Deg	Rad	Sin	Cos	Tan	Cot	Sec	Csc		
15.0	0.2618	0.258 8	0.9659	0.267 9	3.732	1.035	3.864	1.3090	75.0
.1	0.2635	0.260 5	0.9655	0.269 8	3.706	1.036	3.839	1.3073	.9
.2	0.2653	0.262 2	0.9650	0.271 7	3.681	1.036	3.814	1.3055	.8
.3	0.2670	0.263 9	0.9646	0.273 6	3.655	1.037	3.790	1.3038	.7
.4	0.2688	0.265 6	0.9641	0.275 4	3.630	1.037	3.766	1.3020	.6
.5	0.2705	0.267 2	0.9636	0.277 3	3.606	1.038	3.742	1.3003	.5
.6	0.2723	0.268 9	0.9632	0.279 2	3.582	1.038	3.719	1.2985	.4
.7	0.2740	0.270 6	0.9627	0.281 1	3.558	1.039	3.695	1.2968	.3
.8	0.2758	0.272 3	0.9622	0.283 0	3.534	1.039	3.673	1.2950	.2
.9	0.2775	0.274 0	0.9617	0.284 9	3.511	1.040	3.650	1.2933	.1
16.0	0.2793	0.275 6	0.9613	0.286 7	3.487	1.040	3.628	1.2915	74.0
.1	0.2810	0.277 3	0.9608	0.288 6	3.465	1.041	3.606	1.2898	.9
.2	0.2827	0.279 0	0.9603	0.290 5	3.442	1.041	3.584	1.2881	.8
.3	0.2845	0.280 7	0.9598	0.292 4	3.420	1.042	3.563	1.2863	.7
.4	0.2862	0.282 3	0.9593	0.294 3	3.398	1.042	3.542	1.2846	.6
.5	0.2880	0.284 0	0.9588	0.296 2	3.376	1.043	3.521	1.2828	.5
.6	0.2897	0.285 7	0.9583	0.298 1	3.354	1.043	3.500	1.2811	.4
.7	0.2915	0.287 4	0.9578	0.300 0	3.333	1.044	3.480	1.2793	.3
.8	0.2932	0.289 0	0.9573	0.301 9	3.312	1.045	3.460	1.2776	.2
.9	0.2950	0.290 7	0.9568	0.303 8	3.291	1.045	3.440	1.2758	.1
17.0	0.2967	0.292 4	0.9563	0.305 7	3.271	1.046	3.420	1.2741	73.0
.1	0.2985	0.294 0	0.9558	0.307 6	3.251	1.046	3.401	1.2723	.9
.2	0.3002	0.295 7	0.9553	0.309 6	3.230	1.047	3.382	1.2706	.8
.3	0.3019	0.297 4	0.9548	0.311 5	3.211	1.047	3.363	1.2689	.7
.4	0.3037	0.299 0	0.9542	0.313 4	3.191	1.048	3.344	1.2671	.6
.5	0.3054	0.300 7	0.9537	0.315 3	3.172	1.049	3.326	1.2654	.5
.6	0.3072	0.302 4	0.9532	0.317 2	3.152	1.049	3.307	1.2636	.4
.7	0.3089	0.304 0	0.9527	0.319 1	3.133	1.050	3.289	1.2619	.3
.8	0.3107	0.305 7	0.9521	0.321 1	3.115	1.050	3.271	1.2601	.2
.9	0.3124	0.307 4	0.9516	0.323 0	3.096	1.051	3.254	1.2584	.1
18.0	0.3142	0.309 0	0.9511	0.324 9	3.078	1.051	3.236	1.2566	72.0
.1	0.3159	0.310 7	0.9505	0.326 9	3.060	1.052	3.219	1.2549	.9
.2	0.3177	0.312 3	0.9500	0.328 8	3.042	1.053	3.202	1.2531	.8
.3	0.3194	0.314 0	0.9494	0.330 7	3.024	1.053	3.185	1.2514	.7
.4	0.3211	0.315 6	0.9489	0.332 7	3.006	1.054	3.168	1.2497	.6
.5	0.3229	0.317 3	0.9483	0.334 6	2.989	1.054	3.152	1.2479	.5
.6	0.3246	0.319 0	0.9478	0.336 5	2.971	1.055	3.135	1.2462	.4
.7	0.3264	0.320 6	0.9472	0.338 5	2.954	1.056	3.119	1.2444	.3
.8	0.3281	0.322 3	0.9466	0.340 4	2.937	1.056	3.103	1.2427	.2
.9	0.3299	0.323 9	0.9461	0.342 4	2.921	1.057	3.087	1.2409	.1
19.0	0.3316	0.325 6	0.9455	0.344 3	2.904	1.058	3.072	1.2392	71.0
.1	0.3334	0.327 2	0.9449	0.346 3	2.888	1.058	3.056	1.2374	.9
.2	0.3351	0.328 9	0.9444	0.348 2	2.872	1.059	3.041	1.2357	.8
.3	0.3368	0.330 5	0.9438	0.350 2	2.856	1.060	3.026	1.2339	.7
.4	0.3386	0.332 2	0.9432	0.352 2	2.840	1.060	3.011	1.2322	.6
.5	0.3403	0.333 8	0.9426	0.354 1	2.824	1.061	2.996	1.2305	.5
.6	0.3421	0.335 5	0.9421	0.356 1	2.808	1.062	2.981	1.2287	.4
.7	0.3438	0.337 1	0.9415	0.358 1	2.793	1.062	2.967	1.2270	.3
.8	0.3456	0.338 7	0.9409	0.360 0	2.778	1.063	2.952	1.2252	.2
.9	0.3473	0.340 4	0.9403	0.362 0	2.762	1.064	2.938	1.2235	.1
20.0	0.3491	0.342 0	0.9397	0.364 0	2.747	1.064	2.924	1.2217	70.0
.1	0.3508	0.343 7	0.9391	0.365 9	2.733	1.065	2.910	1.2200	.9
.2	0.3526	0.345 3	0.9385	0.367 9	2.718	1.066	2.896	1.2182	.8
.3	0.3543	0.346 9	0.9379	0.369 9	2.703	1.066	2.882	1.2165	.7
.4	0.3560	0.348 6	0.9373	0.371 9	2.689	1.067	2.869	1.2147	.6
.5	0.3578	0.350 2	0.9367	0.373 9	2.675	1.068	2.855	1.2130	.5
.6	0.3595	0.351 8	0.9361	0.375 9	2.660	1.068	2.842	1.2113	.4
.7	0.3613	0.353 5	0.9354	0.377 9	2.646	1.069	2.829	1.2095	.3
.8	0.3630	0.355 1	0.9348	0.379 9	2.633	1.070	2.816	1.2078	.2
.9	0.3648	0.356 7	0.9342	0.381 9	2.619	1.070	2.803	1.2060	.1
		Cos	Sin	Cot	Tan	Csc	Sec	Rad	Deg

Deg	Rad	Sin	Cos	Tan	Cot	Sec	Csc		
21.0	0.3665	0.358 4	0.9336	0.383 9	2.605	1.071	2.790	1.2043	69.0
.1	0.3683	0.360 0	0.9330	0.385 9	2.592	1.072	2.778	1.2025	.9
.2	0.3700	0.361 6	0.9323	0.387 9	2.578	1.073	2.765	1.2008	.8
.3	0.3718	0.363 3	0.9317	0.389 9	2.565	1.073	2.753	1.1990	.7
.4	0.3735	0.364 9	0.9311	0.391 9	2.552	1.074	2.741	1.1973	.6
.5	0.3752	0.366 5	0.9304	0.393 9	2.539	1.075	2.729	1.1956	.5
.6	0.3770	0.368 1	0.9298	0.395 9	2.526	1.076	2.716	1.1938	.4
.7	0.3787	0.369 7	0.9291	0.397 9	2.513	1.076	2.705	1.1921	.3
.8	0.3805	0.371 4	0.9285	0.400 0	2.500	1.077	2.693	1.1903	.2
.9	0.3822	0.373 0	0.9278	0.402 0	2.488	1.078	2.681	1.1886	.1
22.0	0.3840	0.374 6	0.9272	0.404 0	2.475	1.079	2.669	1.1868	68.0
.1	0.3857	0.376 2	0.9265	0.406 1	2.463	1.079	2.658	1.1851	.9
.2	0.3875	0.377 8	0.9259	0.408 1	2.450	1.080	2.647	1.1833	.8
.3	0.3892	0.379 5	0.9252	0.410 1	2.438	1.081	2.635	1.1816	.7
.4	0.3910	0.381 1	0.9245	0.412 2	2.426	1.082	2.624	1.1798	.6
.5	0.3927	0.382 7	0.9239	0.414 2	2.414	1.082	2.613	1.1781	.5
.6	0.3944	0.384 3	0.9232	0.416 3	2.402	1.083	2.602	1.1764	.4
.7	0.3962	0.385 9	0.9225	0.418 3	2.391	1.084	2.591	1.1746	.3
.8	0.3979	0.387 5	0.9219	0.420 4	2.379	1.085	2.581	1.1729	.2
.9	0.3997	0.389 1	0.9212	0.422 4	2.367	1.086	2.570	1.1711	.1
23.0	0.4014	0.390 7	0.9205	0.424 5	2.356	1.086	2.559	1.1694	67.0
.1	0.4032	0.392 3	0.9198	0.426 5	2.344	1.087	2.549	1.1676	.9
.2	0.4049	0.393 9	0.9191	0.428 6	2.333	1.088	2.538	1.1659	.8
.3	0.4067	0.395 5	0.9184	0.430 7	2.322	1.089	2.528	1.1641	.7
.4	0.4084	0.397 1	0.9178	0.432 7	2.311	1.090	2.518	1.1624	.6
.5	0.4102	0.398 7	0.9171	0.434 8	2.300	1.090	2.508	1.1606	.5
.6	0.4119	0.400 3	0.9164	0.436 9	2.289	1.091	2.498	1.1589	.4
.7	0.4136	0.401 9	0.9157	0.439 0	2.278	1.092	2.488	1.1572	.3
.8	0.4154	0.403 5	0.9150	0.441 1	2.267	1.093	2.478	1.1554	.2
.9	0.4171	0.405 1	0.9143	0.443 1	2.257	1.094	2.468	1.1537	.1
24.0	0.4189	0.406 7	0.9135	0.445 2	2.246	1.095	2.459	1.1519	66.0
.1	0.4206	0.408 3	0.9128	0.447 3	2.236	1.095	2.449	1.1502	.9
.2	0.4224	0.409 9	0.9121	0.449 4	2.225	1.096	2.439	1.1484	.8
.3	0.4241	0.411 5	0.9114	0.451 5	2.215	1.097	2.430	1.1467	.7
.4	0.4259	0.413 1	0.9107	0.453 6	2.204	1.098	2.421	1.1449	.6
.5	0.4276	0.414 7	0.9100	0.455 7	2.194	1.099	2.411	1.1432	.5
.6	0.4294	0.416 3	0.9092	0.457 8	2.184	1.100	2.402	1.1414	.4
.7	0.4311	0.417 9	0.9085	0.459 9	2.174	1.101	2.393	1.1397	.3
.8	0.4328	0.419 5	0.9078	0.462 1	2.164	1.102	2.384	1.1379	2
.9	0.4346	0.421 0	0.9070	0.464 2	2.154	1.102	2.375	1.1362	.1
25.0	0.4363	0.422 6	0.9063	0.466 3	2.145	1.103	2.366	1.1345	65.0
.1	0.4381	0.424 2	0.9056	0.468 4	2.135	1.104	2.357	1.1327	.9
.2	0.4398	0.425 8	0.9048	0.470 6	2.125	1.105	2.349	1.1310	.8
.3	0.4416	0.427 4	0.9041	0.472 7	2.116	1.106	2.340	1.1292	.7
.4	0.4433	0.428 9	0.9033	0.474 8	2.106	1.107	2.331	1.1275	.6
.5	0.4451	0.430 5	0.9026	0.477 0	2.097	1.108	2.323	1.1257	.5
.6	0.4468	0.432 1	0.9018	0.479 1	2.087	1.109	2.314	1.1240	.4
.7	0.4485	0.433 7	0.9011	0.481 3	2.078	1.110	2.306	1.1222	.3
.8	0.4503	0.435 2	0.9003	0.483 4	2.069	1.111	2.298	1.1205	.2
.9	0.4520	0.436 8	0.8996	0.485 6	2.059	1.112	2.289	1.1188	.1
26.0	0.4538	0.438 4	0.8988	0.487 7	2.050	1.113	2.281	1.1170	64.0
.1	0.4555	0.439 9	0.8980	0.489 9	2.041	1.114	2.273	1.1153	.9
.2	0.4573	0.441 5	0.8973	0.492 1	2.032	1.115	2.265	1.1135	.8
.3	0.4590	0.443 1	0.8965	0.494 2	2.023	1.115	2.257	1.1118	.7
.4	0.4608	0.444 6	0.8957	0.496 4	2.014	1.116	2.249	1.1100	.6
.5	0.4625	0.446 2	0.8949	0.498 6	2.006	1.117	2.241	1.1083	.5
.6	0.4643	0.447 8	0.8942	0.500 8	1.997	1.118	2.233	1.1065	.4
.7	0.4660	0.449 3	0.8934	0.502 9	1.988	1.119	2.226	1.1048	.3
.8	0.4677	0.450 9	0.8926	0.505 1	1.980	1.120	2.218	1.1030	.2
.9	0.4695	0.452 4	0.8918	0.507 3	1.971	1.121	2.210	1.1013	.1
		Cos	Sin	Cot	Tan	Csc	Sec	Rad	Deg

Trigonometric Functions *(Continued)*

Deg	Rad	Sin	Cos	Tan	Cot	Sec	Csc		
27.0	0.4712	0.454 0	0.8910	0.509 5	1.963	1.122	2.203	1.0996	63.0
.1	0.4730	0.455 5	0.8902	0.511 7	1.954	1.123	2.195	1.0978	.9
.2	0.4747	0.457 1	0.8894	0.513 9	1.946	1.124	2.188	1.0961	.8
.3	0.4765	0.458 6	0.8886	0.516 1	1.937	1.125	2.180	1.0943	.7
.4	0.4782	0.460 2	0.8878	0.518 4	1.929	1.126	2.173	1.0926	.6
.5	0.4800	0.461 7	0.8870	0.520 6	1.921	1.127	2.166	1.0908	.5
.6	0.4817	0.463 3	0.8862	0.522 8	1.913	1.128	2.158	1.0891	.4
.7	0.4835	0.464 8	0.8854	0.525 0	1.905	1.129	2.151	1.0873	.3
.8	0.4852	0.466 4	0.8846	0.527 2	1.897	1.130	2.144	1.0856	.2
.9	0.4869	0.467 9	0.8838	0.529 5	1.889	1.132	2.137	1.0838	.1
28.0	0.4887	0.469 5	0.8829	0.531 7	1.881	1.133	2.130	1.0821	62.0
.1	0.4904	0.471 0	0.8821	0.534 0	1.873	1.134	2.123	1.0804	.9
.2	0.4922	0.472 6	0.8813	0.536 2	1.865	1.135	2.116	1.0786	.8
.3	0.4939	0.474 1	0.8805	0.538 4	1.857	1.136	2.109	1.0769	.7
.4	0.4957	0.475 6	0.8796	0.540 7	1.849	1.137	2.103	1.0751	.6
.5	0.4974	0.477 2	0.8788	0.543 0	1.842	1.138	2.096	1.0734	.5
.6	0.4992	0.478 7	0.8780	0.545 2	1.834	1.139	2.089	1.0716	.4
.7	0.5009	0.480 2	0.8771	0.547 5	1.827	1.140	2.082	1.0699	.3
.8	0.5027	0.481 8	0.8763	0.549 8	1.819	1.141	2.076	1.0681	.2
.9	0.5044	0.483 3	0.8755	0.552 0	1.811	1.142	2.069	1.0664	.1
29.0	0.5061	0.484 8	0.8746	0.554 3	1.804	1.143	2.063	1.0647	61.0
.1	0.5079	0.486 3	0.8738	0.556 6	1.797	1.144	2.056	1.0629	.9
.2	0.5096	0.487 9	0.8729	0.558 9	1.789	1.146	2.050	1.0612	.8
.3	0.5114	0.489 4	0.8721	0.561 2	1.782	1.147	2.043	1.0594	.7
.4	0.5131	0.490 9	0.8712	0.563 5	1.775	1.148	2.037	1.0577	.6
.5	0.5149	0.492 4	0.8704	0.565 8	1.767	1.149	2.031	1.0559	.5
.6	0.5166	0.493 9	0.8695	0.568 1	1.760	1.150	2.025	1.0542	.4
.7	0.5184	0.495 5	0.8686	0.570 4	1.753	1.151	2.018	1.0524	.3
.8	0.5201	0.497 0	0.8678	0.572 7	1.746	1.152	2.012	1.0507	.2
.9	0.5219	0.498 5	0.8669	0.575 0	1.739	1.154	2.006	1.0489	.1
30.0	0.5236	0.500 0	0.8660	0.577 4	1.732	1.155	2.000	1.0472	60.0
.1	0.5253	0.501 5	0.8652	0.579 7	1.725	1.156	1.994	1.0455	.9
.2	0.5271	0.503 0	0.8643	0.582 0	1.718	1.157	1.988	1.0437	.8
.3	0.5288	0.504 5	0.8634	0.584 4	1.711	1.158	1.982	1.0420	.7
.4	0.5306	0.506 0	0.8625	0.586 7	1.704	1.159	1.976	1.0402	.6
.5	0.5323	0.507 5	0.8616	0.589 0	1.698	1.161	1.970	1.0385	.5
.6	0.5341	0.509 0	0.8607	0.591 4	1.691	1.162	1.964	1.0367	.4
.7	0.5358	0.510 5	0.8599	0.593 8	1.684	1.163	1.959	1.0350	.3
.8	0.5376	0.512 0	0.8590	0.596 1	1.678	1.164	1.953	1.0332	.2
.9	0.5393	0.513 5	0.8581	0.598 5	1.671	1.165	1.947	1.0315	.1
31.0	0.5411	0.515 0	0.8572	0.600 9	1.664	1.167	1.942	1.0297	59.0
.1	0.5428	0.516 5	0.8563	0.603 2	1.658	1.168	1.936	1.0280	.9
.2	0.5445	0.518 0	0.8554	0.605 6	1.651	1.169	1.930	1.0263	.8
.3	0.5463	0.519 5	0.8545	0.608 0	1.645	1.170	1.925	1.0245	.7
.4	0.5480	0.521 0	0.8536	0.610 4	1.638	1.172	1.919	1.0228	.6
.5	0.5498	0.522 5	0.8526	0.612 8	1.632	1.173	1.914	1.0210	.5
.6	0.5515	0.524 0	0.8517	0.615 2	1.625	1.174	1.908	1.0193	.4
.7	0.5533	0.525 5	0.8508	0.617 6	1.619	1.175	1.903	1.0175	.3
.8	0.5550	0.527 0	0.8499	0.620 0	1.613	1.177	1.898	1.0158	.2
.9	0.5568	0.528 4	0.8490	0.622 4	1.607	1.178	1.892	1.0140	.1
32.0	0.5585	0.529 9	0.8480	0.624 9	1.600	1.179	1.887	1.0123	58.0
.1	0.5603	0.531 4	0.8471	0.627 3	1.594	1.180	1.882	1.0105	.9
.2	0.5620	0.532 9	0.8462	0.629 7	1.588	1.182	1.877	1.0088	.8
.3	0.5637	0.534 4	0.8453	0.632 2	1.582	1.183	1.871	1.0071	.7
.4	0.5655	0.535 8	0.8443	0.634 6	1.576	1.184	1.866	1.0053	.6
.5	0.5672	0.537 3	0.8434	0.637 1	1.570	1.186	1.861	1.0036	.5
.6	0.5690	0.538 8	0.8425	0.639 5	1.564	1.187	1.856	1.0018	.4
.7	0.5707	0.540 2	0.8415	0.642 0	1.558	1.188	1.851	1.0000	.3
.8	0.5725	0.541 7	0.8406	0.644 5	1.552	1.190	1.846	0.9983	.2
.9	0.5742	0.543 2	0.8396	0.646 9	1.546	1.191	1.841	0.9966	.1
		Cos	Sin	Cot	Tan	Csc	Sec	Rad	Deg

Deg	Rad	Sin	Cos	Tan	Cot	Sec	Csc		Deg
33.0	0.5760	0.544 6	0.8387	0.649 4	1.540	1.192	1.836	0.9948	57.0
.1	0.5777	0.546 1	0.8377	0.651 9	1.534	1.194	1.831	0.9931	.9
.2	0.5794	0.547 6	0.8368	0.654 4	1.528	1.195	1.826	0.9913	.8
.3	0.5812	0.549 0	0.8358	0.656 9	1.522	1.196	1.821	0.9896	.7
.4	0.5829	0.550 5	0.8348	0.659 4	1.517	1.817	1.817	0.9879	.6
.5	0.5847	0.551 9	0.8339	0.661 9	1.511	1.199	1.812	0.9861	.5
.6	0.5864	0.553 4	0.8329	0.664 4	1.505	1.201	1.807	0.9844	.4
.7	0.5882	0.554 8	0.8320	0.666 9	1.499	1.202	1.802	0.9826	.3
.8	0.5889	0.556 3	0.8310	0.669 4	1.494	1.203	1.798	0.9809	.2
.9	0.5917	0.557 7	0.8300	0.672 0	1.488	1.205	1.793	0.9791	.1
34.0	0.5934	0.559 2	0.8290	0.674 5	1.483	1.206	1.788	0.9774	56.0
.1	0.5952	0.560 6	0.8281	0.677 1	1.477	1.208	1.784	0.9756	.9
.2	0.5969	0.562 1	0.8271	0.679 6	1.471	1.209	1.779	0.9739	.8
.3	0.5986	0.563 5	0.8261	0.682 2	1.466	1.211	1.775	0.9721	.7
.4	0.6004	0.565 0	0.8251	0.684 7	1.460	1.212	1.770	0.9704	.6
.5	0.6021	0.566 4	0.8241	0.687 3	1.455	1.213	1.766	0.9687	.5
.6	0.6039	0.567 8	0.8231	0.689 9	1.450	1.215	1.761	0.9669	.4
.7	0.6056	0.569 3	0.8221	0.692 4	1.444	1.216	1.757	0.9652	.3
.8	0.6074	0.570 7	0.8211	0.695 0	1.439	1.218	1.752	0.9634	.2
.9	0.6091	0.572 1	0.8202	0.697 6	1.433	1.219	1.748	0.9617	.1
35.0	0.6109	0.573 6	0.8192	0.700 2	1.428	1.221	1.743	0.9599	55.0
.1	0.6126	0.575 0	0.8181	0.702 8	1.423	1.222	1.739	0.9852	.9
.2	0.6144	0.576 4	0.8171	0.705 4	1.418	1.224	1.735	0.9564	.8
.3	0.6161	0.577 9	0.8161	0.708 0	1.412	1.225	1.731	0.9547	.7
.4	0.6178	0.579 3	0.8151	0.710 7	1.407	1.227	1.726	0.9530	.6
.5	0.6196	0.580 7	0.8141	0.713 3	1.402	1.228	1.722	0.9512	.5
.6	0.6213	0.582 1	0.8131	0.715 9	1.397	1.230	1.718	0.9495	.4
.7	0.6231	0.583 5	0.8121	0.718 6	1.392	1.231	1.714	0.9477	.3
.8	0.6248	0.585 0	0.8111	0.721 2	1.387	1.233	1.710	0.9460	.2
.9	0.6266	0.586 4	0.8100	0.723 9	1.381	1.235	1.705	.9442	.1
36.0	0.6283	0.587 8	0.8090	0.726 5	1.376	1.236	1.701	0.9425	54.0
.1	0.6301	0.589 2	0.8080	0.729 2	1.371	1.238	1.697	0.9407	.9
.2	0.6318	0.590 6	0.8070	0.731 9	1.366	1.239	1.693	0.9390	.8
.3	0.6336	0.592 0	0.8059	0.734 6	1.361	1.241	1.689	0.9372	.7
.4	0.6353	0.593 4	0.8049	0.737 3	1.356	1.242	1.685	0.9355	.6
.5	0.6370	0.594 8	0.8039	0.740 0	1.351	1.244	1.681	0.9338	.5
.6	0.6388	0.596 2	0.8028	0.742 7	1.347	1.246	1.677	0.9320	.4
.7	0.6405	0.597 6	0.8018	0.745 4	1.342	1.247	1.673	0.9303	.3
.8	0.6423	0.599 0	0.8007	0.748 1	1.337	1.249	1.669	0.9285	.2
.9	0.6440	0.600 4	0.7997	0.750 8	1.332	1.250	1.666	0.9269	.1
37.0	0.6458	0.601 8	0.7986	0.753 6	1.327	1.252	1.662	0.9250	53.0
.1	0.6475	0.603 2	0.7976	0.756 3	1.322	1.254	1.658	0.9233	.9
.2	0.6493	0.604 6	0.7965	0.759 0	1.317	1.255	1.654	0.9215	.8
.3	0.6510	0.606 0	0.7955	0.761 8	1.313	1.257	1.650	0.9198	.7
.4	0.6528	0.607 4	0.7944	0.764 6	1.308	1.259	1.646	0.9180	.6
.5	0.6545	0.608 8	0.7934	0.767 3	1.303	1.260	1.643	0.9163	.5
.6	0.6562	0.610 1	0.7923	0.770 1	1.299	1.262	1.639	0.9146	.4
.7	0.6580	0.611 5	0.7912	0.772 9	1.294	1.264	1.635	0.9128	.3
.8	0.6597	0.612 9	0.7902	0.775 7	1.289	1.266	1.632	0.9111	2
.9	0.6615	0.614 3	0.7891	0.778 5	1.285	1.267	1.628	0.9093	.1
38.0	0.6632	0.615 7	0.7880	0.781 3	1.280	1.269	1.624	0.9076	52.0
.1	0.6650	0.617 0	0.7869	0.784 1	1.275	1.271	1.621	0.9058	.9
.2	0.6667	0.618 4	0.7859	0.786 9	1.271	1.272	1.617	0.9041	.8
.3	0.6685	0.619 8	0.7848	0.789 8	1.266	1.274	1.613	0.9023	.7
.4	0.6702	0.621 1	0.7837	0.792 6	1.262	1.276	1.610	0.9006	.6
.5	0.6720	0.622 5	0.7826	0.795 4	1.257	1.278	1.606	0.8988	.5
.6	0.6737	0.623 9	0.7815	0.798 3	1.253	1.280	1.603	0.8971	.4
.7	0.6754	0.625 2	0.7804	0.810 2	1.248	1.281	1.599	0.8954	.3
.8	0.6772	0.626 6	0.7793	0.804 0	1.244	1.283	1.596	0.8936	.2
.9	0.6789	0.628 0	0.7782	0.806 9	1.239	1.285	1.592	0.8919	.1
		Cos	Sin	Cot	Tan	Csc	Sec	Rad	Deg

Deg	Rad	Sin	Cos	Tan	Cot	Sec	Csc		
39.0	0.6807	0.629 3	0.7771	0.809 8	1.235	1.287	1.589	0.8901	51.0
.1	0.6824	0.630 7	0.7760	0.812 9	1.230	1.289	1.586	0.8884	.9
.2	0.6842	0.632 0	0.7749	0.815 6	1.226	1.290	1.582	0.8866	.8
.3	0.6859	0.633 4	0.7738	0.818 5	1.222	1.292	1.579	0.8849	.7
.4	0.6877	0.634 7	0.7727	0.821 4	1.217	1.294	1.575	0.8831	.6
.5	0.6894	0.636 1	0.7716	0.824 3	1.213	1.296	1.572	0.8814	.5
.6	0.6912	0.637 4	0.7705	0.827 3	1.209	1.298	1.569	0.8796	.4
.7	0.6929	0.638 8	0.7694	0.830 2	1.205	1.300	1.566	0.8779	.3
.8	0.6946	0.640 1	0.7683	0.833 2	1.200	1.302	1.562	0.8762	.2
.9	0.6964	0.641 4	0.7672	0.836 1	1.196	1.304	1.559	0.8744	.1
40.0	0.6981	0.642 8	0.7660	0.839 1	1.192	1.305	1.556	0.8727	50.0
.1	0.6999	0.644 1	0.7649	0.842 1	1.188	1.307	1.552	0.8709	.9
.2	0.7016	0.645 5	0.7639	0.845 1	1.183	1.309	1.549	0.8691	.8
.3	0.7034	0.646 8	0.7627	0.848 1	1.179	1.311	1.546	0.8674	.7
.4	0.7051	0.648 1	0.7615	0.851 1	1.175	1.313	1.543	0.8656	.6
.5	0.7069	0.649 4	0.7604	0.854 1	1.171	1.315	1.540	0.8639	.5
.6	0.7086	0.650 8	0.7593	0.857 1	1.167	1.317	1.537	0.8622	.4
.7	0.7103	0.652 1	0.7581	0.860 1	1.163	1.319	1.534	0.8604	.3
.8	0.7121	0.653 4	0.7570	0.863 2	1.159	1.321	1.530	0.8587	.2
.9	0.7138	0.654 7	0.7559	0.866 2	1.154	1.323	1.527	0.8570	.1
41.0	0.7156	0.656 1	0.7547	0.869 3	1.150	1.325	1.524	0.8552	49.0
.1	0.7173	0.657 4	0.7536	0.872 4	1.146	1.327	1.521	0.8535	.9
.2	0.7191	0.658 7	0.7524	0.875 4	1.142	1.329	1.518	0.8517	.8
.3	0.7208	0.660 0	0.7513	0.878 5	1.138	1.331	1.515	0.8500	.7
.4	0.7226	0.661 3	0.7501	0.881 6	1.134	1.333	1.512	0.8482	.6
.5	0.7243	0.662 6	0.7490	0.884 7	1.130	1.335	1.509	0.8465	.5
.6	0.7261	0.663 9	0.7478	0.887 8	1.126	1.337	1.506	0.8447	.4
.7	0.7278	0.665 2	0.7466	0.891 0	1.122	1.339	1.503	0.8430	.3
.8	0.7295	0.666 5	0.7455	0.894 1	1.118	1.341	1.500	0.8412	.2
.9	0.7313	0.667 8	0.7443	0.897 2	1.115	1.344	1.497	0.8395	.1
42.0	0.7330	0.669 1	0.7431	0.900 4	1.111	1.346	1.494	0.8378	48.0
.1	0.7348	0.670 4	0.7420	0.903 6	1.107	1.348	1.492	0.8360	.9
.2	0.7365	0.671 7	0.7408	0.906 7	1.103	1.350	1.489	0.8343	.8
.3	0.7383	0.673 0	0.7396	0.909 9	1.099	1.352	1.486	0.8325	.7
.4	0.7400	0.674 3	0.7385	0.913 1	1.095	1.354	1.483	0.8308	.6
.5	0.7418	0.675 6	0.7373	0.916 3	1.091	1.356	1.480	0.8290	.5
.6	0.7435	0.676 9	0.7361	0.919 5	1.087	1.359	1.477	0.8273	.4
.7	0.7453	0.678 2	0.7349	0.922 8	1.084	1.361	1.475	0.8255	.3
.8	0.7470	0.679 4	0.7337	0.926 0	1.080	1.363	1.472	0.8238	.2
.9	0.7487	0.680 7	0.7325	0.929 3	1.076	1.365	1.469	0.8221	.1
43.0	0.7505	0.682 0	0.7314	0.932 5	1.072	1.367	1.466	0.8203	47.0
.1	0.7522	0.683 3	0.7301	0.935 8	1.069	1.370	1.464	0.8186	.9
.2	0.7540	0.684 5	0.7290	0.939 1	1.065	1.372	1.461	0.8168	.8
.3	0.7558	0.685 8	0.7278	0.942 4	1.061	1.374	1.458	0.8151	.7
.4	0.7575	0.687 1	0.7266	0.945 7	1.057	1.376	1.455	0.8133	.6
.5	0.7592	0.688 4	0.7254	0.949 0	1.054	1.379	1.453	0.8116	.5
.6	0.7610	0.689 6	0.7242	0.952 3	1.050	1.381	1.450	0.8098	.4
.7	0.7627	0.690 9	0.7230	0.955 6	1.046	1.383	1.447	0.8081	.3
.8	0.7645	0.692 1	0.7218	0.959 0	1.043	1.386	1.445	0.8063	.2
.9	0.7662	0.693 4	0.7206	0.962 3	1.039	1.388	1.442	0.8046	.1
44.0	0.7679	0.694 7	0.7193	0.965 7	1.036	1.390	1.440	0.8029	46.0
.1	0.7697	0.695 9	0.7181	0.969 1	1.032	1.393	1.437	0.8011	.9
.2	0.7714	0.697 2	0.7169	0.972 5	1.028	1.395	1.434	0.7994	.8
.3	0.7732	0.698 4	0.7157	0.975 9	1.025	1.397	1.432	0.7976	.7
.4	0.7749	0.699 7	0.7145	0.979 3	1.021	1.400	1.429	0.7959	.6
.5	0.7767	0.700 9	0.7133	0.982 7	1.018	1.402	1.427	0.7941	.5
.6	0.7784	0.702 2	0.7120	0.986 1	1.014	1.404	1.424	0.7924	.4
.7	0.7802	0.703 4	0.7108	0.989 6	1.011	1.407	1.422	0.7907	.3
.8	0.7819	0.704 6	0.7096	0.993 0	1.007	1.409	1.419	0.7889	.2
.9	0.7837	0.705 9	0.7083	0.996 5	1.003	1.412	1.417	0.7871	.1
45.0	0.7854	0.707 1	0.7071	1.000 0	1.000	1.414	1.414	0.7854	45.0
		Cos	Sin	Cot	Tan	Csc	Sec	Rad	Deg

ABBREVIATIONS, SYMBOLS, AND ACRONYMS

The abbreviations, symbols, and acronyms listed here are many of those likely to be encountered by aviation technicians and others involved in the operation and maintenance of aircraft. Additional abbreviations are constantly being issued by various agencies; therefore technicians will find many others in use during the course of their activities in aviation.

A	Amperes, area
AA	The Aluminum Association
AAIP	Approved Aircraft Inspection Program
ABC	After bottom center (piston)
ABDC	After bottom dead center
AC	Air Corps; Advisory Circular
ac	Alternating current
AD	Airworthiness Directive
ADF	Automatic direction finder
ADMA	Aviation Distributors and Manufacturers Association
AF	Air Force
AIAA	American Institute of Aeronautics and Astronautics
AISI	American Iron and Steel Institute
AMS	Aeronauticals Materials Specificaton
AN	Air Force–Navy
AND	Air Force–Navy Design
A&P	Airframe and Powerplant
APU	Auxiliary power unit
AR	Aspect ratio
AS	Aeronautical Standard
ASTM	American Society for Testing Materials
ATA	Air Transport Association of America
ATC	After top center
ATDC	After top dead center
avdp	Avoirdupois pound
b	Wing span
BBC	Before bottom center
BBDC	Before bottom dead center
BC	Bottom center
BDC	Bottom dead center
BF	Buoyant force
bhp	Brake horsepower
BITE	Built-in test equipment
bmep	Brake mean effective pressure
bsfc	Brake specific fuel consumption
BTC	Before top center
BTU	British Thermal Unit
c	Average wing chord
C	Celsius; center; centigrade
CAB	Civil Aeronautics Board
CAS	Calibrated airspeed
CAT	Carburetor air temperature
C_D	Coefficient of drag
C_L	Coefficient of lift
C_M	Coefficient of pitching moment
C-D	Converging-diverging nozzle (jet engine)
CDP	Compressor discharge pressure
CDT	Compressor discharge temperature
CG	Center of gravity
CHT	Cylinder head temperature
CIP	Compressor inlet pressure
CIT	Compressor inlet temperature
₵	Center line
CP	Center of pressure

CPR	Compressor discharge (pressure) ratio
CRS	Certified Repair Station
c_T	Tip chord
CSD	Constant-speed drive
c_s	Root chord
D	Drag
dB	Decibel
dm	Decimeter
DME	Distance-measuring equipment compatible with TACAN
dyn	Dyne
E	Electromotive force (voltage)
EAS	Equivalent airspeed
EC	Exhaust valve closes
EGT	Exhaust gas temperature
EPR	Engine pressure ratio
eshp	Equivalent shaft horsepower
EVC	Engine vane control
EW	Empty weight
EWCG	Empty weight center of gravity
F	Fahrenheit; force; thrust
FAA	Federal Aviation Administration
FAR	Federal Aviation Regulation
FE	Flight environment
FED	Federal
Fg	Gross thrust
FM	Fan marker for ILS; frequency modulation
F	Net thrust
F	Ram drag of engine airflow
FSDO	Flight Standards District Office (FAA)
g	Gram
GADO	General Aviation District Office (FAA)
GAMA	General Aviation Manufacturers Association
GAW	General-Aircraft-Wing
GS	Glide slope of ILS
Hz	Hertz
Hg	Mercury
IA	Inspection Authorization (FAA)
I	Electric current (amperage)
ICAO	Internatonal Civil Aviation Organization
IC	Intake valve closes
IFR	Instrument flight rules
IGV	Inlet guide vanes
ILS	Instrument landing system
IO	Intake valve opens
IPC	Illustrated parts catalog
K	Kelvin
Kn	Knot
kPa	Kilopascals
L	Lift
L/D	Lift/drag ratio
LDA	Localizer-type directional aid
LE	Leading edge
LF	Low frequency
LOC	Localizer of ILS: location
LORAN	Long-range navigation equipment
M	Moment
M	Mach number
MA	Mechanical advantage
MAC	Mean aerodynamic chord

MAP	Manifold pressure (absolute)	UHF	Ultra-high frequency
M or D	Malfunction or Defect Report		
MEL	Minimum Equipment List	V	Velocity; volts; volume
MEK	Methyl-ethyl-ketone	V_A	Design maneuvering speed (aircraft operation)
METO	Maximum except takeoff (power)	V_B	Design speed for maximum gust intensity
MIL	Military	V_C	Design cruising speed
MM	Middle marker for ILS	V_D	Design diving speed
MS	Material standard; Military Standard	V_{DF}/M_{DF}	Demonstrated flight diving speed
MSL	Mean sea level	V_F	Design flap speed
MWCL	Main-wheel center line	V_{FC}/M_{FC}	Maximum speed for stability characteristics
		V_{FE}	Maximum flap extended speed
N	Newton	VFR	Visual flight rules
NACA	National Advisory Committee for Aeronautics (superseded)	V_H	Maximum speed in level flight with maximum continuous power
NAF	Naval Aircraft Factory	VHF	Very high frequency
NAS	National Aerospace Standard; Navy Aircraft Standard	VIGV	Variable inlet guide vanes
		V_{LE}	Maximum landing gear extended speed
NASA	National Aeronautics and Space Administration	V_{LO}	Maximum landing gear operating speed
		V_{LOF}	Lift-off speed
NDB	Nondirectional beacon (ADF)	V_{MC}	Minimum control speed with critical engine inoperative
NPRM	Notice of Proposed Rule Making		
NTC	Negative torque control	V_{MO}/M_{MO}	Maximum operating limit speed
NTS	Negative torque signal	V_{NE}	Never exceed speed
		VOR	Very-high-frequency omnirange station
OAT	Outside air temperature	VORTAC	Collocated VOR and TACAN
OBAWS	On-board aircraft weighing system	V_R	Rotation speed (aircraft takeoff)
OGV	Outlet guide vanes	V_S	Stalling speed or minimum steady flight speed at which the airplane is controllable
OM	Outer marker for ILS		
		V_{SO}	Stalling speed or the minimum steady flight speed in the landing configuration
P	Pressure		
PALS	Precision aircraft landing system	V_{SI}	Stalling speed or the minimum steady flight speed obtained in a specific configuration
PMA	Parts Manufacturing Authorization		
PVA	Polyvinyl alcohol	VTOL	Vertical takeoff and landing
PVC	Polyvinyl chloride	VV	Variable viscosity
p	Density	V_X	Speed for best angle of climb
psi	Per square inch	V_Y	Speed for best rate of climb
		V_1	Critical engine failure speed
Q	Dynamic pressure	V_2	Takeoff safety speed
R	Rankine; Reynolds number	W	Weight
RBN	Radio beacon		
rpm	Revolutions per minute		
S	Airfoil area		
SAE	Society of Automotive Engineers		
sfc	Specific fuel consumption		
SPEC	Specification		
STC	Supplemental Type Certificate		
STD	Standard		
Σ	The sum of		
T	Temperature		
TACAN	Ultra-high frequency tactical air navigation aid		
TAS	True airspeed		
TAFI	Turn-around fault isolation		
TC	Top center; Type Certificate		
TCDS	Type Certificate Data Sheet		
TDC	Top dead center		
TET	Turbine exhaust temperature		
T/F	Turbofan		
TIP	Turbine inlet pressure		
TIT	Turbine inlet temperature		
T/J	Turbojet		
TOP	Turbine outlet pressure		
TOT	Turbine outlet temperature		
T/S	Turboshaft		
tsfc	Thrust specific fuel consumption		
TSO	Technical Standard Order (FAA)		
TSS	Thrust sensitive signal		

● U.S. DEPARTMENT OF TRANSPORTATION

Federal Aviation Administration Regional Offices

ALASKAN REGION
701 C Street, Box 14
Anchorage, Alaska 99513
Tel. 907-271-5645

Area: Alaska

GREAT LAKES REGION
O'Hare Lake Office Center
2300 East Devon Avenue
Des Plaines, Illinois 60018
Tel. 312-694-7000

Area: Illinois, Indiana, Michigan, Minnesota, North Dakota, Ohio, South Dakota, Wisconsin

SOUTHERN REGION
3400 Norman Berry Drive
East Point, Georgia 30344
Tel. 404-763-7222
Mail: P.O. Box 20636
Atlanta, Georgia 30320

Area: Alabama, Florida, Georgia, Kentucky, Mississippi, North Carolina, Ohio, Puerto Rico, Republic of Panama, South Carolina, Tennessee, Virgin Islands

EUROPEAN OFFICE
15, Rue de la Loi (3rd floor)
B-1040 Brussels, Belgium
U.S. mailing address:
   c/o American Embassy
      APO, New York 09667
Tel. AUT 314-548-2225

Area: Europe, Africa, Middle East

CENTRAL REGION
601 East 12 Street
Federal Building
Kansas City, Missouri 64106
Tel. 816-374-5626

Area: Iowa, Kansas, Missouri, Nebraska

NEW ENGLAND REGION
12 New England Executive Park
Burlington, Massachusetts 01803
Tel. 617-273-7244

Area: Connecticut, Maine, Massachusetts, New Hampshire, New Jersey, New York, Pennsylvania, Rhode Island, Vermont

SOUTHWEST REGION
4400 Blue Mound Road
Fort Worth, Texas 76101
Tel. 817-624-4911
   Mail: P.O. Box 1689

Area: Arkansas, Louisiana, New Mexico, Oklahoma, Texas

EASTERN REGION
JFK International Airport
Federal Building
Jamaica, New York 11430
Tel. 718-995-2801

Area: Delaware, District of Columbia, Maryland, New Jersey, New York, Pennsylvania, Virginia, and West Virginia

NORTHWEST MOUNTAIN REGION
1790 Pacific Hwy South, C-68966
Seattle, Washington 98168
Tel. 206-764-7010

Area: Alaska, California, Colorado, Idaho, Montana, Oregon, Utah, Washington, Wyoming

WESTERN-PACIFIC REGION
15000 Aviation Boulevard
Hawthorne, California 90261
Tel. 213-536-6427
   Mail: P.O. Box 92007
   Worldway Postal Center
   Los Angeles, California 90009

Area: American Samoa, Arizona, California, Guam, Hawaii, Japan, Nevada, and Marshall Islands

FAA District Offices

Unless otherwise noted, these are General Aviation District Offices:

Alaskan Region

Anchorage, Alaska. (Flight Standards District Office), 6601 South Air Park Place, Suite 216, 99502; 907-243-1902

Fairbanks, Alaska. (Flight Standards District Office), 3788 University Avenue, 99701; 907-452-1276

Juneau, Alaska. (Flight Standards District Office), Post Office Box 2118, 99803; 907-789-0231

Central Region

Des Moines, Iowa. (Flight Standards District Office), 3021 Army Post Road, 50321; 515-285-9895

Kansas City, Missouri. (Flight Standards District Office), 525 Mexico City Avenue, 64152; 816-243-3800

Wichita, Kansas. (Flight Standards District Office), FAA Building, Room 103, 1801 Airport Road, 67209; 316-946-4462

St. Louis, Missouri. (Flight Standards District Office), 9275 Genaire Drive, Berkeley, Missouri 63134; 314-425-7102

Lincoln, Nebraska. (Flight Standards District Office), General Aviation Building, Lincoln Municipal Airport, 68521; 402-471-5485

Eastern Region

Washington, D.C. (Flight Standards District Office), Post Office Box 17325, 600 West Service Road, Washington Dulles International Airport, Washington, D.C. 20041; 703-557-5360

Baltimore, Maryland. Baltimore-Washington International Airport, North Administration Building, Elm Road, 21240; 301-859-5780

Teterboro, New Jersey. (Flight Standards District Office), 150 Riser Road, Teterboro Airport, 07608; 201-288-1745

Albany, New York. Albany County Airport, CFR&M Building, 12211; 518-869-8482

Farmingdale, New York. Administration Building, 2d Floor, Republic Airport, 11735; 516-694-5530

Rochester, New York. Rochester-Monroe County Airport, 1295 Scottsville Road, 14624; 716-263-5880

Allentown, Pennsylvania. Allentown-Bethlehem-Easton Airport, RAS Aviation Center Building, 18103; 215-264-2888

New Cumberland, Pennsylvania. Capital City Airport, Room 201, Administration Building, 17070, 717-782-4528

Philadelphia, Pennsylvania. (Flight Standards District Office), Philadelphia International Airport, Building 1, Cargo Area, Room 15, 19153; 215-596-0673

Pittsburg, Pennsylvania. Allegheny County Airport, Administration Building, Room 213, West Mifflin, Pennsylvania, 15122; 412-462-5507

Richmond, Virginia. Byrd International Airport, Executive Terminal Building, 2d Floor, Sandston, Virginia 23150; 804-222-7494

Charleston, West Virginia. Kanawha Airport, 301 Eagle Mountain Road, 25311; 304-343-4689

Great Lakes Region

Chicago, Illinois. DuPage Airport, West Chicago, Illinois 60185; 312-377-4500

Springfield, Illinois. No. 3 North Airport Drive, North Quadrant, Capital Airport, 62708; 217-492-4238

Indianapolis, Indiana. Indianapolis International Airport, 6801 Pierson Drive, 46241; 317-247-2491

South Bend, Indiana. Michiana Regional Airport, 1843 Commerce Drive, 46628; 219-236-8480

Detroit, Michigan. Willow Run Airport, East Side, 8800 Beck Road, Belleville, 48111; 313-485-2550

Grand Rapids, Michigan. 5500 44th Street S.E., Kent County International Airport, 49508; 616-456-2427

Minneapolis, Minnesota. Minneapolis–St. Paul International Airport, 6201 34th Avenue South–Room 201, 55450; 612-725-3341

Fargo, North Dakota. (Flight Standards District Office), Post Office Box 5496, 58105; 701-232-8949

Cincinnati, Ohio. Lunken Airport Executive Building, Ground Floor, 4242 Airport Road, 45226; 513-533-8110

Cleveland, Ohio. Federal Facilities Building, Cleveland-Hopkins International Airport, 44135; 216-267-0220

Columbus, Ohio. (Flight Standards District Office), 65 Columbus Branch, 4393 East 17th Avenue, Room 164, Lane Aviation Building, Port Columbus International Airport, 43219; 614-469-7476

Rapid City, South Dakota. (Flight Standards District Office), RR 2, Box 4750, 57701; 605-343-2403

Milwaukee, Wisconsin. (Flight Standards District Office), General Mitchell Field, 5300 S. Howell Avenue, FAA/WB Building, 53207; 414-747-5531

New England Region

Portland, Maine. (Flight Standards District Office-65), Portland International Jetport, General Aviation Terminal, 04102; 207-774-4484

Bedford, Massachusetts. DOTFAA (Flight Standards District Office-61), 2d Floor, Civil Terminal Building, L. G. Hanscom Field, 01730; 617-274-7130

Westfield, Massachusetts. (Flight Standards District Office-63), Barnes Municipal Airport, 1st Floor, Administration Building, 01085; 413-568-3121

Northwest Mountain Region

Denver, Colorado. (Flight Standards District Office), Jefferson County Airport, FAA Building 1, Broomfield, Colorado, 80020; 303-466-7326

Boise, Idaho. (Flight Standards Field Office), 3975 Rickenbacker Street, Boise Airport, 83705; 208-334-1238

Billings, Montana. (Flight Standards Field Office), Room 205A, Administration Building, Billings Logan International Airport; 59105; 406-657-6200

Helena, Montana. (Flight Standards District Office), Room 3, FAA Building, Helena Regional Airport, 59601; 406-449-5270

Eugene, Oregon. (Flight Standards Field Office), Mahlon-Sweet Airport, 90606 Greenhill Road, 97402; 503-688-9721

Portland, Oregon. (Flight Standards District Office), Portland-Hillsboro Airport, 3355 N.E. Cornell Road, Hillsboro, Oregon 97124; 503-221-2104

Salt Lake City, Utah. (Flight Standards District Office), 116 North 2400 West; 84116; 801-524-4247

Seattle, Washington. (Flight Standards District Office), NM-FSDO-61, 7300 Perimeter Road, 98108; 206-431-2738

Spokane, Washington. (Flight Standards Field Office), Felts Field Airport, Administration Building, Post Office Box 11649, 99211-1649; 509-456-4618

Casper, Wyoming. (Flight Standards Field Office), FAA/WB Building, Natrona County International Airport, 1187 Fuller Street, 82601; 307-234-8959

Southern Region

Birmingham, Alabama. (Flight Standards District Office), Municipal Airport, FSS/WB Building, 6500 43rd Avenue, North, 35206; 205-254-1557

Jacksonville, Florida. (Flight Standards District Office), 855 St. John's Bluff Road, Craig Airport, FAA Building, 32211; 904-641-7311

Miami, Florida. (Flight Standards District Office), 5600 N.W. 36th Street, Post Office Box 592015; Miami, Florida 33159; 305-526-2572

St. Petersburg, Florida. (Flight Standards District Office), St. Petersburg-Clearwater Airport, Terminal Building, West Wing, 33520; 813-531-1434

Atlanta, Georgia. (Mid-South Flight Standards District Office), 3420 Norman Berry Drive, Suite 430, 30354; 404-763-7265

Louisville, Kentucky. Bowman Field, FAA Building, 40205; 502-582-6116

Jackson, Mississippi. Post Office Box 6273, Pearl Branch, 39208; 601-960-4633

Charlotte, North Carolina. Municipal Airport, 5318 Morris Field Drive, 28208; 704-392-3214

Raleigh, North Carolina. Route 1, Box 486A, Morrisville, North Carolina, 27560; 919-856-4240

San Juan, Puerto Rico. (Flight Standards District Office), Luis Munoz Marin International Airport, Room 203A, 00913; 809-791-5050

Columbia, South Carolina. 2819 Aviation Way, West Columbia, 29169; 803-765-5931

Memphis, Tennessee. (Flight Standards District Office), International Airport, General Aviation Building, Room 137, 2488 Winchester, 38116; 901-521-3820

Nashville, Tennessee. (Flight Standards District Office), 322 Knapp Boulevard, Room 101, Nashville Metropolitan Airport, 37217; 615-251-5661

Southwest Region

Little Rock, Arkansas. FAA Building, Room 201, Adams Field, 72202; 501-378-5565

Baton Rouge, Louisiana. 9191 Plank Road, 70811; 504-356-5701

Albuquerque, New Mexico. (Flight Standards District Office-61), 2402 Kirtland Drive, S.E., 87106; 505-247-0156

Oklahoma City, Oklahoma. (Flight Standards District Office), Room 111, FAA Building, Wiley Post Airport, Bethany, Oklahoma, 73008; 405-789-5220

Dallas, Texas. 8032 Aviation Place, Love Field Airport, 75235; 214-357-0142

Fort Worth, Texas. (Flight Standards District Office), General Aviation Group, Room 222, Meacham Field, 76106; 817-877-2690

Houston, Texas. (Flight Standards District Office), Hobby Airport, 8800 Paul B. Koonce Drive, Room 152, 77061; 713-643-6504

Lubbock, Texas. (Flight Standards District Office), Route 3, Box 51, 79401; 806-762-0335

McAllen, Texas. Miller International Airport, Terminal Building, 2600 South Main Street, 78503; 512-682-4812

San Antonio, Texas. (Flight Standards District Office), International Airport, 1115 Paul Wilkins Road, Room 201, 78216; 512-824-9535

Western-Pacific Region

Scottsdale, Arizona. Municipal Airport, 15041 North Airport Drive, 85260; 602-241-2561

Fresno, California. 4955 East Anderson, Suite 110, 93727; 209-487-5306

Long Beach, California. (Flight Standards District Office), Long Beach Airport, 2815 East Sprint Street, 90806; 213-426-7134

Los Angeles, California. (Flight Standards District Office), 5885 West Imperial Highway, 90045; 213-215-2150

Oakland, California. (Flight Standards District Office), Post Office Box 2397, Airport Station, 94614; 415-273-7155

Riverside California. (Flight Standards District Office-8), Riverside Municipal Airport, 6961 Flight Road, 92504; 714-351-6701

Sacramento, California. Executive Airport, 6107 Freeport Boulevard, 95822; 916-551-1721

San Diego, California. 8665 Gibbs Drive, Suite 110, 92123: 619-293-5281

San Jose, California. (Flight Standards District Office), San Jose Municipal Airport, 1387 Airport Boulevard, 95110; 408-291-7681

Van Nuys, California. (Flight Standards District Office), Van Nuys Airport, 7120 Hayvenhurst Avenue, Suite 316, 91406; 818-904-6291

Honolulu, Hawaii. (Flight Standards District Office), Honolulu International Airport, Air Service Corporation Building, 218 Lagoon Drive, 96819; 808-836-0615

Las Vegas, Nevada. (Flight Standards District Office), 241 East Reno Avenue, Suite 200, 89119; 702-388-6482

Reno, Nevada. (Flight Standards District Office), 601 South Rock Boulevard, Suite D-102, 89502; 702-784-5321

GLOSSARY

The following list of words and terms represents those that are often encountered in the field of aviation. Additional terms are defined in the text; however, there are still other words or terms that technicians hear or see and for which they may want a definition. For a more complete list of definitions, technicians should consult a good aviation dictionary.

Absolute altimeter. An altimeter designed to give accurate indications of absolute altitude.

Absolute altitude. Actual altitude above the surface of the earth, either land or water. Also terrain clearance.

Adhesive. A substance used to bond two materials together by chemical means.

Adiabatic. Denotes change in volume or pressure without gain or loss of heat.

Aerodynamics. The science relating to the effects of air or other gases in motion.

Aerostat. An aircraft that obtains all or most of its lift by virtue of a confined air or gas lighter than the surrounding air. A balloon or dirigible.

Aileron. A moveable control surface attached to the trailing edge of a wing to control an airplane in roll.

Airscrew. A propeller.

Airworthiness. The state or quality of an aircraft or of an aircraft component which will enable safe performance according to specifications.

Airworthiness directive. A directive issued by the FAA requiring that certain inspections and/or repairs be performed on specific makes and models of aircraft, engines, propellers, rotors, or appliances and setting forth time limits for such operations.

Alloy. A solid solution consisting of two or more metallic constituents. The alloy usually contains one predominant metal to which are added small amounts of other metals to improve strength and heat resistance.

Alloy steel. A steel that contains metallic elements other than those found in carbon steels.

Ambient. Surrounding. Ambient conditions are those conditions existing in the surrounding area.

Amphibian. A fixed-wing or rotary-wing aircraft designed or equipped to take off from and alight upon both land and water.

Aneroid. A thin, disk-shaped box or capsule, usually metallic, that is partially evacuated of air and sealed and that expands or contracts with changes in the pressure of the surrounding air or gas.

Angle of attack. The acute angle between the chord line of a wi̶͟g̶ a̶nd the direction that the air strikes the wing (re-
l̶...

... The ratio of the span of the wing (length) to
... of the wing (width). The wing span squared,

...tary-wing aircraft whose rotor is turned
...t by air forces resulting from the motion
...the air.

Autorotation. A rotorcraft flight condition in which the lifting rotor is driven entirely by action of the air as the rotorcraft is in motion.

Azimuth. Horizontal direction or bearing, measured in degrees from a zero reference point such as north.

Balance. A tab so linked that when the control surface to which it is attached is deflected, the tab is deflected in an opposite direction, thus creating a force that aids in moving the larger surface.

Bearing (direction). The horizontal direction of an object or point, usually measured clockwise from a reference line or direction through 360°. (See also azimuth.)

Bearing (mechanical). A part of a machine that supports a journal, pivot, or pin that rotates, oscillates, or slides.

Blind rivet. Special rivet designed to be used where only one side is accessible and it is not possible to use a bucking bar for riveting.

Bond. An attachment of one material to another or the finish to the metal or fabric.

Bonded structure. A structure whose parts are joined together by chemical rather than mechanical methods.

Bonding agent. An adhesive used to bond structure parts together.

Brake horsepower. The power produced by an engine and available for work through the propeller shaft. It is usually measured as a force on a brake drum or equivalent device and is abbreviated bhp.

Brazing. A method of joining two pieces of metal by wetting their surface with a molten alloy of copper, zinc, or tin.

Bulkhead. A wall, partition, or similar member or structure in an airplane or missile fuselage at right angles to the longitudinal axis of the body and serving to strengthen, divide, or help give shape to the body.

Cabane. A pyramidal arrangement of struts used to support a wing above the fuselage of an airplane or to provide a point of attachment for the inner ends of half axles in some types of landing gear.

Calibrated airspeed. An airspeed value derived when corrections have been applied to an indicated airspeed to compensate for installation errors, instrument errors, errors in the pitot-static system, and aircraft attitude errors.

Camber. Curvature of ∠e medium line of an airfoil section.

Canard. An aircraft or aircraft configuration having its horizontal stabilizing and control surfaces in front of the wing or wings.

Cantilever. A beam or member supported at or near one end only, without external bracing.

Catalyst. A material which is used to bring about a change but does not actually enter into the change itself.

Cathedral. The downward angle of a wing as measured from horizontal on the lateral axis. The outer end of the lower surface of the wing is lower than the root end.

Center of gravity. The point within an aircraft on which, for balance purposes, the total force due to gravity is concentrated.

Center of pressure. The point between the leading and trailing edges of an airfoil at which the aerodynamic forces acting on the airfoil are considered to be concentrated.

Chord. The reference line from which the upper and lower contours of an airfoil are measured. A straight line directly across an airfoil from the leading edge to the trailing edge.

Cohesion The act or process of holding tightly together.

Compressibility burble. A region of disturbed flow produced by and aft of a shock wave.

Compression rib. A rib, more strongly built than others, designed to resist compression forces.

Conduction. The transfer of energy through a conductor by means of molecular activity without any external motion.

Console. A control panel, pedestal, or stand in an airplane or test cell.

Convection. The process by which heat is transferred by movement.

Crack. A partial separation of material usually caused by excessive internal stresses resulting from overloading, vibration fatigue, temperature changes, or defective assemblies.

Dash numbers. Numbers following, and separated from, a part number by a dash. The number usually used in identifying either the components or size of the part.

Datum. An imaginary vertical plane from which all horizontal measurements are taken.

Declination. The angular distance of a body from the celestial equator, measured along the hour circle passing through the body and named north or south according to the direction of the body from the celestial equator.

Demand oxygen system. A system in which oxygen is supplied in proportion to demand during inhalation, the regulator being operated by pressure changes occurring during the breathing cycle.

Density. The weight per unit volume of any substance.

Density altitude. Pressure altitude corrected for free air temperature.

Deviation. The deflection of a compass needle or indicator from magnetic north as a result of local magnetic conditions in aircraft.

Dihedral. The upward angle of a wing as measured from horizontal on the lateral axis. The outer end of the lower surface of the wing is higher than the root end.

Directional stability. Stability of an aircraft with respect to the vertical axis, that is, the tendency of an aircraft not to turn unless induced to do so with rudder control.

Drag. The retarding force acting upon a body in motion.

Drift. The lateral divergence or movement of the flight path of a flying vehicle from the direction of its heading, measured between the heading and track, usually caused by a crosswind.

Eddy. A region of undirected or swirling flow, as in the flow of air about or behind a body; a vortex.

Effective span. The span of an airfoil less corrections for tip loss.

Elevator. A horizontal, hinged control surface, attached to the trailing edge of a horizontal stabilizer of an airplane, designed to control the airplane about the lateral axis.

Elevon. A control surface that functions as both an elevator and an aileron.

Empennage. The assembly of stabilizing and control surfaces at the tail of an aircraft.

Equivalent airspeed. A calibrated airspeed corrected for the effect of compression of air in the pitot system.

Fairing. A part of structure having a smooth, streamlined outline, used to cover a nonstreamlined object or to smooth a junction.

Fan-jet engine. A gas-turbine engine that employs a fan to accelerate a large volume of air through a bypass duct to increase thrust and engine efficiency.

Fastener. A device such as a rivet or bolt used to fasten two objects together.

Fatigue strength. The measured resistance of a body to failure caused by repeated applications of stress.

Ferrous. The term describing metal that is derived from an iron base.

Fill. The direction across the width of a fabric.

Fill threads. Threads running across the width of a piece of fabric.

Fin. A common term for the vertical stabilizer.

Flange. The formed or widened portion at the edges of an I beam or U-channel-shaped part or around a lightening hole. Flanges are formed to add stiffness.

Flutter. A vibration or oscillation of definite period set up in an aileron, wing, or other surface of aerodynamic forces and maintained by the aerodynamic forces and by the elastic inertial forces of the object itself.

Gantry. A frame structure mounted on side supports to span a vehicle, usually traveling on rails, used for erecting and servicing large, vertically launched missiles or spacecraft.

G force. An accelerating force.

Grip range. The difference between the maximum and minimum thickness of material that may be joined by a fastener.

Ground speed. The speed of an aircraft relative to the earth's surface.

Gust. A sudden and brief change of wind speed or direction.

Gyrodyne. A rotorcraft whose rotors are normally engine-driven for takeoff, hovering, and landing and for forward flight through part of its speed range and whose means of propulsion, consisting usually of conventional propellers, is independent of the rotor system.

Gyroplane. A rotorcraft whose rotors are not engine-driven except for initial starting but are made to rotate by action of the air when the rotorcraft is moving, and whose means of propulsion, consisting usually of conventional propellers, is independent of the rotor system.

Gyroscopic precession. Results in a gyroscope reacting to an applied force as though the force were applied 90° in the direction of rotation from the point where the force was actually applied.

Heading. The horizontal direction in which a craft points as it flies through the air, usually expressed as an angle measured clockwise from north to the longitudinal axis of the craft.

Heat sink. In thermodynamic theory, a means by which heat is stored or dissipated in or transferred from the system under consideration. The thick metal shield on the nose cone of a rocket is often designed as a heat sink to absorb heat generated by air friction.

Helicopter. A rotorcraft that, for its horizontal motion, depends principally on its engine-driven rotors.

Honeycomb. A hexagonal cellular material made of thin metal, paper, or plastic used as core material for sandwich structure. Named after the honeybee's honeycomb because of its appearance.

Hypersonic flow. Flow at very high supersonic speed.

Ideal angle of attack. The angle of attack of an airfoil at which the airflow meets the leading edge smoothly resulting in a zero pressure differential across the leading edge.

IFR. The abbreviation of instrument flight rules. An airplane flying IFR is flying according to instrument flight rules.

ILS. The abbreviation for instrument landing system.

Impact pressure. That pressure of a moving fluid brought to rest which is in excess of the pressure the fluid possesses when it does not flow. Also, it is the dynamic pressure lower flow speeds.

Indicated airspeed. The airspeed measurement shown by an airspeed indicator.

Induced drag. That part of the drag of an airfoil caused by the lift, that is, the change in the direction of the airflow.

Installation error. An error in a pitot-static system due to the location of the responsive element with respect to other aircraft components, thus affecting the measurements of the pitot-static instruments.

Interference drag. Drag due to the interference of the airflow around aircraft components close to one another.

Jet aircraft. Aircraft powered by one or more air-breathing jet engines.

Jet engine. Any engine that ejects a jet or stream of gas or fluid, obtaining all or most of its thrust by reaction to the ejection.

Jet stream. A strong, narrow band of wind or winds in the upper troposphere or stratosphere, moving in a general direction from west to east and often reaching velocities of hundreds of miles per hour.

Keel. A longitudinal member or ridge along the center bottom of a seaplane float or hull.

Knot. A rate of speed equivalent to 1 nmi/h (6076.1033 ft/h) [1852 m/h].

Laminated. Composed of thin layers of material firmly bonded or united together.

Laminated structure. A structure of an aircraft made up of layers of material bonded together to form complex shapes.

Lapse rate. The rate of change of temperature, pressure, or some other meteorological phenomenon with altitude, usually the rate of decrease of temperature with increased height.

Lateral stability. The tendency of an aircraft to resist rolling.

Load. The ratio of a specified load to the total weight of the aircraft. Load factors are expressed in terms of G units.

Longitudinal stability. The stability of an aircraft with respect to pitching motions.

Lox. A short term for liquid oxygen.

Mach number. A ratio of the true airspeed of the aircraft divided by the speed of sound in the air through which the aircraft is flying at the time.

Magnetic bearing. The bearing measured relative to magnetic north.

Magnetic heading. The heading measured relative to magnetic north.

Manometer. A gage for measuring the pressure of gases or vapors having a sensing device consisting of a column of liquid in a glass tube.

Mass. The quantity of matter in a body as measured in its relation to inertia.

MIL spec. Standard specifications for materials and parts to ensure compliance with quality and performance standards. The specifications were originally for military purposes; hence the abbreviation MIL.

Negative dihedral. A downward inclination of a wing or other surface.

Parasite drag. The drag produced by air flowing over surfaces not involved in producing lift.

Positive g. A force acting on a body undergoing positive acceleration.

Power ⸺ading. The ratio of the gross weight of an air-pl⸺ ⸺ower.

⸺ ⸺ltitude. Altitude above standard sea level— ⸺ ⸺01.31 kPa]—measured with a barometric

⸺ ⸺on of air or other fluid shoving its way into ⸺ ⸺duct owing to the motion of the intake or ⸺ ⸺id.

Ram effect. The pressure resulting from ram.

Rate gyroscope. A gyroscope sensitive to the rate of angular motion.

Reaction propulsion. Propulsion by reaction to a jet or jets ejected from one or more reaction engines.

Rigid rotor. A helicopter main rotor that is not hinged at the hub to permit independent movement of each blade. The blades have freedom of movement only along the longitudinal axis to permit the changing of pitch.

Rotary-wing aircraft (rotor craft). A type of aircraft that is supported in the air wholly or in part by wings or blades rotating about a vertical axis.

Rudder. A vertical control surface hinged to the tail post, aft of the vertical stabilizer, designed to apply yawing moments to the airplane.

SAE number. Any of a series of numbers established as standard by the Society of Automotive Engineers for grading materials, components, and other products.

Sandwich construction. A type of construction in which two sheets, sides, or plates are separated by a core of stiffening material, such as honeycomb or balsa wood.

Slug. A unit of mass having a value of approximately 32.175 lb [14.594 kg] under standard conditions of gravity.

Speed of sound. The speed of propagation of sound waves. At sea level, in the standard atmosphere, the speed of sound is 761 mph or 661 kn.

Stabilator. A horizontal all-movable tail surface.

Stability. The property of an aircraft or other body to resist displacement and, if displaced, to develop forces that will tend to restore the original condition (straight and level flight).

Stress raiser. A scratch, groove, rivet hole, forgoing defect, or other structural discontinuity causing concentration of stress.

Tandem airplane. An airplane with two or more cockpits or seats, one behind the other.

Thermal. A rising current of warm air.

Thrust. A reaction force measured in pounds.

Torquemeter. A meter for measuring torque, as in the shaft of an aircraft engine.

Torque nose. A mechanism or apparatus at the nose section of the engine that senses the engine torque and activates a torquemeter.

Transonic speed. The speed of a body relative to the surrounding fluid at which the flow is in some places subsonic and in other places supersonic: usually from Mach 0.8 to 1.2.

Troposphere. The lowest layer of the earth's atmosphere. Characterized by relatively steady temperature lapse rate, varying humidity, and turbulence.

Velocity. A vector quantity equal to speed in a given direction.

Vertical-speed indicator. A rate-of-climb indicator.

Vertical stabilizer. A vertical airfoil fixed approximately parallel to the plane of symmetry of an airplane; also called the fin.

Viscosity. The resistance of a fluid to flow.

Washin. A permanent warp or twist given a wing such that some specified angle of attack is greater at the tip than at the root.

Washout. The opposite of washin; that is, a wing warp or twist in which the tip angle is less than the root angle.

Wing loading. The ratio of the total gross weight of the aircraft divided by the total wing area. Wing loading is expressed in pounds per square feet.

Yaw. The movement of an aircraft about the vertical axis.

Zero-lift angle of attack. The angle of attack at which no lift is created.

INDEX

A.T.A. Specification No. 100, 304
  system codes 359–362
Abbreviations and symbols, 375–376
  on drawings, 140
Absolute zero, 49
AC fluid-line fitting, 266
AC (Air Corps) standards, 211
Acceleration, 40–41
Acrobatic category, 96
Acronyms, 375–376
Acrylic resins, 187
Action and reaction, 41
Adding equipment to aircraft, 154–156
Addition, 2
  algebraic, 14
Adiabatic lapse rate, 61
Adjacent-position line, 134
Adjustable trim tabs, 103–104
Adjustable wrenches, 244
Advancing blade concept, 119
Adverse forward loading, 159–160
Adverse rearward loading, 159–160
Adverse yaw, 102
Advisory Circulars, FAA, 291
Aerodynamics, basic, 59–73
Age hardening of aluminum, 193
Ailerons, 101–102
Air, properties of, 59–62
Air ducts, 269
Air pressure, 59–60
  relation with density, 62
Air supply, ground, 322–323
Air temperature, 61
Air Transport Association of America
  (A.T.A.), 304
Aircraft:
  ground handling and servicing,
    307–326
  safety, 307–310
  servicing, 339–343
  standards for hardware, 211–213
Aircraft control, 101–104
Aircraft flight, high-speed, 69–73
Aircraft hardware, 211–232
Aircraft Listings, 297
Aircraft materials, 169–190
Aircraft rivets, 222–224
Aircraft screws, 220–222
Aircraft Specifications, 213, 297
Aircraft stability, 97–100
Airflow, 63–65
  laminar, 65, 90
  turbulent, 65
Airfoil profiles, 75–76
  chord, 64, 74
  Clark Y, 75
  section, 75
  station, 75
  thickness, 74
Airfoil terminology, 74
Airfoils, 64, 74–92
  characteristic curves, 77–78
  characteristics of, 76–77
  dimensions of, 85–86
  NACA reports, 75
  NASA reports, 75
  selection of, 92
  symbols relating to, 76
  taper, 88–89
  wing planform, 85
Airplane in flight, forces on, 94
Airworthiness Alerts, 292
Airworthiness Certificate, 338
Airworthiness Directives, 292–296

AISI (American Iron and Steel
  Institute), 212
Algebra, 13–17
Algebraic signs, 13
Allen wrench, 244
Allowance, drawings, 137
Alterations, classifications of, 353–354
Alternate-position line, 134
Alternative number systems, 27–28
Aluminum and its alloys, 177–180
  aging, 179
  cast, 177
  cold working, 179
  corrosion, 179
  designation codes, 177–178
  hardness and temper designations, 178
  heat treatment of, 193–195
  solution heat treating, 179
  uses for aircraft, 179
  workability, 179
  wrought, 177
Aluminum bronze, 184
Amplitude of vibrations, 55
AMS (Aeronautical Materials
  Specifications), 212
AN fluid-line fittings, 265–268
AN (Air Force and Navy)
  standards, 211
AND (Air Force–Navy Aeronautical
  Design) standards, 212
Angle of attack, 65–66, 105
  critical, 68
Angle of incidence, 98
Angle of maximum lift, 69
Angle of minimum speed, 69
Angles, geometric, definitions of, 17
Annealing, 177
  aluminum, 194
  steel, 195
  titanium, 198
Annual inspections, 328–332, 336–345
Annual rings, wood, 174
Anodizing, 209
ANSI (American National Standards
  Institute), 212
Antitorque pedals, helicopter, 116
Antitorque rotor, helicopter, 115
Approved aircraft inspection program,
  334–335
Archimedes' principle, 52
Area:
  measurements of, 37
  wing, 86
Arithmetic, 1–13
Arm, moment, 144
Articulated rotor, helicopter, 111–114
Artificial aging, aluminum, 179, 194
AS (Aeronautical Standard), 212
Aspect ratio, airfoils, 86
  effects of, 87
Assembly drawings, 122
ASTM (American Society for Testing and
  Materials), 212
Atmosphere, definition of, 59
Atmosphere, standard, 61–62, 363–364
Atmospheric pressure, 59–60
Austenite, 195
Austenitic corrosion-resistant steels, 181
Automatic weight-and-balance
  system, 163
Autorotation feature, helicopter, 117
Aviation snips (shears), 254
Avoirdupois weights, 37–38
Axes, aircraft, 96–97

Bailment, 356
Balance:
  aircraft, 147–148
  control-surface, 104–105
Balance computer, 163
Balancing tab, 104
Ball peen hammer, 251, 252
Ballast, 160
Bank, aircraft turns, 103
Bar (unit of pressure), 60
Bar graph, 28
Barcol hardness tester, 202
Barometric pressure, 60
Basic aerodynamics, 59–73
Basic hole system, drawings, 137
Batteries, servicing, 343
Battery cart, 322
Beading, tube, 278
Bend radius, tubing, 275
Bending stress, 170, 171
Bending tools:
  hand tube benders, 276
  production tube benders, 276
Bending tubing, 275–278
Bernoulli's principle, 63–64
Beryllium, 184
Bidirectional weave, cloth, 189
Binary number system, 28
Biplanes, 108–110
Blade flapping, helicopter, 111–113
Blind bolts, 224
Blind rivets, 224
Block diagram, 123
Blueprint, 122
Bolts, 215–217
  AN general-purpose, 215
  blind, 224
  clevis, 217
  close-tolerance, 216
  drilled-head engine, 216
  lock, 227
Boundary layer, 65, 90
  control of, 65, 90
Box-end wrenches, 243
Boyle's law, 54, 363
Brackets for algebraic terms, 14
Brake and hydraulic systems
  service, 340
Brass, 184
Break line, drawings, 132
Brinell hardness test, 198
British thermal unit (Btu), 49
Brittleness of materials, 170
Broken-line graph, 28
Broken material line, drawings, 132
Bronze, 184
Buckling, compression, 171
Buoyancy, 52
Burble, streamline airflow, 68

Cable fittings, 229–232
Cadmium plating, 209
Caliper:
  micrometer, 236
  slide, 235
  spring, 235
  vernier, 240
Calorie (cal) (thermal unit), 49
Camber, airfoil, 74
Canards, 106
Capacity, measurements of, 37

Carburizing steel, 196
Case-hardening treatments, 195–197
Category:
  acrobatic, 96
  normal, 96
  utility, 96
Cellulose acetate, 187
Celsius (°C) temperature scale, 49
Cementite, 195
Center line drawings, 129
Center of gravity (CG), 142, 148
  CG envelope, 161
    loading charts for, 161
  CG limits, 147–148
  CG location, 142–143
    correction of, 160
    determining, 148
  empty-weight (EWCG) range,
    145, 157
Center of pressure (CP), 67
Center-of-pressure coefficient, 77
Center-of-pressure travel, 67–68
Centimeter (cm) (unit of length), 37
Centrifugal force, 42, 95
Centripetal force, 42
Ceramic cloth, composite material, 189
Channel-lock pliers, 250
Charles' law, 54, 363
Charts, uses of, 28
Checks, wood, 174
Chemical properties, 173
Cherrylock rivets, 224
Cherrymax rivets, 224, 225
Chisels, 252
Chord, airfoil, 64, 74
Chromium-molybdenum steel, 180
Circle(s):
  area of, 21
  geometric, definitions of, 18
Circular (ple) graph, 28
Clark Y airfoil, 75
Cleaning of aircraft, 344–345
  agents for, 344
Clevis bolts, 217
Clevis pins, 222
Clutch-head screwdrivers, 249
Coaxial rotors, helicopter, 119
Coefficient, 14
  center-of-pressure (CP), 77
  of drag, 77
  of expansion, 50
  of lift, 76–78
Cold working of metals, 177
Collective pitch control, helicopter, 116
Color coding for plumbing lines, 286
Combination square, 233–234, 242
Combination wrench, 244
Composite fabrics:
  ceramic, 189
  E glass, 189
  graphite, 189
  Kevlar, 189
  S glass, 189
Composite material, 170, 188
Composition of forces, 43
Compressed gas, safety, 307
Compressibility, 69
Compression stress, 170
Compres____ ood, 174
Co____ ____ eight and balance, 163

____ opter, 117–121
____ or, 114–115
____ 265–269
____ law of, 46

Constructions, geometric, 23–25
Continuous airworthiness inspection
    program, 334
Continuous-line graph, 28
Control locks, 317
Control surfaces, 101–104
Controllability, 101
Controllable trim tabs, 104
Controls:
  aircraft, 101–104
  flight, 101–104
  helicopter, 116–117
Convection, 51
Conventional breaks, drawings, 134
Conversion factors, 365–367
Coordinated turn, flight, 103
Copper, 184
Copper-based alloys:
  aluminum bronze, 184
  beryllium, 184
  brass, 184
  bronze, 184
  manganese bronze, 184
  silicon bronze, 184
Coriolis effect, helicopter, 111–114
Corroded parts, repair of, 207
Corrosion of metals, 177
  aluminum, 179
  titanium, 182
Corrosion control, 206
Corrosion-resistant (stainless) steels
    (CRES), 181
  AISI identification codes, 180
  austenitic, 181
  ferritic, 181
  heat treatment of, 198
  martensitic, 181
Cosine(s), trigonometric, 25–27
Cotter pins, 220
Counterbore, 257
Countersink, 257
Covering materials, aircraft fabric, 175
Cowling fasteners, 228–230
Critical angle of attack, 68
Cross-point screwdrivers, 248
Crowfoot weave, cloth, 189
Crowsfoot wrench, 246
Crystalline structure, metal, 175
Curved surfaces, drawings, 135
Customer relations, 357
Cutting-plane line, drawings, 132
Cutting tools, 252–256
  aviation snips, 254
  chisels, 252
  files, 254
  hacksaws, 253
Cyaniding steel, 196
Cyclic-pitch control, helicopter rotor,
    113, 116–117

Dacron cloth, 175
Daily inspections, aircraft, 335
Dash numbers, drawings, 138
Datum, 145
Decalage, 109–110
Decimal number system, 1
Decimals:
  arithmetic, 1, 7–9
  rounding off, 7
Density:
  of materials, 173
  measurements of, 38
  air, 61–62
Department of Transportation, 376
Depth gage micrometer, 238
Detail drawings, 122

Detail view, drawings, 136
Diagonal-cutting pliers, 251
Dies, thread, 260
Digits, arithmetic, 1
Dihedral, wings, 98–99
Dimension lines, drawings, 132
Dimensioning drawings, 136
Dimensioning holes, drawings, 137
Dimensions:
  location, 136
  size, 136
Directional stability, 100
Discrepancy report, 338
Displacement:
  liquids, 52
  vibration, 55
Dissimilar-metals corrosion, 206
Dissymmetry of lift, helicopter, 111–112
Distance, measurement of, 36–37
Divider, 235
Division, 4
  algebraic, 15
Doppler effect, 58
Double flare, tubing, 266, 278
Downwash, airflow, 64
Drafting room manual, 138
Drafting techniques, 126–141
Drag, 66–67
  coefficient of, 77
  equation for, 77
  induced, 66, 77
  interference, 66
  lift and drag components, 66–67
  parasite, 66, 77
  profile, 77
Drawfiling, 256
Drawing number, 138
Drawings, 122–141
  assembly, 122
  detail, 122
  dimensioning, 136
  installation, 122
  lines, types of, 129–135
  notes, 140
  numbering, 138
  perspective, 126
  production, 122
  revisions, 139
  scale, 139
  symbols, 140
  types of, 122–126
  views and projections, 126–128
  working, 122
Drift, helicopter, 115
Drill-point angles, 257, 259
Drill sizes, 257
  chart, 258
Drill speed and feed, 257
  chart, 259
Drills, 256
Duckbill pliers, 251
Ductility, 170
Ducts, air, 269
Dye-penetrant inspection, 205
Dynamic balance, control surfaces,
    104–105
Dynamic pressure, 63
Dynamic stability, 97–98
Dyne (dyn) (unit of force), 40

E glass cloth, composite material, 189
Ear protection, 261
Eddy-current inspection, 205
Elastic limit, materials, 170
Elastic range, materials, 172
Elasticity of materials, 170

Electrical conductivity, 173
Electrical power supplies, 322
Electrical system drawings, 125
Elevators, aircraft control, 102
Elevons, aircraft control, 108
Employer-technician relations, 357
Empty weight (EW), aircraft, 145
Empty-weight center of gravity
    (EWCG), 145
  computing of, 151–153
  range, 145, 157
Energy:
  kinetic, 45–46
  measurements of, 45–46
  potential, 45–46
Engine(s):
  gas-turbine, starting, 313
  hand cranking, 312
  oil service, 339
  starting, 311–313
    large, 312
    small, 311
English system of measurements, 36–37
Envelope, loading, 161
Epoxy resins, 185
Equations, algebraic, 13
Equilibrium, 144
Erg (measurement of work), 45
Exfoliation, 206
Expansion:
  coefficient of, 50
  thermal, 50
Expansion shock waves, 71–72
Exponent, algebraic, 14
Extension line, drawings, 132
Extreme weight and balance conditions,
    159–160
Eye protection, 261

FAA (see Federal Aviation Administration)
FAA Form 337, major repairs and
    alterations, 347
FAA specifications, hardware, 213
Fabric, aircraft:
  aircraft linen, 175
  airplane cloth, 175
  Dacron cloth, 175
  fill, 175
  finishing tape, 175
  glider cloth, 175
  grade A cloth, 175
  reinforcing tape, 175
  selvage edge, 175
  surface tape, 175
  thread count, 175
  warp threads, 175
Fahrenheit (°F) temperature scale, 49
False start, turbine engine, 313
Federal Aviation Administration (FAA),
    288–290
  district offices of, 377–379
  history of, 288
  organization of, 289
  regional offices of, 376–377
Federal Aviation Regulations (FARs),
    290–291
  publications and forms, 290–302
Ferrite, 195
Ferritic corrosion-resistant steels, 181
Ferrous metals, 180
  heat treating, 195–197
File card (tool), 256
Files (tool), 254–256
Finish, surface roughness, 207
Fire extinguishers, 308–309
Fire safety, 308–309

Fittings:
  cable, 229–232
  hose, 272–274
    installation of, 281–283
  tube, 265–269
Flaperon, 108
Flaps, wing, 80–85, 101
  leading-edge, 85
Flare-nut wrench, 244
Flared tube fittings, 265–268
Flareless tube fittings, 268
  installation of, 279
Flaring tubing, 278
Flight, aircraft, high-speed, 69–73
Flight controls:
  aircraft, 101–104
  helicopter, 116–117
  unusual, 108
Flight line safety, 309–310
Flight manual, FAA approved, 305
Fluid lines:
  fabrication and installation of, 274–285
  fittings, 264–265
  flexible hose, 264
  inspection and maintenance of, 285–286
  rigid, 264
  semirigid, 264
Fluids, 52–53
Fluorescent-penetrant inspection, 205
Foot (ft) (unit of length and distance), 37
Force, 40
  centrifugal, 42, 95
  centripetal, 42
  measurements of, 40
Forces:
  on airplane in flight, 94
  composition of, 43–44
  resolution of, 43–44
Foreshortening, perspective
    drawings, 126
Formulas, geometric, 21–23
Forward swept wing, 106–107
Fractions, arithmetic, 5–9
Frequency of vibration, 55
Fretting corrosion, 207
Friction, skin, 64
Fuel contamination, 323–324
Fueling aircraft, 323–326
  precautions and procedures for,
    325–327
Fulcrum, 47
Functions, trigonometric, 25–27, 367–374
Fusion, heat of, 50

Gages, 240–241
  radius and fillet, 241
  screw-pitch, 241
  small-hole, 240
  telescoping, 240
  thickness, 241
Gallon (gal) (unit of volume or
    capacity), 37
Gap, 109
Gap/chord ratio, 109
Gap/span ratio, 109
Gases, laws of, 54–55, 363
Gasoline grades, 324
GAW (general-aircraft wing) airfoil
    profiles, 79
Gears, 47
General Aviation Airworthiness
    Alerts, 292
General Aviation Manufacturers
    Association (GAMA), 305
General gas law, 55, 363
Geometric constructions, 23–25

Geometric definitions:
  angle, 17
  line(s), 17
  plane, 17
  point, 17
  solid, 17
  surface, 17
Geometric formulas:
  area of a circle, 21
  area of a parallelogram, 22
  area of a regular polygon, 21
  area of a trapezoid, 22
  volume, 23
Geometric shapes:
  circles, 18
  hexagons, 21
  polygons, 19
  quadrilaterals, 20
  trlangles, 19
Geometry, 17–25
Glass-fiber laminates, 186
Glider cloth, 175
Grade A cotton, 175
Grain, wood, 174
Grain size, metal, 177
Gram (g) (unit of weight), 38
Graphical solution equations, 30, 34
Graphite cloth, composite material, 189
Graphs, types and uses of, 28–33
  bar, 28
  broken-line, 28
  circle (pie), 28
  continuous line, 28
  mathematical equations, 30, 34
  nomograms, 30
Gravitation, law of, 38
Gravity, 38, 142
  specific, 38–39
Greek alphabet, 364
Ground air supply, 322–323
Ground effect, helicopter, 115
Ground-support equipment, 320–323
Gust locks, 317
Gyroscopic precession, 113–114

Hacksaws, 253
Hammers, 251
Hand cranking engines, 312
Hand signals for taxiing or towing,
    315–316
Hardening:
  aluminum, 193
  steel, 195
Hardness of materials, 170
Hardness tester calibration, 201
Hardness testing, 198–203
Hardness values (chart), 200
Harmonic motion, 55
Heat, 49–52
  effects of, 49–50
    on aircraft materials, 51–52
  of fusion, 50
  specific, 50
Heat transfer, 50–51
Heat-treating techniques, 191–198
  aluminum, 193–195
  ferrous metals, 195–197
  magnesium, 198
  stainless steel, 198
  steel, 197
  titanium, 197
Hectare (unit of area measurement), 37
Helicopter, 111–121
  configurations, 117–121
  controls for, 116–117
  flight of, 111–113

Helicopter (Cont.):
  hoisting, 320
  power train, 117
  weighing, 166–167
  weight and balance for, 163–168
Hertz (Hz) (unit of frequency), 55
Hexadecimal number system, 28
Hexagon, geometric, definition of, 21
Hi-Lok bolt, 225–227
Hi-Shear fasteners, 225
Hidden line, drawings, 129
High-lift devices, 80–85
High-speed flight, 69–73
High-speed stalls, 105
High-temperature alloys, 184–186
Hole dimensions, drawings, 137
Honeycomb core materials, 188
Hooke's law, 170
Horizontal stability, 99
Horizontal stabilizer, 102
Horsepower (hp) (unit of power), 46
Hose, aircraft, 270–272
  extra-high-pressure, 272
  fire sleeve, 274
  fittings for, 272–274
    installation of, 281–283
  high-pressure, 272
  identification, 272
  installation practices, 284
  low-pressure, 271
  maintenance practices, 283
  medium-pressure, 271
  size, 272
  storage of, 285
Hot-air ducts, 269
Hot start, gas-turbine engine, 313
Hovering, helicopter, 111
Humidity, 62
  effect on density of air, 62
Hydraulic system service, 340
Hydraulics, principles of, 52–54
Hydrometer, 39
Hypersonic flight, 72–73

ICAO (International Civil Aviation
  Organization) standards, air, 61–62
Illustrated parts catalog, 304, 356
Inch (in) (unit of measure), 36
Inclined plane, 48
Index unit, weight and balance, 161–162
Induced drag, 66, 77
Industry standards, 212
Inertia, defined, 40
Inspection Authorization (IA) rating,
  328, 353
Inspection record entries, 346
Inspection reminder, 347
Inspection systems, repair station, 352
Inspections, 328–348, 352
  altimeter, 336
  annual, 328–332, 336–345
  approved aircraft, 334–335
  continuous airworthiness, 334
    daily, 335
  large aircraft programs, 333–335
  light plane, 328–332, 336–345
  manufacturer's recommended, 335
  n____ve testing, 203–206
  ____–332, 336–345
  ____3
  ____ed, 335
  ____–333

Integers, arithmetic, 1
Interference drag, 66
Integranular corrosion, 206
International System of Units (SI metric),
  36–38, 362–363
Interplane interference, 108
Inventory control, 356
Inversion, air temperature and, 61
Invisible outline, drawings, 129
Iron and steel, 180
  SAE identification code, 180
Isometric projection, drawings, 128

Jacking and hoisting aircraft, 317–321
Jo-bolt, 225
Job estimating, 356
Joules (J) (measurement of work), 45

Kelvin (°K) temperature scale, 49
Kevlar cloth, composite material, 189
Kilogram (kg) (unit of weight), 38
Kilometer (km) (unit of length and
  distance), 36
Kilopascal (kPa) (unit of pressure), 53
Kilowatt (kW) (unit of power), 46
Kinetic energy, 45–46
Knot (kn) (rate of speed), 36
Knots in wood, 174

Laminar airflow, 65, 90
Laminated wood, 175
Laminates, 188
Landing wires, biplane wing, 110
Lateral axis, aircraft, 96–97
Lateral stability, aircraft, 99
Lathe filing, 255
Law(s):
  Boyle's, 54, 363
  Charles', 54, 363
  conservation of energy, 46
  of gases, 54–55, 363
  of the lever, 47, 144
  Newton's, 40–41, 64
  Pascal's, 53
  of thermodynamics, 50
Leading-edge flaps, 85
Legal responsibilities of technician,
  356–357
Length, measurements of, 36–37
Leveling means:
  aircraft, 145
  helicopter, 164
Lever, 47
  law of the, 144
Liability of technician, 356
Lift, 66–69
  angle of maximum, 69
  coefficient of, 76–78
  curve for, 78
  effect of air density on, 69
  effect of area on, 68
  effect of velocity on, 68
  equation for, 76
  helicopter, 111–115
Lift and drag components, 66–67
Lift/drag ratio, 67, 77
Lifting and jacking aircraft, 317–321
Limits, tolerance, and allowance, 137
Linear expansion, 50
Lines, drawings, 129–135
Liquid lock, reciprocating engines, 312
Liquids, 52–54
Literal numbers, algebra, 14

Load adjuster, 163
Load cells for weighing aircraft, 148
Load factors:
  design, 96
  flight, 95–96
Loading aircraft, 157–161
Loading chart, 161–162
Loading envelope, 161
Loads:
  aircraft in flight, 96–97
  biplane, 110
Location dimensions, 136
Lock bolts, 227
Logic circuitry for electronic
  systems, 126
Longitudinal axis, 96–97
Longitudinal waves, 56
Low-drag boundary-layer control, 65
Lowest common denominator (LCD),
  arithmetic, 5
Lubrication, 339

MAC (mean aerodynamic chord), 89–90,
  145, 148
Mach number, 70
Machine screw thread, 213
Machine screws, 220–222
  designations, 221
  sizes, 214
Machines, simple, 46–48
Magnaflux, 204
Magnesium, 181
  heat treatment of, 198
Magnetic inspection, 203–205
Main-wheel center line (MWCL), 145
Maintenance, types of, 352–355
Maintenance records, 345–347
Maintenance releases, 355
Major alterations, 353–354
Major repair and alteration Form FAA
  337, 347
Major repairs, 355–356
Malfunction or Defect Report
  (M or D report), 292, 347
Malleability, 170
Maneuverability, 101
Manganese bronze, 184
Manufacturer's publications:
  maintenance, 302–306
  operating, 305–306
Manufacturer's recommended inspection
  program, 335
Martensite, 195
Martensitic corrosion-resistant steels, 181
Mass, 38
Material listing, drawings, 140
Material symbols, drawings, 140
Maximum lift, angle of, 69
Maximum weight, aircraft, 145
Mean aerodynamic chord (MAC), 89–90,
  145, 148
Mean camber line, airfoil, 74
Measurement and layout, 233–242
Measurements and units of measure,
  36–40
  SI, 362–363
Mechanical advantage, 47–48
Mechanical properties of materials, 170
Mechanic's liens, 357
Metal alloys, 177
Metal tubing materials, 265
Metals, 175–185
  heat treating, 191–198
  shapes, 191
  surface corrosion, 206
  surface treatment, 209
Meter (m) (unit of length and distance), 37

Metric system, SI, 36–38, 362–363
Microfiche, 305
Micrometer, 236–240
  caliper, 234
  depth gage, 238
  inside, 237
  vernier, 238
Mile (mi) (unit of length and distance), 36
Military specifications, 213
Mill products, 191
Millibar (mbar) (unit of pressure), 60
Millimeter (mm) (unit of length and distance), 37
Minimum Equipment List (MEL), 305
Minimum fuel, 145
Minimum speed, angle of, 69
Mobile electric power units, 322
Modulus of elasticity, 172
Molecule, gas, 54
Moment, aircraft, 144–145
Moment arm, 144
Momentum, 42
Monel metal, 184
Mooring aircraft, 315
Motion, 40–41
  harmonic, 55
MS (Military Standard), 211
MS fittings, 268
Multiplication, 2
  algebraic, 15

NACA (National Advisory Committee for Aeronautics) airfoil, 75
NAF (Naval Aircraft Factory) standards, 212
NAS (National Aerospace Standard), 211
NASA (National Aeronautics and Space Administration) airfoil, 75
Nautical mile (nm) (navigation), 36
Newton (N) (unit of force), 40
Newton's laws of motion, 40–41. 64
Nicropress sleeves, 230
Nitriding steel, 196
Nomograph, 30
Nondestructive testing, 203–206
Nonthreaded fasteners, 222–228
Normal category, 96
Normal shock waves, 71
Normalizing, steel, 195
Notes, drawings, 140
Number(s), 1
  drawing, 138
  Mach, 70
Number systems, 27–28
  binary, 28
  decimal, 1
  hexadecimal, 28
  octal, 28
Numerals, 1
Nuts, 217
  castellated, 217
  plate, 218
  self-locking, 218

Oblique shock waves, 71
Oblique views, drawings, 128
Octal number system, 28
Offset screwdrivers, 249
Oil service, engine, 320, 339
On-board aircraft weighing system (OBAWS), 163
100-hour inspections, 328–332, 336–345

Open-end wrenches, 242
Operation specifications, 306
Operational inspection, 343
Operator-developed inspection program, 335
Order of operations, algebraic, 16
Orthographic projection, drawings, 128
Ounce (oz) (unit of weight), 37
Overloading, aircraft, 157
Overwing fueling, 326
Oxygen-system service, 342

Panel and cowling fasteners, 228-230
Parallelogram:
  area of, 22
  geometric, definition of, 20
Parasite drag, 66, 77
Parentheses for algebraic terms, 14
Part number, drawings, 138
Parts catalog, illustrated, 304, 356
Pascal (Pa) (unit of pressure), 53
Pascal's law, 53
Pearlite, 195
Pedals, antitorque, helicopter, 116
Percentage, arithmetic, 9
Periodic maintenance, 352
Personnel, repair station, 352
  assignment of, 355
Perspective drawings, 126
Perspective view, 125
Phantom lines, drawings, 129
Phenolic resin, 185
Phillips screwdriver, 249
Physical properties, 173
  conductivity, electrical, 173
  conductivity, thermal, 173
  density, 173
  specific gravity, 173
  thermal expansion, 173
Pictorial drawing, 122
Pins:
  clevis, 222
  cotter, 220
  roll, 222
  taper, 222
Pint (pt) (unit of volume or capacity), 37
Pipe, 264
Pipe schedule numbers, 264
Pipe threads, 264
Pipe-to-tube connectors, 267
Pipe wrench, 246
Pitch, aircraft, 97, 102
Pitching moment, aircraft control, 102
Plane, inclined, 48
Plane or plane surface, geometric, definition of, 17
Plastic range, 172
Plasticity, 170
Plastics, 169, 185–188
Pliers, 250
  channel-lock, 250
  diagonal-cutting, 251
  duckbill, 251
  needle-nose, 251
  safety-wire twisters, 251
  slip-joint, 250
  Vise-Grip, 250
  water-pump, 250
Plumbing systems:
  color coding for lines, 286
  materials for, 264–265
  (See also Fittings; Fluid lines)
Plywood, 175
Point, geometric, definition of, 17
Polyester resins, 186
Polyethylene resins, 187

Polygon(s):
  geometric, definition of, 19
  regular, area of, 21
Polytetrafluoroethylene (Teflon), 187
Polyurethane resins, 186
Portable electronic weighing system, 149
Potential energy, 45–46
Pound (lb) (unit of mass and weight), 37
Power, measurement of, 46
Power train, helicopter, 117
Powers and roots, arithmetic, 10
Precession, gyroscopic, 113–114
Precipitation heat treatment, aluminum, 194
Preflight inspections, 335
Pressure, 52–53, 59–63
  air, relation with density, 62
  atmospheric, 59–60
  barometric, 60
  dynamic, 63
  measurement of, 52–53
  static, 59, 63
Pressure fueling, 326
Pressure interference, biplane, 108
Preventive maintenance, 353
Process code, drawings, 141
Production drawings, 122
Production tube bender, 276
Profile, airfoil (see Airfoil profiles)
Profile drag, 76–77
Progressive inspections, 328–333
Progressive maintenance, 352
Projection drawings:
  isometric, 128
  orthographic, 128
Propeller torque, 98
Proportional limit, 170
Proportional range, 172
Protractor, 242
Psia (pounds per square inch absolute), 60
Pulleys, 47
Punches, 260
Purchasing, 355–356

Quadrilateral, geometric, definition of, 20
Quart (qt) (unit of volume or capacity), 37
Quick-disconnect plumbing fittings, 269

Radiation, 51
Radiography, 205
Radius gage, 241
Rankline (°R) temperature scale, 49
Ratio and proportion, arithmetic, 10
Reamers, 257–259
Receiving inspection, 349
Record entries, inspection, 346
Records:
  maintenance, 345–347
  maintenance and repair, 355
Recrystallization or annealing:
  of aluminum, 194
  of metals, 177
Rectangle, geometric, definition of, 20
Reed & Prince screwdriver, 249
Reference datum, aircraft balance, 145
Regulations (see Federal Aviation Regulations)
Relative wind, 65
Removing equipment from aircraft, 157
Repair stations, 349–358
  certification, 351
  classification of repairs, 354–355
  inspection systems, 352

Repair stations (Cont.)
  management, 355–356
  organization, 349–351
  personnel, 352, 355
  purchasing, 355–356
  ratings, 351
  records, maintenance and repair, 355
  requirements for, 352
Repairs, classification of, 354–355
Resins, 185-187
Resolution of forces, 43–44
Resonance, 57–58
Respirators, 261
Responsibilities of technician, 357
Resultant (composition and resolution of
    forces), 43
Reusable hose fittings, 272–274
Revisions, drawings, 139
Reynolds number, 65
Right triangle, solution of, 26
Rivets, 222–224
  blind, 224
  Cherrylock, 224
  Cherrymax, 224
  designation code, 223
  rivnuts, 224
Rockwell hardness test, 198
Roll, aircraft, 97, 101–102
Roll pin, 222
Root-mean-square (rms) system of surface
    finish designation, 207
Ropes, tie-down, 315–317
Rotary wing, 111
Rotor, helicopter, 111–115
  antitorque, 115
  articulated, 111–114
  blade flapping, 113
  coning, 114–115
  cyclic-pitch control, 113, 116–117
  gyroscopic precession, 113–114
  semirigid, 113
  tandem-rotor, 117–118
  tilt, 119
Rudder, 102–103
Ruddervator, 108
Rules, measuring, 233–235

S glass cloth, composite material, 189
Safety:
  aircraft, 307–310
  compressed gas, 307
  fire, 308–309
  flight line, 309–310
Safety belts, 231
Safety equipment, 261
  ear protection, 261
  eye protection, 261
  respirators, 261
Safety wire, 220
Safety-wire pliers, 251
Sandwich material:
  honeycomb core, 188
  solid-core, 188
Scale, drawing, 139
Scales, measuring and drawing, 233–235
Schematic diagrams, 123
  electrical, 126
Science fundamentals, 36–58
Scientific notation, 11–13
Scleroscope hardness test, 200
Screw-pitch gage, 241
Screwdrivers, 247–250
  clutch-head, 249
  cross-point, 248
  offset, 249
  Phillips, 248
  plain, 247

Screwdrivers (Cont.):
  power, 250
  Reed & Prince, 249
  spiral-rachet, 249
  stubby, 249
Scriber, 261
Sea-level standard conditions, 60
Section, airfoil, 75
Section lines, 134
Sectional view, drawings, 136
Self-tapping screws, 221
Selvage edge, fabric, 175
Semirigid rotor, helicopter, 113
Service bulletins, 302
Servicing aircraft, 339–343
Servicing batteries, 343
Servo tab, 104
Shakes in wood, 174
Shear stress, 170
Shock waves, 71–72
  expansion, 71–72
  normal, 71
  oblique, 71
Shop management, 355–356
Shop sketches, 125
Shore durometer, 201
SI (Système International) metric system,
    36–38, 362–363
Side-by-side rotors, helicopter, 119
Silicon bronze, 184
Silicone resins, 186
Sine(s), trigonometric, 25
Single-view drawings, 136
Size dimensions, 136
Skid, aircraft, 103
Skin friction, 64
Slats, 82
Sleeves, protective, hose, 274
Slip-joint pliers, 250
Slip planes, metal, 176
Slipping, aircraft, 103
Slots, 82–85
Slug, 62
Small-hole gages, 240
Society of Automotive Engineers (SAE),
    212
Socket wrenches, 244–246
Soft hammers, 252
Solid laminate, 188
Solid wood, 174
Solution heat treatment of metals and
    aluminum, 179, 193–194
Sonic boom, 62
Sound, 55–58
  measurement of, 57
  speed of, 69
  transmission of, 57
Space lattice, metal, 176
Span, airfoil, 86
Spanner wrenches, 247
Special inspections, 336
Specific gravity, 38–39
  of materials, 173
Specific heat, 50
Specifications, aircraft, 213, 297
Speed:
  minimum, angle of, 69
  of sound, 69
  vs. velocity, 40
Spiral-ratchet screwdrivers, 249
Splits, wood, 174
Spontaneous ignition, 308
Square:
  combination, 233–234, 242
  geometric, definition of, 20
Squares, cubes, square roots, and cube
    roots (table), 12
Stabilator, 102
Stability, aircraft, 97–100

Stabilizers, 102–103
Stagger, biplane, 109
Stainless steels, 181
  heat treatment of, 198
Stall:
  airfoil, 105
    high-speed, 105
  wing design, 105–106
Stall strips, 105
Stall warning, 105
Standard atmosphere, 61–62, 363–364
Standards:
  aircraft hardware, 211–213
  English system and SI metric system,
    363
Starting engines, 311–313
Static balance, control surface, 104
Static pressure, 59, 63
Static stability, 97
Station, fuselage, 146
Station numbers on drawings, 140
Stations, airfoils, 75
Straight filing, 255
Strain, 171
  calculation of, 171
Strain hardening of metals, 177
Strength, 170
Stress, 170
  bending, 170, 171
  calculation of, 171
  compression, 170
  normal, 171
  shear, 170
  tension, 170
  torsion, 170
Stress corrosion, 206
Stress relief:
  aluminum, 194
  steel, 195
  titanium, 198
Stress-strain diagram, 172
Stubby screwdrivers, 249
Stump bolts, 227
Subsonic flight, 70
Subtraction, 2
  algebraic, 14
Supercritical wing, 92
Supersonic flight, 71–73
Supplemental Type Certificates (STCs),
    175, 302
Surface, geometric, definition of, 17
Surface area, calculation of, 23
Surface corrosion, 206
Surface roughness, 207–209
Surface treatment for metals, 209
Surfaces, control, 101–104
Swaged fittings:
  cable, 230
  tubing, 269
Swashplate, helicopter, 116–117
Sweep angle, 89
Symbols, airfoil, 76
Symbols and abbreviations, 375–376
  on drawings, 140
Système International (SI) metric system,
    36–38, 362–363

T tail, 107–108
Tabs:
  balancing, 104
  servo, 104
  trim, 103–104
Tail rotor, helicopter, 115
Tandem-rotor helicopters, 117–118
Tangent(s), trigonometric, 25–27
Taper, wings, 88–89
Taper pins, 222

Taps, threads, 260
Tare, weight and balance, 146, 151
Taxiing, 313–315
Technician, responsibility and liability of, 357
Teflon, 187
Telescoping gages, 240
Temper designations, aluminum, 194
Temperature, 49
   of air, 61
   heat-treating, aluminum, 193
   heat-treating, steel, 196
Tempering, steel, 195
Tensile testing, 171
Tension stress, 170, 171
Thermal expansion, 50
   of materials, 173
Thermodynamics, laws of, 50
Thermoplastic plastics, effect of heat on, 51
Thermoplastic resins, 185
Thermosetting resins, 185
Thickness gage, 241
Thread, machine screw, 213
   American national form, 214
   lead, 214
   major diameter, 214
   minor diameter, 214
   pitch, 214
   unified national form, 214
Thread, screw:
   coarse, 213–215
   fine, 213–215
Threaded fasteners, 213–222
Threads:
   pipe fittings, 264
   tube fittings, 268
Thrust, 41
   aircraft, 94
   helicopter, 111
Tie-down of aircraft, 315
Tight fit, drawings, 137
Tilt rotor, helicopter, 119
Tip-path plane, helicopter rotor, 111
Tires:
   cleaning, 344
   servicing, 343
Titanium, 181
   alloys of, 182
   heat treatment of, 197
Title blocks, drawings, 138–140
Tolerance, drawings, 137
Ton (unit of weight), 37
Torricelli's experiment, 60
Torque, 47
   propeller, 98
Torque values, tube fittings, 275
Torque wrenches, 246
Torsion stress, 170, 171
Toughness of materials, 170
Towing aircraft, 310
Transducers, weight indication, 149, 163
Translational lift, helicopter, 115
Transonic flight, 71
Transponder inspection, 336
Transposing equations, 16
Transverse flow effect, helicopter, 115
Transverse sound waves, 56
Trapezoid:
   area of, 22
   geometric, definition of, 20
Triangle(s):
   area of, 22
   geometric, definitions of, 19–20
   right, solution of, 26
Trigonometric functions, 25–27, 367–374
Trigonometry, 25–27
Trim tabs, 103–104
Tube benders, 275–278

Tube cutter, 274
Tube fittings, 265–269
   bulkhead, 267
   flared (AC), 266
   flared (AN), 266
      designation, 268
   flared (AN) to pipe, 267
   flareless (MS), 268
      installation of, 279
   lubrication of, 281
   quick-disconnect, 269
   swaged, 269
   tightening, 281
   universal, 267
Tubing, fluid-line, 264–265
   beading, 278
   bending, 275–278
   cutting, 274
   flaring, 278
   identification code, 286
   size, 265
Turbine engines, starting, 313
Turbine fuels, 324
Turbulent airflow, 65
Turn(s), flight:
   coordinated, 103
   loads in, 96
Turnbuckles, control cables, 231
Two-view drawings, 135
Tying down aircraft, 315
Type Certificate Data Sheet, 296–302

Ultimate tensile strength, 172
Ultrasonic inspection, 205
Undrainable oil, 146
Unidirectional weave, fabric, 189
Universal tube fittings (AN), 267
Unusable fuel, 146
Useful load, aircraft, 146
Utility category, 96

Vectors (composition and resolution of forces), 43–44
Velocity:
   measurements of, 40
   speed vs., 40
Ventilator and hot-air ducts, 269
Venturi (tube), 63
Vernier caliper, 240
Vernier micrometer, 238
Vernier scales, 238
Vertical axis, aircraft, 96–97
Vertical stabilizer, aircraft control, 102–103
Vibration, 55
Vickers hardness tester, 199
Views, drawing, 135
Vinyl resins, 187
Viscosity:
   air, 64–65
   liquid, 52
Vise-Grip pliers, 250
Visible outline, drawings, 129
Volume:
   formula for, 23
   measurements of, 37
Vortex generators, airfoils, 90–91
Vortices, wing-tip, 87

Walk-around inspection, aircraft, 335
Warp, fabric, 175
Washers, 219
Washin, aircraft stability, 98

Washout, aircraft stability, 98
Watt (W) (unit of power), 46
Wave motion, 56–57
Waves:
   longitudinal, 56
   transverse, 56
Weave pattern, fabric:
   bidirectional, 189
   crowfoot, 189
   plain, 189
   unidirectional, 189
Webster hardness tester, 201
Weighing aircraft, 148–153
   equipment for, 148–149
   procedures for, 149–151
Weighing point, 146
Weight, measurements of, 38
Weight and balance:
   aircraft, 142–163
   extreme conditions, 159–160
   helicopter, 163–168
   report, 153
   terminology, 144–147
Wind, relative, 65
Window and windshield maintenance, 187
Wing, supercritical, 92
Wing area, 86
Wing fences, 91
Wing flaps, 80–85, 101
   leading-edge, 85
Wing loading, 97
Wing planform, 85
   for high-speed flight, 92
Wing-tip vortices, 87
Winglets, 88
Wire identification chart, 126
Wires, biplane wing:
   flying, 110
   landing, 110
Wiring diagrams, 125
Witness line, drawings, 132
Wood, aircraft, 174–175
   annual rings, 174
   laminated, 175
   plywood, 175
   radial sawing, 174
Work, measurements of, 44
Work hardening of metals, 177
Work order, 349
Working drawing, 122
Wrench sizes, 242
   metric, 242
Wrenches, 242–247
   adjustable, 244
   Allen, 244
   box-end, 243
   combination, 244
   crowsfoot, 246
   flare-nut, 244
   open-end, 242
   pipe, 246
   socket, 244–246
   spanner, 247
   strap, 247
   torque, 246

X-ray inspections, 205
X wing, helicopter, 120–121

Yard (yd) (mesurement of length), 36
Yaw, aircraft, 97, 102–103
Yawing moment, 103
Yield point or yield stress, 172

Zero, absolute, 49